Materials for a Sustainable Future

Materials for a Sustainable Future

Edited by

Trevor M. Letcher
*School of Chemistry, University of KwaZulu-Natal,
Durban, South Africa
Email: trevor@letcher.eclipse.co.uk*

Janet L. Scott
*Centre for Sustainable Chemical Technologies,
University of Bath, Bath, UK*

RSCPublishing

ISBN: 978-1-84973-407-3

A catalogue record for this book is available from the British Library

Published by The Royal Society of Chemistry,
Thomas Graham House, Science Park, Milton Road,
Cambridge CB4 0WF, UK

Registered Charity Number 207890

For further information see our web site at www.rsc.org

Printed in the United Kingdom by Henry Ling Limited, at the Dorset Press,
Dorchester DT1 1HD

Preface

The terms 'sustainable' and 'sustainability' are becoming more and more familiar, with our large industries publicly proclaiming new initiatives linked to sustainable processes and 'greener' ways of doing things through their advertising campaigns and annual reports. These terms are often linked in the public's mind to food and energy production, but there is another large issue to be considered: that of material use and reuse. This book, *Materials for a Sustainable Future*, deals with: the use and reuse of chemicals such as CO_2 and CH_4; new possibilities for using biomass; how much there is of the materials that will one day be in short supply; and how to develop a future society in more sustainable ways. We hope that the book will demonstrate the close connection between sustainability and the materials which are in jeopardy and those that can help to achieve our goal.

The book is largely about materials and hence much of the focus is on chemistry. However, as sustainability relates to ways of living, many other disciplines, such as economics, engineering, politics and government policies, agriculture, biochemistry, sociology, the science of climate change and biology are woven into the chapters of *Materials for a Sustainable Future*.

We are slowly exhausting certain substances derived from our planet; oil, phosphates, helium, gold and even elements such as rare earths are becoming scarce. This book highlights some of the raw materials we need if we are to live in a sustainable way and make it

Materials for a Sustainable Future
Edited by Trevor M. Letcher and Janet L. Scott
© The Royal Society of Chemistry 2012
Published by the Royal Society of Chemistry, www.rsc.org

possible for future generations to be able to live with the comforts and advantages enjoyed by our present generation.

Society will no doubt change considerably over the next 100 years, just as it has done over the past 100 years. The past century has seen the advent of motor transport, domestic electricity, electronics, computers, television, air travel, plastics, and an increase in leisure time for all. The next century will no doubt see dramatic changes as we attempt to curb our excessive exploitation of materials and control our demands on the environment. We must be prepared for change. The topics in this book give guidance to those planning for a more sustainable future. It is not a textbook but a source book, showing another way of looking at materials.

With a rapidly increasing world population and with the acceleration of the destruction of the environment, this book is a clarion call to say 'enough is enough'. We must adjust our needs to the availability of materials in the environment and we will have to shape the environment to fit our needs. We must learn to be good global citizens, so that those who come after us have the same opportunities as we had. We must learn new rules and ways of doing things and this book will go some of the way to introduce us to a new sustainable world.

Amidst the gloom caused by the human exploitation of our planet and with a sustainable future looking ever more distant, there is one comforting thought: that in the past humans have always invented new ways of coping with and solving environmental problems. Disasters such as 'pea-soup' fogs in London led to new rules on burning coal; horse power was replaced by gas and petrol power; and some epidemics have been banished by the invention of new drugs or inoculations. Let us hope that the same will happen to counter our over-zealous exploitation of the planet's resources.

This book provides an up-to-date account of some of the new materials needed for a sustainable existence as we strive, as our race has always done, to maintain longevity and a comfortable way of life. To give one example: over the past 75 years, we have become very dependent on plastics with their thousands of uses. Nearly all our plastics are made from chemicals which are distilled from fossil oil. When the oil runs out or becomes too difficult, or too expensive, to pump out of the ground, what are we going to do? One possibility is to use biomass as the raw material. Much research has gone into exploiting plant material for this purpose and, using only biomass, it does appear possible that we can replace oil as a feedstock for plastic production. Although it is possible to use biomass to produce some transport fuel, it is unlikely that we can replace all our petrol and diesel requirements. There just is not enough land available on Earth to grow the necessary quantities of plant material for motor transport and a further solution needs to be found. However, biomass is going to be an important raw material in the future and is the focus of a number of chapters in this book.

One of the great advantages of this book is that the chapters have been written by scientists or engineers who are experts in their field and include up-to-date statistics, recent research, and references to the latest work. We have chosen to write the book in a very readable format, with the hope that all readers, whether scientists or not, will gain much in knowledge and appreciation of the concept of sustainable living. To help readers, each chapter begins with an Introduction and finishes with a Conclusion written with lay-people in mind. A second advantage of this book is that it brings together many apparently disparate topics, but all are related to sustainability, so that comparisons can be made, synergies developed and issues put into perspective. The book should encourage more and more people to investigate ways of ensuring the survival of future generations.

The audience that we hope to reach includes: industrialists and investors looking at future developments with an eye on suitable investments; policy makers in local and central governments who need to become acquainted with the latest thinking on issues related to sustainability; students, teachers, researchers and professors and scientists and engineers working in the field, who need information, direction and references to new developments; and editors, journalists and the general public who need information on the increasingly popular concepts of sustainable living.

The International System of Quantities is reflected in the book with the use of SI units where ever possible. Also, the International Union of Pure and Applied Chemistry (IUPAC) recommendations on notation and spelling have been used throughout. The book has been supported by IUPAC.

The Editors have attempted to give the book an international flavour with authors coming from 12 different countries; Australia,

Brazil, Canada, England, Finland, Germany, Italy, Scotland, Singapore, South Africa, Sweden and the United States of America.

We wish to express our gratitude to all the authors who have so graciously given time and effort to writing their chapters. We also thank the publication team of the Royal Society of Chemistry, especially Alice Toby Brant and Sue Humphreys, for their assistance in assembling this volume.

Trevor M. Letcher
Janet L. Scott

About the Authors

Ifty Ahmed (Chapter 22) completed his doctoral studies at University College London in the area of biomaterials and tissue engineering. He is currently working as a Senior Research Fellow at the University of Nottingham in the area of biocomposites. His research interests span glass chemistry with special interest in structural biomaterials and synthetic and natural fibres and their composites. He has published more than 40 scientific journal articles and has also published in many scientific conference proceedings. Dr Ahmed is currently project managing a Department of Health-funded project investigating resorbable bone fracture fixation devices made from resorbable composites.

Naa Lamkai Ampofo-Anti (Chapter 21) holds a Bachelor of Science: Design (RIBA Part 1) degree and a Post Graduate Diploma: Architecture (RIBA Part 2) degree from the University of Science and Technology, Kumasi, Ghana. Born in Accra, Ghana, she practised as an architect in Ghana, Nigeria and Cameroon before emigrating to South Africa in 1990 to work for the North West Department of Public Works, Mafikeng. She left her position as Chief Architect of that department to join the CSIR Built Environment (Building Science and Technology Competence Area) in June 2005. She is currently working on a Master of Science in Architecture (by research) degree at the University of Pretoria. Her research speciality is the use of life cycle assessment (LCA)

Materials for a Sustainable Future
Edited by Trevor M. Letcher and Janet L. Scott
© The Royal Society of Chemistry 2012
Published by the Royal Society of Chemistry, www.rsc.org

methodology to enhance the environmental performance of construction products (and services). Starting from 2007, she has written a number of conference papers and client and CSIR in-house reports on the subject of LCA and its applications in the built environment. Since the launch of the *South African Green Building Handbook* in 2008, she has contributed a chapter a year on the topic of LCA.

Michele Aresta (Chapter 13) has a doctorate in industrial chemistry (University of Milan) and is Professor of Chemistry at the University of Bari, Italy, where he is teaching catalysis. He is President of the Inorganic Chemistry Division of EuCheMS and is involved in several international projects and programmes. His fields of interest are metallorganic chemistry and catalysis, chemistry of small molecules, chemistry of CO_2, conversion of biomass and biorefining. He is author of over 250 peer-reviewed publications in international journals, owner of several national and international patents, author of several chapters in books and editor of five books on the chemistry and biochemistry of CO_2. He has received several national and international awards.

Renbi Bai (Chapter 19) is an Associate Professor in the Department of Civil and Environmental Engineering at the National University of Singapore. His research involves functional membranes and membrane processes, adsorption and filtration, photocatalysts and photocatalytic oxidation, surface modification and surface interaction, separation science and technology, water and wastewater treatment. He has published about 80 journal papers and 50 conference papers. His journal papers are well cited and have had more than 1600 SCI citations. He is on the editorial boards of a number of international journals, including *Separation and Purification Technology*.

Natasha A. Birkin (Chapter 15) obtained a Masters degree in chemistry in 2008 from the University of Nottingham, completing a final-year project on the design and synthesis of smart polymers for tissue engineering applications. Since then she has been studying for a PhD with Professor Howdle. The focus of her research is the development of novel CO_2-soluble hydrocarbon surfactants via RAFT polymerisation and the use of supercritical CO_2 as a reaction medium for dispersion polymerisation.

Antonella Colucci (Chapter 8) has a Master's degree in inorganic chemistry on technologies for biodiesel extraction from microalgae. She currently holds a research fellowship within the National Programme 'Extraction of biomass platform molecules and their catalytic conversion' in Italy.

Leroy Cronin (Chapter 18) is the Gardiner Professor of Chemistry at the University of Glasgow. The focus of his work is understanding and controlling self-assembly and self-organisation in chemistry to develop functional molecular and nanomolecular chemical systems, linking architectural design with function and recently engineering system-level functions (*e.g.* coupled catalytic self-assembly, emergence of inorganic materials and fabrication of inorganic cells that allow complex cooperative behaviours). Cronin has won numerous prizes and awards for his research, including the Leverhulme Prize (2007), the Morino Foundation Prize (2008) and the Nexxus young scientist award (2006). He was elected to the Royal Society of Edinburgh in 2009.

Yara Csordas (Chapter 9) graduated from the University of São Paulo, Brazil, with a chemical engineering BSc and a business administration BA. She worked in the technology and innovation areas in Rhodia, Shell, PPG and CP Kelco, acquiring experience in advanced materials and polymers based on petrochemical and renewable sources. Currently, she is Technical Manager of Chem-Trend Southern Hemisphere in Valinhos, Brazil. She has also worked as a consultant to many companies, the University of Campinas and to CGEE, a think-tank supported by the Brazilian Ministry of Science and Technology.

Arno de Klerk (Chapter 11) is a registered professional engineer and holds a PhD in chemical engineering and an MSc in analytical chemistry. After 3 years in forensic science, he moved to industry, where he was involved with industrial Fischer–Tropsch-based coal-to-liquids and gas-to-liquids facilities for 14 years, mainly in design, catalysis and refining. In 2009 he relocated to Canada, where he is the Nexen Professor in Catalytic Reaction Engineering in the Department of Chemical and Materials Engineering at the University of Alberta. His publications include the books *Catalysis in the Refining of Fischer–Tropsch Syncrude* and *Fischer–Tropsch Refining*.

Angela Dibenedetto (Chapter 8) is Associate Professor in the Department of Chemistry at the University of Bari–UNIBA, Italy. Her scientific interests are focused on carbon dioxide utilization in synthetic chemistry, catalysis, coordination chemistry and metallorganic chemistry, green chemistry, marine biomass production by enhanced carbon dioxide fixation, marine biomass as a source of fuels and chemicals applying the biorefinery concept. She is Director of the Interdepartmental Centre on Environmental Methodologies and Technologies (METEA) at UNIBA and Director of the Interuniversity Consortium on Chemical Reactivity and Catalysis (CIRCC).

Paul S. Fennell (Chapter 12) conducted his undergraduate degree and PhD at the Department of Chemical Engineering at the University of Cambridge. His research has centred on the field of pollutant mitigation from power stations, initially working on NO_x destruction for his PhD (under the supervision of Professor Allan Hayhurst), but then working as a PDRA in novel carbon capture and storage methods (specifically, chemical looping technology and the ZECA clean coal process, which hold out the possibility of producing electricity with greater efficiency than current power stations, but with intrinsic separation of CO_2). In 2008 he was appointed to a lectureship in Chemical Engineering at Imperial College London, where he has rapidly built up a large group investigating many aspects of CO_2 capture and bioenergy, with a particular focus on solutions which have high potential for commercial application. He has served on the IEA High Temperature Solid Looping Cycles Network Executive and is in the IChemE Energy Conversion Subject Group. His work has a specific focus on making a difference in the real world and on near-term applicability. He is particularly interested in the mitigation of industrial emissions (such as those from the cement and iron/steel industries) as opposed to focusing entirely on the power sector, since the industrial sector is frequently neglected as a source of CO_2.

Nick H. Florin (Chapter 12) studied chemical engineering at the University of Sydney, Australia, where he also carried out research for his PhD on 'Hydrogen synthesis in biomass gasifiers coupled with carbon dioxide capture using calcium oxide'. He is now an Imperial College Junior Research Fellow and Grantham Institute for Climate Change Researcher. He has 7 years of experience of research and development of calcium looping and is an active member of the International Energy Authority's High-Temperature Solid-Looping Network. He has provided advice to the UK Department for Energy and Climate Change and the Environment Agency on carbon capture and storage and was an invited member of the Norwegian Research Council International Panel of Experts which reviewed CCS and R&D project proposals in 2009, 2010 and 2011.

Ian Forbes (Chapter 17) is Research Fellow in the Northumbria Photovoltaics Application Centre and the University of Northumbria. He has been involved in a wide range of photovoltaics research including III–V-based devices for space and thermophotovoltaic applications, systems performance monitoring, environmental impact assessment and the development of materials for thin-film photovoltaics. He has published over 30 papers in peer-reviewed journals and several invited reviews and a book chapter. His recent research has been focused on thin-film materials in collaboration with other UK groups, most recently as a partner in the first two cycles of the PV Materials for the 21st Century Consortium funded under the EPSRCs SUPERGEN initiative, concerned with the development of sustainable inorganic materials. He is a founder member of the European Kesterite Network that aims to promote research into Cu_2ZnSnS_4. He is currently working on applying photoluminescence and '4C' electrical characterisation techniques to these materials and is also working on environmental impact assessment of thin-film processes.

Fernando Galembeck (Chapter 9) holds BSc and PhD degrees from the University of São Paulo, Brazil. He was a post-doc in the Universities of Colorado and California and a fellow in Unilever Port Sunlight Laboratory. He later joined the faculties of the University of São Paulo (1965–1980) and Campinas (1980–2011), and was Professor of Chemistry since 1987. He has worked in physical chemistry, colloids and surfaces, chemistry of materials, bio- and nanostructured materials, publishing more than 230 papers and authoring 18 patents, seven of which were licensed to industrial companies. He is currently Director of the Brazilian National Nanotechnology Laboratory in Campinas.

Gareth P. Hatch (Chapter 2) is Founding Principal of Technology Metals Research and a co-founder of Innovation Metals. For several years he was Director of Technology at Dexter Magnetic Technologies and holds five patents on a variety of magnetic devices. He is based in Illinois, USA. He has a BEng (Hons) in materials science and technology and a PhD in metallurgy and materials, both from the University of Birmingham. He is a Fellow of the Institute of Materials, Minerals and Mining, a Fellow of the Institution of Engineering and Technology, a Chartered Engineer and a Senior Member of the IEEE.

Kazi M. Zakir Hossain (Chapter 22) obtained his BSc (Hons) and MS (Research) in applied chemistry and chemical technology from the University of Dhaka (Bangladesh). His work experience includes time spent within the pharmaceuticals industry as a quality assurance executive before joining the Daffodil International University as a Lecturer within the Faculty of Engineering in 2009. He is currently completing his PhD at the University of Nottingham, working in the area of biocomposites under the supervision of Dr Ifty Ahmed and Dr Wim Thielemans. His project centres on the processing of biocomposites made using natural fibre-reinforced biopolymers intended for biomedical applications.

Steven M. Howdle (Chapter 15) was born in Rotherham, South Yorkshire, England in 1964. He obtained a first degree in chemistry from Manchester in 1986 and his PhD on 'Spectroscopy in Liquefied Noble Gases' from Nottingham in 1989. His research focuses on sustainable chemistry and in particular on the utilisation of supercritical carbon dioxide for polymer synthesis, polymer processing and preparation of novel polymeric materials for tissue engineering and drug delivery. He has published more than 280 peer-reviewed papers in this field. He holds a chair at the School of Chemistry, University of Nottingham, and prior to this held a Royal Society University Research Fellowship (1991–1999). He has received the Jerwood–Salters' Environment Award for Green Chemistry (2001), the RSC Corday–Morgan Medal and Award (2001), the Royal Society–Wolfson Research Merit Award (2003), the RSC Interdisciplinary Award (2005), the DECHEMA Award of the Max Buchner Research Foundation (2006), the RSC/SCI

Macro Group UK Medal (2008) and the Hanson Medal of the IChemE (2009).

Matthew D. Jones (Chapter 14) is a Lecturer in Chemistry at the University of Bath. He completed his undergraduate studies at the University of Exeter and his PhD at the University of Cambridge. His research is focused on the development of homogeneous and heterogeneous catalysts for sustainable chemical applications, such as polymerisations and the grading up of raw materials.

Andreas Kafizas (Chapter 20) is a postdoctoral worker in Professor Ivan Parkin's research group. He obtained his PhD from University College London in 2011 on combinatorial chemical vapour deposition. He was awarded the Ramsay Medal for his outstanding PhD work. He is the author of some 18 research papers. He has particular research interests in photocatalysis and coatings on glass.

Tanja Kallio (Chapter 16) obtained a DSc (Tech) at Helsinki University of Technology in 2003. She heads a research group at Aalto University focusing on electrochemical characterization and development of materials for electrochemical energy conversion and storage devices. She has investigated polymer electrolytes and electrocatalyst materials for fuel cells for more than 10 years and her latest research interests include lithium ion batteries and supercapacitors. She has published around 30 papers in peer-reviewed journals.

Marianne Labet (Chapter 22) studied paper and printing engineering at the Ecole Française de Papeterie et des Industries Graphiques (INPG) (Grenoble, France) from 2003 to 2006. She obtained her Masters degree in materials science in 2006 working under the supervision of Professor Alain Dufresne and Dr Wim Thielemans. In 2006, she moved to Nottingham (UK) to complete a PhD thesis on the use of starch and cellulose nanocrystals as substrates for the surface-initiated ring-opening polymerisation of lactones under the supervision of Dr Wim Thielemans. Since 2010, she has been working in Professor Chris Breen's team at Sheffield Hallam University on the surface modification of polymers and particles by chromatogenic grafting.

Trevor M. Letcher (Chapter 5) is Emeritus Professor of Chemistry at the University of KwaZulu-Natal, Durban, and a Fellow of the Royal Society of Chemistry. He is a past Director of the International Association of Chemical Thermodynamics and his research involves the thermodynamics of liquid mixtures and energy from landfill. He has published over 250 papers in peer-reviewed journals and edited and co-edited eight books in his research fields. His latest edited and co-edited books are *Heat Capacities* (2010), *Climate Change* (2009) and *Waste* (2011).

Gabriela Alves Macedo (Chapter 9) graduated as a food engineer in 1993 at the University of Campinas, Brazil, and received her MS and PhD in food sciences from the same University in 1997. She spent a few years as a researcher at the Rhodia Research Center in Brazil, joining the Unicamp faculty in the area of food biochemistry and biotechnology. She became an Associate Professor in 2010, with particular interest in microbial enzymes and bioactive compounds.

Niall Mac Dowell (Chapter 12) received his PhD from Imperial College in 2010 and is currently a postdoctoral research associate at the same institution. His background is in the integration of molecular thermodynamics and process modelling and simulation in the context of CO_2 capture. He has published work concerning fundamental molecular thermodynamics in addition to work on CO_2 capture process modelling and the spatio-temporally explicit design of CO_2 transport networks. In 2010 he was awarded the Imperial College Qatar Petroleum Prize in Clean Fossil Fuels for his PhD thesis. He is also a guest lecturer at both Cranfield and Edinburgh Universities on MSc courses in CO_2 capture and transport and is a visiting researcher at the MATGAS Research Centre in Barcelona, Spain.

Geoffrey C. Maitland (Chapter 12) studied chemistry at Oxford University, where he also obtained his doctorate in physical chemistry. After a period as an ICI Research Fellow at Bristol University, he was appointed as Llecturer in Chemical Engineering at Imperial College in 1974. His research focused on molecular interactions and the transport properties of fluids, including the rheology of polymer systems. He spent a secondment with ICI

Plastics Division from 1979 to 1981 and became a Senior Lecturer in 1983. In 1986 he moved to the oil and gas industry with Schlumberger, where he carried out research in oilfield fluids engineering, including the use of colloidal systems for well construction, reservoir stimulation and production enhancement. He held a number of senior technical and research management positions in Cambridge and Paris, most recently as a Research Director. He rejoined Imperial College in September 2005 as Professor of Energy Engineering and his current research covers clean and efficient fossil fuel production with particular emphasis on carbon dioxide mitigation processes, methane hydrate production and energy-related reactor engineering. He is the Director of the Qatar Carbonates and Carbon Storage Research Centre, a long-term research collaboration with Qatar Petroleum and Shell. He was awarded the Hutchison Medal by the Institution of Chemical Engineers in 1998 and served as President of the British Society of Rheology from 2002 to 2005. He was awarded the 2010 IChemE Chemical Engineering Envoy Award for his media work explaining the engineering issues involved in the Gulf of Mexico oil spill.

Joe Mapiravana (Chapter 21) holds a BSC in chemistry and geology from the University of Zimbabwe, an MScTech in the science and technology of materials from the University of Sheffield and a PhD from the University of Leeds. His research fields include the production and application of materials and product and process research and development. He has authored and co-authored a number of technical papers, journal articles, conference papers and other publications. He was won a number of awards for his work.

Gavin M. Mudd (Chapters 1, 4 and 7) has been an active researcher and advocate on the environmental impacts and management of mining for over a decade. He has been involved with many aspects of the mining industry, with a particular specialty in the sustainability of mining, uranium mining and mine rehabilitation. Gavin maintains an independent perspective and has undertaken research for mining companies, community groups and aboriginal organisations. As such, he has developed a unique understanding of the multidisciplinary nature of the environmental aspects of mining and holds a distinctive view on how to quantify an apparent oxymoron – that of 'sustainable mining'. He is currently Senior Lecturer in Environmental Engineering at Monash University in Australia.

Tina-Simone S. Neset (Chapter 6) is an Assistant Professor at the Centre for Climate Science and Policy Research (CSPR) and Department of Water and Environmental Studies at Linköping University. She has an undergraduate background in geography and a PhD from Linköping University focusing on the environmental imprint of human food consumption. Her research areas include material and substance flow analysis studies with a particular focus on resource flows linked to food production and consumption. Within the Nordic Centre of Excellence for Strategic Climate Adaptation Research NORD-STAR, she leads the research on visualization for decision support and data analysis regarding land-use issues and risk assessment in the Nordic Countries. Tina is currently the head of Climate Visualization at CSPR and co-founder of the Global Phosphorus Research Initiative.

Ivan P. Parkin (Chapter 20) is the head of the Chemistry Department at University College London and Professor of Inorganic and Materials Chemistry. He has published over 350 research papers and eight book chapters. His research interests focus on the preparation and functional evaluation of materials, especially thin films on glass formed by chemical vapour deposition. He has other research interests in magnetic hyperthermia, nanoparticle synthesis and hospital-acquired infection.

Lauremce M. Peter (Chapter 17) is Professor of Physical Chemistry at the University of Bath. He has worked for many years on topics related to solar energy conversion, including dye-sensitized solar cells, Earth-abundant materials for thin-film photovoltaics, semiconductor photoelectrochemistry and photobiological systems for energy conversion. He has published around 280 papers in peer-reviewed journals and several book chapters, and has been awarded a number of international prizes for his work. Much of his recent research has involved collaboration with other groups in the UK within the framework of two EPSRC Supergen Consortia: PV Materials for the 21st Century and Excitonic Solar Cells. He currently has a number of international collaborations and splits his time between the University of Bath and the Ludwig Maximillian University in Munich, where he is working on a range of topics including light-driven water splitting and *in situ* microwave measurements on solar cells.

Stefan Pollak (Chapter 15), born in 1978, currently works as a researcher at the Faculty of Mechanical Engineering of the Ruhr University in Bochum, Germany. At the same university he earned his diploma degree in 2005 and his doctoral degree with a thesis on carbon dioxide-based high-pressure spray process for the atomisation of polymers in 2009. In 2011 he spent a period of 4 months at the University of Nottingham in the School of Chemistry working on the CO_2-aided production of polymer foams for tissue engineering. In his current research in Bochum he focuses on the flow behaviour of non-Newtonian fluids under high pressure.

Vinay Prasad (Chapter 11) is a registered professional engineer and holds BTech, MS and PhD degrees in chemical engineering. He has industrial experience in refinery optimisation and in steam methane reforming for hydrogen production. In addition, he has held academic positions in India, the USA and Canada. Since December 2008, he has been on the faculty of the Department of Chemical and Materials Engineering at the University of Alberta, where he conducts research in process monitoring, control and optimisation, and catalysis and reaction engineering.

Justin Salminen (Chapter 16) of European Batteries has wide experience of energy storage systems and developing environmentally friendlier processes in different fields and industries. He holds a DSc (Tech) from Helsinki University of Technology and has been a postdoctoral research fellow at the University of California Berkeley with Professor John M. Prausnitz and Professor John Newman. He has also worked at VTT Technical Research Center of Finland, Lawrence Berkeley National Laboratory and Outotec Research Center. In addition, he has experience in industrial plant start-ups and running plant operations. He is a holder of several patents. He has also given graduate courses on renewable energy technologies, chemical thermodynamics and physical chemistry. He has authored over 100 publications and 150 industrial reports.

Janet L. Scott (Chapter 10) is a Senior Research Fellow in, and Coordinator of, the Centre for Sustainable Chemical Technologies (CSCT), University of Bath. She has previously worked in both industry and academia in three different countries: South Africa (University of Cape Town, 1992–1995; R&D Manager, Fine Chemicals Corporation, 1996–1998); Australia (Monash University, 1999–2006), where she was the Deputy Director of the Centre for Green Chemistry; and the UK, where she held a Marie Curie Senior Transfer of Knowledge Fellowship at Unilever R&D, Port Sunlight (2006–2008). Research interests currently centre on bio-derived chemicals and materials and she works closely with chemical engineers and industrial partners at the CSCT.

Julie Stacey (Chapter 3) has a background in earth and life sciences and has worked operationally and at global policy level in the mining industry for the past two decades. Having been the Sustainable Development Operations Manager for Anglo American plc, she is now an independent consultant on sustainable development. She is an Associate of the Centre for Sustainability in Mining and Industry (CSMI) in the Mining School at the University of the Witwatersrand in South Africa. She is the co-founder of a programme geared to business executives that aims to develop authentic leaders. She is a keen cyclist and enjoys the outdoors.

Paolo Stufano (Chapter 13) is a PhD student at the University of Bari. His interests are in the area of catalysis and reaction mechanism, electrochemistry and CO_2 conversion into chemicals and fuels. He has international collaborations and is active in international and national projects. He is owner of a patent, author of over 10 papers in international journals and of two chapters in books.

Mark D. Symes (Chapter 18) is a Solar Energy Research Fellow at the University of Glasgow in the group of Professor Leroy Cronin. After a PhD in supramolecular and coordination chemistry at the University of Edinburgh under the supervision of Professor David A. Leigh, he took a postdoctoral position at the Massachusetts Institute of Technology with Professor Daniel G. Nocera, examining non-precious metal catalysts for water oxidation. He returned to the UK in 2010 to take up his present position and his current research interests include low-energy routes to water splitting and direct solar-to-fuels energy conversion.

Wim Thielemans (Chapter 22) obtained a Masters degree in chemical engineering from the KU Leuven, Belgium, in 1999 working under the supervision of Professor Jan Mewis and Professor Jan Vermant. He received his PhD in 2004 working with Professor Richard P. Wool (Chemical Engineering, University of Delaware, USA). He subsequently moved to the INPG (Grenoble, France) as a postdoctoral researcher/Marie Curie Research Fellow working with Professor Alain Dufresne and Professor M. Naceur Belgacem. Since August 2006, he has been a lecturer in chemistry and chemical engineering at the University of Nottingham. His research interests include the surface modification of starch and cellulose nanocrystals and their self-assembly.

Gianfranco Unali (Chapter 10) obtained his DPhil on the synthesis of water-soluble polymers from the University of Sussex and has been at Unilever R&D since 2000. His research interests centre mainly on the sustainability of personal care formulations and materials from renewable resources. He leads industrial and publicly funded research projects in this area and is involved in academic research with a number of academic partners.

Llewellyn van Wyk (Chapter 21) holds a Bachelor of Architecture degree from the University of Cape Town. He joined the Council for Scientific and Industrial Research (CSIR) in 2002, where he is a Principal Researcher in the Building Science and Technology Competence Area of the Built Environment Unit. He has delivered several keynote papers at international and national conferences and workshops and is the author of a number of published papers on the subject of architecture, the built environment professions and the constructed environment. He has received a number of awards for his work. He launched the *Digest of South African Architecture* in 1999 and served as the Editor until 2003. He is the current Editor-at-Large of the E-Journal *Greenbuilding South Africa*, which he helped launch in 2007. He is also the Editor of the *Green Building Handbook* (now in its fourth year of publication) and the *Sustainable Transport and Mobility Handbook* (now in its third year of publication). He has contributed book chapters to a number of international publications. He is a Founding Member of the Green Building Council and served as a Director and on its Technical Committee.

Marie Warren (Chapter 15), born in 1986, is currently studying for a PhD in polymer chemistry at the University of Nottingham. Her current research focuses on the supercritical CO_2 processing of polymers for controlled-release drug delivery. She obtained her first degree in chemistry in 2008 from Hull University.

Kin-Ho Wee (Chapter 19) is a former PhD student of Dr Renbi Bai at the National University of Singapore and is currently a research fellow in Dr Bai's research group. He is working in the area of nanomaterials and membrane separation technology.

Zhehan Weng (Chapter 1) graduated from the Environmental Engineering Department at Monash University in 2009 with honours and has worked in research on mining and sustainability, pollution accounting and engineering education research as assistant to Dr Gavin Mudd. Zhehan focuses on global-scale issues and is particularly interested in linking material consumption, environmental impacts and mining together in a holistic manner.

Xiaoying Zhu (Chapter 19) is currently a senior PhD student of Dr Renbi Bai at the National University of Singapore and is working in the area of membrane functionalization.

Contents

Materials for a Sustainable Future
Edited by Trevor M. Letcher and Janet L. Scott
© The Royal Society of Chemistry 2012
Published by the Royal Society of Chemistry, www.rsc.org

Chapter 6
Phosphorus   **163**
Tina-Simone S. Neset

Chapter 7
Uranium   **175**
Gavin M. Mudd

II   Sustainability Related to Biomass

Chapter 8
Aquatic Biomass for the Production of Fuels and Chemicals   **215**
Angela Dibenedetto and Antonella Colucci

IV   Materials Related to Energy Conversion, Storage and Distribution

Chapter 16
Battery and Fuel Cell Materials   **537**
Justin Salminen and Tanja Kallio

Chapter 17
Materials for Photovoltaics   **558**
Ian Forbes and Laurence M. Peter

Chapter 18
Materials for Water Splitting **592**
Mark D. Symes and Leroy Cronin

Introduction

The term 'sustainable development' has its roots in a report issued by the United Nation's World Commission on Environment and Development of 1987, which was chaired by Dr Gro Harlem Brundtland, the then Norwegian Prime Minister. In the paper entitled 'Our Common Future', usually known as the Brundtland Report,[1] an oft-quoted definition of sustainable development appears in this statement: 'Humanity has the ability to make development sustainable to ensure that it meets the needs of the present without compromising the ability of future generations to meet their own needs'. The next two sentences in the report refer directly to resources and to limits that are dependent on the current state of the art with respect to technology: 'The concept of sustainable development does imply limits – not absolute limits but limitations imposed by the present state of technology and social organisation on environmental resources and by the ability of the biosphere to absorb the effects of human activities. But technology and social organisation can be both managed and improved to make way for a new era of economic growth.' Thus the concept of 'needs' (implicitly of the world's poor) and technological development to improve the 'carrying power' of our planet are explicitly dealt with. The final theme of the report is that of economic development, thus the three Ps of sustainability, 'people, planet and profit', are

Materials for a Sustainable Future
Edited by Trevor M. Letcher and Janet L. Scott
© The Royal Society of Chemistry 2012
Published by the Royal Society of Chemistry, www.rsc.org

incorporated. This concept has been reasonably widely accepted by industry as 'triple bottom line' reporting.

In *Materials for a Sustainable Future*, issues of resource depletion and looming shortages in addition to the consequences of resource use are considered. The book is concerned not only with the elements that could be in short supply in the near future, such as phosphorus, helium, and some rare earths, but also with pollutants, such as carbon dioxide and methane, which are being pumped into the atmosphere in sufficiently large quantities to be a threat to our lives on this planet. One approach to such pollutants is to treat them as resources. Furthermore, the book contains chapters concerning chemicals and materials that might soon be required in large quantities to help create a more sustainable way of life, in the light of depletion of fossil resources such as oil. These include: biomass needed to manufacture plastics; special compounds and membranes for water purification, water splitting, photovoltaic cells, batteries and fuel cells; and special materials for buildings, glass technologies and storing hydrogen.

The book is divided into five themes:

- Elements that could soon be in short supply.
- Sustainability related to biomass.
- Sustainability related to the feedstocks: carbon dioxide and methane.
- Materials related to energy conversion, storage and distribution.
- Sustainability related to materials in the urban environment and to water.

Coverage of each of these topics is, of necessity, not exhaustive and there are many areas that are not mentioned. Furthermore, there are other elements besides those mentioned in this book, such as zinc, gallium, germanium, arsenic, indium, hafnium and even silver, the shortage of which many believe will pose a serious threat in the next 100 years (Figure 1).[2] In many ways, this book is a snapshot in time of the present state of strategic elements and compounds on the planet, and of new ideas and processes that could soon bring us closer to a sustainable existence. Some of the topics discussed are moving forward at a rapid rate and new developments appear almost daily, others may only begin to be properly

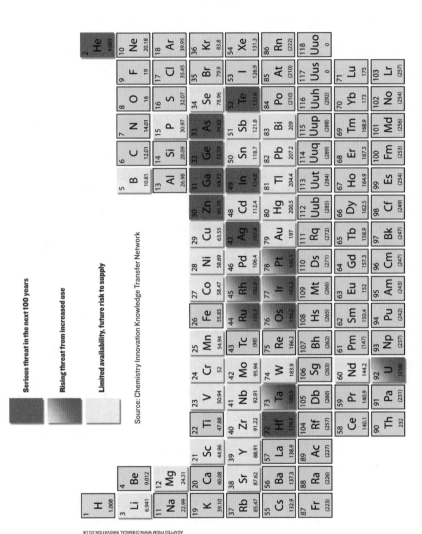

Figure 1 Periodic table showing 'endangered elements', from reference 2 (adapted from Chemistry Innovation Knowledge Transfer Network: www.chemicalinnovation.co.uk).

addressed when prices of desirable or essential goods rise due to impending shortage of supply, or stockpiling by a geopolitical power block.

The developing scarcity of certain elements is becoming widely recognised as having the potential to constrain future technological developments, including those often associated with more sustainable technologies. For example, lithium, widely used in energy storage applications, is predominantly obtained from reserves located in parts of South America; platinum group metals, used in automobile catalytic converters, are concentrated in South Africa and Russia; and the rare earths neodymium and dysprosium are used in the production of light, yet powerful, magnets deployed in the electric motors of hybrid or electric cars and wind turbines, yet these elements are currently mined almost exclusively in China and are increasingly sequestered to service their local market. Clearly, in some cases, actual global scarcity is less of an issue than adequately distributed supply. Rare earths are not particularly rare in the earth's crust and it is possible to reopen or develop new mines to service the need for these materials, while other metals may be in danger of 'running out' in the foreseeable future as global demand rises. [3] In some cases, such as indium, perceived scarcity is challenged, but resources are clearly not infinite.[4] Indium is recovered from the mining of zinc. Cobalt is another metal that could soon be scarce. Growing impacts from mining will most likely cause supply constraints long before some elements run out. This is already happening with the rare earths, with China cutting back for social and environmental reasons.

The strategic issues associated with material supply and control of scarce or unevenly distributed resources have been recognised by nation-states and politico-economic groupings. Japan has developed the 'Element Strategy Initiative' leading to the establishment of the 'Element Strategy Commission'.[5] This initiative is a four pillar approach: substitution, reduction, recycling and regulation. The first of these focuses on replacement of rare *and harmful* elements with abundant, *harmless* ones and is seen as a stimulant to new materials research. The second and third lead to the extension of the lifetime of elements in use and creation of closed materials loops, otherwise known as a 'cradle to cradle' approach as espoused by Braungart and McDonough.[6] The fourth pillar, regulation, often viewed with trepidation by developers and commercialisers of new technologies, is here seen as promoting innovation by ensuring that the other pillars are enacted.[5]

As a resource poor country, Japan focused on these strategically important issues somewhat earlier than many other countries or groupings, but these have followed suit with the publication of 'Critical Raw Materials for the EU'[7] and the US Department of Energy 'Critical Materials Strategy both in 2010 (the latter updated in 2011).[3]

In the cradle to cradle approach, referred to above, the concept of 'technical nutrients'[5] which can be used in continuous cycles without 'downcycling' (loss of integrity or quality) is useful when contemplating elements that are not in particularly short supply, but that humankind is distributing in highly dilute forms about the earth. Phosphorus, ubiquitous in terrestrial and aquatic ecosystems, is not 'rare', in the sense of 'only present on earth in small amounts', but it is rapidly being dispersed in low concentration throughout the ecosphere. Thus high concentrated ores are mined and used directly as fertilisers, or converted into organo-phosphorus compounds in detergents or pesticides,[8] which are then widely distributed with little opportunity for recovery.

Apart from the metals and other elements mentioned above the 'elements of life' are also under threat, albeit of a different kind. The elements of organic compounds, carbon, hydrogen, and oxygen (and to a lesser extent nitrogen, phosphorus and sulfur), form the basis for a huge range of chemicals and materials that we use in growing quantities in the form of polymers, surfactants and fine and commodity chemicals. In addition many organic chemicals serve small, but important roles as preservatives, drugs, flavourants and fragrances. While much of our food is derived from renewable (although not always sustainable) sources a very large percentage of the carbon that is used in organic chemicals and polymers is derived from fossil reserves, mainly oil. While there is little consensus about the date of 'peak oil'[9] all agree that the resource is finite.

Anthropogenically produced carbon dioxide, is widely recognised as a pollutant that is expected to effect large changes in the global climate,[10] yet this may provide a useful carbon building block, as might methane, a significant greenhouse gas.[11] Thus, as well as considering bio-based raw materials as sources of chemicals and materials, we include chapters on the conversion of these two pollutants as sources of chemicals. In such applications, these become a resource rather than (only) a concern.

By far the most rapid means of conversion of useful fossil carbon from hydrocarbons to CO_2 results from combustion in pursuit of energy conversion. Thus, a consideration of materials which allow alternative means of energy harvesting, conversion, storage and transport is essential in a book devoted to *Materials for a Sustainable Future*. While this is a huge topic in itself some of the most rapidly developing topics are covered, though we do not attempt to consider opportunities offered by materials for nuclear fusion or fission.

More than $> 50\%$ of the world's population live in cities and this figure is increasing, with expectations that the urban population will double from 2.6 billion in 2010 to 5.2 billion by 2050.[12] In such a scenario the built environment provides opportunities for energy harvesting via smart materials as well as requiring maximum energy saving, also by application of sophisticated new materials.

Finally, the topic of water and specifically water purification is addressed briefly. Although approximately 70% of the surface of the earth is covered by water, much of this is in a form that is not immediately useful to most terrestrial plants or to humans.[13,14] There are schools of thought that hold that potable water will become the limiting factor in the earth's carrying capacity for human population and that competition for this precious resource may be a future source of global conflict. Once again, the development of smart materials will be critical in accessing fresh water from the seas.

Success in any technical enterprise depends on awareness, decision making and action. We do hope that this book helps in all three areas and that sustainable living can be achieved within the next generation. This is particularly important and necessary as the world's population, currently standing at 6.97 billion,[15] is expected to reach between 7.5 and 10.5 billion by the year 2050.[16]

Trevor M. Letcher
Janet L. Scott

REFERENCES

1. G. H. Brundtland (Chairman), *Our Common Future*, World Commission on Environment and Development, Oxford University Press, Oxford, 1987.

2. For a very readable overview, see: E. Davies, *Chem. World*, 2011, January, 50–54.
3. Various publicly available US Department of Energy reports provide good overviews of elemental scarcity: (a) *US Department of Energy Critical Materials Strategy 2011*, http://energy.gov/pi/office-policy-and-international-affairs/downloads/2011-critical-materials-strategy (last accesed 22 March 2012); (b) *US Department of Energy Critical Materials Strategy 2010*, http://energy.gov/pi/office-policy-and-international-affairs/downloads/2010-critical-materials-strategy (last accesed 22 March 2012).
4. C. Candelise, J. F. Spiers and R. J. K. Gross, *Renew. Sustain. Energy Rev.*, 2011, **15**, 4972–4981.
5. E. Nakamura and K. Sato, *Nat. Mater.*, 2011, **10**, 158.
6. W. McDonough and M. Braungart, *Cradle to Cradle: Changing the Way We Make Things*, North Point Press, New York, 2002.
7. European Commission Enterprise and Industry, *Critical Raw Materials for the EU, Report of the Ad-hoc Working Group on Defining Critical Raw Materials, 2010*, 2010, http://ec.europa.eu/enterprise/policies/raw-materials/files/docs/report-b_en.pdf, (last accessed October 2011). The full report is available from the European Commission Enterprise and Industry: http://ec.europa.eu/enterprise/policies/raw-materials/documents/index_en.htm (last accesed 22 March 2012).
8. H. Diskowski and T. Hofmann, Phosphorus, in *Ullmann's Encyclopedia of Industrial Chemistry*, Wiley-VCH, Weinheim, 2005.
9. B. Gallagher, *Energy Policy*, 2011, **39**, 790–802.
10. M. R. Allen, D. J. Frame, C. Huntingford, C. D. Jones, J. A. Lowe, M. Meinshausen and N. Meinshausen, *Nature*, 2009, **458**, 1163–1166.
11. I. Karakurt, G. Aydin and K. Aydiner, *Renew. Energy*, 2011, **39**, 40–48.
12. United Nations Department of Economic and Social Affairs, Population Division Report, *Population Distribution, Urbanisation, Internal Migration and Development: an International Perspective*, 2011, http://www.un.org/esa/population/ (last accesed 22 March 2012).
13. M. Elimelech, and W. A. Phillip, *Science*, 2011, **333**, 712–717.

14. R. F. Service, *Science*, 2006, **313**, 1088–1090.
15. U.S. Census Bureau, *World POPClock Projection*, http://www.census.gov/population/popclockworld.html (last accessed 15 November 2011).
16. Worldometers, *Real Time World Statistics*, http://www.worldometers.info (last accessed 22 March 2012).

I
Elements That Could Soon
Be in Short Supply

CHAPTER 1

Base Metals

GAVIN M. MUDD* AND ZHEHAN WENG

Environmental Engineering, Monash University, Clayton, VIC 3800,
Australia
*Email: gavin.mudd@monash.edu

1.1 INTRODUCTION

The availability and use of various metals have been so important
to the evolution and development of modern society that historians
name major human epochs after them – namely the Bronze, Iron
and Atomic Ages. Over several millennia, humanity has evolved
from using copper (Cu), gold (Au), bronze (a copper–tin alloy),
silver (Ag), lead (Pb), iron (Fe) and steel, nickel (Ni), aluminium
(Al), zinc (Zn) and uranium (U) to rare earths (to name the most
obvious metals) – with most of the advances occurring since the
dawn of the twentieth century. This growth in the use of metals has
occurred alongside evolution in technology and economic development, especially consumption patterns and most metals now
have active uses and demands.

The natural geological processes which form mineable metal
deposits, however, are extremely slow, effectively making the

Materials for a Sustainable Future
Edited by Trevor M. Letcher and Janet L. Scott
© The Royal Society of Chemistry 2012
Published by the Royal Society of Chemistry, www.rsc.org

deposits a non-renewable resource (within human time scales). Over time, new deposits can be found and exploited, but once depleted, they cannot be replaced. Depending on a metals use pattern, it may be readily recycled, although this is not always straightforward based on current technologies and economics.

Given these basic trends of growing demands and consumption of a non-renewable resource, the obvious issue is: for how long can the current patterns be sustained? This is not a trivial question at all and has been addressed by many scholars over many centuries – all with widely varying views. The German scholar Georgius Agricola in the mid-1500s, arguably the first to document 'modern' metal mining and smelting techniques in his book *De Re Metallica* (1556), recognised that mining often raises opposing views, noting that '... there has always been the greatest disagreement amongst men concerning metals and mining, some praising, others utterly condemning them ...'.[1] Conversely, the famous 'Club of Rome' study called *Limits to Growth* (1972) predicted that modern society could collapse by 2050, in large part due to growing pollution from resource over-exploitation and depletion.[2] Alternatively, the economist John Tilton argued that economic mineral resources are not a stationary, solitary figure, but rather a function of prevailing economic, technological, social and environmental constraints.[3] The debate is ongoing and has been covered further by many scholars.[4–10]

For this chapter, we will focus on the primary trends in base metals mining, specifically Cu, Pb, Zn and Ni. Although Sn is also a base metal, the four chosen are dominant in their economic value and global scale and also contribute to a wide range of critical uses such as pipes, electrical wiring, alloys, galvanising, and chemicals. The principal emphases are twofold – examining trends in economic mineral resources and mine production. These trends for Cu, Pb, Zn and Ni are then examined with respect to crucial sustainability issues in this century, especially energy and greenhouse gas emissions.

Overall, this chapter provides a unique insight into the dominant base metals (Cu, Pb, Zn, Ni), especially economic resources and current mine production, thereby allowing a valuable view of key historical trends, and the current state of affairs and trends that we might expect for coming decades. That is, the core issue of the resource sustainability of the base metals will be assessed.

1.2 THE BASE METALS AND THEIR USES

There is no strict definition or uniformly accepted definition of base metals. In general, the term 'base metals' is commonly related to the chemical behaviour of these particular elements, whereby they can readily oxidise and corrode. In addition, it is used to distinguish them from precious metals such as gold and silver (which do not readily oxidise or corrode) and ferrous metals related to iron (and ferroalloys which include manganese or chromium).

The uses of base metals are varied and include the following:[5]

- *Copper* – pipes and tubing (for gas, water, cables), machine parts (aircraft, automotive, manufacturing), electronics, tools, furniture, coins, jewellery, musical instruments, domestic and industrial appliances, chemicals.
- *Lead* – batteries, cable sheathing, electronics solder, radiation protection, munitions, plumbing, lead crystal.
- *Zinc* – galvanising, alloys, brass, batteries, water treatment, coins, zinc oxide chemicals (*e.g.* pigments, rubber, cosmetics, plastics, pharmaceuticals), zinc sulfide (used in luminous dials, X-ray machines, televisions, fluorescent lights).
- *Nickel* – stainless steel, corrosion-resistant alloys, turbines (gas, jets), plating, coins, catalysts, batteries.

Owing to these wide and varied uses, which are often closely associated with economic activity such as buildings and consumer products, the demand for base metals is commonly closely related to economic conditions. The per capita consumption of metals is also closely linked to economic indicators such as gross domestic product per capita.

1.3 BRIEF ECONOMIC GEOLOGY AND MINING OF BASE METALS

1.3.1 Economic Geology and Major Mineral Deposit Types

For a metal to be worth mining, it needs be present in sufficient concentration and size to be profitable. In general, base metals are not widely abundant in crustal rocks and are present at parts per million (or mg kg^{-1}) levels, but they can be enriched to high

Table 1.1 Average crustal abundance and typical base metal ore grades.

	Cu	*Pb*	*Zn*	*Ni*
Crustal rocks[a] mass/mg kg^{-1}	28	17	67	47
Typical ore grades/mg kg^{-1}	2000–30 000	2000–50 000	2000–150 000	3000–30 000
Typical ore grades/%	0.2–3	0.2–5	0.2–15	0.3–3
Factor	∼70–1070	∼120–2940	∼30–2240	∼60–640

[a]Source: ref. 11.

concentrations (i.e. %) *via* a range of geological processes to form economic orebodies, as shown in Table 1.1.

The processes of orebody formation are a major scientific and professional field, that of economic geology, and the brief outline below is barely an introduction to the topic.[12-15] Another crucial factor to consider is the mineralogy involved, since this strongly influences how easily the metals are extractable from the ore and purified through to a refined metal form (typically >99.9% purity).

In general (but at the risk of unfair over-simplification), the major geological processes which form ore deposits include hydrothermal processes, granite intrusions, magmatic formation, volcanism, metamorphism, sedimentary formations and meteorite-related impacts. A combination of processes can also be important, such as hydrothermal solutions in sedimentary environments or metamorphism of previous mineralisation. It is also common for some metals to occur together, such as Cu–Ni, Pb–Zn or Pb–Zn–Cu (mainly related to their individual elemental geochemistry and the primary processes of orebody formation), along with precious metals, such as Au, Ag and platinum group metals (PGMs), although in widely varying concentrations from 0.1 to 5 g t^{-1} Au, <1 to 500 g t^{-1} Ag and <0.2 to 10 g t^{-1} PGMs.

Copper is dominantly found in mineral deposits broadly classified as porphyry, sediment-hosted and volcanic massive sulfides (VMS) and together these ore types accounted for ∼90% of copper production throughout most of the twentieth century.[16] Porphyry Cu deposits are formed by igneous intrusions which introduce hydrothermal alteration and precipitate Cu, often in association with Au, Ag and/or molybdenum (Mo), with grades often ranging from 0.2 to 1% Cu, but sizes reaching up to billions of tonnes. Sediment-hosted Cu deposits are formed by hydrothermal solutions released into a water reservoir, commonly the ocean, leading to the formation of metal-rich layers. These are also known as sedimentary exhalative (SedEx) deposits, with common grades of 1–3% Cu and sizes from tens to hundreds of millions of tonnes. VMS Cu deposits, as the name implies, are closely related to volcanic activity, whereby hydrothermal solutions derived from volcanism, generally on the ocean floor, lead to the formation of sulfide minerals rich in Cu and commonly associated with Pb, Zn, Au and Ag. Grades can be rich, ranging from 1 to 10% Cu, with highly variable grades for other metals, but sizes are commonly smaller and of the order of tens of millions of tonnes, although there are often many such deposits in a region. Other important types of Cu deposits include iron oxide copper–gold (IOCG), magmatic (or basalt-related) Ni–Cu ores, and also oxide deposits (*i.e.* weathered ores) and native (elemental) Cu ores.

Nickel is dominantly found in komatiitic ore deposits, weathered laterites, magmatic-related deposits and norite (meteorite-related) deposits. Komatiitic deposits are referred to as Kambalda type, named after the field in Western Australia where they were first found in 1966, and are formed from enrichment of Ni sulfide in the komatiitic lava. The grades are often rich, ranging from 2 to 4% Ni, but individual ore lenses may be only several million tonnes, although there may be a large number of these lodes in a region. Grades of Cu and cobalt (Co) are typically low, about 0.2% Cu and 0.02% Co. Laterite Ni deposits are formed from the intense weathering over millions of years of ultramafic rocks, leading to progressive concentration of Ni and Co in surficial soils. Typical grades are 1–2% Ni and 0.02–0.2% Co, with sizes ranging from tens to hundreds of millions of tonnes. The dominant magmatic deposits are related to tectonic rift structures, allowing concordant intrusive sheets to deposit sulfides rich in Ni, Cu, Co and often PGMs. Grades can vary from 1 to 3% Ni, while Cu is more variable at 0.5–5%, Co is about 0.02% but PGMs are extremely variable, ranging from <0.2 up to $15\,\mathrm{g\,t}^{-1}$. The impact of meteorites can lead to severe geological disturbances and significant hydrothermal processes, forming sulfides rich in Ni, Cu and often Fe, Co, PGMs and other metals at similar grades to magmatic ores. The rich PGM ores of South Africa's Bushveld igneous complex and Zimbabwe's Great Dyke often contain low Ni grades containing 0.02–0.1% Ni.

The dominant type of Pb–Zn deposit is the SedEx type, similar to Cu, with VMS and metamorphic (skarn) deposits also playing an important role. Typical grades for all types are 0.2–5% Pb and 0.5–15% Zn, showing that most deposits are zinc dominant, with other important metals including silver at grades of 5–200 g t^{-1} and highly variable Cu at 0.1–3%. Other metals often present include indium (In), germanium (Ge) and cadmium (Cd), although they are typically very low in value compared with Pb–Zn–Ag (although with growing In and Ge use, this is changing).

For most ore types, the base metals are present as sulfide minerals, with the most important economic minerals including chalcopyrite ($CuFeS_2$), chalcocite (Cu_2S), bornite (Cu_5FeS_4), galena (PbS), pentlandite [(Fe,Ni)S] and sphalerite (ZnS). For oxide deposits, the major minerals include garnierite [Ni_3Mg-$Si_6O_{15}(OH)_2 \cdot 6(H_2O)$] and mixed minerals of complex magnesium silicates or oxides (*e.g.* peroditite, limonite) in Ni laterites, malachite [$CuCO_3 \cdot Cu(OH)_2$], azurite [$2CuCO_3 \cdot Cu(OH)_2$] and chrysocolla ($CuSiO_3 \cdot 2H_2O$) in Cu oxide ores, cerussite ($PbCO_3$) and anglesite ($PbSO_4$) for Pb and willemite (Zn_2SiO_4) for Zn. The processing of ores will also depend as much upon the major mineralogy of the host rock or gangue mineralogy, such as carbonates, iron sulfides (*e.g.* pyrite, FeS_2), hematite (Fe_2O_3) and silicates, or other impurities which could interfere with processing or product quality (especially toxic elements such as mercury and arsenic).

1.3.2 Mining and Metal Extraction

In general, there are four primary stages involved in base metals mining and extraction: (1) mining; (2) milling; (3) smelting; and (4) refining. This remains the dominant pathway for most base metal production worldwide, known as the pyrometallurgical route owing to the use of smelting. There are also variations, such as heap leaching, which involves large piles of ore being leached with sulfuric acid and the solutions being processed by solvent extraction and electrowinning, as used in the refining stage, with the overall known process known as hydrometallurgical extraction. A diagrammatic view of a typical modern mine site is shown in Figure 1.1.

The principal mining methods are alluvial, underground and open cut mining. In general, underground and open cut mining are

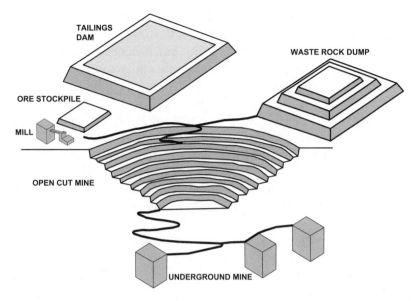

Figure 1.1 Diagrammatic view of a typical modern mining-milling complex.[17]

the predominant forms of mining for base metals, and alluvial mining is rare (and typically practiced only by small-scale or artisanal miners).

The ore is then crushed and ground to a fine powder (a stage known as comminution), which increases the reactive surface area and facilitates more economic extraction in a mill. The dominant milling method is froth flotation, whereby reagents such as xanthates act as 'collectors' and oils or fatty acids as frothing agents are added to a slurry mixture of ground ore and water and air bubbled through it. The base metal sulfides stick to the collector and adhere to the air bubbles, 'floating' to the top with gangue minerals sinking to the bottom – thereby concentrating the sulfides and their contained metals by between 10 and 50 times the original ore grade. Flotation reduces the volume of material for smelting considerably, making smelting more economical due to reduced energy requirements and also lower emissions and pollution loads.[18,19] Perhaps most importantly, however, is that flotation allows the production of individual metal concentrates, especially the separation of Pb and Zn (for most ores).

Smelting involves the thermal treatment of concentrates to produce a metal-rich matte, usually with coking coal. For Cu and

Ni sulfide concentrates, the four major smelter types are the reverberatory furnace, electric furnace, flash smelter and Noranda-type smelter, with the first two being batch operated whereas the last two are continuous and thereby inherently more efficient. In flash smelters, the energy released during the oxidation of the sulfides is used as part of the process and this also generates a high concentration of sulfur dioxide (SO_2) in the exhaust gases, which can be captured and converted to sulfuric acid and sold. For Pb concentrates, the three smelter types include blast furnaces (*i.e.* iron/steel technology), reverberatory furnaces and Imperial smelters. For Zn concentrates, they are generally processed by leaching and electrolysis, with a small quantity also processed by Imperial smelters.

For Ni laterites, either pyro- or hydrometallurgical processing can be used, depending on the laterite ore types being processed (*e.g.* limonite or saprolite). The principal pyrometallurgical process uses rotary kiln electric furnaces (RKEFs), whereas hydrometallurgical processes can be carried out using high-pressure acid leaching (HPAL) or ammonia leaching (Caron process), followed by solvent extraction and electrowinning.

There are two issues of growing importance in the smelting and refining stages – controlling pollution emissions and their localised impacts on the environment (especially SO_2 and particulates), and also the extraction of increasingly valuable by-products such as Cd, Co, Ga, Ge, In, rhenium (Re), tellurium (Te), selenium (Se) and others.

1.3.3   Historical Production Trends

There is a wide variety of data sources for the historical global production of base metals.[20–24] Long-term production is shown in Figure 1.2 with an inset of cumulative metal production, including by country from 1990 to 2010 in Figure 1.3.

The long-term trends for production for all metals are generally increasing, with Cu the fastest. The only anomaly in these trends is Pb, which peaked in the mid-1970s and then declined before beginning to grow again from 1995. This is related to the gradual removal of Pb from petrol, paints and other consumer products (as it posed significant risks to human health and especially children), and also improved recycling programmes for Pb acid batteries.

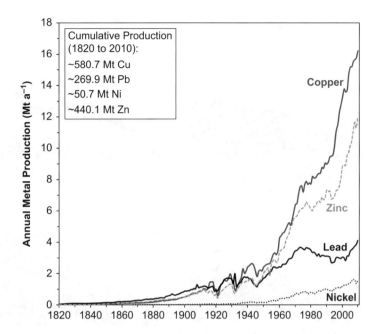

Figure 1.2 Historical global production trends for Cu, Pb, Ni and Zn.

A surprising observation from Figure 1.3 is that although China is a moderate Cu and Ni producer, it is increasingly dominant in Pb and in Zn with 42.7% and 29.2%, respectively, in global production of each metal in 2010. It is a widely held perception that China is primarily an importer with only minor to moderate production – which is clearly not true for Pb–Zn.

1.4 ECONOMIC RESOURCES OF COPPER, LEAD–ZINC AND NICKEL

1.4.1 Formal Reporting of Economic Ore Reserves and Mineral Resources

There are a variety of common terms used to describe or quantify economic mineral resources, including some that have statutory significance. A mineral resource can, at its most basic, be considered as something that can generate inherent value to society, with examples including steel structures, copper electrical wiring, gold jewellery and lithium for energy storage. A mineral resource is identified through geological exploration and, when profitable, can

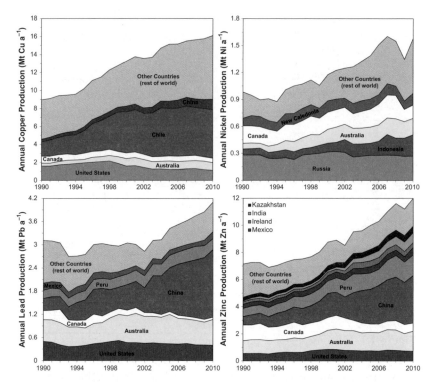

Figure 1.3   Recent global production trends 1990–2010 by country for Cu (top left), Pb (bottom left), Ni (top right) and Zn (bottom right).

be mined to produce a given mineral or metal. The challenge, therefore, is to ascertain and describe which deposits constitute a potentially profitable mineral resource. This determination can vary due to market conditions (*e.g.* price fluctuations), input costs (*e.g.* fuels, labour), ore processability (*e.g.* how easily the minerals can be extracted) or social issues (*e.g.* bans on mining in national parks, opposition to uranium mining).

Given the complexity of defining a mineral resource as profitable and the need to provide clear communication of such results to the public and investors (since most mining companies are publicly listed on their respective national stock exchanges), the Australian mining industry established the Joint Ore Reserves Committee (JORC) Code for reporting mineral resources.[25,26] Any mining company listed on the Australian stock exchange is required by law to use the JORC Code to report on the mineral resources they control. There are equivalent codes in other major mining countries such as Canada (National Instrument 43-101[27]) and South Africa (SAMREC[28]).

The two primary aspects which the JORC and related codes consider are geological and economic probability in claiming a mineral resource as profitable. A range of important 'modifying factors' are compulsory to consider, such as mining, metallurgical, economic, marketing, legal, environmental, social and governmental factors. There are two primary categories used to classify a mineral deposit – ore reserves and mineral resources. The typical distinction is that ore reserves have a very high economic and geological probability of profitable extraction, whereas mineral resources have a reasonable geological probability but are less certain economically. Concise definitions, as used by the JORC Code, are the following:

- *Ore Reserves* – assessments demonstrate at the time of reporting that profitable extraction could reasonably be justified. Ore Reserves are subdivided in order of increasing confidence into Probable Ore Reserves and Proved Ore Reserves.
- *Mineral Resources* – the location, quantity, grade, geological characteristics and continuity of a mineral resource are known such that there are reasonable prospects for eventual economic extraction, although not all modifying factors have been assessed and hence some uncertainty remains. Mineral Resources are subdivided, in order of increasing geological confidence, into Inferred, Indicated and Measured categories.

The United States Geological Survey (USGS) uses the categories of reserves and reserve base (see ref. 21, 2009 Edition). These are broadly similar to JORC's ore reserves and mineral resources, respectively, although the USGS appears to allow for greater inclusion of inferred mineral resources in the reserve base category. It should be noted that reserves base includes reserves, although the USGS stopped reporting reserves base from 2009. An excellent analysis of JORC and its comparison to other systems is referenced here.[29]

For this chapter, an extensive data set of mineral resources by individual project/deposit is compiled for each of the base metals, as reported under the JORC and related codes (using 2010 data or the most recent report). These data sets should be considered a reliable geological estimate (as of ~2010), since all reserves and resources are reported under statutory codes. Whether all projects proceed to production, however, is dependent on economics, mining conditions, processing characteristics and site-specific environmental issues (especially land use, water, mine waste management and energy); amongst others.

1.4.2 Global Status of Base Metal Reserves: USGS Estimates

The USGS presents a global estimate of reserves for most mineral commodities annually in their publication *Minerals Commodity Summaries.*[21] Although the data are approximate, they are sufficient to illustrate countries with significant mineral resources, as summarised in Table 1.2 for base metals.

Table 1.2 Compilation of 2010 USGS global base metal reserves by country (mass metal/Mt), sorted by sum of mass of Cu + Pb + Zn + Ni.

Country	Cu	Pb	Zn	Ni	Sum
Australia	80	27	53	24	184
Chile	150				150
Peru	90	6	23		119
China	30	13	42	3	88
Mexico	38	5.6	15		58.6
USA	35	7	12		54
Russia	30	9.2		6	45.2
Kazakhstan	18		16		34
Indonesia	30			3.9	33.9
Poland	26	1.5			27.5
Zambia	20				20
Canada	8	0.65	6	3.8	18.45
India		2.6	11		13.6
Brazil				8.7	8.7
Bolivia		1.6	6		7.6
New Caledonia				7.1	7.1
Cuba				5.5	5.5
South Africa		0.3		3.7	4
Ireland		0.6	2		2.6
Columbia				1.6	1.6
Madagascar				1.3	1.3
Sweden		1.1			1.1
Philippines				1.1	1.1
Dominican Republic				0.96	0.96
Botswana				0.49	0.49
Venezuela				0.49	0.49
Miscellaneous	80	4	62	4.5	150.5
Total	**~635**	**~80**	**~250**	**~76**	**~1040**

1.4.3 Copper

National estimates of economic Cu resources over time for selected countries are shown in Figure 1.4. A comprehensive compilation of Cu resources by deposits and countries around the world is summarised in Table 1.3, with the 20 largest Cu deposits in Table 1.4.

For Australia, an additional 46.0 Mt Cu (to the figure of 85.6 Mt Cu in Figure 1.4) is reported as sub-economic or marginal resources,[32] giving Australia a total of 131.6 Mt Cu of identified resources. Over time, it is common for identified resources to be upgraded to economic status after further drilling, metallurgical and other studies and be developed for production,[17,33] showing that the higher resource figure is of importance for long-term planning or modelling of Cu production scenarios. The Olympic Dam deposit, at 79.0 Mt Cu, represents ~60% of Australia's identified Cu resources, showing the importance of super-giant deposits.

Chile is clearly in a dominant position in global Cu resources, with ~39% of global Cu resources based on our data and nine of the top 20 deposits; 627.2 Mt Cu also contrasts with the USGS reserves estimate of 150 Mt Cu (and 2008 reserves base estimate of 360 Mt Cu).

The USGS 2010 global estimate of Cu reserves is ~635 Mt Cu, whereas the 2008 reserves base estimate was 1000 Mt Cu, yet the 674 deposits compiled in this study represent 1614.9 Mt Cu, ~60% higher. Our data, however, although certainly comprehensive, are not exhaustive, as they doe not include any Cu deposits in China or

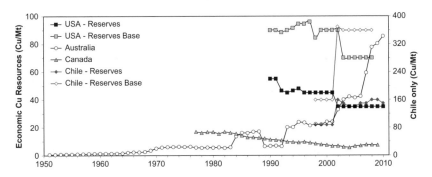

Figure 1.4 Historical trends in economic Cu resources for Australia, Canada, Chile and the USA. Data from refs 17, 21, 30, and 31.

Table 1.3 Compilation of global copper resources by country (Cu mass/Mt).

Country	No. of deposits	Ore/Mt	Cu/%	Cu/Mt
Chile	47	115 102.5	0.54	627.23
Peru	39	27 851.9	0.48	134.47
USA	48	32 785.6	0.41	134.20
Australia	150	20 160.5	0.63	126.71
Democratic Republic of the Congo	21	2288.9	2.34	53.64
Mexico	29	16 442.6	0.32	53.17
Indonesia	5	7458.9	0.67	50.07
Zambia	19	4524.5	1.03	46.74
Canada	78	13 093.7	0.33	42.62
Russia	12	4128.3	0.95	39.13
Mongolia	6	4485.0	0.87	39.09
Kazakhstan	7	6682.0	0.47	31.22
Poland	4	1539.0	2.00	30.77
Papua New Guinea	12	5333.8	0.49	26.08
Philippines	13	5363.3	0.47	25.10
Argentina	6	6451.5	0.39	25.00
Pakistan	1	5867.8	0.41	24.06
Panama	1	6465.0	0.30	19.36
South Africa	39	14 610.3	0.08	11.66
India	14	1394.4	0.82	11.42
Brazil	8	2331.3	0.48	11.29
Iran	1	1200.0	0.7	8.4
Fiji	1	2287.7	0.35	8.01
Botswana	9	1287.7	0.51	6.59
Sweden	18	2607.6	0.23	5.91
Finland	31	2197.7	0.19	4.28
Laos	4	399.8	0.80	3.20
Zimbabwe	4	2138.5	0.11	2.41
Portugal	1	67.2	2.75	1.85
Namibia	9	175.2	0.80	1.41
Spain	3	33.3	3.48	1.16
Saudi Arabia	3	81.3	1.38	1.12
Eritrea	4	94.2	1.13	1.06
Greece	2	192.2	0.55	1.06
Thailand	2	200.0	0.51	1.02
Various	23	1551.4	0.28	4.38
Total	**674**	**318 875**	**0.51**	**1614.9**

other countries with known sizeable Cu resources (*e.g.* the data used for Sarcheshmeh in Iran being several years old, as well as several smaller sites). As of the late 1990s, China had identified 913 Cu deposits containing 73.7 Mt Cu, although only one-third of these resources were considered economic.[34] Furthermore,

Table 1.4 Top 20 copper resources by deposit/project (Cu mass/Mt).

Deposit/project	Status[a]	Ore/Mt	Cu/%	Au/g t^{-1}	Cu/Mt	Au/t	Other metals
Andina, Chile	Op	19 162	0.59		113.63		
El Teniente, Chile	Op	16 756	0.56		93.50		
Olympic Dam, Australia	Op	9075	0.87	0.32	78.95	2904	U–Ag
Collahuasi, Chile	Op	9554	0.81		77.54		
Chuquicamata, Chile	Op	10 497	0.55		57.31		
Escondida, Chile	Op	8509	0.61		52.13		
Grasberg, Indonesia	Op	4946	0.81	0.71	40.21	3493	Ag
Pebble, USA	Dep	10 777	0.34	0.31	36.56	3337	
Taimyr Peninsula, Russia	Op	2188	1.45	0.22	31.74	486	Ni–PGMs
Los Pelambres, Chile	Op	5818	0.53	0.04	30.84	233	Mo
Los Bronces, Chile	Op	6420	0.44		28.39		
Buenavista del Cobre, Mexico	Op	8388	0.33		27.80		
Radomiro Tomic, Chile	Op	7247	0.37		26.67		
Reko Diq, Pakistan	Dep	5868	0.41	0.22	24.06	1291	
Resolution, USA	Dep	1624	1.47		23.87		Mo
Hugo Dummett North, Mongolia	Dev	1426	1.39	0.35	19.81	492	
Cobre Panama, Panama	Dep	6465	0.30	0.05	19.36	342	Ag
Toquepala, Peru	Op	5029	0.36		18.18		
Los Sulfatos, Chile	Dep	1200	1.46		17.52		
Cerro Verde, Peru	Op	4038	0.41		16.72		
Total		**144 987**	**0.58**		**834.8**	**12 578**	

[a]Op, operating; Dev, under development; Dep, deposit.

Canada's national estimate for economic Cu resources in 2009 was 7.29 Mt,[30] yet our data suggest 42.6 Mt Cu. This large discrepancy is mainly related to the fact that Natural Resources Canada (NRC) only includes resources at operating mines or those undergoing development in national estimates and excludes mineral resources at other known deposits. Similarly, the USGS estimates the USA's Cu reserves at 35 Mt and the 2008 reserves base at 70 Mt Cu, whereas our data suggest 134.2 Mt Cu.

An important aspect to note is that of the Cu resources compiled, over half (834.8 Mt Cu) are contained in the 20 largest deposits alone, with these deposits also containing substantial Au (and sometimes Ag and Mo). In addition, most of these deposits are already being mined, meaning that future expansion of Cu production mainly needs to come from expansion at existing mines (*i.e.* brownfields growth), not merely new projects alone (*i.e.* greenfields).

Finally, the trends over time in Figure 1.4 suggest gradual declines in Cu resources for the USA and Canada, compared with strong (almost exponential) growth for Australia. Although it is tempting to label the declines as 'peak copper', when contrasting the data contained in Figure 1.4 with Table 1.3 (and noting the discussion above), it is abundantly clear that the USGS estimates of various national Cu resources are a significant under-estimate of identified Cu resources. For example, the recent (~ 2000) discovery of the giant but deep Resolution Cu deposit in Arizona in an area of intensive historical Cu mining (the Magma Cu field) shows that there remain excellent prospects for exploration to continue to find new deposits, although most likely at greater depth than current and previous Cu mines. Indeed, Australia has continued to find major new deposits over the past 30 years, such as Prominent Hill, Rocklands, Cadia-Ridgeway, Northparkes, Nifty and, most recently, Carrapateena, in addition to ongoing expansion at existing projects (*e.g.* Mt Isa, Mt Lyell, Olympic Dam).

With a global production of ~ 16 Mt Cu in 2010 and economic resources of more than 1600 Mt Cu, this allows for production growth for some decades yet. For comparison, global cumulative production from 1820 to 2010 was ~ 580.7 Mt Cu.

1.4.4   Lead–Zinc

A comprehensive compilation of Pb–Zn resources by deposits and countries around the world is summarised in Table 1.5, with the 20 largest Pb–Zn deposits in Table 1.6. Trends over time in national estimates for some countries are shown in Figure 1.5.

The USGS estimates of global reserves of Pb and Zn are 80 and 250 Mt, respectively, which contrasts with our estimates of 106.8 and 317.8 Mt, respectively. Similarly, Canadian and Australian national estimates of economic Pb/Zn resources are 0.45/4.25 Mt and 34.7/65.2 Mt, respectively, with an additional 25.7/28.8 Mt Pb/Zn of sub-economic resources in Australia, whereas our data suggest 4.63/22.05 and 46.26/79.05 Mt Pb/Zn, respectively.

Table 1.5 Compilation of global lead–zinc–silver resources by country (sorted by Pb + Zn mass/Mt).

Country	No. of deposits	Ore/Mt	Pb/%	Zn/%	Ag/g t⁻¹	Cu/%	Pb/Mt	Zn/Mt	Ag/t	Cu/Mt	(Pb + Zn)/Mt
Australia	91	1528.1	3.03	5.17	51.4	0.20	46.26	79.05	78 569	3.10	125.31
Peru	26	5673.3	0.14	0.66	11.2	0.58	7.70	37.50	63 520	33.05	45.20
India	7	300.8	2.32	9.08	>0.2	—	6.98	27.32	73	0.05	34.29
Mexico	19	3021.3	0.27	0.84	33.8	~0.15	8.03	25.28	10 2148	4.52	33.30
Iran	2	412	1.73	5.24	36.1	—	7.11	21.59	14 886	—	28.70
Canada	36	505.1	0.92	4.36	29.0	0.78	4.63	22.05	14 646	3.96	26.67
Kazakhstan	8	1809.3	0.22	1.06	~6	~0.5	4.02	190.09	10 707	9.46	23.11
USA	8	163.8	2.40	8.71	40.8	0.05	3.94	14.26	6676	0.08	18.20
South Africa	4	291.5	0.88	5.31	0.0	0.15	2.56	15.48	0	0.45	18.04
Portugal	1	186.1	1.00	3.93	52.4	1.26	1.86	7.31	9747	2.34	9.17
Sweden	3	109.1	2.27	5.83	117.7	1.62	2.48	6.36	12 849	1.77	8.84
Finland	3	1589	—	0.50	0.1	0.14	0.00	7.94	151	2.26	7.94
Greenland	6	727.4	0.10	~0.84	0.1	—	0.75	6.11	79	—	6.86
Poland	2	108.0	2.08	3.59	—	—	2.25	3.88	0	—	6.12
Ireland	4	59.8	1.52	8.09	>0.3	—	0.91	4.84	19	—	5.75
Indonesia	1	20.1	8.20	13.70	9.9	—	1.65	2.75	199	—	4.40
Algeria	1	68.6	1.10	4.57	0.0	—	0.75	3.14	0	—	3.89
Greece	2	18.41	4.87	6.52	144.7	—	0.90	1.20	2664	—	2.10
Eritrea	4	94.16	—	2.15	22.3	1.13	—	2.02	2097	1.06	2.02
Various	27	461.3	0.86	2.30	67.6	0.27	3.98	10.63	31 199	1.24	14.61
Total	**255**	**17 147**	**0.62**	**1.85**	**20.4**	**0.37**	**106.8**	**317.8**	**350 200**	**63.3**	**424.5**

Table 1.6 Top 20 lead–zinc–silver resources by deposit/project and country (sorted by Pb + Zn mass/Mt).

Deposit/project, country	Status[a]	Ore/Mt	Pb/%	Zn/%	Ag/g t^{-1}	Cu/%	Pb/Mt	Zn/Mt	Ag/t	Cu/Mt	(Pb + Zn)/Mt
McArthur River, Australia	Op	175.1	4.67	10.66	47		8.18	18.66	8154		26.84
Mehdiabad, Iran	Dep	394	1.6	4.2	36		6.30	16.55	14 184		22.85
Mt Isa–George Fisher North, Australia	Op	154.6	4.27	8.50	68		6.60	13.14	10 442		19.74
Rampura Agucha, India	Op	120.36	1.96	13.83			2.36	16.64			19.01
Peñasquito (mill only), Mexico	Op	1742.8	0.26	0.64	24.6		4.47	11.09	42 935		15.56
Gamsberg North/East, South Africa	Dep	186.4		6.89				12.84			12.84
Red Dog, USA	Op	62.5	4.19	16.21			2.62	10.13			12.75
Shalkiya, Kazakhstan	Op	274	0.76	3.09			2.09	8.47			10.56
Mt Isa–George Fisher South, Australia	Op	71.5	5.85	8.66	126		4.18	6.19	9012		10.38
Neves-Corvo, Portugal	Op	186.13	1.00	3.93	52.4	1.26	1.86	7.31	9747	2.34	9.17
Antamina, Peru	Op	1934.3	0.00	0.44	10.7	0.84	0.00	8.42	20 697	16.29	8.42
El Brocal (San Gregorio), Peru	Op	75.94	2.26	8.06	12.4		1.72	6.12	945		7.84
Talvivaara, Finland	Op	1550		0.49		0.13		7.60		2.08	7.60
Dugald River, Australia	Dep	53.0	1.85	12.46	36.4		0.98	6.60	1929		7.59
Cannington, Australia	Op	72	6.5	3.4	241		4.68	2.45	17 352		7.13
Cerro de Pasco, Peru	Op	165.96	1.12	2.66	98.0	0.12	1.86	4.41	16 271	0.21	6.27
Angouran, Iran	Op	~18	~4.5	~28	~40		0.81	5.04	702		5.85
Sindesar Khurd, India	Op	60.82	3.33	5.26	67.29		2.02	3.20	5625		5.22
East Region, Kazakhstan	Op	83.60	1.22	4.78	35.2	2.34	1.02	4.00	1302	1.96	5.02
Century–Century East, Australia	Op	37.0	1.5	11.8			0.56	4.37			4.93
Total		**7418**	**0.71**	**2.34**	**21.5**	**~0.3**	**52.3**	**173.2**	**159 298**	**22.9**	**225.5**

[a]Op, operating; Dev, under development; Dep, deposit.

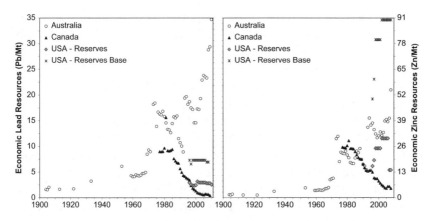

Figure 1.5 Historical trends in economic Pb–Zn resources for Australia, Canada and the USA. Data from refs 17, 21, 30, and 31.

As with Cu, our data virtually exclude China, with only one deposit with publicly reported data (curiously operated by an Australian miner). China has abundant and widespread Pb–Zn resources, with 45.36 and 110.08 Mt Pb and Zn, respectively,[34] with one-quarter and one-third of the Pb and Zn resources considered as economic. Typical ore grades are 0.5–2% Pb and 1–5% Zn, although some higher grade deposits were known, up to 5% Pb or 15% Zn.

For Australia, the strong growth over time is largely due to expansion of known deposits, since most of the deposits currently known were discovered by the 1970s, with few later discoveries. The two exceptions are the large and rich Century (Zn-dominant) and Cannington (Ag–Pb-dominant) deposits in western Queensland, which were both discovered in the early 1990s. Some large deposits remain undeveloped, such as at Dugald River and at Admiral Bay, whereas there are active expansion plans being assessed and/or developed at historic producers such as Broken Hill, McArthur River and Mt Isa (the last could include a large open cut, which might involve relocation of a significant portion of the Mt Isa township).

In contrast, Canada appears to have had very little brownfields or greenfields exploration success, except at Kidd Creek in Ontario and some Cu–Zn fields in Quebec. Canadian Pb–Zn ore resources are typically moderate in size and of lower grade than their international cousins.

The extent of our data for the USA is also very small, with eight resources containing 3.94/14.3 Mt Pb/Zn, compared with USGS data of 7/12 Mt Pb/Zn. The giant and high-grade Red Dog deposit in Alaska is dominant in our data set, with 2.62/10.1 Mt Pb/Zn. Unfortunately, the prolific Missouri Pb–Zn mining district is absent in our data, since the operating companies do not readily report economic resources (and production) data to the same extent as their Canadian and Australian counterparts.

Overall, like Cu, there are certainly a range of large and/or high-grade Pb–Zn resources known globally. Specifically, global production in 2010 was $\sim 4.1/\sim 12$ Mt Pb /Zn, compared with more than 105.9/312.8 Mt Pb/Zn. Therefore, given the lower ratio of resources to annual production, there are more constraints to long-term production growth for Pb–Zn than Cu. For comparison, global cumulative production from 1820 to 2010 was $\sim 269.9/440.1$ Mt Pb/Zn.

1.4.5 Nickel

There are two primary types of nickel deposits, sulfides and laterites, and a detailed review of historical global trends and production has been published.[35] In essence, Ni sulfide ores are easier to treat using pyrometallurgy, whereas Ni laterite ores require more intensive processing through either pyro- or hydrometallurgical techniques. The compiled Ni resources are therefore split between these two dominant ore types, and Ni sulfide resources by deposits and countries around the world are summarised in Table 1.7 and the 20 largest Ni sulfide deposits in Table 1.8, and Ni laterite resources by deposits and countries around the world are summarised in Table 1.9 and the 20 largest Ni laterite deposits in Table 1.10. Trends over time for some countries are shown in Figure 1.6.

The USGS 2010 global estimate of Ni reserves is ~ 76 Mt Ni, whereas the 2008 estimate of the global Ni reserves base was 150 Mt Ni, yet the 366 deposits compiled in this study represent 259.7 Mt Ni. The dominant resource type is clearly laterites, with 169.2 Mt Ni, compared with 90.5 Mt Ni in sulfides. Our data, however, although certainly comprehensive, are not completely exhaustive, as they only include three Ni deposits in China (the large Jinchuan Ni–Cu–Co–PGMs resource (~ 2005[36]) and two very small Mn–Ni–Co deposits). As of the late 1990s, China had

Table 1.7 Compilation of global nickel sulfide resources by country (Ni + Cu mass/kt).

Country	No. of deposits	Ore/Mt	Ni/%	Cu/%	Co/%	Ni/kt	Cu/kt	Co/kt
Russia	7	3502.8	0.67	1.02	0.003	23 423	35 566	119.4
South Africa	25	9657.9	0.17	0.07		16 757	7099	81.5
USA	9	3887.9	0.15	0.45	0.005	5716	17 550	181.5
Canada	44	4603.8	0.35	0.11	0.014	16 265	4984	645.8
Australia	54	2034.8	0.57	0.11	0.014	11 591	2189	275.1
China	3	435.2	1.38	0.87		6002	3800	
Finland	25	2178.4	0.22	0.18	0.018	4774	4029	399.1
Botswana	4	477.2	0.25	0.23		1178	1112	
Brazil	4	303.5	0.60	0.15	0.014	1817	466.7	42.8
Tanzania	2	75.5	2.20	0.30	0.158	1657	223.5	118.9
Zimbabwe	7	326.2	0.26	0.10		854.3	340.1	
Vietnam	1	21.8	0.75	0.13	0.09	163.1	28.3	19.6
Zambia	3	20.6	0.56	0.10	0.121	115.0	21.5	24.9
Spain	1	13.0	0.56	0.46		72.9	59.5	
Total	**200**	**27 561**	**0.33**	**0.28**	**~0.007**	**90 494**	**77 514**	**2025**

identified 83 Ni 'districts' containing 7.77 Mt Ni, with 3.70 Mt Ni considered to be economic.[34] Most of China's Ni resources are magmatic-related sulfide deposits, except for the Mojiang laterite resource (0.53 Mt Ni). In contrast, Canada's 2009 national estimate for economic Ni resources was 3.30 Mt Ni,[30] yet our data suggest 16.3 Mt Ni, the difference being due to the inclusion of operating mines and projects under development only by NRC.

Similarly for Cu, the largest 20 laterite resources account for almost half (47.0%) of the global total, whereas the largest 20 sulfide resources are dominant at 70.6% of the global total. For both ore types, about half of the 20 largest resources are already being mined, while the remainder are split equally between being developed or remaining as a deposit. Hence there is more room for greenfields expansion from new projects than for Cu, and also possible brownfields expansions. The longer term future of the Ni sector will be increasingly dominated by laterites, despite historically being dominated by sulfides (see Section 1.5.3).

Finally, the trends over time in Figure 1.6 suggest gradual declines in Ni resources for Canada, compared with strong (almost exponential) growth for Australia. Although this again could make it tempting to label Canada's decline as 'peak nickel', on contrasting the data contained in Figure 1.6 with Table 1.7 (and noting

Table 1.8 Top 20 nickel sulfide resources by deposit/project (Ni + Cu mass/kt).

Deposit, country	Status[a]	Ore/Mt	Ni/%	Cu/%	Co/%	Pt/g t^{-1}	Pd/g t^{-1}	Ni/kt	Cu/kt	Co/kt
Talnakh-Norilsk, Russia	Op	2188.4	0.77	1.45		1.01	3.72	16783	31740	
Jinchuan, China	Op	432	1.39	0.88	0.025			6000	3800	108.0
Mogalakwena, South Africa	Op	2779.9	0.18	0.11		0.94	1.05	5004	3058	
Nokomis-Duluth, USA	Dev	823.87	0.202	0.637	0.01	0.179	0.398	1667	5245	82.4
Mesaba, USA	Dep	1200	0.09	0.43				1080	5160	
Dumont, Canada	Dep	2105.3	0.27		0.011			5653		222.4
Kola, Russia	Op	546.3	0.63	0.30		0.02	0.04	3427	1633	
Turnagain, Canada	Dep	1841.8	0.21		0.013			3793		241.7
Kingashskoye-Verkhnekingashskoye, Russia	Dep	484.6	0.45	0.21	0.018		0.55	2162	1001	85.0
Talvivaara-Kuusilampi, Finland	Op	890	0.22	0.13	0.02			1958	1157	178.0
Spruce Road, USA	Dep	529.5	0.15	0.43				814	2268	
Sudbury (Vale Inco), Canada	Op	112.3	1.20	1.53	0.04	0.9	1.1	1348	1718	44.9
NorthMet, USA	Dev	807.1	0.078	0.268	0.0067	0.069	0.245	630	2162	53.8
Yakabindie, Australia	Dep	437	0.57					2491		
Talvivaara-Kolmisoppi, Finland	Op	660	0.23	0.14	0.02			1518	924	132.0
Nebo-Babel, Australia	Dep	392	0.3	0.3				1176	1176	
Santa Rita, Brazil	Op	285.5	0.60	0.16	0.015			1716	463	42.8
Sheba's Ridge, South Africa	Dep	764.3	0.19	0.07				1452	535	
Nkomati, South Africa	Op	407.7	0.35	0.13	0.02	0.20	0.25	1440	530	81.5
Kevitsa, Finland	Dev	275	0.31	0.41				843	1120	
Total		**17 963**	**0.34**	**0.35**	**~0.007**			**60 954**	**63 691**	**1273**

[a]Op, operating; Dev, under development; Dep, deposit.

Table 1.9 Global nickel laterite resources by country (Ni + Co mass/kt).

Country	No. of deposits	Ore/Mt	Ni/%	Co/%	Ni/kt	Co/kt
Australia	43	4064.1	0.74	0.052	30 172.0	2111.8
New Caledonia	11	1546.3	1.84	0.045	28 520.9	693.9
Indonesia	14	1615.2	1.61	0.038	25 999.9	613.0
Cuba	7	1287.1	1.20	0.081	15 482.2	1039.4
Brazil	19	1145.2	1.37	0.047	15 664.2	540.8
Philippines	22	1055.6	1.12	0.040	11 787.4	419.9
Cote d'Ivoire	1	293	1.46	0.11	4277.8	322.3
Russia	4	412.9	0.99	0.026	4090.7	107.9
Burundi	3	266.4	1.34	0.066	3579.3	175.9
Greece	3	300.0	1.07		3217.4	
Columbia	1	269	1.08		2903	
Cameroon	2	323.1	0.61	0.211	1977.9	683.0
Guatemala	3	173.4	1.44	0.054	2505.4	93.8
Madagascar	3	232.0	0.99	0.082	2286.8	190.3
Papua New Guinea	2	268	1.03	0.081	2769	218.0
Myanmar (Burma)	2	150	1.45		2169	
India	5	214.8	0.93	0.067	1993.0	143.3
Serbia	3	261.6	0.73	0.004	1910.1	9.4
Solomon Islands	2	160.7	1.04	0.088	1671.8	140.9
Kazakhstan	3	188.3	0.78	0.040	1471.3	75.9
Dominican Republic	2	86.9	1.53		1326.5	
Miscellaneous	11	336.5	1.03	0.043	3471.8	146.2
Total	**166**	**14 650**	**1.16**	**~0.053**	**169 248**	**7726**

the discussion above), it is abundantly clear that the USGS estimates of various national Ni resources are a significant underestimate of identified Ni resources. For example, there are two giant but low-grade Ni–Co sulfide deposits in Canada, Dumont in Quebec and Turnagain in British Columbia, which are being actively investigated for development. The Sudbury Basin of northern Ontario is still providing new discoveries, often PGM rich, although mostly at greater depths, and also new deposits in the Lynn Lake district of northern Manitoba. In Australia, almost all of the growth in Ni resources, including sub-economic resources, has come from ongoing exploration and developments in Ni laterites (discussed in more depth later), whereas sulfide discoveries appear to be merely replacing the Ni mined.[37]

Current global production of ~1.6 Mt Ni in 2010 and economic resources of much more than 260 Mt Ni clearly allow for production growth for some decades yet. For comparison, global cumulative production from 1820 to 2010 was ~50.7 Mt Ni.

Table 1.10   Top 20 nickel laterite resources by deposit/project (Ni + Co mass/kt).

Deposit, country	Status[a]	Ore/Mt	Ni/%	Co/%	Ni/kt	Co/kt
Kalgoorlie Group, Australia[b]	Dep	970.1	0.74	0.045	7218	435.7
Halmahera-Weda Bay, Indonesia	Dep	423.8	1.50	0.073	6343	310.2
Prony, New Caledonia	Op	400	1.4		5600	
Thio, New Caledonia	Op	208	2.4	0.1	4992	208.0
Sangaji, Indonesia[c]	Op	295.2[c]	1.63		4817	
Pinares de Mayari, Cuba	Op	400	1.1	0.1	4400	400.0
Tiuba-Biankouma, Cote d'Ivoire	Dep	293	1.46	0.11	4278	322.3
Punta Gorda, Cuba	Op	310	1.3	0.11	4030	341.0
Jacaré, Brazil	Dep	288.8	1.27	0.13	3660	375.4
Koniambo, New Caledonia	Dev	158.6	2.49		3946	
Doniambo Group, New Caledonia	Op	158.1	2.47		3903	
Gag Island, Indonesia	Dep	214.3	1.73	0.08	3708	171.4
Nakety, New Caledonia	Op	229	1.48	0.12	3389	274.8
San Felipe, Cuba	Dep	234	1.3		3042	
Murrin Murrin, Australia	Op	268	1.01	0.074	2707	198.3
Cerro Matoso, Columbia	Op	269	1.08		2903	
Mindoro, Philippines	Dev	315.1	0.83	0.06	2615	189.1
Sulawesi, Indonesia	Dep	162	1.63	0.08	2636	131.4
Weld Range (Ni), Australia	Dep	330	0.75	0.06	2475	198.0
Pujada, Philippines	Dep	200	1.3		2600	
Total		**6128**	**1.29**	**~0.058**	**79 263**	**3556**

[a]Op, operating; Dev, under development; Dep, deposit.
[b]This project is several deposits reported together (ranging from ~0.1 to 2 Mt Ni each).
[c]Reported as wet ore only.

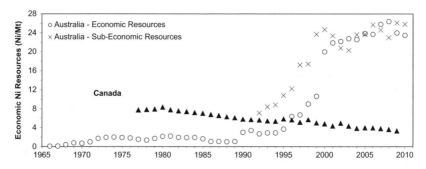

Figure 1.6   Historical trends in economic Ni resources for Australia and Canada. Data from refs 17, 30, 31 and 35.

1.5 STATISTICS OF BASE METAL MINES AROUND THE WORLD

1.5.1 Copper

The fortunes of the Cu sector have continually changed and evolved over time, moving from relatively small scale use (~ 50 kt y^{-1} of Cu around 1850) to large-scale application once the age of electricity began from the late 1800s. Annual Cu production soared to ~ 500 kt y^{-1} by 1900 and ~ 2500 kt y^{-1} by 1950 and reached a new record of 16.2 Mt in 2010 (see Figure 1.2).

Over this period, the industry has moved to predominantly open cut mines, in addition to varying ore types, as shown in Figure 1.7. The rise of the porphyry sulfide ores at the outset of the twentieth century coincided with surging demand for electrical applications, the emergence of some large-scale, low-grade but highly profitable open cut mines (*e.g.* Bingham Canyon in the USA and Mt Lyell in Australia, although both of the original sites are now dwarfed in scale by modern open cut mines) and new flotation technology for sulfide ores.

The trends in ore grade, over approximately the same period, are shown in Figure 1.8, demonstrating a long-term decline. For Australia, the rich ores of the mid-1800s, grading $\sim 15\%$ Cu, were oxide ores and at the surface, leading to easy mining and direct smelting at famous mines such as Burra Burra and Peak Downs. By the 1890s, however, the oxide ores were depleted and the

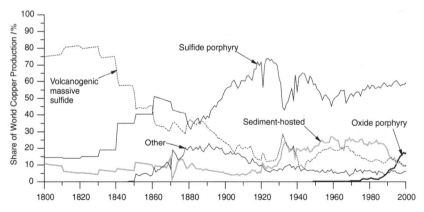

Figure 1.7 Historical trends in Cu ore types (based on global data).[16]

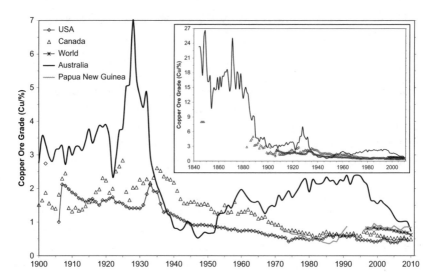

Figure 1.8   Historical trends in Cu ore grades for Australia, Canada, Papua New Guinea, USA and the World. Data updated from refs 17 and 38 and references therein.

remaining resources were dominantly sulfides. Although the sulfides were lower grade at 2–5% Cu, they were considerably larger orebodies and this allowed larger project scales to be developed at sites such as Moonta-Wallaroo, Mt Lyell, Cobar and Mt Morgan. The brief rise in Australia's ore grade in the 1920s was due to the closure of low-grade open cuts in preference for high-grade underground mines to survive the depressed Cu market of the decade. In the 1930s the Cu market again came back to strength and even larger open cuts were developed on yet lower grade ore, shifting the average ore grade down to historical lows (reaching ~0.53% Cu in 1947). The development of the high-grade Mt Isa Cu lodes (~3% Cu) in the 1950s saw the Australian average ore grade rise, and only since the late 1990s, with the development of several large but lower grade projects, has the grade started to decline again (reaching ~0.75% Cu in 2010). For Canada and the USA, Cu ore grades are a function of the increasing scale of porphyry mines, and also declining Cu grades in the Sudbury Ni–Cu field in Canada.

A detailed compilation of Cu mines is given in Table 1.11, representing 12.32 Mt (76%) of global Cu production in 2010. Similarly for resources, about half of global production is from the

Table 1.11 Annual global production statistics for copper mines (2010 data, Cu production/kt y^{-1}).

Deposit/project	Stages[a]	Ore/Mt	Cu/%	Au/g t^{-1}	Cu/kt	Au/t	WR/Mt	OC[a] ore/%	Other metals
Escondida, Chile	OC-F/HL	123.894	0.93		1010.8	5.41	340.43	100	
Codelco Norte, Chile	OC-F	123.220	0.81		903.7		373.91	100	Mo
Grasberg, Indonesia	OC/UG-F	83.122	0.83	0.92	608.4	62.82	344.96	63	Ag
Collahuasi, Chile	OC-F/HL	56.347	1.02		504.0		120.95	100	
Polska-Rudna, Poland	UG-F-S-R	29.3	1.64		425.4	0.776		0	Pb–Ag
El Teniente, Chile	UG-F/HL	47.000	0.97		403.6			0	Mo
South American (Mill)[b]	OC-F	68.9	0.65		385.6			100	Mo–Au
Los Pelambres, Chile	OC-F	58.181	0.76		384.6			100	Mo
Taimyr Peninsula, Russia	UG/OC-F	16.118	2.31		348.4			18	Ni–PGMs
North American (HL)[b]	OC-HL	236.8	0.24		338.4			100	
Antamina, Peru	OC-F	36.507	1.00		301.5		83.53	100	Pb–Zn–Mo
Batu Hijau, Indonesia	OC-F	43.375	0.66	0.65	297.6	22.92	60.11	100	
Bingham Canyon, USA	OC-F-S-R	53.551	0.53	0.38	249.8	14.49	111.92	100	Mo–Ag
Kansanshi, Zambia	OC-F	21.518	1.3		231.1	3.41	23.85	100	
South American (HL)[b]	OC-HL	98.1	0.41		228.6			100	
Los Bronces, Chile	OC-F/HL	62.176	0.54		221.4			100	
Andina, Chile	OC/UG-F	23.830	0.91		188.5		43.10	50.6	Mo
North American (Mill)[b]	OC-F	69.1	0.32		180.5			100	Mo
Spence, Chile	OC-HL	18.300	1.27		178.1		55.17	100	
Zhezkazgan, Kazakhstan	UG/OC-F	23.309	0.82		170.4			8.2	Ag
Toquepala, Peru	OC-F/HL	88.757	0.19		169.4	0.094	90.56	100	Mo
Cuajone, Peru	OC-F	31.429	0.52		165.0	0.115	94.72	100	Mo
Ok Tedi, Papua New Guinea	OC-F	22.191	0.84	0.97	159.8	15.13	26.63	100	
Mt Isa, Australia	UG-F	6.092	2.91		157.7			0	
Lumwana, Zambia	OC-F	18.58	0.86		146.7		82.92	100	

Table 1.11 (Continued)

Deposit/project	Stages[a]	Ore/Mt	Cu/%	Au/g t^{-1}	Cu/kt	Au/t	WR/Mt	OC[a] ore/%	Other metals
Zaldivar, Chile	OC-HL	40.598	0.58		144.2		28.63	100	
Cananea-La Caridad, Mexico	OC-F-S-R	70.069	0.21		144.1	0.326	14.09	100	Mo
Alumbrera, Argentina	OC-F	37.428	0.45	0.46	140.3	12.60	59.12	100	
Olympic Dam, Australia	UG-F-S-R	7.046	1.88	0.54	131.8	2.13	0.56	0	U–Ag
Tenke Fungurume, Democratic Republic of the Congo	OC-F	3.76	3.51		120.4			100	Co
Sossego, Brazil	OC-F	17.4	0.84	0.4	117.0	3.48		100	
Prominent Hill, Australia	OC-F	9.537	1.32	0.82	112.2	6.11	53.35	100	Ag
Ray, USA	OC-F-S	44.899	0.23		105.1			100	
50–100 kt y^{-1} Cu (23 mines)	Mostly OC-F	307.0	0.69	~0.1	1708.7	10.52	254.3	84.24	Ag
<50 kt y^{-1} Cu (98 mines)	Mostly OC-F	354.5	0.43	~0.4	1237.6	94.12	356.3	60.72	Ag
Total (154 mines)		**2352.0**	**0.62**	**~0.2**	**12 320**	**254.46**	**>>2620**	**84.94**	
2010 global production					**~16 200**	**~2700**			

[a]WR, waste rock; OC, open cut; UG, underground; F, flotation; HL, heap leach; S, smelter; R, refinery.
[b]Freeport McMoRan report North and South American mines as groups, only split through heap leach or mill processing; South America includes Cerro Verde in Peru and Candelaria, Ojos del Salado and El Abra in Chile, while North America includes Morenci, Sierrita, Bagdad, Safford and Miami in Arizona, and Tyrone and Chino in New Mexico.

top 20 mines, which have widely varying scales from 170 to 1000 kt y^{-1} of Cu. The dominance of open cut mining is clear, with this method providing 85% of ore processed (and the lack of reporting waste rock data for some mines) and most of the largest projects using open cut mining (the exception being the large underground mines at El Teniente in Chile and Polska-Rudna in Poland). In general, most projects only include a mine and mill (or heap leach), with a minority including a smelter and/or refinery on-site (or nearby). It is worth noting that many mines producing <50 kt y^{-1} of Cu are by-/co-product mines, mostly with Au.

An important aspect of open cut mines is waste rock – the barren rock which has to be excavated to allow access to the ore. Although waste rock has very low (*i.e.* uneconomic) metal concentrations, it may contain sulfides such as pyrite, which can lead to the generation of acid and metalliferous drainage (AMD) due to sulfide oxidation and escape of the acidic and metal-rich solutions.[39–41] Furthermore, waste rock can provide a source of sediment or dust to adjacent ecosystems or communities and also increases the area of land impacted by mining operations. In 1987, the US Environmental Protection Agency (USEPA) is attributed to have stated that:[42]

'... *problems related to mining waste may be rated as second only to global warming and stratospheric ozone depletion in terms of ecological risk. The release to the environment of mining waste can result in profound, generally irreversible destruction of ecosystems.*'

In parallel with the shift to open cut mining in the Cu sector, there has been an even faster growth in waste rock generation, although the data are far from complete since many mines still fail to report annual waste rock data. That is, all mines (even underground mining) should be reporting complete information on waste rock generated, ore processed, ore grades and metals extracted. Typical waste rock-to-ore ratios range from 2 to 5, although values can be as high as 15 or more. As seen in Table 1.11, many mines do not report waste rock (especially small mines). At some of the larger mines, annual waste rock movements now exceed 300 Mt y^{-1}. At Ok Tedi and Grasberg (and previously at the former Bougainville–Panguna mine), waste rock is directly dumped or allowed to erode gradually into the river, and this leads to severe

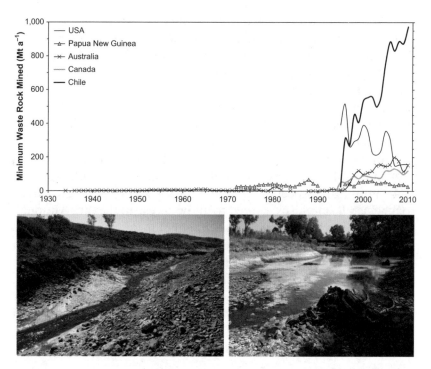

Figure 1.9 Top: *Minimum* waste rock mined annually in some countries. Data updated from refs 17 and 38 and references therein. Bottom: selected images of acid and metalliferous drainage from waste rock (note the green, blue and red colours in both). Left: Cu-rich AMD waste rock dump seepage at the Redbank Cu mine, Australia. Right: the *rehabilitated* waste rock dumps are still causing severe AMD impacts on the Finniss River at the former Rum Jungle U-Cu mine, Australia.

environmental and social impacts (see Figure 1.9). The available *minimum* data for waste rock in some countries are shown in Figure 1.9, including images of AMD at some sites. Although there is now better recognition of the need to proactively assess and manage waste rock in the mining industry and environmental regulators, there is clearly still significant room for improvement in reporting and demonstrating environmental management.[43]

1.5.2 Lead–Zinc

The Pb and Zn sectors are very closely linked, since the metals generally occur together in SedEx or VMS deposits. Until about

1900, each metal was mined separately and it was only after flotation arrived that it became possible to develop Pb–Zn ores and produce separate concentrates (which was preferred for smelting).

Pb enjoyed gradual growth throughout the 1800s, until World War I provided major interruptions in addition to the famous 18 month strike at Broken Hill in Australia during 1919–20. The 1920s–30s proved to be boom–bust times, with a brief but major decline after World War II followed by exceptional growth in annual production to 3.67 Mt Pb in 1973. At this point, however, the growing awareness of Pb's toxicity, especially towards children, led to reduced demand for its use in petrol (eventually being phased out as an additive in most western countries in the 1980s) and also major recycling programmes for Pb acid batteries. With surging demand and internal capacity, China has now come to dominate global Pb production, which reached a new record of ~4.1 Mt Pb in 2010.

In contrast, although Zn production used to be a close second to Pb until the 1950s, since that time annual production has grown substantially, driven by Zn's wide variety of uses. From 1950 to 2010, global Zn production grew from ~2 to ~12 Mt, and it is now triple that of Pb. The higher Zn production is a function of the commonly higher Zn grades in SedEx and VMS deposits and also the larger market and slightly higher prices for Zn (on average ~1.2 times Pb from 1950 to 2009).

Historically, the Pb–Zn sector has used underground mining, although many of the world's largest mines now use open cuts. At some projects, the underground mines are going deeper over time, such as Kidd Creek in Canada (now up to 2.9 km deep), whereas at the Broken Hill field in Australia new projects are being developed in remnant pillars of shallow historical mining, in addition to continuing operations and future plans at depths ranging from 1 to 2 km.

The trends in ore grade over time for Australia and Canada are shown in Figure 1.10, demonstrating a long-term decline, although Zn has not declined as rapidly in recent decades compared with Pb. The erratically low Australian grades in the early years are due to only yields being reported and not true assayed ore grade, with low production leading to low ore grades. For both countries, the data represent almost 100% of the reported production and are therefore an accurate reflection of the sector's history in each country.

A detailed compilation of Pb–Zn mines is given in Table 1.12, representing 1.34 Mt Pb (31.9%) and 5.68 Mt Zn (47.3%) of global

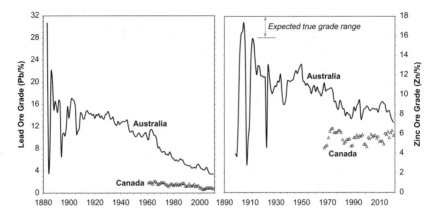

Figure 1.10   Historical trends in Pb–Zn ore grades for Australia and Canada. Data updated from refs 17 and 38 and references therein.

production in 2010. Unfortunately there is only one small mine in China in the data set, with China producing ~ 1.75 Mt Pb (41.7%) and ~ 3.5 Mt Zn (29.2%). Similarly for resources, about half of global production is from the top 20 mines, which are very Zn dominant and have widely varying scales from ~ 80 to 750 kt y^{-1} of Pb + Zn.

The dominance of open cut mining is clear for large producers (and, like Cu, the lack of waste rock data), but others are mostly underground. Another key factor is ore grade, with most of the largest 10 mines having grades of 8–13% Zn. In general, most projects only include a mine and mill, with a minority including a smelter and/or refinery on-site (or nearby). For mines with no Ag data in Table 1.12, this is mainly due to lack of reporting Ag grades and production. The rich Mexican Ag deposits, famously sought by the Spanish Conquistadors, are still very high grade at 200–500 g t^{-1} Ag (data in the 'Miscellaneous' group in Table 1.12). Unfortunately, there are not as many historical data available for waste rock in the Pb–Zn sector, with none presented here (other than in Table 1.12).

1.5.3   Nickel

The mining of Ni was small scale and mostly centred in Europe and New Caledonia until the discovery of the giant Sudbury Basin in

Table 1.12 Annual production for top 25 lead–zinc–silver mines and miscellaneous smaller mines (2010 data, Pb+Zn production/Mt y⁻¹).

Deposit/project, country	Ore/Mt	Pb/%	Zn/%	Ag/g t⁻¹	Cu/%	Pb/kt	Zn/kt	Ag/t	Cu/kt	Pb+Zn/kt	WR[a]/Mt	OC[a] ore/%
RampuraAgucha, India	5.856	2.15	13.09			68.8	677.4			746.2	77.1	100
Red Dog, USA	3.572	5.4	18.2			109.9	538.0			647.9	~5	100
Century, Australia	5.211	1.18	12.13	24.6		32.3	510.6	95.3		542.9	42.5	100
Mt Isa, Australia	8.566	2.7	5.6	53.1		143.7	355.0	210.7		498.7		~90
Antamina, Peru	36.507		1.39		1.00	5.8	386.2	398.7	301.5	392.0	83.5	100
Cannington, Australia	3.075	9.72	3.21	459.3		256.1	59.4	1200.4		315.5		0
Brunswick, Canada	3.068	3.1	8.0	98		90.4	214.0	329.5	8.2	304.4		0
McArthur River, Australia	2.247	4.5	11.0	~40.5		31.6	183.5	45.5		215.2		100
Lisheen, Ireland	1.588	1.9	12.2			20.6	175.1			195.7		0
Tara, Ireland	2.593	1.4	7.0			18.5	167.3			185.8		0
Yauli Group, Peru	3.426	0.67	5.10	111.8	0.16	18.3	159.3	307.4	3.2	177.6		0
East Balkhash, Kazakhstan	4.610		4.76	73.21	2.48		154.4	78.6	85.6	154.4		0
Animón-Chungar, Peru	1.476	1.69	7.48	123.8	0.18	54.8	45.8	238.6		100.6		0
Perseverance, Canada	1.062		14.1		~1.26		139.4		10.0	139.4		0
Skorpion, Namibia	1.358		11.2				138.5			138.5		100
Maleevsky-Zyrianovsk, Kazakhstan	~2.25	~1.13	~5.96	~69.94	~1.79	~19.1	~114.0	~94.4	~20.1	~133.1		0
Garpenberg, Sweden	1.443	2.5	6.6	133	0.1	29.3	86.0	140.1	0.2	115.3		0
Broken Hill, Australia	1.638	3.63	4.40	37.9		51.2	63.6	44.7	4.0	114.8		0
Rosebery, Australia	0.725	4.09	12.67	125.0	0.38	25.4	84.2	78.7	2.1	109.5		0
Cerro de Pasco, Peru	3.217	1.21	4.09	67.5	0.06	24.5	83.1	102.2		107.6		0
Greens Creek, USA	0.726	~3.5	~9.3	414.804		23.0	67.6	224.1		90.6		0

Table 1.12 (Continued)

Deposit/project, country	Ore/Mt	Pb/%	Zn/%	Ag/g t⁻¹	Cu/%	Pb/kt	Zn/kt	Ag/t	Cu/kt	Pb+Zn/kt	WRᵃ/Mt	OCᵃ ore/%
Black Mountain, South Africa	1.379	4.2	3.3		0.3	50.6	36.1		2.5	86.7		0
Kidd Creek, Canada	2.429	~0.49	~4.06	~37.0	2.28	7.7	86.1	59.2	52.6	86.1		0
Golden Grove, Australia	1.597		5.11		2.37		73.3		34.3	81.0		0
Cerro Lindo, Peru	2.534	0.4	3.2	31.1	0.8	6.2	73.6		16.7	79.7		0
Miscellaneous (42 mines)	59.4	0.59	2.50	50.5	0.56	251.6	1010.3	2224.0	247.6	1261.9	>>25.9	43.2
Total (67 mines)	**161.6**	**1.15**	**4.45**	**46.1**	**0.62**	**1339.3**	**5672.1**	**5872.1**	**788.6**	**7021.0**	**>>234**	**54.6**
2010 global production						~4200	~12 300	~22 900	~16 200	~16 500		

ᵃWR, waste rock; OC, open cut.

1883 in northern Ontario, Canada. Although Sudbury was initially considered a rich Cu prospect, initial smelting problems instead led to the realisation that it was a rich Ni–Cu sulfide ore. After considerable effort and pioneering metallurgical research, the Sudbury Basin quickly emerged as the dominant global Ni producer by the start of the twentieth century. Sudbury dominated global Ni production for the next 50 years, although after World War II the scale of the Russian Ni sulfide fields at Nori'lsk-Talnakh on the Taimyr Peninsula in Siberia and on the Kola Peninsula, near Finland, was starting to grow, in addition to the re-emergence of New Caledonian laterites. In the late 1960s, the Western Australian Ni boom leap-frogged Australia into position as a major Ni player. From the 1970s, a range of countries began to exploit their laterite resources, leading to major production in Indonesia, the Philippines and Cuba, amongst others. Hence, although Ni sulfides had dominated throughout most of the twentieth century, by the start of the twenty-first Ni laterites were beginning to dominate, and look set to continue to increase their share given its stronger resource base (further details can be found elsewhere[35]).

From the late 1990s, three major Ni laterite projects were developed in Western Australia, at Cawse, Bulong and Murrin Murrin, using improved technological developments in HPAL. Although they were alleged to provide a paradigm shift and to lower the cost curve for Ni mining, the engineering, construction and operation of HPAL plants proved highly problematic and the Cawse and Bulong projects were quickly financial and technical failures.[17,35] The Murrin Murrin project, also the largest, barely managed to scrape by financially and survive until the boom Ni prices of the late 2000s – annual production levels remain, however, below the original design capacity and are around 30 kt y^{-1} for Ni and 2 kt y^{-1} for Co. Most new laterite projects around the world are now looking to HPAL technology, in addition to potentially heap leaching.

The trends in ore grade for some countries are shown in Figure 1.11, demonstrating a long-term decline. The ore type over time is also included in Figure 1.11, showing the overall dominance of sulfides through the twentieth century but the re-emergence of laterites more recently. The low grades in Australia from $\sim$1995 are due to the development of the Mt Keith large-tonnage, low-grade disseminated Ni sulfide mine ($\sim$0.6% Ni), and also the Ni

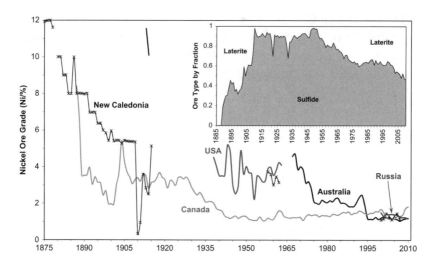

Figure 1.11 Historical trends in Ni ore grades for Australia, Canada, New Caledonia, Russia and the USA, combined with an inset of ore type over time. Data updated from refs 17, 35 and 38 and references therein.

laterite projects from 1999 ($\sim$1% Ni). In Canada, the recent increase in ore grade was driven by the high-grade Voisey's Bay mine (3.2% Ni) and the major workers strike and lower production at Sudbury leading to the need to process higher grade ore. The data represent close to 100% of the reported production (except Russia, which excludes the minor laterite production) and are therefore an accurate reflection of the sector's history in each country.

Detailed compilations of Ni sulfide and Ni laterite mines are presented in Tables 1.13 and 1.14, representing 627.8 kt Ni (39.8%) and 541.5 kt Ni (34.3%) of global production in 2010, respectively. Unfortunately there are no data from China, which produced $\sim$79 kt Ni ($\sim$5%). Similarly to resources, $\sim$60% of global Ni production is from the top five mines by each ore type, showing the dominance of giant producers. The type of mine is variable for Ni sulfides, with examples of large-scale underground, open cut and mixed operations, although virtually all Ni laterites are mined by open cut (the one small underground mine at the Agios Ioannisas group in Greece is the exception). The waste rock data for Ni mines are rarely reported.

Table 1.13 Annual production for top 10 nickel sulfide mines and miscellaneous smaller mines (2010 data, Ni + Cu + Co production/kt y^{-1}).

Deposit, country	Ore/Mt	Ni/%	Cu/%	Co/%	PGMs/g t^{-1}	Ni/kt	Cu/kt	Co/t	OC ore/%
Taimyr Peninsula, Russia	16.118	1.41	2.31		7.51	196.5	348.4		18
Voisey's Bay, Canada	1.510	3.20	2.44	~0.06		42.3	33.0	524	100
Sudbury (Vale Inco), Canada	2.660	1.78	1.53	~0.02	1.1	22.4	34.0	302	0
Kola Peninsula, Russia	8.336	0.67	0.29		0.08	39.0	17.3		88
Sudbury (Xstrata), Canada	1.472	1.40	2.81	~0.04		16.8	36.1	341	0
Leinster, Australia[a]	~2.75[a]	~1.9[a]				~44.4[a]			0[a]
Mt Keith, Australia[a]	~11[a]	~0.55[a]				~42.4[a]			100[a]
Raglan, Canada	1.280	2.45	0.68	~0.06		28.2	7.1	567	0
Thompson, Canada	2.158	1.67	0.12	~0.02		29.8	1.0	189	0
Kambalda, Australia	1.061	3.01	0.22			26.8	1.9	143	0
Miscellaneous (35 mines)	134.2	0.17	0.06		~2.7	139.2	51.5	779	39.8
Total	**182.2**	**0.46**	**0.34**		**~0.4**	**627.8**	**530.4**	**2845**	**41.7**
2010 world production						**1578**	**16 118**	**88 000**	**41.7**

[a]BHP Billiton refuse to report site-based production data, with the data for Leinster and Mt Keith based on older WMC data and approximate Ni production for Western Australia (after taking into account other Ni mines).

Table 1.14 Annual production for top 10 nickel laterite mines and miscellaneous smaller mines (2010 data, Ni + Co production/ kt y^{-1}).

Deposit, country	Ore/Mt	Ni/%	Co/%	Ni/kt	Co/t	OC ore/%
Sorowako, Indonesia	4.129	2.05	0.048[a]	76.0	1198	100
PT Antam ore exports, Indonesia	5.571[b]	1.8[a]		75.2		100
Doniambo, New Caledonia	3.978	1.59[a]		53.7		100
Cerro Matoso, Columbia	5.38	1.08[a]		49.4		100
Moa Bay, Cuba	3.00	1.26[a]	0.13[a]	34.0	3706	100
SMSP Group, New Caledonia	2.327	2.03[a]		35.4		100
Murrin Murrin, Australia	2.640	1.27	0.103	28.4	1976	100
Coral Bay, Philippines	2.555[b]	1.33		25.4		100
Larymna, Greece	2.5[a]	1.05[a]		19.6		95
Pomalaa-Halmahera, Indonesia	1.503[b]	1.85		18.7		100
Miscellaneous (13 mines)	10.7	1.61		125.7		100
Total	**44.3**	**1.55**		**541.5**	**6880**	**99.7**
2010 world production				**1578**	**88 000**	

[a]Ore milled and grade not reported, values estimated based on other company information (*e.g.* reserves. resources).
[b]Wet ore reported, based on conversion to dry ore for consistency.

1.6 SUSTAINABILITY ISSUES, CONSTRAINTS AND THE FUTURE OF BASE METALS

The future of the base metals sector of the global mining industry is by no means assured simply based on the economic resource base documented in this chapter. The future of the mining industry is being challenged by a range of often competing and complex challenges and below is only a brief review of arguably the most critical issues – namely energy intensity and associated greenhouse gas emissions. Other essential issues may include mine waste management (tailings and waste rock), water consumption and impacts and risks to water resources, social licence to operate, rehabilitation performance, pollution risks and management (*e.g.* dust, SO_2, heavy metals in particulates, accidental leaks and spills), economic conditions (markets, supply/demand), wealth sharing and technology policy (*e.g.* electric cars, renewable energy) – clearly the future of base metals will be more constrained by all of these and potentially other factors rather than simply the quantity of resource remaining.

1.6.1 Energy and Greenhouse Gas Emissions Intensity

The production of a refined base metal requires a substantial investment of energy, from mining through milling to smelting and final refining. The energy sources typically include diesel for trucks and mobile machinery, and also various sources of electricity (mainly coal, gas or hydroelectricity). Given the dominance of fossil fuels for most energy systems, this leads to a very close link between energy consumption and greenhouse gas emissions (GGEs), primarily as carbon dioxide (CO_2) and its equivalent warming potential from other GGEs such as nitrous oxide and methane (collectively given the symbol CO_{2e}). GGEs are the primary pollution driver for human-induced climate change.[44]

It is important to account for the energy and GGEs over the full production chain for a refined metal, a method of analysis known as life cycle assessment (LCA), with numerous studies completed for Cu, Pb, Zn and Ni.[45–48] It is critical to ensure that the various pathways for refined metal production are assessed, such as pyrometallurgy or hydrometallurgy, in addition to allowing for regional differences in inputs, especially in electricity sources.

In general, the energy inputs for the smelting and refining stages are relatively constant per unit metal, since the input concentrate grades are also somewhat stable. As ore grades decline, however, the energy intensity in the mining and milling stages begins to increase substantially, since more rock has to be mined and processed to maintain the output – leading to an inverse relationship between ore grade and energy or GGEs intensity. The relationships between ore grade and energy intensity for Cu or GGEs intensity and Ni production for various process routes are shown in Figure 1.12 (the corresponding grade–GGEs graph for Cu and grade–energy graph for Ni are similar to those shown). The relationship between energy and GGEs intensity for some Ni mines, based on operational data from sustainability reports, is shown in Figure 1.13, demonstrating the more intensive processing required for Ni laterites. In addition, for similar energy intensity, the GGEs intensity can be significantly different, demonstrating the importance of electricity source (*e.g.* hydroelectricity or coal). The performance over time for the Kalgoorlie Ni smelter in Western Australia is shown in Figure 1.14, showing a decline in the 1990s

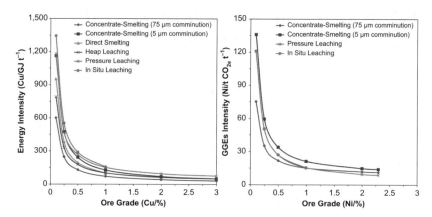

Figure 1.12 Relationship between ore grade and energy intensity of Cu production (left), and ore grade and greenhouse gas emissions intensity of Ni production (right). Redrawn from ref. 49.

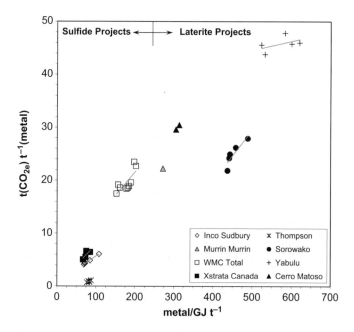

Figure 1.13 Relationship and GGEs intensity for some Ni mines (data from
sustainability reports). Redrawn from ref. 35.

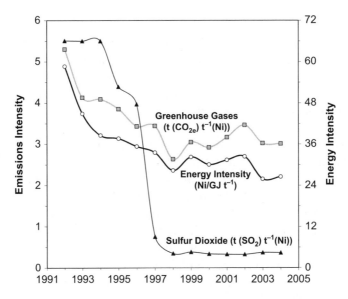

Figure 1.14   Energy, GGEs and sulfur dioxide intensity over time for the Kalgoorlie Ni smelter.[35]

for energy and GGEs intensity but being fairly stable in the 2000s, whereas the SO_2 intensity dropped dramatically in 1997 due to the installation of an SO_2 capture and sulfuric acid plant.

The current ore grades of Cu and Ni mines are mostly 0.3–3% Cu (average ~0.6% Cu), 0.4–3% Ni for sulfides (average ~0.5% Ni) and 1–2% Ni for laterites (average ~1.5% Ni). This shows that Cu and Ni sulfides are rapidly approaching the strong inflection point in the graphs at about 0.5% ore grade, whereas laterites still appear to be some distance from this point.

Collectively, as ore grades decline, the inverse relationship between ore grade and energy or GGEs intensity will place significant upward pressure on energy inputs and GGEs outputs for base metal production. It is very clear that it will be increasingly important to be able to demonstrate mine site performance in the manner shown here, especially as legislation continues to evolve in this area alongside community sentiment and expectations. Energy, GGEs and pollution issues generally will provide real and substantial constraints on foreseeable base metal production rather than simply the available economic resource remaining.

1.6.2   Other Ore Processing Issues

Another major issue facing the future of the base metals industry is how readily existing technology can process a given ore. There are some deposits for which it has taken many years to develop a viable approach to processing, and for others it still has not been solved. For example (amongst others):

- *McArthur River (Australia)* – despite discovery in 1955 and being recognised as a giant deposit, the ore is extremely fine grained and advanced flotation technology able to process the ore was not available until the 1990s. Even after development in 1995 the mine can only produce a bulk Pb–Zn concentrate which very few smelters in the world can process. The mine also remains controversial due to a ~ 6 km diversion of the McArthur River itself, which flows directly over the deposit, to allow conversion from underground to open cut mining from 2006.
- *Armstrong (Australia)* – this Ni sulfide deposit is in the Kambalda field and had long been known but not developed owing to its high arsenic content and low iron to magnesium ratio, making it very difficult to produce a suitable concentrate. Titan Resources, however, developed an open cut mine without addressing these risks and the first tonne of ore was rejected by the Kambalda mill as outside contractually agreed specifications – pushing Titan to near bankruptcy in the process.[50]
- *Mesaba (USA)* – the Cu occurs frequently in the Cu-poor mineral cubanite ($CuFe_2S_3$), which is often interlocked with the Ni mineral pentlandite, making it difficult to produce a high-grade Cu concentrate suitable for smelting.[51]
- *Western Australia Ni laterites* – despite being recognised during the late 1960s Ni boom, laterites were ignored as uneconomical. Although the development of Cawse, Bulong and Murrin Murrin in the late 1990s promised much, they clearly failed to live up to expectations, with the more recent BHP Billiton project at Ravensthorpe also failing to sustain economical production and being closed prematurely (and then quickly sold very cheaply to Canada's First Quantum) – collectively casting a lingering doubt over the future of Ni laterites (Australian at least).

As such, ores are becoming more complex and often require a finer grinding size to allow economical extraction. The presence of impurities is also proving to be more of a challenge, either in processing terms (aka the Armstrong Ni fiasco) or with respect to environmental risks such as mercury or acid and metalliferous drainage (there is also a need to consider the European REACH legislation in the light of maximum contaminant levels).

On the positive side, however, an increasing opportunity for the base metals sector is by-product metal recovery. As noted in the various resources table, many sulfide ores contain precious metals, especially Au and Ag, and many porphyry Cu ores can contain Mo. Historically, metals such as PGMs, Te, Se, Ge or Cd have been extracted from smelter slags or slimes. With the growing sophistication of metals use in modern technology, such as flat screen displays, touch screens, energy efficiency (especially magnets) and specialty alloys, there are rapidly growing expectations that other metals commonly found in base metal ores, such as In and Re, will also require extraction. At present, these metals are rarely included in economic studies of deposits and, in the deposits compiled for this study, only a handful included In or Re in their formal mineral resource. For iron oxide Cu–Au ores, rare earths can commonly be present also, with the super-giant Olympic Dam resource in Australia containing ~ 50 Mt of rare earths alone – worth \$4.5 trillion (\$4.5 \times 10^{12}$) (or the same as BHP Billiton's entire iron ore resources in the Pilbara of Western Australia) compared with the Cu–U–Au–Ag value of $\sim \$1$ trillion (based on January 2012 metal prices). As base metal ore grades decline, there will be growing opportunities to extract additional value from by-product metals contained in the ore.

1.6.3 Ocean Base Metal Resources?

There may be a temptation to believe that since mineral resources are finite, there will be a need in the future to move from land- to ocean-based resources. However, although the ocean does contain significant ore types, such as seafloor massive sulfides (associated with volcanic vents, considered the origin of VMS deposits) and manganese nodules (chemical and sedimentary accretions on the ocean floor), they are invariably at water depths from 1 to 6 km – which is beyond the scope of present technology and economic feasibility.

There has been ongoing research into this area for a few decades now and estimates of potential base metals from nodules alone in the ocean's depths include the following:

- *Clarion-Clipperton zone, western Pacific Ocean* – 340 Mt Ni and 265 Mt Cu.[52]
- *Global estimate of recoverable metals* – 130–215 Mt Ni and 105–175 Mt Cu.[53]
- *Global estimate of in situ geological resources* – 290–1500 Mt Ni and 240–1240 Mt Cu.[54]

There remains significant community concern over potential environmental impacts of deep sea mining, especially poorly understood deep ocean biodiversity, which is only beginning to be investigated. As of 2012, there are yet to be any commercial deep-sea mining projects and, given that formal mineral resource codes are not readily applicable to estimating economic mineral resources in seafloor nodules and mining projects, it is extremely speculative to predict to what extent, if any, the above estimates may become realistically available resources in the future – despite their potential size.

1.7 CONCLUSION

This chapter has provided an extensive review of the primary base metals, copper, lead, nickel and zinc, focusing on economic resources reported by companies and global production trends. These metals can be found in a wide variety of deposit types, dominantly as sulfide deposits for Cu, Pb and Zn, whereas Ni is mainly in laterite ores and numerous sulfide deposits. For all metals, the compiled data significantly exceeded the global reserve estimates produced annually by the USGS. This demonstrates that there still remains a sizeable economic resource base for all Cu, Pb, Zn and Ni, able to sustain production for some decades generally, although Pb and Zn appear to have weaker prospects for sustained long-term production growth than Cu and Ni. It should be noted, however, that exploration is still proving successful but becoming more difficult, although truly comprehensive global data sets on mineral resources remain elusive owing to a lack of regular public reporting by some major countries (especially China and some others). The future of the base metals mining sector will be increasingly dominated by critical issues such as the energy and greenhouse gas emissions intensity of production, and also a range of other competing issues (water, wealth sharing, mine waste management, pollution mitigation, *etc.*). As ore grades gradually decline, the inverse relationship between energy and GGEs intensity and grade will place significant upward pressure on base metals production – making it imperative for mining companies to demonstrate their performance, especially transparently. Given the ongoing rise of China, the impending emergence of India and not to forget other regions of the world such as Africa, South America and Central Asia, there will be a need for base metals for some decades to come to meet reasonable demands for a range of products and services. It is clear that the sustainability of base metals production over coming decades will be governed more by the environmental costs and risks in mining them rather than simply the tonnage remaining – and the journey thus continues.

REFERENCES

1. G. Agricola, *De Re Metallica*, Reprint, ed. H. C. Hoover and L. H. Hoover, Dover Publications, New York, 1950, pp. 4, 1556.
2. D. H. Meadows, D. L. Meadows, J. Randers and W. W. Behrens III, *The Limits to Growth. A Report for the Club of Rome's Project on the Predicament of Mankind*, Universe Books, New York, 1972.
3. J. E. Tilton, *On Borrowed Time? Assessing the Threat of Mineral Depletion*, Resources for the Future, Washington, DC, 2003.
4. W. J. Rankin, *Minerals, Metals and Sustainability – Meeting Future Material Needs*, CSIRO Publishing and CRC Press, Melbourne, 2011.
5. International Institute for Environment and Development and World Business Council for Sustainable Development, *Breaking New Ground: Mining, Minerals and Sustainable Development*, Earthscan, London, 2002.
6. C. D. Da Rosa, J. S. Lyon and P. M. Hocker (eds), *Golden Dreams, Poisoned Streams: How Reckless Mining Pollutes America's Waters and How We Can Stop It*, Mineral Policy Center, Washington, DC, 1997.

7. J. E. Young, *Mining the Earth*, Worldwatch Institute, Washington, DC, 1992.

8. R. Raymond, *Out of the Fiery Furnace: the Impact of Metals on the History of Mankind*, Macmillan, Melbourne, 1984.

9. R. B. Gordon, M. Bertram and T. E. Graedel, *Proc. Natl. Acad. Sci. USA*, 2006, **103**, 1209.

10. R. Eggert, in *Australian Mineral Economics – A Survey of Important Issues,* ed. P. Maxwell, Australasian Institute of Mining and Metallurgy, Melbourne, 2006, pp. 187–194.

11. R. L. Rudnick and S. Gao, in *Treatise on Geochemistry*, ed. H. D. Holland and K. K. Turekian, Elsevier-Pergamon, Oxford, 2003, Chapter 3.01.

12. A. M. Evans, *An Introduction to Economic Geology and Its Environmental Impacts*, Blackwell Science, Oxford, 1997.

13. P. Laznicka, *Giant Metallic Deposits: Future Sources of Industrial Metals*, 2nd edn, Springer, Heidelberg, 2010.

14. W. L. Pohl, *Economic Geology: Principles and Practice*, Wiley-Blackwell, Chichester, 2011.

15. H. G. Dill, *Earth Sci. Rev.*, 2011, **100**, 1.

16. M. D. Gerst, *Econ. Geol.*, 2008, **103**, 615.

17. G. M. Mudd, *The Sustainability of Mining in Australia: Key Production Trends and Their Environmental Implications for the Future*, Department of Civil Engineering, Monash University, and Mineral Policy Institute, Melbourne, October 2007, Revised April 2009.

18. S. M. Bulatovic, *Handbook of Flotation Reagents: Chemistry, Theory and Practice. Flotation of Sulfide Ores*, Vol. 1, Elsevier, Amsterdam, 2007.

19. S. M. Bulatovic, *Handbook of Flotation Reagents: Chemistry, Theory and Practice*, Vol. 2, Elsevier, Amsterdam, 2010.

20. C. J. Schmitz, *World Non-Ferrous Metal Production and Prices, 1700–1976*, Frank Cass, London, 1979.

21. US Geological Survey, *Minerals Commodity Summaries*, US Geological Survey, Reston, VA, Years 1996–2011.

22. Australian Bureau of Agricultural and Resource Economics, *Australian Commodity Statistics*, Canberra, Years 1995–2010 (formerly *Commodity Statistical Bulletin*, 1986–1994).

23. US Bureau of Mines, *Minerals Yearbook*, US Government Printing Office, Washington, DC, Years 1933–1993.

24. US Geological Survey, *Minerals Yearbook*, Reston, Virginia, USA, Years 1994–2009.

25. AusIMM, MCA and AIG, *Australasian Code for Reporting of Exploration Results, Mineral Resources and Ore Reserves: the JORC Code*, Joint Committee of the Australasian Institute of Mining and Metallurgy, Australian Institute of Geoscientists and Minerals Council of Australia, Parkville, VIC, 2004.
26. P. R. Stephenson, *IMM Trans. B, Appl. Earth Sci.*, 2001, **110**, B121.
27. Ontario Securities Commission, *National Instrument 43-101 – Standards of Disclosure for Mineral Projects, Form 43-101F1 and Companion Policy 43-101CP*, Ontario Securities Commission, Toronto, 2011.
28. SAMRCWG, *South African Code for the Reporting of Exploration Results, Mineral Resources and Mineral Reserves (the SAMREC Code)*, South African Mineral Resource Committee Working Group (SAMRCWG), The Southern African Institute of Mining and Metallurgy (SAIMM) and Geological Society of South Africa (GSSA), Johannesburg, 2009.
29. I. Lambert, Y. Meizitis and A. D. McKay, *AusIMM Bull.*, 2009, December, 52.
30. Natural Resources Canada, *Canadian Minerals Yearbook*, Natural Resources Canada, Ottawa, ON, Years 1944–2009.
31. Geoscience Australia, *Australia's Identified Mineral Resources*, Geoscience Australia, Canberra, ACT, Years 1992–2010.
32. Geoscience Australia, *Australia's Identified Mineral Resources – Preliminary Resources Table (online)*, Geoscience Australia, Canberra, ACT, 2011, www.ga.gov.au (last accessed 24 March 2012).
33. G. M. Mudd, *Res. Pol.*, 2010, **35**, 98.
34. X. Zhu (ed.), *Mineral Facts of China*, Science Press, Beijing, 2006.
35. G. M. Mudd, *Ore Geol. Rev.*, 2010, **38**, 9.
36. Jinchuan Group, *Jinchuan Group Ltd – Overview*, 2009, Jinchuan Group, Jinchang, www.jnmc.com (last accessed 10 December 2009).
37. R. Schodde, presented at AMIRA International's 8th Exploration Managers Conference, AMIRA, Yarra Valley, VIC, 2010.
38. G. M. Mudd, in *48th Annual Conference of Metallurgists – Green Technologies for Mining and Metallurgical Industries*, Metallurgical Society, Canadian Institute of Mining, Metallurgy and Petroleum, Sudbury, ON, 2009, p. 273.

39. J. Taylor and S. Pape (eds), *Managing Acid and Metalliferous Drainage*, Leading Practice Sustainable Development Program for the Mining Industry, Commonwealth Department of Industry, Tourism and Resources, Canberra, ACT, 2007.

40. D. W. Blowes, C. J. Ptacek, J. L. Jambor and C. G. Weisener, in *Treatise on Geochemistry*, ed. H. D. Holland and K. K. Turekian, Elsevier-Pergamon, Oxford, 2003, Chapter 9.05.

41. B. Lottermoser, *Mine Wastes – Characterisation, Treatment, Environmental Impacts*, 3rd edn, Springer, Berlin, 2010.

42. European Environmental Bureau, *The Environmental Performance of the Mining Industry and the Action Necessary to Strengthen European Legislation in the Wake of the Tisza-Danube Pollution*, European Environmental Bureau, Brussels, 2000, p. 1.

43. G. M. Mudd, in *SDIMI 2009 – Sustainable Development Indicators in the Minerals Industry Conference*, Australasian Institute of Mining and Metallurgy, Gold Coast, *QLD*, 2009, pp. 377–391.

44. Intergovernmental Panel on Climate Change, *Climate Change 2007: the Physical Science Basis – Summary for Policymakers*, IPPC, c/o World Meteorological Organization, Geneva, February 2007.

45. H.-J. Althaus and M. Classen, *Int. J. Life Cycle Assess.*, 2005, **10**, 43.

46. S. Alvarado, P. Maldonado, A. Barrios and I. Jaques, *Energy*, 2002, **27**, 183–196.

47. D. Lunt, Y. Zhuang and S. R. La Brooy, in *Green Processing 2002: International Conference on the Sustainable Processing of Minerals*, Australasian Institute of Mining and Metallurgy, Cairns, QLD, 2002, p. 185.

48. T. E. Norgate and N. Haque, *J. Cleaner Prod.*, 2010, **18**, 266.

49. T. E. Norgate and S. Jahanshahi, *Min. Eng.*, 2010, **23**, 65.

50. D. Giurco, T. Prior, G. M. Mudd, L. Mason and J. Behirsch, *Peak Minerals in Australia: a Review of Changing Impacts and Benefits*, prepared for CSIRO Minerals Down Under Flagship – Mineral Futures Collaboration Cluster, by the Institute for Sustainable Futures (University of Technology, Sydney) and Department of Civil Engineering (Monash University), March 2010.

51. D. L. Jones, K. Mayhew and L. O'Connor, presented at the 48th Annual Conference of Metallurgists – Green Technologies for Mining and Metallurgical Industries, Metallurgical Society, Canadian Institute of Mining, Metallurgy and Petroleum, Sudbury, ON, 2009.

52. C. L. Morgan, in *Handbook of Marine Mineral Deposits*, ed. D. S. Cronan, CRC Marine Science Series, CRC Press, Boca Raton, FL, 2000, p. 145.

53. A. A. Archer, *Trans. Inst. Min. Metall.*, 1981, **90**, A1–A6.

54. A. A. Archer, in *Manganese Nodules: Dimensions and Perspectives*, ed. United Nations Ocean Economics and Technology Office, Reidel, Dordrecht, 1979, pp. 71–81.

CHAPTER 2

Rare Earths

GARETH P. HATCH

Technology Metals Research LLC, 180 S. Western Avenue #150,
Carpentersville, IL 60110, USA
Email: ghatch@techmetalsresearch.com

2.1 INTRODUCTION

The so-called rare-earth elements (REEs) are a unique group of
chemical elements that exhibit a range of special electronic, magnetic, optical and catalytic properties. These elements are enablers;
their use in components manufactured from a wide range of alloys
and compounds, can have a significant effect on the performance of
complex engineered systems.

The International Union of Pure and Applied Chemistry
(IUPAC) defines the rare-earth metals (*i.e.* the REEs) as the 15
lanthanoid elements in addition to scandium (Sc) and yttrium (Y).[1]
Table 2.1 shows the positions of the REEs in the periodic table of
elements. One of the lanthanoid elements, promethium (Pm), is
radioactive, with no stable or long-lived isotopes. It therefore
occurs in the Earth's crust in vanishingly small amounts, as a result
of the decay of other radioactive elements. Sc, although exhibiting
similar properties to the other REEs, is seldom found together with
the lanthanoids in mineral occurrences and is therefore usually

Materials for a Sustainable Future
Edited by Trevor M. Letcher and Janet L. Scott
© The Royal Society of Chemistry 2012
Published by the Royal Society of Chemistry, www.rsc.org

Table 2.1 The periodic table of the elements.[2]

1	2	3	4	5	6	7	8	9	10	11	12	13	14	15	16	17	18
1 **H** 1.008																	2 **He** 4.003
3 **Li** 6.941	4 **Be** 9.012											5 **B** 10.81	6 **C** 12.01	7 **N** 14.01	8 **O** 16.00	9 **F** 19.00	10 **Ne** 20.18
11 **Na** 22.99	12 **Mg** 24.31											13 **Al** 26.98	14 **Si** 28.09	15 **P** 30.97	16 **S** 32.07	17 **Cl** 35.45	18 **Ar** 39.95
19 **K** 39.10	20 **Ca** 40.08	21 **Sc** 44.96	22 **Ti** 47.87	23 **V** 50.94	24 **Cr** 52.00	25 **Mn** 54.94	26 **Fe** 55.85	27 **Co** 58.93	28 **Ni** 58.69	29 **Cu** 63.55	30 **Zn** 65.38	31 **Ga** 69.72	32 **Ge** 72.64	33 **As** 74.92	34 **Se** 78.96	35 **Br** 79.90	36 **Kr** 83.80
37 **Rb** 85.47	38 **Sr** 87.61	39 **Y** 88.91	40 **Zr** 91.22	41 **Nb** 92.91	42 **Mo** 95.96	43 **Tc**	44 **Ru** 101.1	45 **Rh** 102.9	46 **Pd** 106.4	47 **Ag** 107.9	48 **Cd** 112.4	49 **In** 114.8	50 **Sn** 118.7	51 **Sb** 121.8	52 **Te** 127.6	53 **I** 126.9	54 **Xe** 131.3
55 **Cs** 132.9	56 **Ba** 137.3	57–71	72 **Hf** 178.5	73 **Ta** 180.9	74 **W** 183.9	75 **Re** 186.2	76 **Os** 190.2	77 **Ir** 192.2	78 **Pt** 195.1	79 **Au** 197.0	80 **Hg** 200.6	81 **Tl** 204.4	82 **Pb** 207.2	83 **Bi** 209.0	84 **Po**	85 **At**	86 **Rn**
87 **Fr**	88 **Ra**	89–103	104 **Rf**	105 **Db**	106 **Sg**	107 **Bh**	108 **Hs**	109 **Mt**	110 **Ds**	111 **Rg**	112 **Cn**						

Lanthanoids

57 **La** 138.9	58 **Ce** 140.1	59 **Pr** 140.9	60 **Nd** 144.2	61 **Pm**	62 **Sm** 150.4	63 **Eu** 152.0	64 **Gd** 157.3	65 **Tb** 158.9	66 **Dy** 162.5	67 **Ho** 164.9	68 **Er** 167.3	69 **Tm** 168.9	70 **Yb** 173.1	71 **Lu** 175.0

Actinoids

89 **Ac**	90 **Th** 232.0	91 **Pa** 231.0	92 **U** 238.0	93 **Np**	94 **Pu**	95 **Am**	96 **Cm**	97 **Bk**	98 **Cf**	99 **Es**	100 **Fm**	101 **Md**	102 **No**	103 **Lr**

Table 2.2   The rare-earth elements of most interest to the industrial supply chain.

Element	Symbol	Atomic number	Element	Symbol	Atomic number	Element	Symbol	Atomic number
Lanthanum	La	57	Europium	Eu	63	Erbium	Er	68
Cerium	Ce	58	Gadolinium	Gd	64	Thulium	Tm	69
Praseodymium	Pr	59	Terbium	Tb	65	Ytterbium	Yb	70
Neodymium	Nd	60	Dysprosium	Dy	66	Lutetium	Lu	71
Samarium	Sm	62	Holmium	Ho	67	Yttrium	Y	39

(although not always) not included in discussions pertaining to the rare earths. From the industrial viewpoint, therefore, the rare earths generally of most interest are the lanthanoids (except Pm) + Y – the 15 elements shown in Table 2.2.

REEs can be grouped into the so-called light REEs (LREEs) and heavy REEs (HREEs), using strict definitions based on the specific configurations of 4f electrons within each REE atom. On this basis, the LREEs are La–Ce–Pr–Nd–Sm–Eu–Gd. Moving across the periodic table from La, which has no 4f electrons, to Gd, incremental numbers of unpaired electrons with clockwise spin are present in each successive element.[3] Gadolinium has seven 4f electrons with clockwise spin, leading to a stable, half-filled 4f electron shell. The HREEs then, are Tb–Dy–Ho–Er–Tm–Yb–Lu + Y. Continuing along the row of lanthanoids, starting with Tb, incremental numbers of electrons with anticlockwise spin are present in each successive element, up to Lu. These additional electrons are 'paired' with the electrons with clockwise spin already present. Y is considered to be an HREE because of its similar properties to these elements.[4] It should be noted that for somewhat obscure reasons, many rare-earth exploration and mining companies incorrectly refer to Eu and Gd as HREEs.

A further distinction is frequently made, when an additional grouping, the medium REEs (MREEs), is defined based on the way in which groups of REEs stay together during the processing of rare-earth ores. In this scenario, the LREEs would be La–Ce–Pr–Nd, the MREEs would be Sm–Eu–Gd, with the rest defined as HREEs. Given the inherent confusion that can result from these various naming schemes, the reader should be careful to determine which definitions have been used whenever reviewing REE-related information provided by industry players.

REEs are frequently separated and sold in their oxide form and therefore it is customary to render mineral-deposit data in terms of rare earth oxide (REO) equivalents. The corresponding terms light- (LREO), medium- (MREO) and heavy-rare-earth oxide (HREO) are used as appropriate, particularly when assessing the supply and demand characteristics of the overall rare-earths market. Exploration-stage projects may report REE concentrations as elemental parts-per-million (ppm) values, which will be slightly lower than their oxide equivalents. Total-rare-earth oxide(s) [TREO(s)] refers to the sum total of rare-earth-oxide equivalents present in a deposit.

2.2 APPLICATIONS AND END USES OF RARE EARTHS

Rare earths are used in a variety of chemical forms and in a wide variety of applications, some more complex than others. Table 2.3 shows some examples of end-use applications for individual REEs.

The first broad group of applications of rare earths is the process enablers – end uses that are used as part of the production life cycle of other materials and components, where the REEs do not stay with the processed material. In general, the properties of simple REE compounds are utilised, such as REOs. There are many specific process-enabling applications of REEs; significant examples include the following:

- *Fluid-cracking catalysts:* These are materials used in the petroleum-refining industry, to convert heavy crude oil into gasoline and other valuable products. La and Ce are added to the catalytic compounds to stabilise the overall matrix and to take advantage of their ability to interact with the hydrogen atoms found in the long-chain hydrocarbon molecules in the starting raw material. This interaction aids in the transformation of the crude oil into useful petroleum products.
- *Automotive catalytic converters:* Modern vehicles use catalytic converters to reduce the emission of pollutants that result from the internal combustion process. CeO_2 is the primary rare-earth compound used in this process (with some La_2O_3

Table 2.3 Selected end uses and applications of rare earths.

Rare earth element	Symbol	Selected end uses and applications
Lanthanum	La	Fluid-cracking catalysts, energy-storage batteries, hydrogen-storage materials, glass additives, optical lenses
Cerium	Ce	Glass polishing, glass colorizing and decolorizing, UV-resistant glass, phosphors, catalytic convertors, self-cleaning ovens, precipitation-hardening agent
Praseodymium	Pr	Ceramic pigments, permanent magnets, fiber-optic amplifiers, scintillators, alloying agents, magnetocaloric alloys
Neodymium	Nd	Permanent magnets, glass coloring, lasers, welding goggles, capacitors, enamel colorants
Samarium	Sm	Permanent magnets, lasers, capacitors, chemical reagents, nuclear control rods
Europium	Eu	Phosphors, fluorescent lighting
Gadolinium	Gd	Medical-imaging contrast agents, phosphors, scintillators, nuclear-reactor shielding, microwave components
Terbium	Tb	Phosphors, magneto-optic films, sonar actuators, fuel-cell stabilisers
Dysprosium	Dy	Permanent magnets, capacitors, sonar actuators, lasers, nuclear control rods, dosimeters, high-intensity lighting
Holmium	Ho	Microwave components, nuclear control rods, lasers, glass colorizing, magnetic pole pieces
Erbium	Er	Glass colorizing, optic-fiber amplifiers, lasers, nuclear control rods, cryocoolers
Thulium	Th	Lasers, portable X-ray sources, microwave components
Ytterbium	Yb	Optic-fiber amplifiers, stress gauges, thermal-barrier coatings, lasers
Lutetium	Lu	Phosphors
Yttrium	Y	Thermal-barrier coatings, crucibles, phosphors, sensors, ceramics, microwave components, grain refiners, superconductors

and Nd_2O_3), usually in conjunction with platinum-group metals.

- *Polishing media:* Significant amounts of CeO_2 (with some La_2O_3 and Nd_2O_3) are utilised in the polishing of glass, mirrors, TV screens, computer displays and the wafers used to produce silicon chips. When used in a fine powder form, the REOs react with the surface of the glass to form a softer layer (the so-called 'mechano-chemical' effect), thus making it easier to polish the surface to a high-quality finish.

The second group of end uses for rare earths involve incorporating various REEs into sometimes complex alloys and compounds, which are then used in engineered components. These components might then be utilised in sub-assemblies, which in turn might be used to produce a complex engineered product or device. In some cases, relatively small amounts of REEs are used in the overall product, but their presence is critical for the functionality of the ultimate end application.

There are many specific 'building-block' applications of REEs; significant examples include the following:

- *Permanent magnets:* The use of REEs in certain magnetic alloys has made it possible to produce permanent magnet materials that generate very strong magnetic fields. At the same time, these magnets are able strongly to resist being demagnetised when exposed to other magnetic fields or to elevated temperatures. The REEs present in these alloys, such as Nd, Pr, Sm, Dy and in the past Tb, effectively help to 'channel' the inherent ferromagnetism of transition metals such as iron (Fe) and cobalt (Co).

  These characteristics have revolutionised magnetics design in recent years, most notably in the production of high-performance electric motors, which convert electricity into mechanical motion, and electric generators, which, operating in reverse, convert mechanical motion into electricity. Permanent-magnet motors (PMMs) and generators (PMGs) are used in, for example, Prius-class hybrid electric vehicles (HEVs), to power the vehicle and also to recapture energy associated with braking, respectively. Each Prius HEV requires an estimated 1 kg of Nd + Pr and perhaps 100–200 g of Dy in the various magnets required for operation.[5] PMGs can be used in megawatt-scale, next-generation wind turbines. The magnets in these large turbines contain an estimated 150–200 kg of Nd + Pr, and perhaps 20–35 kg of Dy, per megawatt of generating capacity.[6]

  In addition to being able to produce such electrical machines with higher efficiencies and greater performance, rare-earth-permanent-magnet (REPM) materials have made it possible to miniaturise motors, loudspeakers, hard-disk drives, cordless power tools and other applications that use permanent magnets to operate, while maintaining the same or better output characteristics as other technologies.

The two most important families of REPM materials are those based on compounds of Sm, Co and other elements and those based on compounds of Nd, Fe and B, along with other REEs such as Pr and Dy. The latter family is the more widely used. The HREE Dy is used in Nd-based REPM materials, to increase the ability of the material to resist being demagnetised as a result of elevated temperatures or stray magnetic fields – a characteristic known as coercivity. Increased additions of Dy or Tb to Nd-based materials give them increased coercivity (although Tb is generally not used today for this purpose). A wide range of grades or blends of material, tailored to meet particular specifications, are used today.

- *Energy storage:* Compounds of La and Ni are used to produce battery cells for energy storage. The presence of La permits the absorption of H in the cell and the ease of reversal of this electrochemical process leads to La–Ni–H compounds being particularly suitable for rechargeable-battery applications.

  Although recent developments in battery cells that utilise lithium ion technology are gaining ground in certain applications, batteries based on La–Ni–H are still a very cost-effective and reliable method of storing electricity, for applications including Prius-class HEV battery packs (which use an estimated 2.3 kg of La per vehicle for this application)[5] and others.

- *Phosphors:* Phosphor materials emit light after being exposed to electrons or ultraviolet (UV) radiation. Liquid-crystal displays (LCDs) and plasma screen displays, light-emitting diodes (LEDs) and compact fluorescent lamps (CFLs) all utilise such materials. Compounds containing Eu, Y and Tb are frequently used to produce phosphors and are fined-tuned for particular color outputs. Since much more of the electrical energy is converted into light than with conventional light sources, phosphor materials are significantly more energy-efficient than older technologies, requiring much less electricity to produce the same outputs.

- *Glass additives:* CeO_2 and La_2O_3 are used as additives in the glass industry for a variety of purposes. They are used to remove undesirable coloration in commercial glass by reducing the effects of the presence of Fe within the material. They are also used to reduce UV light penetration, thus protecting the interiors of vehicles and other materials from degradation over time. They can also be used to increase the refractive index of glass lenses. Other REEs such as Nd can be added to glass to provide either decorative or functional coloration or tinting to glass.

2.3 RARE-EARTH MINERALS AND GEOLOGY

In recent years, it has become an oft-repeated cliché to say that 'rare earths' are neither 'rare' nor 'earths'. REEs are metals, but are not found in the metallic form; they are also fairly well distributed within the Earth's crust (Ce, for example, has a crustal abundance of ~43 ppm.[7] What is rare is for these elements to be concentrated enough to result in a deposit of rare earths that can be mined economically. Rarer still are economically mineable deposits containing appreciable amounts of HREEs.

Table 2.4 shows a selection of the minerals that contain REEs. There are well over 220 known REE-bearing minerals,[9] but very few have ever been commercially exploited (*i.e.* bastnaesite, monazite, loparite, xenotime and the ion-adsorption clays). Because the REEs are chemically very similar to each other, they are all found together in their associated minerals, substituting for other ions to varying degrees. Such minerals include silicates, carbonates, oxides, phosphates and halides. The variations in REE content in different

Table 2.4 Selected rare-earth-bearing minerals.[7,8]

Mineral	REO content/mass%	Common formula[a]
Allanite	38	$(Ca,RE)_2(Al,Fe)_3(SiO_4)_3OH$
Apatite	19	$Ca_5(PO_4)_3(F,Cl,OH)$
Bastnaesite	75	$RECO_3F$
Brannerite	9	$(U,Ca,RE)(Ti,Fe)_2O_6$
Eudialyte	9	$Na_{15}Ca_6(Fe,Mn)_3Zr_3(Si,Nb)Si_{25}O_{73}(OH,Cl,H_2O)_5$
Fergusonite	53	$RE(Nb,Ti)O_4$
Loparite	30	$(RE,Na,Ca)(Ti,Nb)O_3$
Monazite	65	$(LRETh)PO_4$
Synchysite	51	$CaRE(CO_3)_2F$
Xenotime	61	$REPO_4$

[a]RE = rare-earth element.

occurrences of the same minerals gives each deposit a unique 'signature' in terms of specific REE content.

The occurrence of REE-bearing minerals is influenced by a variety of rock-forming and hydrothermal processes and they are typically found in a number of distinct rock types.[7] Primary REE-bearing mineral deposits include the following:

- *Carbonatites:* Igneous rocks mostly consisting of carbonate minerals and whose mineralisation is frequently high grade (*i.e.* REEs are present in high concentrations relative to the host rock) and rich in the LREEs.
- *Alkaline igneous rocks:* Rich in the alkali metals, niobium (Nb), titanium (Ti), zirconium (Zr) and the REEs. Mineralisation is frequently rich in HREEs, albeit at lower grades.
- *Hydrothermal occurrences:* Deposits that result from the flow and actions of high-temperature fluids.

Secondary mineral deposits are also an important source of rare-earth minerals.[7] Examples include the following:

- *Placer deposits:* These are occurrences where minerals have been transported by the flow of water, such as rivers, lakes and seas, with a wide range of original occurrences. Mineralisation is usually of very low grade, but because the minerals have been finely divided through natural processes, they are relatively straightforward to process.
- *Residual weathered deposits:* As the name implies, these are deposits that have formed over long periods as a result of weathering processes. Examples include the formation of lateritic clays (a type of soil that forms in tropical climates) and the rare ion-adsorption clays. These materials have a wide range of concentrations, but with very low grades. Again, because of the relative ease of processing, these low grades are not necessarily an impediment to commercialisation, and in the case of the ion-adsorption clays of southern China, these are at present the world's only commercial source of HREEs.

Table 2.5 shows the typical REO distribution for REE-bearing minerals of interest, found in selected locations.

Table 2.5 Typical rare-earth distributions within mineral deposits of interest.[10-14]

REO	REO/mass%						
	Deposit						
	Bayun Obo, Inner Mongolia	Maoniuping, Sichuan	Mountain Pass, California	Bayun Obo, Inner Mongolia	Mount Weld, Western Australia	Longnan, Jiangxi	Karnasurt, Lovozero
	Mineral						
	Bastnaesite	Bastnaesite	Bastnaesite	Monazite	Monazite	Ion-adsorption clay	Loparite
La_2O_3	26.0	37.4	33.2	25.7	25.14	2.18	27.8
CeO_2	50.4	47.1	49.1	50.1	44.97	1.09	57.1
Pr_6O_{11}	5.4	3.61	4.3	5.4	4.90	1.08	3.7
Nd_2O_3	15.8	10.3	12.0	16.5	17.18	3.47	8.7
Sm_2O_3	1.07	0.82	0.79	1.11	2.44	2.34	0.91
Eu_2O_3	0.21	0.10	0.10	0.21	0.56	0.37	0.13
Gd_2O_3	0.52	0.26	0.20	0.52	1.54	5.69	0.21
Tb_4O_7	0.03	0.05	0.06	0.02	0.17	1.13	0.07
Dy_2O_3	0.14	0.07	0.05	0.28	0.58	7.48	0.09
Ho_2O_3	0.02	0.02	0.02	–	0.08	1.60	0.03
Er_2O_3	–	0.04	0.02	–	0.22	4.26	0.07
Tm_2O_3		0.01	0.02		0.02	0.60	0.07
Yb_2O_3	0.06	0.01	0.02	0.05	0.09	3.34	0.29
Lu_2O_3	–		0.02	0.01	0.02	0.47	0.05
Y_2O_3	0.29	0.14	0.10	0.28	2.09	64.90	0.14
Total	**99.94**	**99.93**	**100**	**100.18**	**100**	**100**	**99.36**

Table 2.6 Potential global rare-earth resources.[15]

Country	Potential resources, mass/t
China	55 000 000
Commonwealth of Independent States	19 000 000
USA	13 000 000
India	3 100 000
Australia	1 600 000
Other	22 100 000
Total	**113 800 000**

HREEs are generally much scarcer than LREEs. The majority of REE-bearing mineral deposits are dominated, in tonnage terms, by the presence of LREEs. So-called HREE mineral deposits are therefore generally designated as such on the basis of the potential value of that deposit by virtue of the presence of HREEs, not their overall tonnage, because HREEs tend to be more valuable owing to their scarcity. However, HREEs tend to occur in lower material grades within the mineral resource.

The United States Geological Survey publishes so-called global reserve estimates for rare earths, the details of which are summarised in Table 2.6.[15] Such estimates do not actually describe mineral reserves in the normal regulatory sense (*i.e.* those parts of a mineral resource that are economically viable), but rather potential mineral resources.

2.4 MINING AND PROCESSING OF RARE EARTHS

Most of the global mining of REEs takes place in China, with minor production in Russia and India. Small amounts of previously mined ores are currently processed in the USA. At the Bayun Obo mine in Inner Mongolia, the rare-earth minerals are extracted as a byproduct of iron-ore mining; they are also frequently produced as by-products elsewhere. For other current sources, such as Mountain Pass, California, the rare earths present are the primary product of interest. Obviously the disposition of rare earths as either primary products or by-products affects the economics of the entire operations used in their mining.

Hard-rock rare-earth mines generally use open-pit methods for extraction, which involves the removal of overburden followed by blasting and digging. Placer deposits use scraping or dredging techniques to obtain the ores. In either case, once the ore has been removed from the ground, it is subject to physical beneficiation in order to concentrate the quantities of rare earths. This involves the crushing and grinding of the ore material, prior to initial chemical processing, which produces a slurry containing 25–35% solids. Additional standard processes such as froth flotation, gravitational, magnetic or electrostatic separation may be utilised as appropriate, in order to produce useful rare-earth concentrates containing 40–60% rare earths.

The next stage in processing is to chemically beneficiate the concentrates – using acids and alkalis to subject the rare-earth concentrates to solution and precipitation cycles. The specific chemicals and methods used very much depend on the minerals being processed and the natural distribution of the REEs present. The chemical processes used to chemically beneficiate rare-earth concentrates are optimised for each particular rare-earth deposit being exploited. Figure 2.1 shows the chemical processing routes for a variety of rare-earth minerals.

Ion-adsorption clay deposits are frequently subject to *in situ* leaching and extraction methods, where holes are drilled into the deposit and reagent are poured into the holes, which then percolate through the rocks, bringing with them the rare earths present.[7] The latter technique has been subject to widespread criticism for the resulting negative environmental impact associated with it.

As a result of first the physical and then the chemical beneficiation processes, concentrates containing the equivalent of 60–90 wt% TREOs may be obtained. The final stage of processing these concentrates typically involves the separation of individual REEs via the solvent-extraction (SX) process. Figure 2.2 gives an example of a flow diagram for processing bastnaesite concentrates at Mountain Pass, California, into individual REE products. Purities as high as 99.5–99.99% are not uncommon for the resulting materials; high purification levels are required for certain application, hence materials will be reprocessed across successive SX circuits, to remove impurities (mostly other REEs). SX is a time- and energy-intensive process, but has become the standard method for achieving separated rare-earth products.

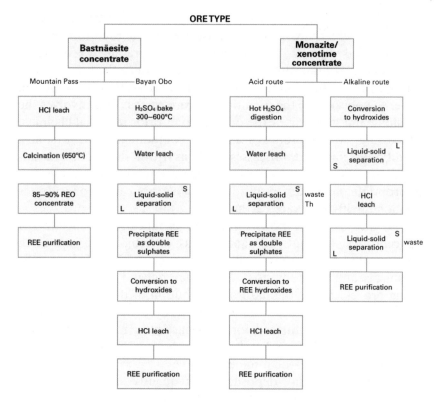

Figure 2.1   Examples of processing routes for the chemical beneficiation of rare-earth elements. Courtesy of the British Geological Survey.[7]

2.5   RARE-EARTH DEMAND DRIVERS

The demand for many of the REEs has increased significantly in recent years, owing to their use in a wide range of applications that underpin the modern technological age. Table 2.7 shows estimates of the end-use demand for REEs in 2010.

The majority of rare earths are used within China, where a downstream industry has been steadily growing to take advantage of low labor rates, in addition to close proximity to the sources of supply. The majority of REEs consumed in the USA are LREEs, used in FCCs. Japan and South Korea still have a strong base of manufacturing processes that require REEs, despite the recent trend for Japanese companies to procure more downstream semi-finished or finished products from China.

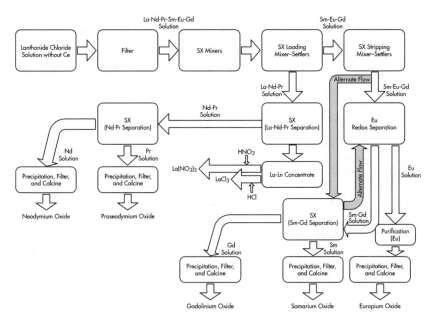

Figure 2.2   Solvent-extraction process for Mountain Pass rare-earth mineral deposit. Courtesy of the Society for Mining, Metallurgy and Exploration.[8]

Table 2.7   Estimated global rare-earth demand in 2010 (REOs ± 15%).[16,17]

Sub-sector	China	Japan and SE Asia, mass/t	USA, mass/t	Others, mass/t	**Total, mass/t**	Market share/%	Market value/%
Permanent magnets	20 500	4000	500	1000	**26 000**	21	39
Catalysts	9000	3000	9000	3500	**24 500**	20	16
Metal alloys	14 500	5500	1000	1000	**22 000**	18	15
Polishing powders	9500	7000	1000	1500	**19 000**	15	10
Glass additives	7000	1500	1000	1500	**11 000**	9	4
Phosphors	5500	2000	500	500	**8500**	7	12
Ceramics	2500	2500	1500	500	**7000**	6	2
Other	4000	2000	500	500	**7000**	6	2
Total	**72 500**	**27 500**	**15 000**	**10 000**	**125 000**	100	100
Market share	**58%**	**22%**	**12%**	**8%**	**100%**		

Table 2.8   Projected future global demand for rare earths.[16–18]

Sub-sector	2010e (± 20%), mass/t	AAGR, 2010–2015, mass/%	2015f (± 20%), mass/t	AAGR, 2015–2020, mass/%	2020f (± 25%), mass/t
Permanent magnets	26 000	8–12	40 000	10–15	67 000
Metal alloys	22 000	8–12	36 000	4–7	47 500
Polishing powders	19 000	4–8	25 000	5–7	34 000
Catalysts	24 500	0–2	25 500	4–6	32 000
Phosphors	8500	6–10	12 500	5–7	17 000
Glass additives	11 000	Decreases	10 000	4–6	12 500
Ceramics	7000	4–6	9000	5–7	12 000
Other	7000	9–13	12 000	20–25	33 000
Total demand	**125 000**	**4–8**	**170 000**	**6–10**	**255 000**

e, estimated; f, forecasted.

Most of the applications shown in Table 2.7 have seen robust demand growth rates in the past decade, a reflection of the use of rare earths in components for electronics and consumer goods, labor-saving devices, appliances, vehicles and a wide range of industrial products.

As economies around the world continue to grow, so the demand for REEs grows with it. Demand has been particularly strong over the past decade for REPMs, ceramic materials for capacitors and for the metal alloys used for energy storage. In some cases, relatively small amounts of REEs are used in each device but, with millions of units being sold each year, the overall consumption becomes significant.

Table 2.8 shows the estimated demand for REEs in 2015 and 2020, suggesting a potential doubling of overall demand by 2020, compared with 2010 levels. In addition to the established drivers associated with general global economic growth, significant additional demand for REEs is expected from the so-called 'green' or 'clean-tech' sector. Three specific sub-sectors will likely drive this demand:

- *HEVs and other electric vehicles (EVs):* Significant quantities of REEs will be required to meet the demand for battery alloys in HEVs and for PMMs and PMGs in HEVs and other types of EVs. Although there is much interest in the use of Li-ion batteries for various types of EVs, the significant unit costs for such batteries at present will likely mean a significant demand for HEVs using La–Ni–H batteries into at least the medium term. Given the superior energy densities of Li-ion batteries, however, if the price of these materials can be reduced, it is likely that in the long term growth in the use of La-based alloys for energy storage will begin to decline.

Increasing market penetration of EVs will mean that the demand for Nd-based permanent magnets in motors and generators will remain strong. Such materials will require additions of HREEs such as Dy, to ensure that the machine performance is maintained even at the elevated temperatures frequently found in these 'under-the-hood' applications. A number of initiatives to replace PMMs with induction motors have been announced recently, primarily due to price increases for rare earths; however, such electrical machines are usually bulkier, less efficient and require additional cooling. It is likely that REE-based PMMs and PMGs will remain the devices of choice for these EV applications.

- *Direct-drive PMG (DDPMG) wind turbines:* One of the largest potential future uses of REPMs, and thus Nd, Dy and Pr, is in 'next-generation' utility-scale wind turbines, which utilise DDPMG systems. Conventional wind turbines utilise sets of electromagnetic coils within the generator, connected to a large mechanical gearbox. This gearbox has historically been a significant source of reliability issues and, by using a PMG in the turbine, the gearbox can be eliminated, thus reducing maintenance issues and potentially reducing the weight at the top of the tower that supports the turbine.

  To date, the market penetration of these DDPMG wind turbines has been modest, despite mainstream-media reports to the contrary. However, more and more wind-turbine manufacturers are introducing DDPMG-based turbines into their product line-ups and all expectations are that if prices for rare earths stabilise, then this will be a strong growth area for REPMs and the REEs required to produce them.

- *Energy-efficient devices and appliances:* There is a growing demand for consumer goods and other products that use energy more efficiently in their operation. Already the requirement for improved energy efficiency is being mandated in a number of jurisdictions around the world, as a means of reducing fossil-fuel usage and to reduce the emission of greenhouse gases. Examples include the replacement of incandescent light bulbs with CFLs and LEDs, which use a variety of HREEs. Increasing numbers of electrical appliances, such as heating and air-conditioning units, washing machines and driers, use REE-based PMMs to meet the demands for greater efficiency.

2.6 RARE EARTH SUPPLY DRIVERS

Until the 1940s, India and Brazil were the principal sources of REEs, producing them in relatively small quantities compared with modern production rates. Australia, Malaysia and South Africa began to supply REEs in the middle part of the twentieth century from monazite placer deposits, and it was at this point that the Mountain Pass bastnaesite mine in California, USA, came on-stream. The latter quickly became the world's largest producer of REEs.

In the 1980s, China began to produce REEs from bastnaesite and monazite in ever-increasing quantities, ultimately dominating the market, as can be seen in Figure 2.3. As demand for HREEs increased, ion-adsorption clays and xenotime were mined in the southern provinces of China from the early 1990s onwards – the only commercially significant producers of HREEs currently in operation. The rise of low-cost production in China put significant pressures on the rest of the global producers and one by one they closed, including the Mountain Pass mine in 2002, leaving only a handful of non-Chinese producers.

As shown in Table 2.9, an estimated 123 kt of REOs were produced worldwide in 2010.[22] Mines in China accounted for over 96% of that production, where the principal sources of LREO production are located in the Inner Mongolia autonomous region and in Sichuan province. HREOs are primarily sourced from Jiangxi province. Other provinces also produce REOs. The Chinese Ministry of Land and Resources allocates rare-earth production quotas to a number of provinces and regions each year. Historically, the actual amount of rare earths produced has been significantly higher than the production quotas allocated. This may

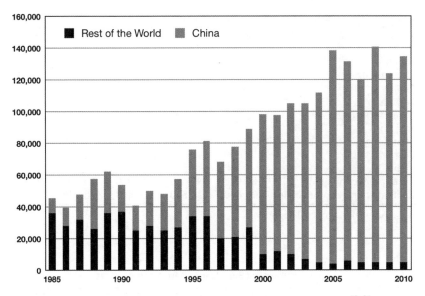

Figure 2.3 Historical global supply of rare-earth oxides (tonnes).[19–21]

Table 2.9 Estimated global rare-earth production in 2010 (mass/t).[22]

Rare earth oxide	China	Rest of world	Total
La_2O_3	33 530	1650	35 180
CeO_2	46 950	2010	48 960
Pr_6O_{11}	6050	180	6230
Nd_2O_3	19 710	510	20 220
Sm_2O_3	2450	40	2490
Eu_2O_3	300	0	300
Gd_2O_3	1670	10	1680
Tb_4O_7	240	0	240
Dy_2O_3	940	10	950
Y_2O_3	5890	40	5930
Other	1180	10	1190
Total	**118 910**	**4460**	**123 370**

have changed in 2011, following stricter enforcement by the authorities. Table 2.10 shows the statistics from recent years.

As the world's largest producer of REEs, China's actions and initiatives tend to have the largest effects on the rare-earth sector. In recent years, the authorities there have imposed export quotas on the shipment of REEs out of China, as detailed in Figure 2.4. A variety of reasons have been suggested for these restrictions,

Table 2.10   Chinese rare-earth production-quota allocations (mass/t).[22]

Province/region	2007	2008e	2009	2010	2011
Fujian	220	320	720	1500	2000
Guangdong	700	1000	1500	2000	2200
Guangxi	200	200	200	2000	2500
Hunan	100	300	800	1500	2000
Inner Mongolia	46 000	46 000	46 000	50 000	50 000
Jiangxi	7400	7400	7400	8500	9000
Shandong	1200	1200	1500	1500	1500
Sichuan	31 000	31 000	24 000	22 000	24 400
Yunnan	200	200	200	200	200
Total production quota	**87 020**	**87 620**	**82 320**	**89 200**	**93, 800**
Estimated actual production	**120 800**	**124 500**	**129 400**	**118 900**	**TBD**

e, estimated; TBD, to be determined.

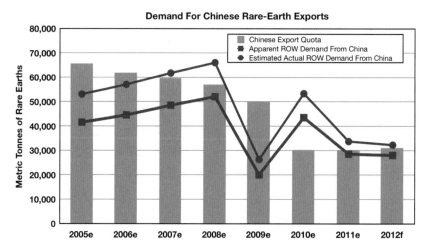

Figure 2.4   Demand for Chinese rare-earth exports.[16,17,23]

ranging from a desire to encourage downstream industries to migrate to China (bringing jobs and revenues to that country) to recent initiatives relating to the shutdown and closure of inefficient polluting mines in order to allow for environmental remediation. The latter would, of course, be more effective if restrictions were placed on overall production and not just on exports. China has also had ongoing issues relating to uncontrolled and illegal operations, which by some estimates supply up to 15–20% of the world's REE demand (this is the difference between the two 'rest-of-the-world demand' lines in Figure 2.4). China has made a number of policy decisions relating to cleaning up the negative environmental impact of its unregulated REE mining activities.

Despite the imposition of export quotas, until recently there have been few real supply issues for REEs because demand was generally matched by supply. Buffer stocks held by significant end users of these materials compensated for any shortfalls in specific REEs.

This all changed in July 2010, when the Chinese authorities announced a significant reduction in export quotas for the latter half of 2010 – a maximum of ~ 8000 t of REOs for export, bringing the total for 2010 to just over 30 000 t.[24] This was a 40% reduction over 2009 and caused considerable consternation in the rare-earth industry.

The initial results of this action were very significant price increases for the export of LREEs, in some cases by 1000–1500%. The announcement at the end of December 2010 of an REE export quota of $\sim 14\,500$ t for the first part of 2011[25] did little to quell concerns; however, the second announcement for the second half of 2011 brought the quotas to around the same level as in 2010,[26] and prices started to fall from their peak in July 2011. The pricing characteristics of REEs are discussed in further detail in Section 2.8.

Table 2.11 presents a summary of the recent export-quota allocations by the Chinese authorities. It should be noted that the Chinese authorities allocate export quotas to individual trading companies in China, some of which are partly foreign-owned. The quotas prior to 2012 were monolithic, *i.e.* there was no differentiation between specific REEs or REOs within those quotas. Furthermore, the export quotas do *not* apply to semi-finished or

Table 2.11 Chinese rare-earth export-quota allocations (mass/t).[27]

Sub-group	2009 H1	H2	2010 H1	H2	2011 H1	H2	2012e H1	H2p
Domestic	15 043	18 257	16 304	6208	10 762	12 221	17 926	TBD
Foreign JV	6685	10 160	5978	1768	3746	3517	6978	TBD
Sub-total	**21 728**	**28 417**	**22 282**	**7976**	**14 508**	**15 738**	**24 904**	**6226**
Total	**50 145**		**30 258**		**30 246**		**31 130**	

e, estimated; p, projected; JV, joint venture; TBD, to be determined.

finished goods, such as permanent magnets or magnet alloys, produced in China. At present they apply only to the raw material forms of REEs and simple REE-based compounds, along with some ferroalloys.

At the end of December 2011, the Chinese Ministry of Commerce announced the first round of allocations of rare-earth export quotas for 2012 to individual companies operating in China.[27] The total export quotas allocated during this first round came to 24 904 t of rare earths. The approach taken in this announcement was unusual for three reasons:

- The Ministry issued separate quota allocations for light-(LRE) and medium/heavy-(M/HRE) rare-earth products and not just for rare earths as a whole. Such a breakdown had been anticipated for some time, but 2012 marked the first time that separate allocations had been rolled out.
- Also for the first time, the Ministry clearly telegraphed the intended total export quotas for the entire year, prior to making the usual follow-up allocation announcement later in 2012. They stated that the first round of quota allocations (totaling 24 904 t) would represent 80% of the quota allocations for 2012, which indicates that the total for 2012 would be 31 130 t of rare earths, slightly higher than for 2011.
- The Ministry separated individual companies into two groups – the first group received confirmed quota allocations, whereas the second received only provisional allocations. Companies were placed into one of these groups based on their progress towards implementing new pollution-control regulations, and the latter group will only receive their allocated quotas if they meet the various requirements by July 2012. Companies that fail to meet the new requirements would have their quotas reallocated to other companies.

In recent years, there has been growing concern that China could, for its own reasons, decide unilaterally to restrict the export of REEs to the rest of the world or could possibly consume all of its available domestic supply.[28] The threats to the supply chain are not only of a geopolitical nature, however. A second vulnerability results simply from the geographic 'bottleneck' caused by the concentration of LREE production in the Bayun Obo region of China and of HREE production in southern China. It would require just one moderate earthquake in one of these regions to cause severe problems for the entire global supply chain for these materials.

There is therefore significant interest in developing sources of REE supply outside China. At present, two advanced-stage projects of this type are nearing completion. The first is the revival of the Mountain Pass mine in California, owned by Molycorp Inc. The company plans to be producing at a rate of $19\,000\,t\,a^{-1}$ of separated oxides by the end of 2012, with the possibility of doubling capacity to $40\,000\,t\,a^{-1}$ some time after that.[29] The other near-term project is a mine at Mount Weld in Western Australia, owned by Lynas Corp., coupled with a concentration facility in Australia and a processing and separation facility in Malaysia. The Mount Weld project is scheduled to be producing at a rate of $11\,000\,t\,a^{-1}$ of separated oxides by the end of 2012, with a future expansion to $22\,000\,t\,a^{-1}$ forming part of the company's strategy.[30]

Despite the significant quantities of REEs that will become available outside China in the near future, very little of this production will consist of HREEs. Although they tend to be used in smaller quantities than LREEs, nevertheless, given the growing demand for specific HREEs such as Dy and Tb, the question of HREE supply outside China will remain a significant concern for the foreseeable future. Additional projects with some appreciable HREE content are slated to come on-stream in 2013 and beyond, but most of these are at fairly early stages of execution and face a variety of developmental challenges. In addition, there are numerous exploration projects under way all around the world to find additional resources, for both LREEs and HREEs, some of which are detailed in Section 2.9.

2.7 SPECIFIC SUPPLY AND DEMAND ISSUES

As stated previously, REEs are found together in a variety of mineral deposits. The demand profile and end uses of the individual REEs, however, do not match the physical distribution of the elements in those deposits. For example, the REE content of high-coercivity Nd-based REPMs may contain $(Nd + Pr):Dy$ ratios ranging from 50:1 to 2:1 – ratios that are significantly richer in Dy than would typically be found in naturally occurring LREE mineral deposits such as monazite or bastnaesite, where the ratios might range from 50:1 to 100:1 or greater, depending on the particular deposit.

This means that in order to produce enough of the more scarce (and more valuable) HREEs, in addition to the most important LREE Nd, additional quantities of REE minerals have to be processed. This has the potential to result in surpluses of the more abundant REEs such as La and Ce. For this reason, and because of the inherent potential value of the finished products, REE deposits that are relatively rich in Dy, Tb and other HREEs (and also Nd) are particularly attractive for exploration and development. The REE industry tends to view deposits with HREO:(HREO+ LREO) ratios of ~10% or greater as being 'HREE enriched'. Projects with ratios of less than 10% might also be of interest for their HREE content, but only if the overall REE material grade is significant.

Within the wider technology-metals sector, there have been a number of attempts to assess the criticality of certain elements and compounds from end use, industrial and strategic perspectives. Recent examples include studies by the US National Academy of Sciences (NAS),[31] the European Commission (EC),[32,33] the US Department of Energy (DOE),[34,35] the UK House of Commons,[36] a joint study by the American Physical Society and Materials Research Society[37] and Caltech's Resnick Institute.[38] In each of these studies, rare earths (either as a group or as individual elements) were evaluated for criticality and generally deemed to be critical for a variety of applications.

One of the recent reports by the EC ranked REEs collectively as being at the highest degree of supply risk out of 41 different materials reviewed.[32]

The DOE studies on critical materials were particularly detailed in their approach to evaluating which elements are critical to clean-tech applications.[34,35] Using a format adapted from the earlier NAS study,[31] the DOE assessed the potential criticality of specific elements across short- and medium-time scenarios. The studies identified five elements as being the most critical in the short term (2011–2015) for clean-tech applications such as wind turbines, EVs, photovoltaic cells and fluorescent lighting. 'Criticality' in this context was a measure of the importance of a particular element to the clean-tech sector, combined with the estimated risks of supply disruptions. All five of the elements were REEs, specifically Nd,

Eu, Tb, Dy and Y. In the longer term (2015–2025), all of these five REEs remained designated as 'critical'. Dy was ranked as being at the highest level of criticality in both cases, reflecting its important use in REPM materials.

There is growing concern elsewhere in the USA regarding the vulnerability of the defense and other supply chains to the afore-mentioned issues.[39-41] Although defense applications for rare earths do not necessarily require significant quantities of REEs, the components and devices that do use them are frequently critical to the functionality of a wide variety of defense platforms and other supporting technologies. This has led to significant activity in the US Congress in the past 2 years, in the form of various pieces of proposed legislation, to create new policies that would reduce the dependence on China as a source of supply.

There are indications that the supply and demand projections for the critical REEs mentioned above mean that it could be 2014 before we see a permanent surplus of supply of Nd, 2015–2016 for Eu and Tb, 2016 for Y and as late as 2017 for Dy.[22] Of course, much could change between now and these future dates and these data assume that the projects outlined in Section 2.9 come on-stream on time, but they give some indication of the challenges to come between now and then for this sector.

2.8 RARE-EARTH PRICING

As stated previously, export-control policies in China have had a dramatic effect on recent prices for rare earths. A key inflection point occurred in July 2010, following the announcement of H2-2010 quota allocations. This led to significant price increases for exported materials, particularly for LREEs and LREOs. The root cause of this was the unofficial imposition of a quota 'surcharge' by traders and producers in China. This surcharge essentially placed value on each unit of quota available, in addition to the particular rare earth material itself. For the latter half of 2010, it was increases in this surcharge that led to some dramatic price increases for La, Ce and other LREE-based compounds and a disconnect between internal prices within China and those of exported materials. Table 2.12 shows historical pricing information for selected rare earths. Figures 2.5 and 2.6 show recent pricing for La_2O_3 and Dy_2O_3, respectively.

Table 2.12 Recent average prices[a] for selected rare-earth oxides exported from China (US$ kg^{-1}).

Rare-earth oxide	2008	2009	2010	2011
La_2O_3	7	6	22	102
CeO_2	4	5	21	100
Pr_6O_{11}	25	15	46	196
Nd_2O_3	25	15	47	233
Sm_2O_3	5	5	17	102
Eu_2O_3	433	464	552	2868
Gd_2O_3	9	7	22	147
Tb_4O_7	608	352	534	2349
Dy_2O_3	104	105	228	1441
Y_2O_3	15	13	26	139

[a]Prices derived from Metal-Pages spot prices for 99% REO content FOB China (99.9% for Eu_2O_3 and 99.999% for Y_2O_3).[42]

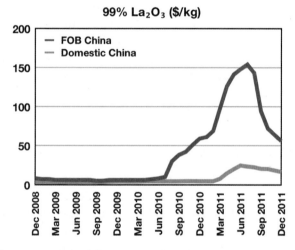

Figure 2.5 Recent Metal-Pages spot prices for FOB (export) and domestic (internal use) La_2O_3 material.[42]

The next key inflection point occurred around February 2011, when prices for rare earths used internally within China also started to increase. This was likely a result of increased speculation and stockpiling of materials inside China. There was also some evidence to suggest that some materials were being siphoned off for export, albeit illegally, in order to capitalise on the arbitrage between internal and external spot prices for these materials.

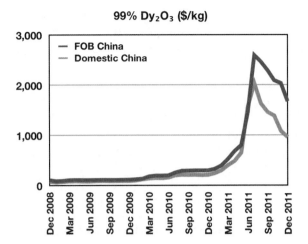

99% Dy₂O₃ ($/kg)

Figure 2.6   Recent Metal-Pages spot prices for FOB (export) and domestic (internal use) Dy₂O₃ material.[42]

Prices for most materials peaked in the summer of 2011, and prices of all of the rare earths have declined since then. The official export channels saw a significant reduction in volumes of materials shipped from China, indicating some degree of demand destruction, particularly for La- and Ce-based materials, which typically constitute 60–70% of export volumes. In the face of the escalating prices, end users of these materials in the FCC and glass-polishing industries made efforts to reduce or to replace these elements, and this was reflected in the volumes.

It should be noted that pricing data for rare earths can be deceptive. These materials are not traded openly on an exchange and so there is no transparency. The so-called spot prices are based on limited datasets, produced for reported transactions to publications such as Metal-Pages[42] and Asian Metal.[43] At the peak of export prices in the summer of 2011, these spot prices were sometimes twice the prices of actual trades taking place, according to anecdotal evidence from within the industry. The trends, however, are likely to be accurate and the unreliability of spot-pricing information has decreased as overall prices have decreased.

As noted above, the recent escalation in prices gave significant impetus to reducing the material intensity, *i.e.* the amount of material used per component or unit. In addition to FCC producers reducing La usage and polishing-powder producers doing the same for Ce, there were indications in 2011 that end users of REPMs started to switch to designs that either used fewer magnets, smaller magnets or different material grades, containing less Dy in particular.

It is important to reiterate, therefore, the dependence of demand projections on the assumed prices for these materials. It is likely that in the medium term, LREO prices will decline further as Mountain Pass and Mount Weld come on-stream, producing significant incremental quantities of LREOs. Price declines for M/ HREOs will likely take longer to occur, since it will be some time before significant new quantities of these rare earths are available.

2.9   FUTURE SOURCES OF RARE-EARTH SUPPLY

Given the recent growth in demand for rare earths and future projected growth associated with clean-tech applications, it is perhaps not surprising that there has been something of an explosion in exploration for new sources of rare earths, in recent years.

At the end of 2011, there were over 400 active exploration and development projects for rare earths, being managed by nearly 250 different companies, in 35 countries, not including India and China.[44] Clearly, not all of these projects are equal – many are based on superficial soil or rock-chip sampling. Some, however, have been under development for several years and are well on their way to production.

Table 2.13 lists the more advanced exploration and development projects; this list includes projects that have a minimum of a defined mineral resource, compliant with relevant technical codes such as Canada's NI-43-101 and Australia's JORC, and/or are currently undergoing engineering and construction work. Figure 2.7 shows their locations, along with the locations of the leading current producers of rare earths and their present state of development.

It can be seen that there are numerous potential new sources of rare-earth supply, given the diversity of locations and project types around the world. However, a key issue is that many companies at the advanced stages of project development plan to produce REE concentrates – intermediate products that will require further processing and separation before they can be used. Unfortunately,

Table 2.13 Selection of rare-earth projects under development outside China.[44]

Project	State/Province	Country	Principal owner
Bear Lodge	Wyoming	USA	Rare Element Resources Ltd
Bokan	Alaska	USA	Ucore Rare Metals Inc.
Clay-Howells	Ontario	Canada	Rare Earth Metals Inc.
Cummins Range	Western Australia	Canada	Navigator Resources Ltd and Kimberley Rare Earths Ltd
Dubbo	New South Wales	Australia	Alkane Resources Ltd
Eco Ridge	Ontario	Canada	Pele Mountain Resources Inc.
Eldor	Quebec	Canada	Commerce Resources Corp.
Hastings	Western Australia	Australia	Hastings Rare Metals Ltd
Hoidas Lake	Saskatchewan	Canada	Great Western Minerals Group Ltd
Kangankunde	Balaka	Malawi	Lynas Corporation Ltd
Kipawa	Quebec	Canada	Matamec Explorations Inc.
Kutessay II	Chui	Kyrgyzstan	Stans Energy Corp.
Kvanefjeld	Kujalleq	Greenland	Greenland Minerals and Energy Ltd
Montviel	Quebec	Canada	Geomega Resource Inc.
Mount Weld	Western Australia	Australia	Lynas Corporation Ltd
Mountain Pass	California	USA	Molycorp Inc.
Nechalacho	Northwest Territories	Canada	Avalon Rare Metals Inc.
Nolans Bore	Northern Territory	Australia	Arafura Resources Ltd
Norra Kärr	Smaland	Sweden	Tasman Metals Ltd
Sarfartoq	Qaasuitsup	Greenland	Hudson Resources Inc.
Steenkampskraal	Western Cape	South Africa	Great Western Minerals Group Ltd
Strange Lake	Quebec	Canada	Quest Rare Minerals Ltd
Two Tom	Labrador	Canada	Rare Earth Metals Inc.
Wigu Hill	Morogoro	Tanzania	Montero Mining and Exploration Ltd
Zandkopsdrift	Northern Cape	South Africa	Frontier Rare Earths Ltd and Korea Resources Corp.

there are currently very few separation facilities located outside China and those that do exist are not independent. Sending concentrates to China for separation is not currently practical, since the re-export of the finished goods back out of China would be subject to restrictions associated with quota allocations. The separation of REEs is a capital-intensive business and therefore the

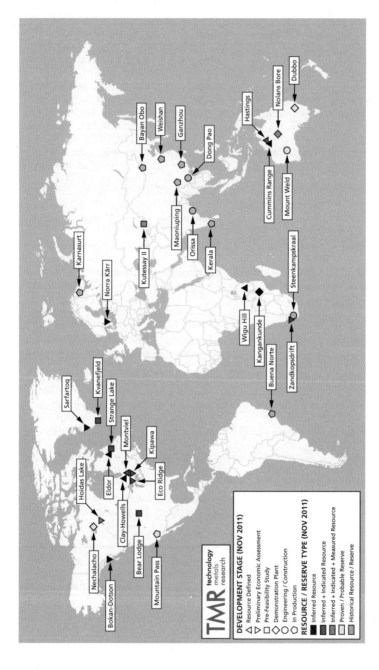

Figure 2.7 Current and potential future sources of rare-earth supply.

processing and separation of rare earths, even if diverse sources of mineral ores can be developed, remains a potential bottleneck for the industry.

In addition to the production of new 'virgin' sources of rare-earth supply, there is increased interest in the recycling of rare-earth materials, from existing components and devices that contain them, in both the public and private sectors.[45] Examples include the use of hydrogen to recycle REPMs[46] and efforts to recycle REE-containing battery materials and phosphors.[47,48] Until recently, such efforts have typically been dismissed as being uneconomic, but in the face of recent price increases for REEs, such efforts are coming under greater scrutiny.

REFERENCES

1. N. G. Connelly, T. Damhus, R. M. Hartshorn and A. T. Hutton, *Nomenclature of Inorganic Chemistry: IUPAC Recommendations 2005*, RSC Publishing, Cambridge, 2005.
2. M. E. Weiser and T. B. Coplen, *Pure Appl. Chem.*, 2011, **83**, 359.
3. J. B. Hedrick, personal communication, 2011.
4. A. H. Daane, in *The Encyclopedia of the Chemical Elements*, ed. C. A. Hampel, Reinhold, New York, 1968, pp. 810–821.
5. J. Lifton, personal communication, 2010.
6. G. P. Hatch, *How Does the Use of Permanent Magnets Make Wind Turbines More Reliable?*, Terra Magnetica, Carpentersville, IL, 2009, http://www.terramagnetica.com/?p=263 (last accessed 29 December 2011).
7. A. Walters and P. Lusty, *Rare Earth Elements*, British Geological Survey, Nottingham, 2011.
8. S. B. Castor and J. B. Hedrick, in *Industrial Minerals and Rocks: Commodities, Markets and Uses*, ed. J. W. Kogel, N. C. Trivedi, J. M. Barker and S. T. Krukowski, Society for Mining, Metallurgy and Exploration, Littleton, CO, 2006, pp. 769–792.
9. A. P. Jones, F. Wall and C. T. Williams, in *Rare Earth Minerals: Chemistry, Origin and Ore Deposits*, ed. A. P. Jones, F. Walland C. T. Williams, Chapman and Hall, London, 1996, pp. 349–356.
10. R. Chi and J. Tian, *Weathered Crust Elution-Deposited Rare Earth Ores*, Nova Science, New York, 2008.

11. C. Wu, Z. Yuan and G. Bai, in *Rare Earth Minerals: Chemistry, Origin and Ore Deposits*, ed. A. P. Jones, F. Wall and C. T. Williams, Chapman and Hall, London, 1996, pp. 281–310.

12. J. B. Hedrick, in *U.S. Geological Survey Minerals Yearbook–2002*, United States Geological Survey, Reston, VA, 2002, pp. 61.1–61.16.

13. N. Curtis, *Rare Earths: We Touch Them Everyday*, Roskill International Rare Earths Conference Presentation, November 2011, Lynas Corporation, Sydney, 2011.

14. C. K. Gupta and N. Krishnamurthy, *Extractive Metallurgy of Rare Earths*, CRC Press, Boca Raton, FL, 2005.

15. D. J. Cordier, *Mineral Commodity Summaries: Rare Earths*, United States Geological Survey, Reston, VA, 2011.

16. D. J. Kingsnorth, *Rare Earths: Reducing Our Dependence upon China*, IMCOA, Mount Claremont, Western Australia, 2011.

17. G. P. Hatch, *Internal Report #1*, Technology Metals Research, Carpentersville, IL, 2011.

18. D. J. Kingsnorth, personal communication, 2011.

19. G. P. Hatch, *A Summary Overview of the Rare-Earths Market*, Technology Metals Research, Carpentersville, IL, 2011.

20. Z. Chen, personal communication, 2011.

21. J. Chegwidden, *Rare Earths: an Evaluation of Current and Future Supply*, China Rare Earth Summit 2010, Roskill Information Services, London, 2010.

22. G. P. Hatch, *Critical Rare Earths: Global Supply and Demand Projections and the Leading Contenders for New Sources of Supply*, Technology Metals Research, Carpentersville, IL, 2011.

23. G. P. Hatch, *On Rare Earths, Quotas and Embargoes*, Technology Metals Research, Carpentersville, IL, 2010, http://www.techmetalsresearch.com/?p=2304 (last accessed 29 December 2011).

24. G. P. Hatch, *China's Rare Earths Game Plan: Part 1 – The Effects of Reduced Export Quotas*, Technology Metals Research, Carpentersville, IL, 2010, http://www.techmetalsresearch.com/?p=1263 (last accessed 29 December 2011).

25. G. P. Hatch, *How The H1–2011 Rare-Earth Quotas Were Allocated*, Technology Metals Research, Carpentersville, 2010, http://www.techmetalsresearch.com/?p=2583 (last accessed 29 December 2011).

26. G. P. Hatch, *Chinese Rare-Earth Export Quotas for H2–2011*, Technology Metals Research, Carpentersville, IL, 2011, http://www.techmetalsresearch.com/?p=3865 (last accessed 29 December 2011).

27. G. P. Hatch, *The First Round Of Chinese Rare-Earth Export-Quota Allocations for 2012*, Technology Metals Research, Carpentersville, IL, 2011, http://www.techmetalsresearch.com/?p=4744 (last accessed 29 December 2011).

28. J. Lifton, *The New 'Great Game'*, Technology Metals Research, Carpentersville, IL, 2009, http://www.techmetalsresearch.com/?p=158 (last accessed 29 December 2011).

29. Molycorp Inc., *Form S-1: Registration Statement*, United States Securities and Exchange Commission, Washington, DC, 2010, http://www.sec.gov/Archives/edgar/data/1489137/000095012310035593/d70469sv1.htm (last accessed 29 December 2011).

30. N. Curtis, *Rare Earths: We Touch Them Everyday*, Investor Presentation, May 2011, Lynas Corporation, Sydney, 2011.

31. R. D. Eggert, A. S. Carpenter, S. W. Freiman, T. E. Graedel, D. A. Meyer, T. P. McNulty, B. M. Moudgil, M. M. Poulton, L. J. Surges, E. A. Eide and N. D. Rogers, *Minerals, Critical Minerals and the U.S. Economy*, National Academies Press, Washington, DC, 2008.

32. M. Catinat (Chair), *Critical Raw Materials for the EU: Report of the Ad-hoc Working Group on Defining Critical Raw Materials*, European Commission Directorate-General for Enterprise and Industry, Brussels, 2010.

33. R. L. Moss, E. Tzimas, H. Kara, P. Willis and J. Kooroshy, *Critical Metals in Strategic Energy Technologies: Assessing Rare Metals as Supply-Chain Bottlenecks in Low-Carbon Energy Technologies*, European Commission Joint Research Center, European Union, Luxembourg, 2011.

34. D. Bauer, D. Diamond, J. Li, D. Sandalow, P. Telleen and B. Wanner, *Critical Materials Strategy: December 2010*, US Department of Energy, Washington, DC, 2010.

35. D. Bauer, D. Diamond, J. Li, M. McKittrick, D. Sandalow and P. Telleen, *Critical Materials Strategy: December 2011*, US Department of Energy, Washington, DC, 2011.

36. Science and Technology Committee, *Strategically Important Metals: HC726*, House of Commons, London, 2011.

37. R. Jaffe, J. Price, G. Ceder, R. Eggert, T. Graedel, K. Gschneidner, M. Hitzman, F. Houle, A. Hurd, R. Kelley, A. King, D. Milliron, B. Skinner, F. Slakey and J. Russo, *Energy Critical Elements: Securing Materials for Emerging Technologies*, American Physical Society and Materials Research Society, Washington, DC, 2011.
38. Resnick Institute, *Report on Critical Materials for Sustainable Energy Applications*, ed. N. Fromer, R. G. Eggert and J. Lifton, California Institute of Technology, Pasadena, CA, 2011.
39. C. Hurst, *China's Rare Earth Elements Industry: What Can the West Learn?*, Institute for the Analysis of Global Security, Washington, DC, 2010.
40. J. Neumann, J. Kim, E. Carson, B. Corby, M. Ahearn, B. El Osta and M. Delaney Ramaker, *GAO-10-617R: Rare Earth Materials in the Defense Supply Chain*, United States Government Accountability Office, Washington, DC, 2010.
41. M. Humphries, *Rare Earth Elements: the Global Supply Chain*, Congressional Research Service, Washington, DC, 2010.
42. Metal-Pages, *Metal Prices – Rare Earths*, Metal-Pages, Teddington, 2011, http://www.metal-pages.com/metalprices/rareearths/ (last accessed 2 January 2012).
43. Asian Metal, *AM Rare Earth Prices*, Asian Metal, Pittsburgh, PA, 2011, http://www.asianmetal.com/price/initPriceListEn.am?priceFlag=8 (last accessed January 2, 2012).
44. G. P. Hatch, *TMR Advanced Rare-Earth Projects Index*, Technology Metals Research, Carpentersville, IL, 2011, http://www.rareearths.org (last accessed 29 December 2011).
45. T. G. Goonan, *Rare Earth Elements – End Use and Recyclability: Scientific Investigations Report 2011–5094*, United States Geological Survey, Reston, VA, 2011.
46. M. Zakotnik, I. R. Harris and A. J. Williams, *J. Alloys Compd.*, 2008, **450**, 525.
47. D. Schüler, M. Buchert, R. Liu, S. Dittrich and C. Merz, *Study on Rare Earths and Their Recycling: Final Report for the Greens/EFA Group in the European Parliament*, Öko-Institut, Freiburg, 2011.
48. H. Kara, A. Chapman, T. Crichton, P. Willis and N. Morley, *Lanthanide Resources and Alternatives*, Oakdene Hollins, Aylesbury, 2010.

CHAPTER 3

Gold

JULIE L. STACEY

Centre for Sustainability in Mining and Industry (CSMI), University of the Witwatersrand, Private Bag 3, WITS, 2050, Johannesburg, South Africa
Email: julie@envalution.com

3.1   INTRODUCTION AND HISTORY

Gold. This most evocative of words carries with it myriad implications and symbolism: beauty, value, purity, worth, superiority, desirability, solidity, certainty, reliability. From gold credit cards, to gold stars, the Gold Standard, to golden years, anything associated with the colour is held in most societies to be something that should be pursued, worked towards and even envied.

The history of civilisation and of gold is intimately linked,[1] and the attainment of gold is used to symbolise progress. Although estimates vary, it is reported that gold was probably first used over 6000 years ago for ornamental objects.[2] Gold smelting was recorded in Egypt about 3600 years ago and there is evidence of gold jewellery from Mesopotamia 2600 years BC.[1] The earliest records in Africa point to gold smelting in 1000 AD, from artefacts found at Mapungubwe in the Limpopo Province in South Africa. People's desire for gold is evident in that its discovery has sparked a number of 'gold rushes' across the world, including in Australia in 1851

Materials for a Sustainable Future
Edited by Trevor M. Letcher and Janet L. Scott
© The Royal Society of Chemistry 2012
Published by the Royal Society of Chemistry, www.rsc.org

(although gold had been discovered earlier), South Africa in 1886 and the so-called Klondike Gold Rush in Canada in 1896.

Gold is most renowned for its use in jewellery and as a basis for modern currency. Used along with silver for military funding and reward in Japan around the sixteenth and seventeenth centuries,[3] gold as a basis for currency extended to the western world in the late nineteenth century.

As with all mining commodities, the price and supply of gold have fluctuated dramatically in a complex interplay of supply, demand, cost and perceived value. Any discussion of the sustainability of gold must address the concept of sustainable development and address the use of a seemingly finite and non-renewable resource.

3.1.1 Sustainable Development

For the purposes of this discussion, the sustainability of gold is framed within the Five Capitals model of sustainable development. In this hierarchical model of dependencies, the five 'asset bases' of natural, social, human, financial and manufactured capital are required to sustain life and to achieve and maintain a certain quality thereof. The Five Capitals model talks to stocks and flows of capital between these areas and the need to ensure that no capital is drawn below an undefined threshold level, beyond which the capital is no longer able to serve its functions. Natural capital – the environment – is the basis for the existence and functionality of all other capitals. Minerals are part of the environment: being deemed a finite resource,[4] mining must subscribe to the concept of weak sustainability. In this framework, natural capital can be used up as long as it is converted into another capital of equal value. Ideally – in a strong sustainable environment – no capital must be used up beyond its tipping point. The problem with the weak sustainability concept is that once the specific element of natural capital is exhausted – in this case minerals – any aspects of the other capitals that were dependent on minerals will also collapse. Given the extensive use of minerals to support current technology, infrastructure and products in daily use, a collapse of mineral resources would have profound ramifications for modern life.

This model is used to present the sustainability of gold in two perspectives: (i) if gold reserves are indeed finite, how can access to the available resource be prolonged, and (ii) to what use can gold be put with respect to contributing to economic, social and environmental aspects of sustainable development? Such a discussion requires an understanding of the properties of gold.

3.2 WHY IS GOLD SPECIAL?

Unique properties of gold make it desirable and useful. Its desirability relates to its appearance and its scarcity.[5] One of the 10 rarest elements, gold is found at ~0.001–0.004 grams per tonne of earth (g t^{-1}; 1–4 parts per billion, ppb). To put this in context, copper exists at just over 50 ppm and iron at 56 000 ppm.[6] This natural scarcity is exacerbated by the artificial scarcity created by those who have historically owned it[5] and the uses to which it has been put. The high social value placed on gold means that its use and ownership have always been restricted.

Contributing to its exclusivity, unlike the other precious metals – silver, platinum and palladium – gold off-take is largely driven by geography and extremely high degrees of discretionary spend.[7]

The unique physical and chemical properties of gold (Table 3.1)[8] enhance its desirability. Given its position in the periodic table, gold has an unexpectedly small atomic radius, contributing to its unique properties.[9] To all practical ends, gold is indestructible.[10] It is highly ductile (able to be drawn into a thin wire) and malleable (able to be beaten into different shapes without breaking). In fact,

Table 3.1 The properties of gold.[8]

Property	Value
Atomic number	79
Atomic mass	196.9655 g mol^{-1}
Electronegativity according to Pauling	2.4
Density	19.3 g cm^{-3} at 20 °C
Melting point	1062 °C
Boiling point	2000 °C
Van der Waals radius	0.144 nm
Ionic radius	0.137 nm (+1)
Isotopes	7
Electronic shell	[Xe] $4f^{14}5d^{10}6s^1$
Energy of first ionisation	888 kJ mol^{-1}
Energy of second ionisation	1974.6 kJ mol^{-1}
Standard potential	+ 1.68 V (Au$^+$/Au)

Table 3.2 The colours and constituents of gold alloys.[2]

	Content (%)			
Colour reference	Gold	Silver	Copper	Palladium
Yellow gold	75	16	9	–
White gold	75	4	4	17
Rose gold	75	25	–	–
Red gold	50	50	–	–

28 g (1 ounce) of gold can be beaten into a sheet of nearly 28 m^2 or drawn into a 5 μm thick wire 80 km long.[8]

In its massive form, gold is almost inert (forming few chemical compounds) and one of the least reactive elements, being one of the few metals that do not corrode in the presence of oxygen – it therefore does not tarnish. Gold has only one naturally occurring and stable isotope, ^{197}Au. Counterintuitively, in nanoparticle form, gold is an effective catalyst.

Gold is a very good conductor of heat and electricity and can be alloyed with other metals such as copper, silver and platinum. Being very soft, gold is easy to work and such alloying increases its durability. With a melting point of 1064 °C, gold can be melted and cast. Most jewellery is an alloy of gold.[2]

'Pure gold', which is not alloyed with other metals, is referred to as 24 karat gold; the standard of gold trade is known as karatage. An alloy of 50% gold by weight is 12 karat and 75% gold by weight is 18 karat.[2] Understandably, alloys of gold have a lower value per unit of weight than pure gold, are harder and are less tarnish resistant.

Gold is typically referred to by its colour, which is a function of its alloy mix (Table 3.2).[2,11]

3.3 THE USES OF GOLD

Modern life is built on mineral products.[12] Gold has many uses in society and its unique properties mean that it is likely to maintain a leading place in research for sustainable development (see Section 3.6).

Gold has been used in decoration, as a symbol of status and wealth and in medical applications for thousands of years. The

Chinese are reported to have used gold for medical treatments from 2500 BC.[13] The Bible reflects that drinking gold was used as a punishment for wrongdoing.[14] The Romans used a colloidal solution of gold to create a mauve or ruby tint for glass and applied it in the manufacture of goblets and stained-glass windows.[15]

In 2010, it was estimated that the global historical production of gold amounted to 168.3 kt,[10] of which 75% has been mined since 1910. In modern times, gold has been applied as the basis for economic exchange and is used for electronics, engineering, industrial, dental and medical purposes, beauty products and photography. Applications in some of these fields are discussed below.

3.3.1 Economics

Gold plays a significant practical and psychological role in global economics, having been used as the physical basis for currency, in addition to being a medium for risk management, investment for wealth preservation and as a mechanism to diversify economic value.[16]

From the mid-1800s in the UK, until 1971 in the USA, gold was the basis of global currency: this system was termed the Gold Standard. Switzerland was the last country to separate gold from its currency, in 2000. The Gold Standard meant that paper money of particular value was supported in reality by a pre-set quantity of gold and could be exchanged for that fixed mass of gold. Today, paper money is backed up by what is termed the *fiat* system, which is in effect an arbitrary guarantee of the value of the money. Under the *fiat* system, money is no longer underpinned by gold and it has no intrinsic value, which is taken entirely on faith. In unstable economic or political times, the markets buy up gold that is in reserve; it therefore still supports the economy, although indirectly. A very recent example of this is that following the Greek economic crisis in late 2011, as the Eurozone debt issues took hold in Italy, gold recovered in value. This separation of gold from currencies has benefit in as far as both the full service value of gold and the transactional value of money can be exercised simultaneously.[17]

The central banks of many countries hold gold in reserve, because of its perceived stability of value (Figure 3.1). The current importance of gold in backing up financial efforts is highlighted by

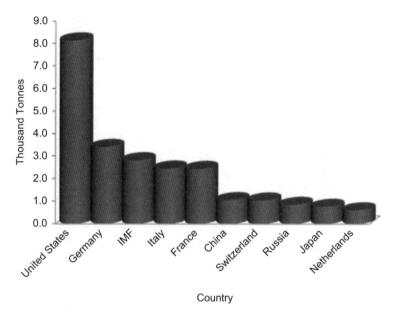

Figure 3.1   Tonnes of gold held by country.[18]

the fact that as of November 2011, the USA, Germany, Italy and
France all retained more than 70% of their foreign reserves in the
form of gold. Portugal recorded the highest such proportion at
88.9% of foreign reserves.[18]

3.3.2   Engineering

The applications of gold in engineering are widespread. Its resistance to corrosion and tarnishing, low volatility, durability and low
maintenance requirements make gold a dependable commodity.
Some of these properties make gold ideal for aerospace applications. As very thin sheets, gold reflects infrared light, being used in
radiant heat shields in space suits and on sun visors for space
helmets,[19] to protect both eyes and skin from the extreme radiation
levels outside the Earth's protective atmosphere. Similarly, sheets
of polyester film coated with gold are also used to stabilise spacecraft temperatures. Thin sheets of gold serve as lubricants between
moving parts of spacecraft, where organic lubricants would break
down.[2] Gold is also used extensively in aerospace circuitry, given

its properties as a reliable connector and conductor and the need to endure harsh conditions in space with no opportunity for maintenance.

3.3.3   Industrial and Electronics

The most important industrial use of gold is in the manufacture of electronic equipment. Its capacity for effective use in minute quantities, while retaining its corrosion-free and rapid conductive and connective qualities, makes it ideal for micro-applications in modern electronic devices. Gold is used variously in bonding wires in semiconductor packages, hybrid circuits, soldered coatings and solders for printed circuit boards, contact pads and as an electro-plating chemical.[20] Every mobile phone manufactured today – approximately one billion per annum – contains roughly 50 US cents worth of gold. The reliance of modern-day communications, healthcare and economic systems, and also education efforts and the entertainment industry, on computers highlights the importance of the application of gold in electronics in supporting and elevating standards and quality of life.

Other industrial applications of gold include in glassmaking. Minute amounts of gold can be suspended or dispersed in glass. Used in construction, the gold-containing glass acts as a thermoregulator, reflecting solar radiation outwards and internal heat inwards. The amounts of gold required are very small, making this a cost-effective and energy-saving approach to climate control for buildings.[2]

Gold is also used in traditional photographic development solutions, producing a rich blue–black image. Such applications result in longer duration images, due to gold's non-reactivity.

3.3.4   Dental and Medical

Gold – mainly as alloys – is used in dentistry for bridges and crowns. It is chemically inert, easy to work and non-allergenic,[2] and the amounts used are very small, making it financially viable.

Gold has been used medically for thousands of years. Current applications include direct treatment or therapeutic options for a variety of conditions, and also in medical research and supportive technologies.

Gold is commonly used in the form of radioisotopes or salts to treat liver disease, psychiatric disorders, cancer, tuberculosis[14] and autoimmune diseases such as arthritis.[21] For example, rheumatoid arthritis can be treated by injections of a weak solution of sodium aurothiomalate or aurothioglucose.[2] Both the chemical and physical properties of gold are medically applicable. Lagophthalmos, in which a person's eye cannot close properly, is treated with small amounts of gold implanted in the eyelid, to weigh the lid down.

Colloidal gold – a suspension of gold nanoparticles – with its intense red colour is used in diagnostics. Gold nanoparticles absorb protein molecules on their surface and this is exploited for use with electron microscopy. Immunogold labelling makes use of this characteristic to identify the presence of particular antigens,[22] which take the form of 'dense round spots' in the electron microscope tissue sample.[23] Gold and its alloys are also used to coat biological specimens for viewing in electron microscopy.[24] The properties that allow this application include gold's high electrical conductivity, high density and high output of secondary electrons.

Finally, because of its high conductivity, non-reactivity and reliability, gold is used in many surgical instruments, electronic medical equipment and life-support devices.

3.3.5 Jewellery

As at the end of 2010, over 84 kt of gold existed in the form of jewellery.[10] representing $\sim 50\%$ of the total available above-ground stocks.[25,26]

3.4 HOW MUCH GOLD DO WE USE?

The use of gold is unique, in as far as the bulk of production is applied to discretionary spending, which is defined as the purchase of gold for the joy of ownership and not needed. Until the recent financial crisis, nearly 80% of spending on gold was for jewellery.[7] The production of gold therefore largely serves societal wants rather than their needs, which has particular implications for its sustainable use.

The supply and demand of gold are complex. Simplistically:

$$\text{total supply of gold} = \text{total mine supply} \\ + \text{official sector sales} + \text{recycled gold} \quad (3.1)$$

where

$$\text{total mine supply} = \text{total mine production} \\ + \text{net producer hedging} \quad (3.2)$$

For 2010, the equation reads as follows:[26]

$$\text{total supply of gold}/t = [2686 + (-108)] + (-77) + (1651) \quad (3.3)$$

Therefore, with 2686 t produced through mining and with a demand of $\sim 4000\,$t,[26] 2010 saw the highest ever recorded annual production and demand figures for gold. In the 2010 calendar year, jewellery used 2017 t, bar and coin investment 1150 t, exchange traded funds (ETFs) and similar 368 t and technology applications 466 t.[†] The financial market for gold includes private investment and official sector holdings: central banks hold most of these reserves. As can be seen from the above equations, the shortfall between mining production and demand is addressed through official sector sales and recycling.

Demand patterns for gold have shifted in the past two decades, not least due to the 2008 global financial crisis. From 1997 to 2002, on average 81% of gold produced was used for jewellery, 8% for retail investment and 10% for industrial applications. From 2007 to 2009, on average 52% was used for jewellery, 38% for retail investment and 11% for industrial applications.[27] This highlights the role of gold in financial risk management. In 2010, the top consumers of gold for the purposes of jewellery were India (746 t), China (400 t) and the USA (129 t).[28] In a 2006 it was estimated that 50 000 t of unmined reserves of gold remain.[29]

The industrial demand split (electronics and other) has remained fairly constant for the past 5 years, with a dip following the

---

[†]A reliable breakdown of gold use statistics is difficult to present, as there is no common terminology with respect to usage. For example, one source text may refer to technological applications, which include medical, dental, industrial and electronic. Another text may separate out industrial and electronic applications or medical and dental. Nonetheless, the trend is evident in the available statistics reproduced here.

financial crisis.[29] Total industrial demand in 2011 is estimated to reach a little less than 120 t.[16]

3.5 WHERE AND HOW IS GOLD MINED?

3.5.1 Gold Production

In 2009, the top five global gold producers were China (320 t), the USA (223 t), Australia (222 t), Russia (205 t) and South Africa (197 t), with the global production estimates ranging from 2460 t[30] to 2689 t.[25]

3.5.2 Mining Methods and Processes

The feasibility of any single mine depends on the interaction of a series of economic and physical factors, including ore grade, size and nature of the deposit, relative amount of overburden and associated geology.

Gold is found (i) as alluvial gold (in sediments, often in a water course), (ii) in hard-rock deposits (at concentrations of less than 10 g t^{-1} or 0.0001% by mass),[31] often associated with base metals, silver or uranium, and (iii) in very low concentrations in oceans (0.00005 ppm). If it is close to the surface, gold can be mined through open-pit methods. Currently, most gold mines are underground. The deepest gold mine, which is also the deepest mine in the world, is in South Africa. Mponeng, located near the town of Carltonville, 65 km south-west of Johannesburg, is owned by AngloGold Ashanti.[32] In September 2010, Mponeng was 3778 m below datum, 3502 m below surface and 949 m below sea level. In 2011, Mponeng was reported as being nearly 4 km deep.

Alluvial gold can be won through panning or using washing tables, relying on gravity concentration, with the addition of mercury to enhance recovery. This method is still largely in use by the world's informal mining sector. Metallic mercury is a liquid at room temperature, has a specific gravity of 13.5, with a boiling point of 357 °C, and forms an amalgam with gold. Mercury also has a low vapour pressure and evaporates at room temperature. Mercury vapour has severe environmental toxicity and human health impacts.

Gold is recovered from hard rock through a combination of comminution (crushing, grinding and milling) and mineral processing. Of particular concern for the sustainability of gold production is the use of mercury (largely for informal mining) and cyanide (in the formal mining sector) to recover gold.

Where the gold is finely dispersed in the ore, it is liberated typically through cyanide salt leaching (the main method used in the industry for gold recovery in China, Australia, USA, Russia, South Africa, Peru, Canada, Indonesia, Uzbekistan, Papua New Guinea, Ghana, Tanzania, Brazil, Mali and Chile),[33] followed by recovery from the leach solution, through adsorption on activated carbon. Gold is then recovered through stripping and/or electrolysis. If the gold is closely associated with sulfide minerals, froth flotation, followed by roasting or wet pressure oxidation, is used.

Of the 1.1 Mt of hydrogen cyanide produced annually, only 66 000 t (or 6% of global production) is used in gold recovery.[34] The highly complex gold cyanidation process is described in a reaction known as Elsner's equation:

$$4Au + 8CN^- + O_2 + 2H_2O \rightarrow 4Au(CN)^{2-} + 4OH^- \qquad (3.4)$$

The amount of cyanide required for dissolution may be as low as 350 mg L^{-1} (as NaCN).

The cyanide leaching process is achieved through either heap/dump leaching or agitated pulp leaching. Heap leaching involves the percolation of a cyanide solution through piles of ore located on pads lined with impermeable membranes. The gold leaches into the cyanide solution, which is directed into storage for further treatment. With low capital costs, heap leaching is nonetheless slow and inefficient, with a gold recovery rate only 50–70%.

In an agitated leaching circuit, milled ore is contained in tanks in which it is agitated either mechanically or through air injection. The greater exposure of the gold-bearing 'slurry' to the cyanide solution that is achieved through agitation improves the efficiency of the gold recovery.

The gold–cyanide complex in solution is then exposed to activated carbon, either through carbon-in-leach (CIL) processes or carbon-in-pulp (CIP). The activated carbon is introduced directly into the CIL tanks or into separate CIP tanks after leaching. Carbon treatment recovers more than 99.5% of the gold in solution.

The gold-loaded carbon and pulp are separated by air drying or hydrodynamic screening. Alternatively, elution makes use of hot caustic aqueous cyanide solution to separate the gold from the carbon. Carbon is then regenerated and reused in the adsorption process, while zinc cementation or electrowinning produces the gold concentrate from the eluate. The gold concentrate is then refined to bullion through stages of smelting, chlorination and electro-refining to a purity of up to 99.999%.

The use of mercury and cyanide for gold recovery carries significant human and environmental health concerns. These concerns, and also issues relating to recovery efficiencies in mildly refractive gold ores, have led to the examination of alternatives to the cyanidation process, including that of thiosulfate leaching.[35] Thiosulfate forms stable anionic bonds with gold and silver and is cheaper and less toxic than cyanide. The chemistry of thiosulfate leaching is, however, even more complex and less stable than that of cyanide and requires the use of large volumes of reagents.

$$4Au + 8S_2O_3^{2-} + O_2 + 2H_2O \rightarrow 4Au(S_2O_3)_2^{3-} + 4OH^- \qquad (3.5)$$

The main chemical components of the leaching process are ammonium thiosulfate and ammonium sulfate, both commonly used as fertilisers. Thiosulfate processing negates the use of CIP or CIL processes; recovery through resin-in-pulp technology and the cementation of gold on copper metal powder are being explored.[36]

3.6 THE SUSTAINABILITY OF GOLD

Traditional views of sustainable development focus on environmental protection and socio-economic relations.[37] There is a difference between the sustainability of the gold product and that of the process of producing gold, notably mining and processing. Certain groups argue that there is no net benefit from gold mining and, indeed, that there is a net detriment.[38,39] The net benefit of the application of gold (as a product) to sustainable development efforts looks much more promising, however.

Ultimately, the sustainability of gold will be determined by a complex interplay of economic feasibility, rates of supply, rates of demand, sources of gold and the social acceptance (or rejection) of the full environmental and social costs of gold mining.

3.6.1 Issues of Sustainability in Gold Production

A 3.8% increase in gold production (all attributable to mining) was recorded for 2010,[33] the bulk of which went into investment to shore up fears following the economic crisis of 2008, rather than into physical applications.

Gold mining is linked with severe negative environmental, social and economic impacts in developed and developing countries.[40] From an environmental perspective, the bulk of sustainability issues linked to gold mining at site level relate to (i) the use of highly hazardous chemicals such as cyanide and mercury, (ii) the enormous amounts of waste produced, (iii) the scale of landscape and biodiversity damage especially in open cut operations, (iv) large-scale use of other natural resources such as water and (v) the carbon intensity of specifically diesel use and the energy intensity of deepening mines. Gold ore grades are declining[38] and resource intensity will increase as declining ore grades require ever increasing inputs of raw materials and energy for mining and recovery processes, at higher cost, with increasing proportions of waste generated. Conservative waste statistics estimate that for a 'typical' 0.333 g, 18 karat gold ring, over 20 t of waste is produced. Phrased differently, for every ounce of gold mined, over 76 t of waste is generated.[41] It is estimated that for every tonne of gold mined, 143 GJ of energy and 691 kL of water are used and 11.5 t of CO_2 is generated.[4]

Because of its mining methods, gold mining records the highest rates of fatalities in the mining industry.[42] Furthermore, it is estimated that over 25% of ex-gold mine workers in South Africa suffer from silicosis. It is also prevalent in Canada.[41] Silicosis results from inhaling silica, which is associated with the ore in which gold is found, and results in lung scarring, which ultimately leads to death.

The negative environmental and social costs of gold mining[40] differ between developed and developing countries and between the formal and informal sectors.[43] In developing countries, impacts recorded include deforestation, acid mine drainage, noise, dust, air and water pollution from arsenic, cyanide and mercury, social disorganisation, a loss of livelihoods, mass displacement[39] and reduced access of rural households to natural and social capital resources.[44] About 3.4 kt of mercury is released into the environment annually worldwide by small-scale gold mining, with major ecosystem functioning, health and socio-economic impacts.[45,46]

Acid mine drainage (AMD) is potentially one of the worst environmental consequences in the long term related to the mining of gold. AMD occurs when pyrite (iron disulfide) oxidises in the presence of oxygen, usually in water. The process is two-stage, both of which produce sulfuric acid,[4] resulting in a decrease in pH of the receiving water body. The overall reaction governing AMD is[48]

$$4FeS_2 + 14H_2O + 15O_2 \rightarrow 4Fe(OH)_3 + 8SO_4^{2-} + 16H^+ \qquad (3.6)$$

The process of oxidation occurs in normal weathering, but with mining, the process is greatly accelerated and is too fast for natural neutralisation processes. The acceleration occurs as mining dramatically increases the surface area of the rock fragments.

The low pH (high acidity) increases the solubility of aluminium and other heavy metals present, increasing the metal content of the water, the sulfate content and the suspended solids. The overall impact of AMD is highly dependent on many variables, including the rate of release. Catastrophic incidents have been reported globally; the US Environmental Protection Agency (EPA) has on record 66 such events.[49] Typical effects of AMD include fish kills, decreased water quality, toxicity and increased erosivity of the water. Currently, AMD is threatening the Gauteng and North West provinces of South Africa, as a result of over 100 years of gold mining in the Witwatersrand. The declining gold mining industry in the area means that the pumping that kept the low-pH water under control is tailing off. As a result, the level of the (largely acidic) water, which was contained in the mined out areas, is rising, threatening the economic centre of South Africa.

Civil society rejection of perceived poor business ethics, human rights abuses, especially by security firms associated with gold mines,[41] and the social and environmental consequences of gold mining is increasing markedly, with global campaigns such as the 'No Dirty Gold' collaborative programme created in 2004. Within the past 5 years, mutual funds have divested from gold mining companies or have put the option to do so to the shareholder vote.[29]

3.6.1.1 Site-level Contributions to Sustainable Development. The contribution of gold mining to economic development is significant. Its largest benefit has been found to be its contribution to foreign direct investment in developing countries.[43] That this investment is not always equitably distributed is beyond the control of mining houses; collective global efforts such as the Extractive Industries Transparency Initiative (EITI) are addressing these concerns.

Moves to internalise the social and environmental costs of gold mining are necessary but not sufficient to contribute to sustainable development at site level. This approach will slow unsustainable patterns but does not fundamentally change them.

The management of gold waste must maximise the inert nature of tailings and rock waste, while enhancing the stability of these structures, physically and/or chemically. As relative proportions of waste increase, better technological options for waste disposal such as thickened tailings, dry stacking and paste backfill are already in use.[50] It is estimated that to truly internalise waste management costs, gold processing plants should set aside at least one-third of their capital for this purpose.[33]

Gold mines must decrease energy use, for environmental and economic reasons. One of the highest sources of energy consumption in gold mines is for underground cooling, to reduce the working temperature from that of the virgin rock (up to 60 °C) to legally permissible levels of less than 30 °C. Demand side management[‡51] and harnessing the gravitational energy released by the water falling to the bottom of the mine as part of underground cooling systems have been shown to reduce energy consumption by up to 2.3%.[52]

From a safety perspective, technology is being used to improve matters. For example, a robot is being used to survey new mining areas that are too dangerous for people to access. Another advantage of these robots is that they can do the work twice as fast as humans.[53]

3.6.2 Issues of Sustainability in Gold Use

Gold is not 'consumed' when it is used and is largely available for re-use.[54] Despite this, the gold price does not reflect the fact that gold is both non-renewable and essentially irreplaceable, nor does the price include the environmental and social costs of gold

---

‡Demand side management (DSM) occurs when a utility intervenes in the marketplace to alter the energy use profile of consumers. DSM can take the form of a number of activities, including improving end-user efficiency, use of energy-efficient technologies and pricing strategies based on peak and off-peak demand profiles.

mining.[55] Gold as a product however, could potentially be separated from the full cost of mining, from a sustainable development perspective.

Approximately 62% of gold ever produced is in private hands – ~ 104 kt is in the form of bullion, coin and jewellery.[56] Other than perceived retention of wealth and risk management applications, the majority of gold is used for non-essential purposes. There is two-thirds more gold above ground (having been mined) than in reserve below it (in other words, currently unmined).[29] If these above-ground gold sources were released into circulation and if society reduced the non-essential demand, there could be no further need to mine. However, it would be nonsensical to suggest that mining should stop immediately, given the socio-economic dependencies on the industry. A shift in policy to recycle above-ground stocks, rather than mining new sources, could be the answer to sustaining the supply of gold and to addressing the many environmental and social concerns associated with its production processes. Given that historically over 80% of gold is used for jewellery, the power to sustain gold supplies is largely in the hands of consumers.[57]

Over the past 5 years, mined production has accounted for 59% of the supply, recycled gold accounts for 36% of annual supply (largely from jewellery in the developing world) and official sector supplies 6%. Recycled gold is gold that has been recovered from fabricated products, melted, refined and recast into bullion for resale.[10] An innovative suggestion from Richards[55] is a deposit and lease scheme for metals to address their indestructibility and irreplaceability.

3.6.3 How Can Gold Contribute to Sustainable Development?

Although largely inert in its massive form, research has discovered that gold becomes reactive at nanoparticle size. The use of gold in nanotechnology is dramatically increasing its application in addressing issues of sustainability. The understanding of the unique characteristics of gold at nanoparticle size as compared with its bulk form is fairly recent; research into gold-based chemistry is therefore relatively underdeveloped.[58] The perception of the cost of gold has also hindered research advances. The understanding of the properties of nanoparticle gold is fuelling enthusiasm and funding and new applications continue to emerge. The minute quantities of gold used in nanoparticle science and engineering make such applications relatively inexpensive and a boost to sustainable technologies, many of which are deemed to lack commercial viability and/or be excessively expensive.[58]

3.6.3.1 Gold and Environmental Protection. The current high-priority global environmental challenges include reducing reliance on fossil fuel energy sources and reducing greenhouse gas emissions (which two factors are directly linked), control of hazardous air pollutants and the reduction of pollution. Research is demonstrating the practical application of gold in addressing some of these pressing issues.

As long as energy remains largely fossil-fuel based, reducing energy use is a direct way to reduce greenhouse gas emissions. In extreme climates, energy consumption for the thermoregulation of buildings is considerable. One such application of gold is found on the Royal Bank Plaza building in Toronto, Canada. Its 14 000 windows are coated with pure gold, whose infrared optical reflectivity reduces heating and ventilation needs. About 70 kg of gold was used, worth ~ US\$4 million. The reduced energy costs make this an economically viable option.[58]

Energy is likely to be a limiting factor for economic development in the future. Gold is being investigated for use in fuel cells (separator plate technology), which is expected to be highly significant as a future transport power source. With hydrogen fuel cells, the expected dominant technology, it is believed that gold has a role to play in hydrogen generation and purification and as a hydrogen/oxygen catalyst.[58] Lithium–air batteries using gold alloy catalysts have shown significantly improved efficiency.[15] Gold nanoparticles are also under investigation for use in solar cells.

Platinum is used in catalytic converters for automobiles for pollution and emission control. Gold is being investigated as an alternative metal, but its ability to withstand high temperatures is problematic. Diesel engines are, however, increasingly popular, due not least to their lower fuel consumption. Catalytic converter systems for diesel engines operate at lower temperatures, making gold a possible viable alternative to platinum. Conversely, gold nanoparticles are highly reactive at low temperatures and this could have significant application in controlling start-up emissions of cold vehicle engines. In another application of the lower temperature functionality of gold catalysts, a prototype air purification unit was designed in 2002 which removes carbon monoxide from the air at room temperature for application in commercial and industrial settings.[9]

Other pollution control measures are making use of gold nanoparticles. Gold is being used to detect and trap mercury (arising from coal-fired boilers and linked to human health issue such as Alzheimer's disease) and carbon monoxide (often from poorly ventilated boilers), with catalytic activity taking place even at very low temperatures.[15] Gold nanoparticles also have a unique ability to break down harmful chlorinated pollutants and to determine pesticide concentrations in water supplies.

3.6.3.2 Industrial and Engineering Applications of Gold for Sustainability. There are increasing numbers of new applications for gold,[20] including in the industrial sector, although this has constituted only 12% of historical demand for above-ground stocks – around 19.8 kt to date.[59] While growth in demand is anticipated for new industrial uses, its characteristics at the nanoparticle level suggest that growth in gold demand will be in the variety of applications, rather than in significant quantities.[58]

An intriguing development for the contribution of gold to sustainable development is the substance known as graphene. Graphene is microscopically thin, consisting of a single atom-thick sheet of carbon atoms, arranged in a hexagonal lattice.[60] It has a remarkable range of properties: it is very strong, transparent and highly conductive. This means that it could be used for a whole range of applications, including flexible electronics that can be worn, fast broadband, high-performance computing and lightweight components for airplanes and other machines. This has implications for healthcare, improved efficiency (and therefore reduced cost) of the manufacture of electronic devices, energy use and social development – including education.

Graphene is produced by exposure to carbon-rich vapours at temperatures in excess of 1000 °C, in the presence of a thin catalyst film on which the graphene assembles, typically nickel or copper.

These high temperatures preclude in-manufacturing circuit integration, as many substances used in electronics cannot withstand the heat. By including less than 1% of gold in the nickel film, the graphene growth occurs at a temperature of 450 °C. The presence of the gold also improves the conductivity of the graphene, as it affects its growth pattern, allowing larger flakes to develop before joining. Thus high-quality graphene at lower temperatures may have distinct electronic applications, although commercial viability is some way off.

Current technologies can also benefit from the application of gold, including conductive inks, touch-sensitive screen technology and high-density data storage.[15]

Nanoelectronics is an expanding field of technology, harnessing material characteristics that are fundamentally different from macroscopic properties for the processing, transmission and storage of data.[61] From molecules that act as transistors to quantum dots, where the spin of electrons is used to process information, nanoelectronics has the potential for self-assembly of mechanical devices, to reduce power consumption and to reduce the weight of electronic devices. Researchers at the University of Delaware have grown gold microwires from the application of dielectrophoresis in a gold nanoparticle solution. The wires of less than 1 μm in diameter self-assembled up to 5 mm in length.[9]

Other nanoelectronic research is moving towards high-density data storage. For example, a memory chip is under investigation that is projected to be capable of storing over one terabyte (10^{12} bytes) of memory per square inch.[62]

3.6.3.3 Medical Applications. The application of gold to new and current priority health issues is expanding. Gold is being applied to HIV/AIDS diagnostics and treatment. Antiretroviral (ARV) drugs were a breakthrough that changed the face of HIV/AIDS treatment globally. An ARV drug based on gold has been developed in South Africa.[63] More recently, two HIV diagnostic devices have been developed that make use of colloidal gold nanoparticles to measure CD4-positive white blood cells in HIV patients.[15] CD4 counts are indicative of the progress of HIV to full-blown AIDS. The devices are simple and robust and are durable in the harsh African climate where the pandemic is rife.

The use of gold in dentistry declined in the 1970s owing to its cost. However, concerns about the health effects of the use of mercury amalgams is resulting in an upswing in the popularity of gold in dentistry.[12]

The incidence of cancer is increasing dramatically. In 2008, cancer caused 13% of all deaths globally, at 7.6 million deaths,[64] and is predicted to rise to 15 million new cases per year by 2020. Gold nanoparticles have high surface reactivity and are highly biocompatible.[65] The high surface area and size relationships of nanoparticles to cells allow for targeted specificity in treatments. Unlike generalised chemotherapy, which attacks all cells, gold nanoparticles that are carrying specific treatments can be focused on cancer cells. Not only is the treatment thus targeted in terms of which cells to attack, but also highly specific doses of treatment can be sent to the tumour.[15]

Similar targeted cancer treatment using gold nanoparticles is being applied in a completely different way. Radiation treatment is being refined through using radiofrequency-heated gold nano-particles, with their unique physical and optical properties, to destroy cancer tissues using thermal abrasion technology that does not damage surrounding cells.[15]

In addition to their application in therapeutic uses, gold nano-particles are being used in diagnostics. Ideally, diagnostic tech-nology must be highly precise, sensitive and inexpensive. The protein-binding properties of gold nanoparticles and their stability, durability, consistent high quality, non-toxic nature and relatively low cost make them ideal for quick and reliable testing. Applica-tions of nanoparticle gold in diagnostics include pregnancy testing and the detection of food-borne pathogens including *Salmonella*, *Escherichia coli* and *Campylobacter*.

Pfizer is currently researching a needle-free vaccine-delivery system, where vaccine-coated gold nanoparticles are delivered into the epidermis (the outer layers of the skin) by a burst of pressurised helium.

Because of their protein-specific binding properties, gold nano-particles are being used in the early diagnosis of prostate cancer. Prostate cancer has few early symptoms, but the earlier it is detected, the higher is the rate of full recovery. If prostate cancer proteins are present, when gold nanoparticles are injected into a patient the cancer proteins cluster around the gold nanoparticles in

the blood, and these clusters are then detectable by photon correlation spectroscopy.

Gold nanoparticles are also being applied in infection control,
making use of its substantial antimicrobial properties. It is thought
that this technology may be applicable in the increasingly difficult
fight against drug-resistant bacteria.[15] Silver – long used for its
antimicrobial properties – decays with time. Used in conjunction
with gold, the antimicrobial properties are enhanced and last
longer. This technology is even being considered for use in soaps
and detergents.

3.7 CONCLUSION

Our perceptions of gold as being too expensive, too rare and of use
only in making beautiful but non-essential things are being challenged. Gold as a product is perhaps one of the most sustainable
elements currently available to us. The processes by which we
obtain gold – through mining and associated activities – come with
high environmental and social costs that significantly limit its
sustainability.

Of the 168 kt of gold ever mined, 104 kt (62%) remains in private
hands. Of this, jewellery comprises 84 kt and the rest is bullion or
coinage. It is estimated that only a further 50 kt remain available in
mineable reserves underground. Reducing the demand for gold for
jewellery would enhance the sustainable prospects for the metal
considerably. Rethinking the psychology of our economy and
reducing our reliance on gold as a perceived risk management
commodity could facilitate the release of some of the extensive
above-ground stocks, which far outweigh the remaining underground reserves. Such a transition would have to be managed with
due care and full cognisance of the socio-economic dependencies
within the gold mining industry. The move would have to include
policy responses which broaden and deepen backward and forward
linkages in the economy, to improve the sustainability of the
industry's contribution, through growth and spreading of the
benefits to the wider population.[66]

The real value of gold lies in its properties at nanoparticle size
and it is in this form that it holds the most promise in contributing
to a sustainable future. Its applications in addressing global issue
such as climate change and energy use, pollution abatement, fuel

cell technology and myriad health-related issues make it one of the most valuable metals for the future. Funding for gold nanoparticle-related research has been relatively limited, but is increasing and must continue to do so to commercialise the promising laboratory results to date.

Its truly unique properties reduce the likelihood of alternatives being found for gold applications. The current social and environmental costs of producing gold are unsustainable. Delinking our use of gold from its current mode of supply through improved recycling, release of above-ground stocks and its increasing application in addressing genuinely beneficial social and environmental needs could be the future of this remarkable metal.

REFERENCES

1. World Gold Council, *About Gold*, 2011, http://www.gold.org/about_gold/story_of_gold/heritage/ (last accessed 13 September 2011).
2. Geology.com, *The Many Uses of Gold*, 2005–2011, http://geology.com/minerals/gold/uses-of-gold.shtml (last accessed 13 September 2011).
3. A. Kobata, *Econ. History Rev.*, 1965, **18**, 245.
4. G. Mudd, *Gold Mining and Sustainabilty: a Critical Reflection*, Encyclopeadia of Earth, http://www.eoearth.org/article/Gold_mining_and_sustainability:_A_critical_reflection (last accessed 8 November 2011).
5. E. Schoenberger, *Cultural Geogr.*, 2010, **18**, 3.
6. Jefferson Lab, *It's Elemental*, http://education.jlab.org/itselemental/index.html (last accessed 24 November 2011).
7. R. O'Connell, *What Sets the Precious Metals Apart from Other Commodities?*, World Gold Council, London, 2005.
8. Lenntech, *Gold – Au*, http://www.lenntech.com/periodic/elements/au.htm (last accessed 13 September 2011).
9. C. Corti, R. Holliday and D. Thompson, *Gold Bull.*, 2002, **35**, 111.
10. A. Bhatia, N. Dempster and G. Milling-Stanley, *Liquidity in the Global Gold Market*, World Gold Council, London, 2011.
11. J. Emsley, *Nature's Building Blocks: an A–Z Guide to the Elements*. Oxford University Press, Oxford, 2008.
12. BGS, *Minerals in Our Lives, Mineral Matters*, No. 7, British Geological Society, Nottingham, 2004.

13. S. Fricker, *Gold Bull.*, 1996, **29**, 53.
14. D. Lazzeri, S. Lazzeri, M. Figus, M. Nardi and M. Pantaloni. *BMJ*, letter to editor, 2010, http://www.bmj.com/rapid-response/ 2011/11/02/%E2%80%9Caurum-potable%E2%80%9D-%E2% 80%9Cdrinkable-gold-alchemists%E2%80%9D (last accessed 2 May 2012).
15. T. Keel, R. Holliday and T. Harper, *Gold for Good. Gold and Nanotechnology in the Age of Innovation*, World Gold Council and Cientifica, London, 2010.
16. World Gold Council, *Gold Investment Digest*, World Gold Council, London, Second Quarter 2011.
17. T. Quint and M. Shubik, *Cowles Foundation Discussion Paper*, No. 1814, 1 August 2011.
18. World Gold Council, *World Official Gold Holdings: International Financial Statistics*, World Gold Council, London, 2011.
19. L. Mallan, *Suiting Up for Space: the Evolution of the Space Suit*, John Day, New York, 1971.
20. P. Goodman, *Gold Bull.*, 2002, **35**, 21–26.
21. L. Messori and G. Marcon, in *Metal Ions and Their Complexes in Medication*, ed. A. Sigel, CRC Press, Boca Raton, FL, 2004, p. 280.
22. W. Faulk and G. Taylor, *Immunochemistry*, 1971, **8**, 1081.
23. J. Roth, M. Bendayan and L. Orci, *J. Histochem. Cytochem.*, 1980, **28**, 55.
24. J. Bozzola and L. Russell, *Electron Microscopy: Principles and Techniques for Biologists*, Jones and Bartlett Learning, Burlington, MA, 1999.
25. P. Klapwijk, *Gold Survey 2011*, GFMS Ltd, http://www.gfms. co.uk/Presentations/2011/Gold-Survey-2011-Presentation-London.pdf (last accessed 26 November 2011).
26. World Gold Council, *Gold Demand Trends Third Quarter 2011*, http://www.gold.org/download/pub_archive/pdf/GDT_Q3_ 2011.pdf (last accessed 26 November 2011).
27. S. Janse van Rensburg, *Gold: National and International Background*, Mintek, Johannesburg, 2010.
28. World Gold Council, *Jewellery Key Markets*, http://www.gold. org/jewellery/markets/ (last accessed 26 November 2011).
29. S. Ali, *J. Cleaner Prod.*, 2006, **14**, 455.
30. BGS, *World Mineral Production 2005–2009*. British Geological Survey, Nottingham, 2011.

31. ICMI, *International Cyanide Management Code for the Manufacture, Transport and Use of Cyanide in the Production of Gold*, International Cyanide Management Institute, Washington, DC, 2011.
32. A. Parsons, Environmental Policy Advisor, AngloGold Ashanti, interviewed by J. Stacey, 7 November 2011.
33. S. Janse van Rensburg and S. Zang, presented at the Conference of Metallurgists, Montreal, 2 October 2011.
34. ICMI, *Use of Cyanide in the Gold Industry, International Cyanide Management Code for the Gold Mining Industry*, http://www.cyanidecode.org/cyanide_use.php (last accessed 13 September 2011).
35. W. Yen, G. Deschenes and M. Aylmore, presented at the 33rd Annual Operator's Conference of the Canadian Mineral Processors, Ottawa, 2001.
36. SGS, *Metallurgical Processes, Thiosulfate Leaching*, http://www.met.sgs.com/thiosulfate-leaching-an-alternative-to-cyanidation (last accessed 18 November 2011).
37. G. Hilson, *Miner. Resources Eng.*, 2001, **10**, 397.
38. G. Mudd, *Resources Policy*, 2007, **32**, 42.
39. A. Kumah, *J. Cleaner Prod.*, 2006, **14**, 315.
40. Harvard Law School, *All That Glitters: Gold Mining in Guyana. The Failure of Government Oversight And the Human Rights of Amerindian Communities*, Human Rights Program, Harvard Law School, Cambridge, MA, 2007.
41. Earthworks and Oxfam America, *Golden Rules, Making the Case for Responsible Mining*, Earthworks and Oxfam America, Washington, DC, 2007.
42. Chamber of Mines of South Africa, *Facts and Figures 2010*, Chamber of Mines of South Africa, Johannesburg, 2011.
43. World Gold Council, *The Golden Building Block: Gold Mining and the Transformation of Developing Economies, with an Economic Life-cycle Assessment of Tanzanian Gold Production*, World Gold Council, London, 2009.
44. J. Bury, *Geogr. J.*, 2004, **170**, 78.
45. WWF Guianas, *Gold Mining Pollution Abatement*, http://www.wwfguianas.org/our_work/goldmining/ (last accessed 8 November 2011).
46. W. Basson, *Sciencescope*, 2009, **4**, 55.
47. T. S. McCarthy, *South African Journal of Science*, 2011, **107**.

48. M. Williams, *Acid Mine Drainage*, University of Colorado, Boulder, CO, http://snobear.colorado.edu/Markw/Mountains/07/AcidMineDrainage/amd_daniel.ppt (last accessed 23 November 2011).
49. S. Jennings, D. Neuman and P. Blicker, *Acid Mine Drainage and Effects on Fish Health and Ecology: a Review*, Reclamation Research Group, Bozeman, MT, 2008.
50. D. Franks, D. Boger, C. Cote and D. Mulligan, *Resources Policy*, 2011, **36**, 114.
51. A. Schutte, R. Pelzer and M. Kleingeld, *Demand-Side Energy Management of a Cascade Mine Surface Refrigeration System*, Cape Peninsula University of Technology, http://active.cput.ac.za/energy/web/ICUE/DOCS/330/Paper%20-%20Schutte%20A.pdf (last accessed 8 November 2011).
52. C. Thomaz, *Mining Weekly*, http://www.miningweekly.com/article/energy-efficient-ice-technology-used-at-gold-majors-operations-2009-03-20 (last accessed 2 May 2012).
53. N. Eyriey, *Business Cornwall*, http://www.businesscornwall.co.uk/news-by-industry/manufacturing-in-cornwall/worlds-deepest-gold-mine-survey-123 (last accessed on 2 May 2012).
54. Gold Recycling Association of Canada, *85% of Existing Gold Can Be Easily Recycled*, http://goldrecycling.ca/gold-recycling/ (last accessed 22 November 2011).
55. J. Richards, *J. Cleaner Prod.*, 2006, **14**, 324.
56. M. George, *USGS 2007 Minerals Yearbook: Gold*, US Geological Survey, http://minerals.usgs.gov/minerals/pubs/commodity/gold/myb1-2007-gold.pdf (last accessed 22 November 2011).
57. R. Sarin, *J. Cleaner Prod.*, 2006, **14**, 305.
58. C. Corti and R. Holliday, *Gold Bull.*, 2004, **37**, 20.
59. World Gold Council, *The Importance of Gold in Reserve Asset Management*, World Gold Council, London, 2010.
60. R. Weatherup, B. Bayer, R. Blume, C. Ducati, C. Baehtz, R. Schlogl and S. Hofmann, *Nano Lett.*, 2011, **11**, 4154.
61. McGill University, *Nanoelectronics*, http://www.physics.mcgill.ca/~peter/nanoelectronics.htm (last accessed 23 November 2011).
62. UnderstandingNano.com, *Nanotechnology in Electronics (Nanoelectronics)*, http://www.understandingnano.com/nanotechnology-electronics.html (last accessed 23 November 2011).

63. M. Upton, *Safeguarding Workplace and Community Health. How Gold Mining Companies are Fighting HIV/AIDS, Tuberculosis and Malaria*, World Gold Council, London, 2008.
64. World Health Organization, *Fact Sheet No. 297: Cancer*, http://www.who.int/mediacentre/factsheets/fs297/en/ (last accessed 23 November 2011).
65. Nano.com, *Development of Biocompatible Gold and Silver Nanoparticles for Medical Applications*, http://www.azonano.com/news.aspx?newsID=3788 (last accessed 4 November 2011).
66. R. Bloch and G. Owusu, *MMCP Discussion Paper*, No. 1, University of Cape Town and Open University, 2011.

CHAPTER 4

Platinum Group Metals

GAVIN M. MUDD

Environmental Engineering, Monash University, Clayton, VIC 3800, Australia
Email: gavin.mudd@monash.edu

4.1 INTRODUCTION

Platinum group (PGMs) possess a range of unique chemical and physical properties. PGMs are increasingly finding important uses in a variety of environmentally related technologies, such as chemical process catalysts (especially in oil refineries), catalytic converters for vehicle exhaust control, hydrogen fuel cells, electronic components and a variety of specialty medical uses, amongst others. Given the need to expand almost all of these uses to meet this century's environmental and technological challenges, demand growth for PGMs can reasonably be expected to be sustained long into the future.

Global production of PGMs is dominated by South Africa owing to its large resources in the Bushveld Complex, while other countries such as Russia, Canada, Zimbabwe and the USA play a minor but important role. Concerns are being raised, however, about the long-term ability to supply PGMs to meet future needs,[1,2] and also allegations of significant environmental and

Materials for a Sustainable Future
Edited by Trevor M. Letcher and Janet L. Scott
© The Royal Society of Chemistry 2012
Published by the Royal Society of Chemistry, www.rsc.org

social impacts such as water pollution, unfair village relocation, economic disparity and compensation issues.[3–5]

This chapter presents a detailed review of the PGMs sector, including detailed mine and mineral resource information. A range of detailed data sets are compiled, focusing on annual production statistics and ore reserves and mineral resources by principal ore types. Given the concerns over possible future PGM supply shortages, this chapter provides an authoritative case study on resource sustainability for a group of metals that are uniquely concentrated in a select few regions of the Earth.

4.2 THE PLATINUM GROUP METALS AND THEIR USES

The uses of PGMs are wide and varied. Platinum's most common uses are in catalytic converters for exhaust control in transport vehicles ($\sim 50\%$), jewellery ($\sim 30\%$) and minor uses spread across chemicals, electrical components, glass, financial investment and petroleum process catalysts, as shown in Figure 4.1. Demand for PGMs in catalytic converters is showing strong growth in recent years, nearly tripling since 1990. The introduction of fuel cell technologies in hydrogen applications and vehicles could be another major demand for PGMs in the medium to long term.

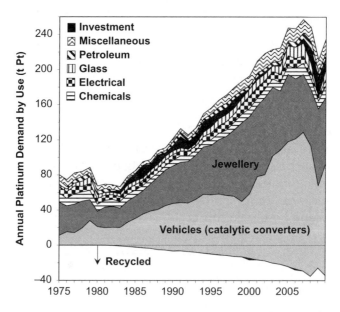

Figure 4.1 Platinum demand by application (1975–2010).[6]

4.3 BRIEF GEOLOGY OF PLATINUM GROUP METALS

4.3.1 Principal Ore Types

The platinum group metals comprise a heavy sub-group of platinum (Pt), iridium (Ir) and osmium (Os) and a light sub-group of palladium (Pd), rhodium (Rh) and ruthenium (Ru). Gold (Au) is often present in PGM ores, although mainly at low grades. There are broadly considered to be four economic types of PGM ores:[7,8]

- *Stratiform deposits:* Where PGMs occur in large Precambrian mafic to ultramafic layered intrusions; major examples include the Merensky and Upper Group 2 (UG2) reefs of the Bushveld Complex, South Africa, the Great Dyke reefs in Zimbabwe and the Stillwater reef complex in Montana, USA. These are usually considered primary owing to their size (~ 10–1000 Mt) and grade (3–10 g t^{-1} PGMs, ~ 0.1–1% Ni + Cu).
- *Norite intrusions:* Where meteoritic impact is considered to have been instrumental in allowing PGM emplacement; the major example is the Sudbury Irruptive Complex in Ontario, Canada (~ 10–1000 Mt, 1–3 g t^{-1} PGMs, ~ 2–3% Ni + Cu).
- *Ni-Cu bearing sills:* sub-volcanic sills related to deep seated structural lineaments; with examples being the Noril'sk–Talnakh District (or Taimyr Peninsula) in Russia and the Jinchuan deposits in China (~ 10–1000 Mt, 2–15 g/t PGMs, ~ 3–5% Ni + Cu).
- *Placer deposits:* Alluvial sedimentary deposits containing coarse PGMs (mainly Pt; generally very small in scale).

4.3.2 The Bushveld Complex, South Africa

The north-eastern region of South Africa hosts the Bushveld Complex, a large igneous complex with dimensions of ~ 370 km east–west and up to 240 km north–south.[8] It consists of multiple mafic layers formed during the intrusion of the Bushveld granites, giving rise to stratiform reefs up to 1.4 km in total thickness and reaching 9 km in depth. Owing to cover by younger sediments, outcrop of the Bushveld complex occurs in two bracket-like lobes on the west and east sides plus a linear lobe to the north. A regional and geological map and cross-section are given in Figure 4.2. Further details on the geology were given by Vermaak[8] and Cawthorn *et al.*[9,10]

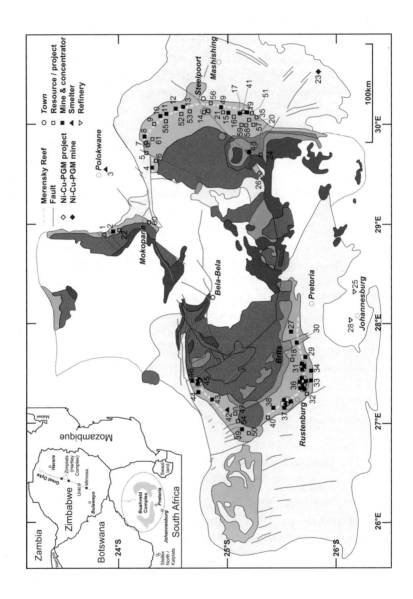

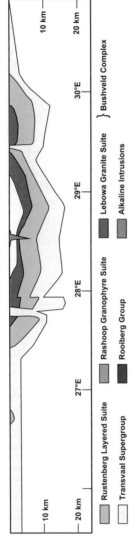

Figure 4.2 Geology of the Bushveld Complex, South Africa, and operating platinum group metal mines, smelters, refineries and projects (inset of southeastern Africa). All details are approximate only, based on 2009 data and most recent names. Names: 1 Boikgantsho, 2 Mogalakwena, 3 Polokwane, 4 Limpopo, 5 Mphahlele, 6 Zondernaam, 7 Liger, 8 Bokoni, 9 Ga-Phasha, 10 Twickenham, 11 Marula, 12 Smokey Hills, 13 Modikwa,14 Kennedy's Vale-Spitzkop, 15 Mototolo, 16 Der Brochen, 17 Everest North-Vygenhoek, 18 Leeuwkop (Afplats), 19 Everest South-Hoogland, 20 Booysendal, 21 Tamboti, 22 Akanani, 23 Nkomati, 24 Loskop, 25 Impala Springs, 26 Sheba's Ridge, 27 Eland, 28 Lonmin, 29 Pandora JV, 30 Crocodile River, 31 Marikana Group (Lonmin), 32 Townlands, 33 Kroondal, 34 Marikana PSA (AqP-AAP), 35 Berg, 36 Bathopele-Khomanani–Thembelani–Khuseleka–Siphumelele mines, Waterval Smelter, Rustenburg Refineries (former RPM Rustenberg Group), 37 Impala Group, 38 Bafokeng-Rasimone, 39 Pilanesberg, 40 Frischgewaagd (3-4-11), 41 Mareesburg, 42 Mortimer, 43 Union, 44 Tumela, 45 Northam, 46 Dishaba, 47 Magazynskraal, 48 Blue Ridge, 49 Two Rivers, 50 Platmin Other, 51 Klipriver, 52 Garatau, 53 Tubatse, 54 Rooderand-Ruighoek, 55 Tjate, 56 Grootboom, 57 Everest West-Sterkfontein, 58 Chieftains Plain, 59 Walhalla, 60 Oorlogsfontein (War Springs), 61 Lesego. Main map adapted from Scoates and Friedman.[11]

Although platinum was first reported in 1906 from the Bushveld region, it was not until 1924 that geologist Hans Merensky made the first discovery of economic amounts of platinum in the Bushveld – with this PGM layer then called the Merensky reef. Another major reef layer is the Upper Group 2 (UG2) chromitite reef (first noted in 1906). The Eastern and Western Bushveld contain both the Merensky and UG2 reefs, and the Northern Bushveld contains the Platreef. The Merensky and UG2 reefs are remarkably continuous over tens to hundreds of kilometres, with the PGMs mineralogically associated with base metal sulfides or chromites.[8,10] Most PGM reefs also contain low-grade nickel (Ni) and copper (Cu).

The Merensky reef was the first to be successfully mined commercially, whereas the UG2 proved more challenging to process until technology became available from the 1980s (pioneered by Lonmin). The very thin nature of the Merensky and UG2 reefs (~ 1 m) requires narrow mining techniques. The Platreef is much thicker (ranging from ~ 4 to ~ 200 m) and is mined by open cut at Mogalakwane due to the shallower reef depth. The depth of underground mines can range from 100 to 2000 m, with most currently active up to several hundred metres deep.

4.4 ECONOMIC RESOURCES OF PLATINUM GROUP METALS

4.4.1 Reported Ore Reserves and Mineral Resources

The metals contained in PGM ores are often reported as '4E' (Pt + Pd + Rh + Au) or '6E' (Pt + Pd + Rh + Ru + Ir + Au) grades, with osmium extremely rarely reported (3E is Pt + Pd + Au). A compiled summary of economic PGM reserves and resources by mine and ore type for 2010 are given in Tables 4.1, 4.2 and 4.3 for the Eastern, Western and Northern Bushveld Complex, respectively, Great Dyke reserves and resources in Table 4.4, other global PGM resources in Table 4.5 and Ni–Cu–PGMs resources in Table 4.6. All data are derived from company annual reports (or other technical studies) based on statutory mineral resource reporting codes (*e.g.* Australia's JORC Code;[12] South Africa's SAMREC[13] and Canada's NI-43-101[14]). A summary is presented in Table 4.7, giving a global resource of ~ 90 400 t PGMs (4E or 2.91 billion oz).

Table 4.1 PGM ore reserves and resources – Eastern Bushveld Complex, South Africa (2010 data).

Mine/project	Principal companies	Reserves Merensky Ore/Mt	4E/g t⁻¹	Reserves UG2 Ore/Mt	4E/g t⁻¹	Resources[a] Merensky Ore/Mt	4E/g t⁻¹	% Cu	% Ni	Resources[a] UG2 Ore/Mt	4E/g t⁻¹	% Cu	% Ni	PGMs (4E)/t
Bokoni	Anooraq Resources/Anglo American Platinum	31.2	4.14	39.0	5.25	198.4	4.93	0.09	0.21	328.6	6.40	0.06	0.17	3414.7
Booysendal	Mvelaphanda Resources/Northamerican Platinum					255.5	4.69	0.121	0.265	580.5	3.47	0.009	0.080	3212.6
GaPhasha	Anooraq Resources/Anglo American Platinum					236.1	5.20	0.08	0.22	311.0	6.14	0.04	0.15	3137.5
Der Brochen	Anglo American Platinum					177.6	4.38	0.064	0.125	319.9	4.73	0.02	0.09	2291.0
Spitzkop-Kennedy's Vale	Eastern Platinum					343.3	2.93	0.12	0.26	262.5	4.73	0.017	0.054	2249.0
Twickenham	Anglo American Platinum			73.4	5.37	161.4	5.04	0.05	0.14	171.3	5.76	0.02	0.14	2194.3
Modikwa	Anglo American Platinum/African Rainbow Minerals			56.0	4.86	208.8	2.70	0.12	0.29	216.6	5.99	0.02	0.13	2133.4

Table 4.1   (*Continued*)

Mine/project	Principal companies	Reserves Merensky Ore/Mt	Merensky 4E/g t⁻¹	UG2 Ore/Mt	UG2 4E/g t⁻¹	Resources[a] Merensky Ore/Mt	Merensky 4E/g t⁻¹	% Cu	% Ni	UG2 Ore/Mt	UG2 4E/g t⁻¹	% Cu	% Ni	PGMs (4E)/t
Walhalla	Aquarius Platinum					135.0	4.30			185.0	5.70			1635.0
Tamboti	Impala Platinum/Kameni					141.1	3.81			177.6	5.58			1528.2
Tubatse	Nkwe Platinum					153.2	4.45			105.8	7.18			1440.9
Lesego	Village Main Reef					77.2	6.75			119.5	6.52	0.068	0.182	1300.1
Chieftains Plain	Aquarius Platinum					85.0	4.30			115.0	5.70			1021.0
Garatau	Nkwe Platinum					69.7	4.40			78.6	5.35			727.2
Tjate	Jubilee Platinum					56.3	4.66	0.13	0.20	76.2	5.67	0.039	0.079	694.3
Marula	Impala Platinum					52.8	4.23	0.11	0.17	52.6	8.57			674.0
Limpopo JV[b]	Lonmin					49.6	4.16	0.11	0.17	107.3	4.49	0.09	0.14	688.9
Mphahlele[c]	Platmin									121.4	3.93	0.092	0.132	476.6
Limpopo (Dwaalkop)	Mvelaphanda Resources/Lonmin					76.1	2.98			37.6	4.35	0.09	0.14	390.1
Two Rivers	African Rainbow Minerals/Impala Platinum					22.6	3.31			56.7	3.91			296.3
Blue Ridge	Aquarius Platinum/Imbani Platinum									81.4	3.19			259.2

Project	Company												Contained
Mototolo	Anglo American Platinum/XK Platinum Partnership	15.8	3.82						27.6	4.36			180.7
Everest	Aquarius Platinum								34.2	3.38			115.9
Loskop JV[d]	Lonmin/Platmin								24.6	4.04[d]	0.011		99.5
Grootboom[c]	Platmin			2.9	8.61				21.6	4.28		0.028	92.4
Anglo Pt Other Projects	Anglo American Platinum								10.3	5.87			85.4
Mareesburg	Eastern Platinum								15.9	3.92	0.026		62.3
Kliprivier	Nkwe Platinum								24.4	2.3			56.2
Berg[c]	Platfields								23.6	2.02			47.8
Everest West-Sterkfontein	Aquarius Platinum								13.0	3.50			45.5
Liger (Leeuwkop-Tigerpoort)	Platfields								9.6	4.75		0.070	45.4
Everest South-Hoogland	Aquarius Platinum								6.6	2.88			19.0
Smokey Hills	Platinum Australia					0.140	0.287		3.99	4.72			18.8
Everest North-Vygenhoek	Aquarius Platinum								2.8	5.11			14.2
Sub-totals		**31.2**	**4.14**	**184.2**	**5.06**	**2502.6**	**4.23**	**~0.094**	**~0.207**	**3723.2**	**5.11**	**~0.034**	**~0.115**	**30 647.4**

[a]Resources are in addition to reserves.
[b]Includes Baobab-Doornvlei-Zebedeila.
[c]Ore type not specified, UG2 assumed.
[d]3E grade only.

Table 4.2   PGM ore reserves and resources – Western Bushveld Complex, South Africa (2010 data).

| Mine/project | Principal companies | Reserves | | | | Resources[a] | | | | | | | | PGMs (4E)/t |
| | | Merensky | | UG2 | | Merensky | | | | UG2 | | | | |
		Ore/Mt	4E/g t^{-1}	Ore/Mt	4E/g t^{-1}	Ore/Mt	4E/g t^{-1}	% Cu	% Ni	Ore/Mt	4E/g t^{-1}	% Cu	% Ni	
Marikana (Lonmin)	Lonmin					342.8	4.99	0.10	0.18	559.6	5.08	0.01	0.03	4553.5
Tumela	Anglo American Platinum	29.9	5.80	149.4	4.72	141.0	8.12	0.10	0.26	216.2	5.65	0.01	0.11	3245.0
Impala	Impala Platinum					278.3	5.74			234.4	6.72			3173.0
Bafokeng-Rasimone	Royal Bafokeng/ Anglo American Platinum	78.2	4.28	58.2	3.93	118.8	6.79	0.11	0.26	150.9	5.29	0.01	0.10	2168.2
Afplats-Leeuwkop Group	Impala Platinum									358.8	5.09			1826.3
Union	Anglo American Platinum	0.2	5.64	56.2	4.11	81.9	6.21	0.07	0.26	122.8	5.71	0.01	0.11	1442.3
Dishaba	Anglo American Platinum	16.0	5.25	84.3	4.58	32.2	7.79	0.10	0.23	51.9	5.71	0.01	0.12	1017.3
Zondereinde (Northam)	Northamerican Platinum					68.44	7.09			62.6	5.06			802.1
Eland	Xstrata			14.8	3.98					153.1	4.31			660.4
Pandora JV	Anglo American Platinum/ Lonmin									139.3	4.25	0.01	0.02	651.0
Siphumelele	Anglo American Platinum	13.2	5.36	23.8	3.79	13.8	6.93	0.12	0.24	73.8	5.29	0.01	0.10	647.0

Project / Operator													
Magazynskraal 3JQ — Bakgatla/Pallinghurst/Anglo American Platinum					47.0	6.58			63.5	4.65			604.5
Frischgewaagd 3-4-11 (WBJV) — Wesizwe/Anglo American Platinum					45.1	6.35			56.4	4.54			542.5
Thembelani — Anglo American Platinum	22.6	5.05	14.0	3.66	5.3	5.53	0.10	0.24	61.5	5.24	0.01	0.10	516.9
Rustenburg Other — Anglo American Platinum					22.6	6.40	0.10	0.22	57.8	5.33	0.01	0.10	452.7
Khomanani — Anglo American Platinum	10.6	5.01	15.6	3.74	12.7	6.61	0.10	0.23	35.7	5.26	0.01	0.10	383.2
Pilanesburg (T-R JV)[b,c] — Platmin/Moepi									109.3	3.10	0.024	0.087	339.0
Khuseleka — Anglo American Platinum	6.7	4.69	56.1	3.95	3.7	5.87	0.10	0.21	9.3	5.42	0.01	0.10	325.1
Projects 1–3 (WBJV) — Platinum Group Metals/Wesizwe Platinum					23.5	7.27			32.6	4.50			318.0
Crocodile River Gubevu — Eastern Platinum/Gubevu									63.2	4.05			256.0
Kroondal — Anglo American Platinum/Aquarius Platinum									20.1	5.55			111.5
Marikana (PSA) — Anglo American Platinum/Aquarius Platinum									34.9	5.02			175.4

Table 4.2 (Continued)

| | | Reserves | | | | Resources[a] | | | | | | | | |
| | | Merensky | | UG2 | | Merensky | | | | UG2 | | | | |
Mine/project	Principal companies	Ore/ Mt	4E/ g t⁻¹	Ore/ Mt	4E/ g t⁻¹	Ore/ Mt	4E/ g t⁻¹	% Cu	% Ni	Ore/ Mt	4E/ g t⁻¹	% Cu	% Ni	PGMs (4E)/t
Bathopele	Anglo American Platinum			43.8	2.94	1.5	4.18			1.1	3.59	0.01	0.10	139.0
Ledig 1–2	Wesizwe Platinum					10.7	6.00			13.23	4.56			124.7
Townlands	Aquarius Platinum									16.3	6.29			102.7
Rooderand[b]	Nkwe Platinum/ Platinum Australia									24.8	3.8	0.027	0.051	94.2
Schietfontein	Xstrata									22.90	2.29			52.4
Zilkaatsnek	Xstrata									4.34	2.50			10.8
Frischge- waagd 1	Wesizwe Platinum					0.28	6.66			0.34	4.30			3.3
Sub-totals		**177.4**	**4.86**	**516.2**	**4.18**	**1249.7**	**6.18**	**~0.099**	**~0.221**	**2750.9**	**5.09**	**~0.011**	**~0.070**	**24 738.1**

[a]Resources are in addition to reserves.
[b]Ore type not specified, UG2 assumed.
[c]Includes Tuschenkomst-Ruighoek ('T-R').

Table 4.3 PGM ore reserves and resources – Northern Bushveld Complex, South Africa (2010 data).

Mine/project	Principal companies	Reserves		Resources[a]				
		Ore/Mt	4E/g t^{-1}	Ore/Mt	4E/g t^{-1}	% Cu	% Ni	PGMs (4E)/t
Mogalakwene	Anglo American Platinum	609.4	2.82	2170.5	2.03	0.11	0.18	6124.6
Akanani	Lonmin/Incwala Resources			291.9	3.84	0.12	0.22	1121.4
Boikgantsho	Anooraq Resources/Anglo American Platinum			280.8	1.31	0.08	0.13	367.9
Oorlogsfontein (War Springs)[b]	Platmin			47.0	1.11[b]	0.10	0.13	52.1
Rooipoort	Caledonia Mining Corp.			18.1	1.28	0.11	0.19	23.3
Sub-totals		**609.4**	**2.82**	**2808.3**	**2.13**	**0.11**	**0.18**	**7689.3**

[a]Resources are in addition to reserves.
[b]3E grade only.

Table 4.4 PGM ore reserves and resources – Great Dyke, Zimbabwe (2010 data).

Mine/project	Principal companies	Reserves		Resources[a]				PGMs (4E)/t
		Ore/Mt	4E/g t^{-1}	Ore/Mt	4E/g t^{-1}	% Cu	% Ni	
Zimplats	Impala Platinum			1879.0	3.63	0.11	0.13	6818.5
Serui[b]	Amari Resources/Zimbabwe Mining Development Corp.			137.8[b]	4[b]			551.1
Bokai	Eurasian Natural Resources Corp./Zimbabwe Mining Development Corp.			91.6	3.62	0.17		331.6
Mimosa - South Hill	Aquarius Platinum/Impala Platinum			86.6	3.74	0.12	0.14	323.6
Unki	Anglo American Platinum	41.7	3.78	77.6	4.13	0.15	0.22	320.5
Mimosa - North Hill	Aquarius Platinum/Impala Platinum			48.6	3.64	0.11	0.14	177.0
Sub-totals		**41.7**	**3.78**	**2321.2**	**3.67**	**0.11**	**0.12**	**8679.9**

[a]Resources are in addition to reserves.
[b]Only contained PGMs reported (17.14 Moz or 551.1 t), ore tonnage and grade assumed.

Table 4.5 PGM ore reserves and resources – miscellaneous worldwide (2010 data).

Mine/project	Principal companies	Ore/Mt	$3E/g\ t^{-1}$	% Cu	% Ni	PGMs $(4E)/t$
Skaergaard (Greenland)	Platina Resources	1520.0	0.86			1307.2
Stillwater (USA)	Stillwater/Norilsk	37.01	16.8	0.07	0.1	621.7
Nokomis-Duluth (USA)	Wallbridge/Antofagasta	823.9	0.67	0.64	0.20	550.6
NorthMet (USA)	Polymet Mining	1029.4	0.292	0.238	0.067	300.9
Arctic-Ahmavaara (Finland)[a]	Gold Fields	187.77	1.22	0.175	0.069	229.3
Kalplats (South Africa)	African Rainbow Minerals/Platinum Australia	137.4	1.53			210.2
Lac des Iles (Canada)	North American Palladium	35.5	3.99	0.07	0.09	141.5
Marathon (Canada)	Marathon PGM	121.0	1.05	0.236		127.6
Arctic-Konttijärvi (Finland)[a]	Gold Fields	75.24	1.30	0.097	0.046	97.6
Eagle's Nest (Canada)	Noront Resources	21.6	4.44	0.98	1.39	95.9
Nunavik (Canada)	Jilin Jien Nickel/Goldbrook Ventures	27.15	2.83	1.10	0.89	76.7
Panton (Australia)[b]	Platinum Australia	14.3	5.20	0.075	0.27	74.5
MunniMunni (Australia)[b]	Platina Resources	23.6	2.94[b]	0.15	0.09	69.4
River Valley (Canada)	Anglo American Platinum/Pacific North West	33.0	1.36	0.10	0.02	44.9
Pedra Blanca (Brazil)	Anglo American Platinum/Solitario	12.9	2.27	0.03	0.23	29.4
Thunder Bay North-Current Lake (Canada)	Magma Metals	10.79	2.13			23.0
Weld Range (Australia)	Atomaer Holdings	6.3	1.7			10.7
Broken Hammer (Canada)	Wallbridge Mining	0.251	3.79	1	0.10	1.0
Parkin (Canada)	Wallbridge Mining	0.351	1.93	0.69	0.59	0.7
Sub-totals		**4117.4**	**0.97**	**0.37**	**0.13**	**4012.6**

[a]2007 data.
[b]4E grade.

Table 4.6 Nickel–copper–PGM ore resources – miscellaneous worldwide (2010 data).

Mine/project	Principal companies	Ore/Mt	$3E/g \ t^{-1}$	% Cu	% Ni	PGMs $(3E)/t$
Kingashskoye-Verkhnekingashshskoye (Russia)	Unknown	484.55	0.63	0.45	0.21	305.5
Kola Peninsula (Russia)	Norilsk Nickel	546.3	0.08	0.30	0.63	42.6
Maslovskoy (Russia)	Norilsk Nickel	215	6.53	0.51	0.33	1404.0
Mokopane (South Africa)[a]	Blackthorn Resources	39.7	0.55	0.085	0.146	21.9
Nkomati (South Africa)[a]	Norilsk Nickel	407.7	0.87[a]	0.13	0.35	356.0
Nor'ilsk-Talnakh (Russia)	Norilsk Nickel	2188.4	4.95	1.45	0.77	10 843.2
Sheba's Ridge (South Africa)	Aquarius Platinum/Anglo American Platinum	764.3	0.90	0.07	0.19	687.9
Shebandowan West (Canada)	North American Palladium	1.46	1.20	0.62	0.93	1.8
Tati-Selkirk (Botswana)[a]	Norilsk Nickel	135.3	0.57[a]	0.27	0.23	77.0
Vale Inco Sudbury (Canada)	Vale Inco	112.3	2.33	1.53	1.20	262.2
Kabanga (Tanzania)	Xstrata/Barrick Gold	58.2	0.58	0.33	2.63	33.8
Jinchuan (China)	Jinchuan	432	0.2	0.88	1.39	86.4
Kevitsa (Finland)	First Quantum Minerals	275	0.87	0.41	0.31	239.3
Maturi (USA)	Franconia Minerals	119.9	0.38	0.67	0.25	45.6
Birch Lake (USA)	Franconia Minerals	216.81	0.85	0.52	0.17	183.8
Shakespeare (Canada)	Ursa Major Minerals	14.02	0.88	0.36	0.34	12.3
Sudbury-Fraser Morgan (Canada)	Xstrata	8.65	0.26	0.55	1.85	2.3
Eagle (USA)	Rio Tinto	4.8	1.2	2.40	2.82	5.8
Sudbury-Onaping Depth (Canada)	Xstrata	15.7	0.97	1.25	2.74	15.3
Denison (Canada)	Vale Inco/Lonmin	0.7	6.30	0.96	0.55	4.4
Kuhmo Group (Finland)	Altona Mining	9.819	0.41	0.08	0.38	4.0
Werner Lake-Big Zone (Canada)	Puget Ventures	0.172	1.21	0.26	0.62	0.2
Wildara-Horn (Australia)	Breakaway Resources	0.6	0.5	0.3	1.39	0.3
Sub-totals		**6051.5**	**2.42**	**0.78**	**0.60**	**14 635.4**

[a]4E grade.

Table 4.7 Summary of global PGM resources (2010 data).

Mine/project	Ore/Mt	4E/g t⁻¹	% Cu	% Ni	PGMs (4E)/t
Eastern Bushveld Complex, South Africa	6441.2	4.76	~0.06	~0.15	30 647.4
Western Bushveld Complex, South Africa	4694.2	5.27	~0.04	~0.12	24 738.1
Northern Bushveld Complex, South Africa	3417.7	2.25	0.11	0.18	7689.3
Great Dyke, Zimbabwe	2362.9	3.67	0.11	0.12	8679.9
Miscellaneous, global	4117.4	0.97	0.37	0.13	4012.6
Nickel–copper–PGMs, global	6051.5	2.42	0.78	0.60	14 635.4
Totals	**27 085**	**3.34**	**0.28**	**0.24**	**90 400**

This should be considered a reliable geological estimate (as of 2010), since all reserves and resources are reported under statutory codes, although whether all projects proceed to production is dependent on economics, mining conditions, processing characteristics, energy, site-specific environmental issues (especially land use, water and mine waste management), social issues (especially governance and economic empowerment issues) and so on. The Bushveld Complex clearly dominates, with 70.9% of reported global PGM resources (Table 4.7) – only the Norilsk-Taimyr Peninsula resource comes close in size to the Bushveld, with the Great Dyke just behind.

Another important aspect of PGM resources is the proportion of individual metals. Given that only a minority of companies report PGM resource grades as 6E, only 4E (or 3E) was included in Tables 4.1–4.6. For the Bushveld, based on available 6E reporting, 6E grades are typically about 1.16 times 4E grade. An approximate breakdown by individual metal is given in Table 4.8, based on ore-weighted grades where 4E or 6E split data are reported. Although Ru and Ir can be highly variable between ore types, the UG2 reef clearly has higher Ru and Ir grades than Merensky and Great Dyke ore.

The evolution of reported reserves and resources over time for some of the major mines in the Bushveld and Great Dyke is shown in Figure 4.3, with a comparison of 2010 reserves and resources with cumulative production from 2000 to 2010 in Table 4.9. This shows that most major producers have continued to increase total resources over time.

Table 4.8 Approximate PGM grades by individual metal and ore type (weighted average of available 2010 data; number of data points in parentheses).

Mine/project	All grades/g t^{-1}						
	Pt	Pd	Rh	Ru	Ir	Os	Au
Eastern Bushveld – Merensky Reef	2.47 (13)	1.29 (13)	0.13 (13)	~0.25 (4)	~0.04 (4)	~0.2[a]	0.28 (13)
Western Bushveld – Merensky Reef	3.78 (13)	1.69 (13)	0.29 (13)	~0.56 (1)	~0.12 (1)	~0.1[a]	0.27 (13)
Eastern Bushveld – UG2 Reef	2.42 (19)	2.08 (19)	0.44 (19)	~0.84 (6)	~0.20 (6)	~0.2[a]	0.08 (19)
Western Bushveld – UG2 Reef	2.97 (5)	1.47 (5)	0.56 (5)	~1.04 (2)	~0.26 (2)[b]	~0.1[a]	0.04 (5)
Platreef	0.99 (4)	1.12 (4)	0.07 (2)	–	–	–	0.15 (3)
Great Dyke	1.81 (5)	1.35 (4)	0.16 (4)	0.14 (3)	0.07 (3)	0.016[c]	0.27 (5)
Miscellaneous	0.16 (17)	0.66 (17)	0.05 (4)	–	–	–	0.12 (12)
Nickel–copper–PGMs	0.75 (14)	2.49 (14)	–	–	–	–	0.17 (14)

[a]Approximate only, based on data from refs 9,15 and 16.
[b]Consistent with ref. 17.
[c]Data based on personal communication (Zimplats) (also consistent with ref. 17).

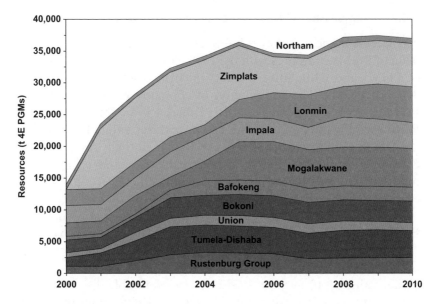

Figure 4.3 Reported reserves and resources over time for some of the major mines in the Bushveld and Great Dyke.

Table 4.9 Comparison of 2010 reserves and resources with cumulative production from 2000 to 2010 for some major PGM mines.

Mine/project	Reserves and resources (2010)			Cumulative production (2000 to 2010)		
	Ore/Mt	4E/g t^{-1}	PGMs (4E)/t	Ore/Mt	4E/g t^{-1}	PGMs (4E)/t
Rustenburg Group	505.0	4.88	2463.5	113.3	4.36	414.2
Tumela-Dishaba	720.8	5.91	4261.4	72.2	5.32	310.0
Union	222.0	5.52	1225.9	59.7	3.85	168.1
Bokoni	597.1	5.72	3414.7	15.0	4.45	54.1
Bafokeng-Rasimone	406.1	5.34	2168.2	25.3	4.41	89.8
Mogalakwane	2779.9	2.20	6124.6	62.4	3.41	144.1
Impala	535.9	7.57	4056.7	168.0	4.80	575.5
Lonmin	1212.0	4.65	5635.8	138.9	4.97	411.6
Northam	131.1	6.12	802.0	22.9	5.36	108.6
Zimplats	1879.0	3.64	6839.6	19.8	3.21	53.0

4.4.2 Comparison of Reserve Resource Estimates

The US Geological Survey (USGS) 2010 global reserves value of 66 000 t of PGMs compares with the 2009 global production of ∼465 t of PGMs and cumulative production from 1900 to 2009 of ∼13 300 t of PGMs. The compilation of company-reported economic PGM reserves + resources in 2010 (Tables 4.1–4.6) of ∼90 400 t of PGMs significantly exceeds the USGS reserves value. Historically, the USGS used to report 80 000 t of PGMs in their 'reserves base' category (*e.g.* USGS, 2009), although this category is no longer reported (from 2010) owing to the significant uncertainty in this estimate (reserves base included reserves).

The USGS 2010 reserves value for South Africa is 63 000 t of PGMs, which compares favourably with the total from Tables 4.1–4.6 of 64 351 t, in contrast, for Russia the USGS estimates reserves of 1100 t of PGMs whereas the data in Table 4.6 show a much higher value of 12 595 t of PGMs – a considerable discrepancy.

Tentative concerns were raised in the late 1960s about the future availability of Pt, with reasonably confident assessments pointing to extractable resources of at least 6220 t of Pt from 1970 to 2000[18] – over this same period, South African production was 2245 t of Pt (noting present resources in Tables 5.1–5.6). The extent of recoverable resources was linked to ore grades and mining depth, with considerable potential for continuing to increase reserves for some time[18] – a prediction strongly confirmed by Figure 4.3.

Although Gordon *et al.*[1] suggested that Pt could be in short supply within a few decades as demand continues to grow, this proposition is not supported by the available reserves–resources data. In the Bushveld Complex, most resources are only estimated to a mining depth of 2 km (the current economic mining limit), with considerable potential for additional PGMs known at greater depths and also in other lower grade reefs (1–3 g t^{-1}) not currently exploited – potential PGMs could reach 311 000 t in the Bushveld alone.[10,19,20] Hence the critical sustainability issue in the future is not resource size but the associated environmental and social costs.[21]

4.5 PLATINUM GROUP METAL MINES AROUND THE WORLD

4.5.1 Overview

The mining of PGM ores is through conventional underground or open cut techniques. The next stage is grinding and gravity-based (or dense media) separation, followed by flotation to produce a PGM-rich concentrate. The run-of-mine ore grades are typically several g t^{-1}, whereas concentrates are some hundreds of g t^{-1}.[8] Concentrate is then smelted to produce a PGM-rich Ni–Cu matte, with the PGMs extracted and purified at a precious metals refinery (including Ni–Cu by-products). The processing is therefore more analogous to base metals than Au, which relies on cyanide leaching and hydrometallurgy. Smelting of concentrates from Ni–Cu mining can also be a moderate source of PGMs (*e.g.* Russia, Canada; see later tables). Further details on PGM ore processing were given by Vermaak,[8] Merkle and McKenzie[22] and Cole and Ferron.[23]

4.5.2 Production Trends

The historical global production of PGMs by country is shown in Figure 4.4, with nominal and real prices over time shown in the inset. The boom–bust cycle of nominal prices is evident, although in real terms PGM prices have been stable on average over the long-term. Demand by use was shown in Figure 4.1. The severe impact of the recent global financial crisis is evident, as demand declined substantially from 2007 to 2009 (mainly due to the

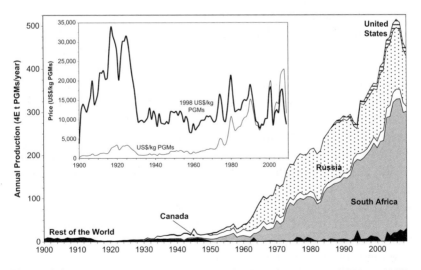

Figure 4.4 Historical global production of PGMs by country (1900 to 2009), including nominal and real price data in the inset (data updated from ref. 21).

collapse of autocatalyst demand for new vehicles), although it began to recover in 2010.

A detailed compilation of PGM production statistics for 2010 in the Bushveld Complex is given in Table 4.10, other PGM projects in Table 4.11 and Ni–Cu-PGM projects in Table 4.12. As shown, the overall average PGM ore grade is ~ 3.8 g t^{-1} (4E) in the Bushveld, although higher for Stillwater in the USA. For Ni–Cu–PGM projects, the ore grade is highly variable (only yield is commonly reported and not assay grade), with the relatively high Pd content of the Taimyr Peninsula ore standing out, showing similar grades to current mines (allowing for a somewhat lower extraction efficiency). The recent trends in PGM ore grades by company are shown in Figure 4.5, demonstrating a gradual decline for most companies, with only a couple showing stable grades.

In comparing production versus resources (*i.e.* Tables 4.10–4.12 *versus* Tables 4.1–4.6), ore grades can be expected to decline in the future to a minor extent, but this will more critically depend on the balance of ore types developed and processed, such as a preference for shallower low grade Platreef over deeper Merensky/UG2 projects.

A major trend in the Bushveld over the past few decades has been the emergence of UG2 ore as the dominant source. The processing of UG2 ore was more difficult than that of Merensky ore, owing its high chrome content, and the ability to treat UG2 ore was pioneered by Lonmin in the 1980s. By the 1990s, the process technology had been adapted and proven and numerous projects began to mine the UG2 reef in conjunction within existing mines (*e.g.* Bokoni, Rustenburg Group), or new PGM projects were developed based solely on UG2 resources (*e.g.* Kroondal, Twickenham). Furthermore, the Mogalakwane project was developed on the northern Platreef by Anglo American Platinum in the mid-1990s, becoming the first commercial mine in this region. The Mogalakwane project is mined by large-scale open cut owing to the shallower nature of the reef, although it is lower in grade than the Eastern or Western Bushveld. The recent proportion of ore types for South Africa and Zimbabwe is shown in Figure 4.6.

In general, most Bushveld projects use underground mining techniques ($\sim 84\%$ of ore, based on Tables 4.10 and 4.11), although there are some open cut mines where the reefs outcrop or subcrop. For many recent projects, a small-scale open cut is used

Table 4.10 Major platinum group metal producers – Bushveld Complex, South Africa (2010 data).[a]

Mine/project	Ore milled/kt	4E/g t⁻¹	Pt/kg	Pd/kg	Rh/kg	Ru/kg	Ir/kg	Au/kg	Ni/t	Cu/t	Ore source	Mine type	Principal owners
Bathopele[b]	3107	3.02	4403.8	2544.0	768.2	1346.6		43.5	300	100	UG2 100%	UG 100%	Anglo Platinum 100%
Khomanani[b]	1317	4.38	3144.2	1467.9	301.7	391.9		124.4	700	400	UG2 28.3%, Mer. 71.7%	UG 100%	Anglo Platinum 100%
Thembelani[b]	1447	4.23	3035.4	1620.3	438.5	755.7		62.2	500	200	UG2 80.2%, Mer. 19.8%	UG 100%	Anglo Platinum 100%
Khuseleka[b]	1967	3.97	4095.9	2021.5	472.7	715.3		130.6	900	500	UG2 55%, Mer. 45%	UG 100%	Anglo Platinum 100%
Siphumelele[b]	1032	5.09	2991.8	1306.2	223.9	211.5		143.1	700	500	UG2 0.1%, Mer. 99.9%	UG 100%	Anglo Platinum 100%
Tumela[c]	4488	4.02	9423.3	4378.9	1427.5	2233.0		140.0	1000	500	UG2 80.5%, Mer. 19.5%	UG 100%	Anglo Platinum 100%
Dishaba[c]	1908	4.79	4864.0	2233.0	600.2	833.5		115.1	800	400	UG2 43.7%, Mer. 56.3%	UG 100%	Anglo Platinum 100%
Union	5543	3.37	9454.4	4183.0	1449.3	2407.1		108.9	800	300	UG2 64.7%, Mer. 35.3%	UG 100%	Anglo Platinum 85%
Bafokeng	2407	4.31	5826.9	2392.7	417.5	722.3	139.5	323.9	2251	1425	UG2 3.2%, Mer. 96.8%	UG 100%	AngloPt 33%, RBH 67%
Bokoni[d]	1044.1	4.12	1953.1	1309.3	195.9	234.4		112.0	700	400	UG2 32%, Mer. 68%	UG 100%	AngloPt 49%, AnRes 51%
Mogalakwane[e]	10 380	2.60	8468.5	8807.5	513.2	531.8		901.9	8500	5600	Platreef 100%	OC 100%	Anglo Platinum 100%
Twickenham	58	4.20	112.0	99.5	18.7	34.2		3.1	nd	nd	UG2 100%	UG 100%	Anglo Platinum 100%
Modikwa	2270	5.25	4089.7	4007.6	849.0	1211.4	299.3	105.2	663	410	UG2 100%	UG 100%	AngloPt 50%, ARM 41.5%
Mototolo JV	2262	3.33	3436.6	2021.5	581.6	1125.8		46.7	300	100	UG2 100%	UG 100%	AngloPt 50%, Xstrata 50%
Kroondal JV	6180	2.59	7477.7	3780.9	1385.0	2330.5	552.7	62.9	400	100	UG2 100%	UG 100%	AngloPt 50%, AqPt 50%
Marikana JV	2230	2.65	2566.5	1188.8	431.1	879.8	204.7	25.1	100	100	UG2 100%	UG 100%	AngloPt 50%, AqPt 50%
Everest	150	3.09	158.5	82.6	20.5	nd	nd	2.6	nd	nd	UG2 100%	UG 100%	Aquarius Platinum 100%
Two Rivers	2920	3.95	4382.2	2537.4	735.0	1220.2	295.0	59.4	nd	nd	UG2 100%	UG 100%	ARM 55%, Implats 45%
Northam	2038	5.19	6099.4	2999.7	799.9	1352.0	293.0	100.0	1752	957	UG2 50.8%, Mer. 49.2%	UG 100%	Northam Platinum 100%

Table 4.10 (Continued)

Mine/project	Ore milled/kt	4E/g t^{-1}	Pt/kg	Pd/kg	Rh/kg	Ru/kg	Ir/kg	Au/kg	Ni/t	Cu/t	Ore source	Mine type	Principal owners
Implats	13 531	4.60	27 100.5	14 284.2	384.9	8157.5		nd	4900	nd	UG2 60.2%, Mer. 39.8%	UG 100%	Implats 100%
Marula	1545	4.36	2180.1	2257.9	457.2	845.9		nd	216.6	nd	UG2 100%	UG 100%	Implats 73%
Crocodile River	1861	4.10	1957.1	851.8	332.5	742.7		14.0	nd	nd	UG2 100%	UG 100%	Eastplats 87.5%
Marikana (Lonmin)	11 175	4.64	21 595.1	10 129.2	3019.2	4700.5	1009.9	471.1	2972	1824	UG2 75.6%, Mer. 24.4%	UG 98.8%, OC 1.2%	Lonmin 82%
Eland	900[g]	4.5[h]	1940.9	842.3	318.8	nd	nd	nd	nd	nd	UG2 100%	OC 100%	Xstrata 74%, Ngazana 26%
Blue Ridge	1082	2.39	1094.1	539.0	171.8	231.0	47.7	17.8	nd	nd	UG2 ~15%, Mer. ~85%	UG 100%	AqPt 50%, Imbani 50%
Pilanesberg	2906	1.72	1149.2[i]	528.6[i]	138.7[i]	nd	nd	57.1[i]	nd	nd	UG2 100%	OC 100%	Platmin 72.4%
Smokey Hills	398	3.58	430.3[i]	442.5[i]	88.8[i]	nd	nd	36.6[i]	nd	nd	UG2 100%	UG ~50%, OC ~50%	Platinum Australia 79%
Total	**86.15 Mt**	**3.80**	**143 431**	**78 858**	**19 941**	**>28 000**		**>3207**	**>28 500**	**>14 000**	UG2 ~65.6%, Mer. ~22.3%	UG 85.4%, OC 14.6%	

[a] Abbreviations: JV, Joint Venture; UG2, upper group 2 reef; Mer., Merensky reef; OC, open cut; UG, underground; AngloPt, Anglo Platinum; AqPt, Aquarius Platinum; RBH, Royal Bafokeng Holdings; AnRes, Anooraq Resources; ARM, African Rainbow Resources; nd, no data.
[b] Formerly part of the RPM Rustenburg group.
[c] Formerly part of the RPM Amandelbult group.
[d] Formerly known as Lebowa.
[e] Formerly known as Potgietersrust.
[f] Ore processed and PGMs production includes some tailings reprocessing (about 30% of material milled in 2009).
[g] approximate estimate only.
[h] Ore grade assumed from reported resources.
[i] Production based on total 4E split from resources.

Table 4.11 Major platinum group metal producers – Great Dyke, Zimbabwe and Stillwater, USA (2010 data).[a]

Mine/project	Ore milled/kt	4E/g t⁻¹	Pt/kg	Pd/kg	Rh/kg	Ru/kg	Ir/kg	Au/kg	Ni/t	Cu/t	Ore source	Mine type	Principal owners
Mimosa (Zimbabwe)	2300	3.59	3149	2382	251	190	98	426	2539	2074	Great Dyke[100%]	UG[100%]	AqPt[50%], Implats[50%]
Zimplats (Zimbabwe)	4095	3.00	5408	4391	480			602	3131	2231	Great Dyke[100%]	UG[100%]	Implats[87%]
East Boulder (USA)	363	12.7	933	3203	62			280	525	374	Stillwater[100%]	UG[100%]	Norilsk[51.5%]
Stillwater (USA)	709	17.1	2519	8428							Stillwater[100%]	UG[100%]	Norilsk[51.5%]
Total	**7466 Mt**	**4.99**	**12 009**	**18 404**	**793**	**>190**	**>98**	**1308**	**6195**	**4679**			

[a]Abbreviations as in Table 4.10.

Table 4.12 Major nickel–copper–platinum group metal producers (2010 data).[a]

Mine/project	Ore milled/kt	% Ni	% Cu	Ni/kt	Cu/kt	PGMs/ g t⁻¹	Pt/kg	Pd/kg	Au/kg	PGMs/kg	Mine type	Owners
Nkomati	5058	0.39	0.17	13.0	5.7	0.80				1910	OC, UG	Norilsk[50%], ARM[50%]
Tati	8380	0.23	0.23	12.0	11.9	~0.5[b]	435	2612		3047	OC	Norilsk[85%], Botswana[15%]
Vale Sudbury	2660	1.78	1.53	22.4	34.0	1.60[b]	1089	1866	1306	4261	UG	Vale Inco[100%]
Taimyr Peninsula	16 118	1.41	2.41	196.5	348.4	7.51	20 619	84 654		105 273	UG, OC	Norilsk[100%]
Kola Peninsula[c]	8336	0.67	0.29	39.0	17.3	0.08					UG, OC	Norilsk[100%]
Kambalda[d,e]	~1061	~3.01	~0.22	~26.8	~1.9	~0.8[b]				781	UG	BHP Billiton[100%,e]
Total	**~41.6 Mt**	**~0.96**	**~1.16**	**~310**	**~419**	**~3.1**	**>22 143**	**>90 132**	**>1306**	**>115 272**		

[a]Abbreviations as in Table 4.10.
[b]Yield not assay grade.
[c]PGMs production included in Taimyr Peninsula.
[d]Approximate data only (adapted from Mudd, 2010).
[e]All Western Australian Pt–Pd assumed to be from Kambalda only.

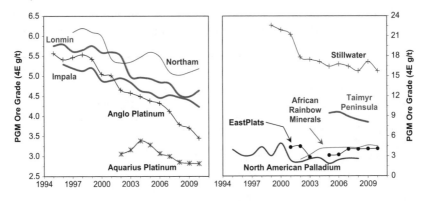

Figure 4.5 Recent PGM ore grades by company during processing.

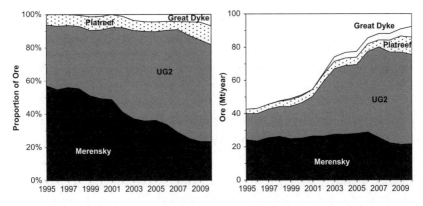

Figure 4.6 Proportion of PGM ore by type for South Africa and Zimbabwe.

initially to facilitate the development of underground mining (*e.g.* to make decline construction more economic). A major trend in the Bushveld is the increasing depth of underground mines, with current underground and open cut mine depths (where known) compiled in Table 4.13. It is worth noting that Impala Platinum only report resources to a maximum depth of 2350 m at present (although some gold mines in the Witwatersrand Basin to the south of the Bushveld now reach more than 4 km in depth). The trend of increasing mine depth is also strongly evident in the Sudbury Basin Ni–Cu–PGM field in Canada, where mines now reach a maximum depth of ~2.5 km.[24]

Table 4.13 Approximate maximum depths for underground and open cut PGM mines, Bushveld Complex, Great Dyke and Stillwater.

Western Bushveld	maximum Depth/m	Eastern Bushveld	maximum Depth/m	Open cut Project	maximum Depth/m
Bathopele	284	Twickenham	110	Mogalakwane	281
Khomanani	1181	Modikwa	628	Marikana JV	200
Thembelani	1055	Mototolo JV	190	Eland	105
Khuseleka	783	Bokoni	500	Pilanesberg	180
Siphumelele	1402	Everest	250		
Tumela	1400	Blue Ridge	275		
Dishaba	1347				
Union	1700	**Others**			
Bafokeng	491	Mimosa (Zimbabwe)	200		
Kroondal JV	400	Zimplats (Zimbabwe)	'Shallow'		
Northam	2200	East Boulder (USA)	~500		
Impala	1000	Stillwater (USA)	595		

4.6 SUSTAINABILITY ISSUES, CONSTRAINTS AND THE FUTURE OF PLATINUM GROUP METALS

At an annual production of ~465 t of PGMs (2009), the 90 400 t of PGMs would last nearly two centuries. Alternatively, assuming a linear trend line of annual PGMs production from 1960 to 2009 (1960 is when production began to increase strongly) gives a correlation coefficient of 96.0%, as shown in Figure 4.7, with cumulative production of ~80 400 t of PGMs from 2010 to 2100. As presented in this chapter, there are clearly sufficient resources of PGMs known at present to meet growing demands for some decades – meaning that other factors will be much more critical in determining PGM supply.

The production of PGMs requires a range of inputs, such as electricity, diesel, water and other chemicals, and these are commonly reported by companies in annual sustainability reports.[25] Given these data, it is possible to estimate a variety of crucial metrics to assess the environmental costs of PGM production – such as energy, water and greenhouse gas emissions (GGEs) intensity; a recent study of the PGMs sector was given by Mudd.[26]

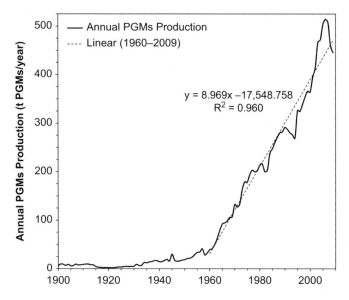

Figure 4.7 Linear trend line of global PGM production (1960 to 2009).

On average, the energy intensity is 222 GJ kg^{-1} PGMs and the water intensity is 800 m^3 kg^{-1} PGMs, with a GGEs intensity of 51.2 t CO$_{2\text{-e}}$ kg^{-1} PGMs. The energy intensity shows a negative correlation with ore grade (*i.e.* as ore grade declines, energy intensity increases); current resources are similar in ore grade to ore being processed, meaning that ore grade is not likely to be a major factor with respect to energy and GGEs. Total GGEs will therefore mainly be a function of production levels. Water is more complex, however, with some mines showing major savings in annual water consumption and intensity whereas others show increases. In addition, water is critical for surrounding communities and there remains ongoing controversies regarding impacts on water resources from mining in South Africa and linked social impacts and issues.[3–5]

In the Bushveld, PGM production is dominated by electricity, mainly related to the high proportion of narrow underground mining techniques and also intensive processing (*i.e.* mining, milling, smelting and refining). Furthermore, South African electricity is dominated by coal sources, which also gives rise to a strong greenhouse gas emissions footprint associated with electricity consumption.[26] It is also clear that energy efficiency alone will not be sufficient to address the need to reduce GGEs intensity (made especially urgent with the ongoing electricity supply crisis in South Africa). As production grows in the medium to longer term, however, if renewable energy sources were used this would lower the GGEs intensity of PGM production considerably. Although the viability of different renewable energy sources is outside the scope of this chapter, given South Africa's geography and climate, options such as baseload solar thermal power (*i.e.* solar thermal with heat storage) would appear to be well placed to meet this challenge.

Finally, it must be remembered that there are a range of complex social and governance issues affecting South Africa and Zimbabwe.[4,5] In the post-apartheid era, the South African government has set policy and legislative frameworks for black economic empowerment (BEE), whereby ownership of mines and minerals is to be shared with local communities. In theory, BEE is intended to help provide the impetus for addressing historical socio-economic disadvantage through a joint industry–government–community partnership. A similar approach is, in theory, being adopted in Zimbabwe. The reality, however, is meandering and slow – and it is difficult to predict the ongoing impacts of complex social and governance issues on the PGMs sector in South Africa and Zimbabwe. It is clear that they cannot be ignored and will form an integral part of the potential of the Bushveld and Great Dyke to meet global demands for PGMs.

4.7 CONCLUSION

This chapter has reviewed the platinum group metals, focusing particularly on current economic resources and production trends. Most of the world's PGMs are found in South Africa, with moderate amounts in Russia and Zimbabwe and minor amounts in Canada and other countries. Based on a detailed compilation of reported resources for 2010, there are ~90 400 t of PGMs globally, compared with current annual production of about 465 t of PGMs. It is clear that there are abundant PGMs to meet growing global demands for some decades. The primary constraints which are already proving challenging for the PGMs sector are environmental and social in nature – such as economic benefits sharing, energy and emissions intensity, water consumption and impacts on water resources. Overall, PGMs can certainly provide a range of metals for sustainable materials and technologies into the future, but they will come at increasing environmental and social costs which need to be carefully assessed and managed to ensure the most sustainable outcomes for the countries involved and also global issues such as greenhouse gas emissions and climate change risks.

REFERENCES

1. R. B. Gordon, M. Bertram and T. E. Graedel, *Proc. Natl. Acad. Sci. USA*, 2006, **103**, 1209.
2. C.-J. Yang, *Energy Policy*, 2009, **37**, 1805.
3. M. Curtis, *Precious Metal – The Impact of Anglo Platinum on Poor Communities in Limpopo, South Africa*, ActionAid, Johannesburg, 2008.
4. D. Rajak, *Res. Econ. Anthropol*, 2008, **28**, 297.
5. S. C. Mnwana and W. Akpan, Platinum Wealth, Community Participation and Social Inequality in South Africa's Royal Bafokeng community – A Paradox of Plenty?, in *SDIMI 2009–4th International Conference on Sustainable Development Indicators in the Minerals Industry*. Australasian Institute of Mining and Metallurgy: Gold Coast, Queensland, 2009, p. 283.
6. Johnson Matthey, *Market Data Tables – Platinum Supply and Demand*, www.platinum.matthey.com/publications/market-data-tables/ (last accessed 16 April 2011).
7. L. J. Cabri (ed.), *The Geology, Geochemistry, Mineralogy and Mineral Beneficiation of the Platinum-Group Elements*, Canadian Institute of Mining, Metallurgy and Petroleum, Montreal, 2002.
8. C. F. Vermaak, *The Platinum-Group Metals – A Global Perspective*, Mintek, Randburg, South Africa, 1995.
9. R. G. Cawthorn, R. K. W. Merkle and M. J. Viljoen, in *The Geology, Geochemistry, Mineralogy and Mineral Beneficiation of the Platinum-Group Elements*, ed. L. J. Cabri, Canadian Institute of Mining, Metallurgy and Petroleum, Montreal, 2002, p. 389.
10. R. G. Cawthorn, *Platinum Met. Rev.*, 2010, **54**, 205.
11. J. S. Scoates and R. M. Friedman, *Econ. Geol.*, 2008, **103**, 465.

12. AusIMM, MCA and AIG, *Australasian Code for Reporting of Exploration Results, Mineral Resources and Ore Reserves: the JORC Code*, Joint Committee of the Australasian Institute of Mining and Metallurgy, Australian Institute of Geoscientists and Minerals Council of Australia, Parkville, VIC, 2004.

13. SAMRCWG, *South African Code for the Reporting of Exploration Results, Mineral Resources and Mineral Reserves (the SAMREC Code)*, South African Mineral Resource Committee Working Group (SAMRCWG), The Southern African Institute of Mining and Metallurgy (SAIMM) and Geological Society of South Africa (GSSA), Johannesburg, 2009, www.samcode.co.za (last accessed 24 March 2012).

14. Ontario Securities Commission, *National Instrument 43-101 – Standards of Disclosure for Mineral Projects, Form 43-101F1 and Companion Policy 43-101CP*, Ontario Securities Commission, Toronto, 2011.

15. S.-J. Barnes and W. D. Maier, in *The Geology, Geochemistry, Mineralogy and Mineral Beneficiation of the Platinum-Group Elements*, ed. L. J. Cabri, Canadian Institute of Mining, Metallurgy and Petroleum, Montreal, 2002, p 431.

16. B. Godel, S.-J. Barnes and W. D. Maier, *J. Petrol.*, 2007, **48**, 1569.

17. T. Oberthür, in *The Geology, Geochemistry, Mineralogy and Mineral Beneficiation of the Platinum-Group Elements*, ed. L. J. Cabri, Canadian Institute of Mining, Metallurgy and Petroleum, Montreal, 2002, p. 483.

18. L. B. Hunt and F. M. Lever, *Platinum Met. Rev.*, 1969, **13**, 126.

19. R. G. Cawthorn, *S. Afr. J. Sci.*, 1999, **95**, 481.

20. R. G. Cawthorn, *New Sci.*, 2007, **196**, doi: 10.1016/S0262-4079(07)62720-1 (online only).

21. B. J. Glaisterand and G. M. Mudd, *Miner. Eng.*, 2010, **23**, 438.

22. R. K. W. Merkle and A. D. McKenzie, in *The Geology, Geochemistry, Mineralogy and Mineral Beneficiation of the Platinum-Group Elements*, ed. L.J. Cabri, Canadian Institute of Mining, Metallurgy and Petroleum, Montreal, 2002, p. 793.

23. S. Cole and C. J. Ferron, in *The Geology, Geochemistry, Mineralogy and Mineral Beneficiation of the Platinum-Group Elements*, ed. L. J. Cabri, Canadian Institute of Mining, Metallurgy and Petroleum, Montreal, 2002, p. 811.

24. G. M. Mudd, *Ore Geol. Rev.*, 2010, **38**, 9.
25. G. M. Mudd, in *SDIMI 2009 – Sustainable Development Indicators in the Minerals Industry Conference*, Australasian Institute of Mining and Metallurgy, Gold Coast, Australia, 2009, p. 377.
26. G. M. Mudd, *Platinum Met. Rev.*, 2012, **56**, 2.

CHAPTER 5

Helium

TREVOR M. LETCHER

School of Chemistry, University of KwaZulu-Natal, Durban 4041,
South Africa
Email: trevor@letcher.eclipse.co.uk

5.1   INTRODUCTION

Helium is a strategic element in that its cryogenic properties are
unique and for many applications there are no substitutes. It is
found in gas wells, and once the gas wells have been depleted there
will be no helium in significant and winnable quantities on Earth. It
is possible that this might happen within the next 100 years.[1]

Helium is a monatomic noble gas of atomic number 2, is
colourless, odourless, and non-toxic, is the least reactive of all
elements and has the lowest boiling and melting points of all the
elements. It has an exceptionally high thermal conductivity. It is
the second lightest element, next to hydrogen, and also the second
most abundant element in the observable universe (again after
hydrogen), making up about 24% of the total elemental mass in the
observable universe and in our Sun.[2] Most of the helium-4 in the
universe is believed to have been formed in the Big Bang. Our
atmosphere contains only a tiny fraction of helium but none of it
originates from the Big Bang. Our terrestrial helium, including that

Materials for a Sustainable Future
Edited by Trevor M. Letcher and Janet L. Scott
© The Royal Society of Chemistry 2012
Published by the Royal Society of Chemistry, www.rsc.org

found in the atmosphere, has been created by the natural radio-active decay of some heavy elements. This helium has found its way into natural gas wells, from which it diffuses into the atmosphere if not contained by impervious rock. Its low molecular mass and velocity enable it to escape from the gravitational pull of the Earth. The rate at which it does escape is about equal to the rate at which it is formed in the atmosphere, hence its relatively constant atmospheric concentration.

Helium was discovered in 1868 during a solar eclipse by a Frenchman, Jules Janssen, and an Englishman, Norman Lockyer, the first Editor and founder of the journal *Nature*. It was detected as a yellow spectral line in sunlight and named helium, after the Greek word for sun, by Lockyer. He noticed the unknown wave-length at 587.6 nm which was slightly less than the wavelengths of the well-known sodium lines.[3] Atmospheric helium was found about 25 years later by William Ramsay[4] and in 1903 it was dis-covered in large reserves in the natural gas fields in the USA.

5.2 PROPERTIES[5,6]

Table 5.1 gives the main properties of helium.

Below a pressure of 2.5 MPa, helium is a liquid even at absolute zero. Above this pressure, helium solidifies at about 1.1 K (melting point) Below the lambda point of 2.18 K, the properties of helium change radically; viscosity virtually disappears and heat transfer rates increase beyond that of any known substance.

Table 5.1 Properties of helium.

Property	Value
Density	0.145 g cm^{-3} for the liquid at its melting point or 0.178 kg m^{-3} for the gas at standard temperature (273.15 K) and pressure (101 325 Pa)
Atomic mass	4.002602 g mol^{-1}
Melting point	0.95 K (at 2.5 MPa)
Boiling point	4.22 K
Critical point	5.19 K (at 0.227 MPa)
Heat of fusion	0.138 kJ mol^{-1}
Heat of vaporisation	0.0829 kJ mol^{-1}
Molar heat capacity	20.786 J mol^{-1} K^{-1}
Lambda point	2.18 K
Speed of sound	972 m s^{-1}
Thermal conductivity	0.1513 W m^{-1} K^{-1}

Below its boiling point of 4.22 K and above its lambda point (2.18 K), helium-4 is called helium I and it behaves like an ordinary cryogenic liquid; boiling when heated and contracting when cooled. However, below its lambda point, helium-4 is known as helium II; it does not boil on heating (because of its very high conductivity) and it expands on cooling. These weird properties are linked to its quantum states and helium I and II are known as quantum fluids. Helium II has other strange properties, such an apparent anti-gravitational property which allows it to creep out of containers. It is because of these strange quantum mechanical properties that helium has been so extensively studied.

5.3 ISOTOPES

Helium-3 and helium-4 are the two stable isotopes of helium.[7] The ratio of ^{3}He to ^{4}He depends on the radioactive origin, which makes it possible to distinguish between the helium formed during the Big Bang and the helium formed through heavy element decay on Earth. Helium-4 is the most common isotope and is produced on Earth by alpha particle decay ($^4_2$He^{2+}). Helium-3 is produced by the beta decay of tritium. The unusually high stability of helium-4 is due to the facts that its nucleons are arranged in a complete shell and that it has a high binding energy per nucleon, which is much higher than the binding energies of lithium, beryllium and boron, the next three elements in the periodic table. This stability and the large binding energy are the reasons for its being such a stable product of nuclear fusion (as in the Sun) and of radioactive decay.[5]

There are at least six other isotopes of helium, but none of them is stable.

5.4 APPLICATIONS

The unique properties of helium are its low boiling point, low solubility, high thermal conductivity and inertness, and these dictate its uses.

5.4.1 Cryogenics and MRI Scanners

The largest single use of helium is for cryogenic purposes (24% by volume)[8] and most of that is for medical magnetic resonance imaging (MRI) scanners and nuclear magnetic resonance (NMR) machines in research laboratories. Helium at 4.2 K and also at lower temperatures is used to cool the superconducting magnets. The supercooled wires, with resistances of close to zero, conduct much larger electric currents than wires at room temperature. In this way, intense magnetic fields can be created. It is possible to create similar magnetic fields with very special electromagnets but the problem then become one of removing the heat dissipated in the coil windings. In any event, superconducting magnets are cheaper to operate. The coil winding need only be cooled below its critical point (the temperature at which the winding coils loose their electrical resistance and become superconducting). The critical temperature is usually much higher than liquid helium temperature but the superconducting coils work more efficiently the lower the temperature. The helium for the superconducting magnets is usually kept in efficient dewar vessels, surrounded by liquid nitrogen at 77 K to reduce the evaporation of the helium.[9]

It is also used as a cooling gas in nuclear plants as helium, with its stable nucleus, will not become radioactive in a radioactive environment.

It is very much in demand in science research laboratories and especially for physics research experiments where it is used in bubble chambers, particle accelerators and plasma confinement for fusion research. One of the largest research users of helium is the Large Hadron Collider at CERN which requires 160 t of helium $(1.0 \times 10^6 \text{ m}^3)^\dagger$ over the next few years to cool the accelerator and magnets to 1.8 K to ensure superconducting conditions.[10] Helium is also used to cool substances such as hydrogen and oxygen fuel used in space rockets.[11]

5.4.2 Pressurised Purging and the Space Programme

Helium plays a vital role in purging and pressurising rocket propulsion systems. The propellant for most rockets nowadays is liquid hydrogen and oxygen and to purge and, pressurise and fill the tanks while the hydrogen and oxygen are being consumed, the purging gas must not freeze at the temperature of liquid fuel

---

†The volume units, used by the gas industry, refer to standard cubic metres (SCM) at 101 325 Pa and 293.15 K. These units are used for helium gas volumes in this chapter, unless stated otherwise.

(hydrogen boils at 20.28 K). It must also have a low solubility in the liquid fuel and be inert. The only substance that fits these requirements is helium. The next possible inert substance is neon, but its boiling point is too high (24.5 K), so it cannot be used. By way of example, the Saturn V booster rocket used in the Apollo missions needed about 370 000 m^3 for each launch.[11]The helium used each year in the space programme amounts to about 20% of the total helium.[12]

5.4.3   Shield Gas for Welding

Helium is used as a shield gas for arc welding as the gas is inert. It acts as a protection from aerial oxidation and nitrogen reactions. It is especially important for welding metals with high heat conductivities such as aluminium and copper. In this respect, it is preferred to the cheaper argon. The helium used for this purpose amounts to 18% of the total available each year.[12]

5.4.4   Controlled Atmosphere

Helium is also used as a protective atmosphere in the growing of crystals such as silicon and germanium for the production of wafers in the semiconductor industry. It is a common carrier gas in chromatography as a result of its inertness. The controlled atmosphere application takes up 16% of the total helium available.[12]

5.4.5   Leak Detection

Helium is used to detect leaks in high-vacuum and high-pressure equipment vessels. Owing to its small atomic size and rapid diffusion rates, it will more readily diffuse from a small crack in a container than almost any other gas. Its inertness plays an important role in this application, which uses about 6% of the total amount of helium available each year.[12]

5.4.6   Breathing Mixtures

Helium is added to oxygen tanks so that divers can breathe at depths greater than 120 m below sea-level. The helium acts like nitrogen in the atmosphere in that it dilutes the oxygen. Nitrogen cannot be used as it is soluble in the fatty and nervous tissues of the body and leads to decompression sickness when divers ascend too fast to the surface. Here, both the low solubility of helium (in body tissues) and its inertness (it does not react with other gases in the tank or with the material of the tank) play important roles.[12] This application takes up about 3% of the annual helium available.

5.4.7   Airships and Balloons

Being very much lighter than air, helium can be used to fill balloons and airships to give them lift. For example, 'aerostats' or tethered blimps or balloons are used in advertising.[12] Its inertness is again important in this application, but diffusion through the balloon and airship material must be overcome. Hydrogen gas is more buoyant (by about 7%), but it is flammable, which helium is not.

5.4.8   Cooling and Condensing Other Gases to Liquids

Helium is used in the liquefaction of hydrogen and oxygen for rocket fuel.[13]

5.4.9   Lasers

Helium has been extensively used in helium–neon lasers and these have found a ready market as barcode readers.[12,14] The red He–Ne laser has also used to read optical disk devices and has been used widely in laboratories. These lasers have recently been replaced by cheaper diode lasers[15] Helium lasers are used in eye surgery.

5.4.10   Party Balloons and the Donald Duck Voice

Helium, being lighter than air and a safe gas, has long been associated with party balloons. Also, the Donald Duck voice is a very common party trick, based on the high value of the speed of sound in helium. The speed of sound in air (343 m s^{-1}) is very much lower. Talking normally, in an atmosphere of helium, produces a sound with a higher frequency than a normal voice [almost three times higher as the frequency of the sound is directly proportional to the speed of sound ($c = \lambda v$)].

5.4.11 Other Uses

The unique properties of helium are also put to good use in the following:

- supersonic wind tunnels;
- resuscitation pumps in heart surgery;
- isotopic dating of uranium and thorium (using the helium isotope ratios);
- high-speed 'push gas' in air-to-air guided missiles;
- academic research for cooling and for studying particle collisions in particle accelerator laboratories. Because helium is a very effective conductor of heat, it is used to transfer energy from exothermic particle reactions (fracturing) in the field of particle physics, as in the quest for the Higgs boson particle.

5.5 SUBSTITUTES FOR HELIUM

The unique properties of helium make it very difficult to replace in many applications. There is no substitute for cryogenic applications that require low gas temperatures of less than 20 K, especially if the gas has to be inert, as in the purging of space rocket fuel tanks.

The inertness of helium is not unique and in some cases it can be replaced by argon, which is cheaper. Argon is used in large quantities for welding.

Helium's low density and hence excellent lifting power is not unique and helium can be replaced by hydrogen for filling blimps and balloons if the flammability of hydrogen is not a problem.

5.6 EXTRACTION OF HELIUM FROM OIL WELLS

High levels of helium are found in old rock formations and strata with the helium trapped together with natural gas by impermeable rock caps or domes.[8] The Amarillo Fields Office,[8] attached to the US strategic repository for helium, has analysed 22 000 wells from 26 countries, including the USA. It was found that more than half of all the natural gas discovered to date contains non-commercially viable helium ($<0.1\%$). About 10% have economically attractive helium contents of $>0.4\%$ and can be as high as 2.8%, as are found in the Rocky Mountains and High Plains of the USA. Because helium is always the minor constituent of a gas well, the wells are always exploited first for their hydrocarbon content. About 66% of the known accumulations of helium have already been tapped for their hydrocarbon content, and most of the helium has been dissipated into the atmosphere and space. Only a few percent of that was exploited for helium. New fields are continually being discovered but unfortunately they usually belong to a newer geological age and hence do not contain winnable quantities of helium.[16]Helium is currently separated from natural gas in the USA, Poland, Algeria, Russia, Qatar, Canada and China and, more recently, in Australia, where the first separation plant opened on 3 March 2010.[17] The USA has the largest known helium–natural gas fields in the world, with 10 wells producing helium, four in Kansas and one each in Oklahoma, Colorado, New Mexico, Utah, Wyoming and Texas.

The USA has a Federal Helium Reserve, an underground repository for helium, at the Cliffside Storage Facility in a natural gas storage formation in a geological area called the Bush Dome near Amarillo in Texas. The repository was started in 1925 because of the perceived strategic nature of helium, and it gained ground in the 1950s as a result of the 'cold war'. A decade ago, however, with the reserve in debt, it was agreed to sell off the stored helium slowly.[8,18]

In all, there are at least 17 natural gas wells in the world producing helium and eight more are to be commissioned between 2011 and 2017, with two of the new wells in the USA.

Helium is separated from natural gas by successive liquefaction and removal of the other components, namely hydrocarbons, methane, nitrogen and neon. Activated charcoal is used in the final purification step, resulting in 99.995% pure Grade A helium. In the final production step, helium is produced in a liquefied form, ready for export of transportation in efficient dewar vessels.

There is a small amount of helium in the atmosphere (0.00052% by volume), but it is not commercially viable to extract it. However, it is not impossible to extract it, as Kamerlingh Onnes showed in 1908 when he used liquid air, then liquid hydrogen and finally the Joule–Thomson effect to liquefy helium from the atmosphere.[16]

Table 5.2   US Statistics for helium for 2006–2010.[8]

Source	Volumes per year[a]				
	2006	*2007*	*2008*	*2009*	*2010*
Extracted from natural gas	79	77	80	78	77
Withdrawn from storage	58	61	50	40	48
Consumption	75	74	60	47	54
Exports	62	64	70	71	71

[a]The volumes are in units of millions of standard cubic metres (SCM) ($10^6 \, m^3$).

5.7   WORLD SUPPLY AND DEMAND

The statistics for helium in the USA are given in Table 5.2.

In 2008,[19] approximately $130 \times 10^6 \, m^3$ of helium were extracted from wells or withdrawn from storage in the USA. This amounted to 78% of the world's total extraction, which was $169 \times 10^6 \, m^3$ for that year. The helium reserve contained over $1.2 \times 10^9 \, m^3$ of helium in 1980 but, with the recent directive to sell off the reserves, it stood at $0.67 \times 10^9 \, m^3$ in 2010.

In 2006, the total helium potential reserves and resources of the USA were estimated to be $20.6 \times 10^9 \, m^3$ and the estimate of the amount in the rest of the world was $31.3 \times 10^9 \, m^3$. This includes all possible helium from wells know to have winnable amounts of helium.

The total US usage in 2010 was $54 \times 10^6 \, m^3$ and the corresponding figure for the whole world was $180 \times 10^6 \, m^3$. With total possible reserves of over $50 \times 10^9 \, m^3$, if the usage remains the same and if all the possible reserves can be tapped, the world has only 270 years of helium left. However, if reserves fall and consumption increases, the figure will be more like 100 years.

5.8   CONCLUSION

Helium is a very special element with a set of unique properties that dictate its many uses. Most of these uses relate to its cryogenic properties. For many applications, there are no alternatives and it is indeed a strategic element.

Commercially viable quantities of helium are found only in natural gas wells and in particular those which are geologically old and are capped with a dome or layer of impervious rock. Most of the gas wells in the world have been shown to have little or no helium. The strategic nature of the element is highlighted when one considers that once the natural gas wells containing helium have been exhausted, there will be no extractable helium on Earth. This scenario might take between 100 and 270 years.

REFERENCES

1. *RSC News*, April 2011, pp. 8–11, and RSC's Roadmap *Chemistry for Tomorrow's World*, Royal Socety of Chemistry, Cambridge.
2. H. Hasen, *Helium*, Rosen Publishers, New York, 2006.
3. J. B. Hearnshaw, *The Analysis of Starlight*, Cambridge University Press, Cambridge, 1986.
4. W. Ramsay, *Proc. R. Soc. London*, 1895, **59**, 325.
5. K. H. Benneman and J. B. Ketterson (eds), *The Physics of Liquid and Solid Helium, Part I*, Wiley, New York, 1978.
6. R. C. Weast and M. J. Astle (eds), *CRC Handbook of Chemistry and Physics*, CRC Press, Boca Raton, FL, 1981–2.
7. C. A. Hampel, *The Encyclopedia of the Chemical Elements*, Van Nostrand Reinhold, New York, 1968, pp. 256–268.
8. K. Salazar and M. K. McNutt (eds), *Helium: Minerals Commodity Survey – 2011*, US Geological Survey, Reston, VA, http://minerals.usgs.gov/minerals/pubs/commodity/helium/ (last accessed 11 November 2011).
9. S. W. van Sciver, *Helium Cryogenics*, Plenum Press, New York, 1986.
10. http://www.gasworld.com/messer-to-provide-helium-for-theproject/ 2387.article by Rob Cockerill, 23 January 2008.
11. T. J. Kelly, *Moon Lander: How We Developed the Apollo Lunar Module*, Smithsonian Institute Press, Washington, DC, 2001.
12. Inter-American Corporation, *Helium Facts: Uses for Helium*, http://www.helium-corp.com/facts/heliumusers.html (last accessed 15 November 2011).
13. J. Wilks and D. S. Betts, *An Introduction to Liquid Helium*, Clarendon Press, Oxford, 1987.
14. C. S. Willett, *An Introduction to Gas Lasers*, Pergamon Press, Oxford, 1974, p. 407.

15. J. Emsley, *Nature's Building Blocks*, Oxford University Press, Oxford, 2008.
16. B. M. Abraham, *Helium, McGraw-Hill Encyclopaedia of Science and Technology*, McGraw-Hill, New York, 2002, p. 443.
17. *ABC News*, 3 March 2010.
18. E. M. Smith, T. W. Goodwin and J. Schillinger, *Adv. Cryog. Eng*, 2003, **49A**, 119.
19. K. Salazar and M. K. McNutt (eds), *Helium. Mineral Commodity Summaries 2008*, US Geological Survey, Reston, VA, http://minerals.usgs.gov/minerals/pubs/commodity/helium/mcs-2009-heliu.pdf (last accessed 15 November 2011).

CHAPTER 6

Phosphorus

TINA-SIMONE S. NESET

Centre for Climate Science and Policy Research, Water and
Environmental Studies, Linköping University, 581 83 Linköping, Sweden
Email: tina.schmid.neset@liu.se

6.1 INTRODUCTION

Phosphorus (P) is the eleventh most abundant element in the
Earth's crust and an essential nutrient for all organisms and of
particular importance for global food security. Agricultural production depends on phosphorus for maintaining high yield levels
that are currently supplying food to seven billion people. However,
the natural phosphorus cycle through which phosphorus is accu-
mulated in the sediments of sea beds through organic deposition
over millions of years has, since the late nineteenth century, been
increasingly disturbed. The increasing global demand for food led
to the mining of phosphate rock in many areas of the world where
large, easily mined, low-cost deposits were found. Global consumption has increased significantly since the mid-twentieth century and the sustainability of long-term consumption of this fossil
resource has been discussed amongst scholars for several years.
Still, the lack of institutional structures, transparent data and
information regarding this resource has created great uncertainty

Materials for a Sustainable Future
Edited by Trevor M. Letcher and Janet L. Scott
© The Royal Society of Chemistry 2012
Published by the Royal Society of Chemistry, www.rsc.org

for estimating the timeline and long-term access of phosphorus for global food production.

Phosphorus was first discovered by the German alchemist Henning Brand in the late seventeenth century,[1] who distilled and boiled 50 buckets of his own urine in his pursuit of the philosopher's stone. The ability to glow in the dark led to its name derived from the Greek words *phôs* (light) and *phoros* (bearer).[2] Since then, phosphorus has found uses in many areas. White phosphorus is highly reactive and flammable and explosive when exposed to air, which has led to its use in a number of military applications.[2] Phosphorus is also very important in many industrial applications including matches and detergents.

6.2 PROPERTIES AND NATURAL CYCLES

Phosphorus is a non-metallic chemical element with the atomic number 15. It is highly reactive and does not appear as a basic element in Nature, but always in combination with other elements (*e.g.* Mg, Al, Ca, Na, F, Cl, Cd, U and O). Phosphorus has, however, several allotropes, with white, red and black phosphorus being the most common. It is one of the elements that is found in all forms of life on Earth. About 1% of the human body consists of phosphorus; this is mainly found in the bones and teeth (calcium phosphates) and plays an important role in energy transport and storage through ATP (adenosine triphosphate).

The bio-geochemical cycle of phosphorus has, unlike the other main elements of life such as C, N, H, O and S, no atmospheric form. Phosphorus contained in bedrock is eroded and washed out into water courses, lakes and the ocean and finally deposits as calcium phosphates ($CaPO_x$) in the sediments. Over millions of years, this sedimentary bedrock has been changed by tectonic movements and uplifting and also by weathering. In its aquatic stage, phosphorus has a minor cycle involving uptake by aquatic organisms, which in turn are eaten by larger organisms or decay and become part of the sediment. On land, phosphorus is cycled from the soil to plants, to animals and, through decay, back to the soil. These smaller cycles have a significantly shorter time frame, but are part of the main phosphorus cycle. Anthropogenic activities, in particular from the late nineteenth century onwards, have impacted significantly on this natural phosphorus cycle. Leakage

Table 6.1 Properties of phosphorus.

Property	Value
Density at 20 °C/g cm^{-3}	(White) 1.82
Atomic number	15
Atomic mass/g mol^{-1}	30.9738
Melting point/°C	(White) 44.1
Boiling point/°C	(White) 280

from extensive fertilisation of agricultural soils has contributed to the eutrophication of surface waters and coastal areas, resulting in algal blooms and dead zones across the globe. While these diffuse emissions of phosphorus are due to structural changes in agriculture, the sources of phosphorus emissions from urban settlements and industrial activities have been mitigated through technically improved wastewater treatment. While these measures were mainly driven from a pollution perspective, the recovery of phosphorus from wastewater streams is still a challenge, given the pollution in mixed wastewater streams and the reluctance to reuse human excreta and sewage sludge in food production.

The basic properties of phosphorus are given in Table 6.1.

6.3 APPLICATIONS

Phosphorus has a vast array of applications, ranging from industrial use in detergents to nutrient additives in animal feed; however, only a small share of the globally mined phosphorus is used in its elemental form. The main use of phosphorus today is for food production, which covers ~90% of the globally mined phosphate rock.[3]

6.3.1 Fertilisers

Phosphorus is an essential nutrient for food production. As the natural fertility and nutrient content of soils vary, farmers have over centuries added phosphorus to their cropland in the form of manure, human excreta, guano and other concentrated sources of phosphorus either to compensate for the nutrients that have been removed at harvest or to increase the productivity of the soil. Other common approaches to secure the fertility of agricultural soil include shifting cultivations or regular flooding.[4] The introduction of mineral fertiliser in the late nineteenth century made the replacement of nutrients to the soil easier, and also more regular, and decreased the share of recycled and recovered phosphorus in the food system. Since the mid-twentieth century, the consumption of phosphorus from fossil phosphate rock has increased significantly, in line with both the increase in global population and the intensification of global agricultural production.

Globally, 82% of mined phosphate rock is used for the production of chemical fertilisers, which are applied at levels from over 10 kg per hectare in Asia and the USA to less than 2 kg per hectare in sub-Saharan Africa.[3]

6.3.2 Feed Additives

Another application area that is predominantly linked to food production is the addition of phosphorus to animal feed. Non-ruminants (*e.g.* pigs and poultry) in particular have a limited ability to absorb phosphorus. In order to avoid productivity losses or malnutrition, phosphorus salts are added to their feed.[3] Globally, the share of animal feed additives is 7% of the total amount of mined phosphate rock.[5]

6.3.3 Food Additives

Different food products contain differing amounts of phosphorus and foods of animal origin, nuts and cereals have higher concentrations of phosphorus than, for instance, vegetables and fruit. Furthermore, an increasing number of food products contain phosphorus as a food additive, more for its functionality than for its nutritional value. In baking products, phosphates are applied for leavening using an acid-based reaction with bicarbonate for the production of carbon dioxide. In beverages, it has multiple functions: as an acidulant, as a clouding agent, and as a stabiliser for protein in dairy drinks. Also in other dairy products, phosphates are used as an emulsifying agent, for protein stabilisation, for coagulation and to inhibit bacteriophages. In meat and seafood products, phosphates are applied to increase moisture and to stabilise emulsions.

Phosphate additives can also be found in egg products, fruits and vegetables, pasta and pet food.[6]

6.3.4 Detergents

One of the most intense application areas is the use of phosphates as detergents and cleaning products. Detergent products containing phosphates have historically accounted for at most 10% of the global phosphate rock use. Detergents containing phosphates have raised significant environmental concerns as a result of the pollution of surface waters and have been banned in many European countries.[7] There are several alternatives to using phosphates in detergents but it has been argued that there is no other 'natural' solution; some alternatives have been the object of debate and linked to environmental tradeoffs.

6.3.5 Pyrotechnic and Military Applications

Phosphorus is used in a number of pyrotechnic applications such as fireworks and matches. Match production was one of the early applications of phosphorus and the initial matches were made from highly toxic white phosphorus; this was later changed to the safer red phosphorus. Phosphorus is used in many military applications, such as phosphate bombs, bomb shells, tracers, grenades and smoke cartridge incendiaries.[1]

6.3.6 Industrial Applications

There is a wide variety of industrial applications of phosphorus and it is used, for example, in flame retardants, water treatment, surface treatment, and corrosion prevention. It is also used in paints (*e.g.* water-based paints), in surface coatings and in the processing of ceramics. Elemental phosphorus is used as an insecticide, oil additive and alloying agent.

6.3.7 Phosphate–Lithium Batteries

Another potentially important application is in phosphate–lithium batteries,[8] which are used, for example, in electric vehicles. If the global demand for these vehicles increases as predicted, this could mean an increased demand on phosphorus.

6.4 GLOBAL PHOSPHORUS RESERVES

In mineral form, phosphorus appears predominantly as apatite, but can be found in nearly 190 different minerals. The outer part of the Earth's crust contains 0.12% of phosphorus. Sedimentary rock deposits can have a P_2O_5 content of up to 12–20%; the currently extracted phosphate rock from sedimentary deposits has a content that varies between 2 and 20%.[9] The concentration in igneous rocks is significantly lower and is estimated to be, on average, below 5%. The global abundance of phosphorus is estimated at 4×10^{15} t,[10] which is significantly larger than the amount of phosphorus that can be considered as a global reserve. The distribution of phosphorus reserves around the world is given in Table 6.2.

Much of the global deposits are expected to be found in the sea bed, as a result of millions of years of erosion of bedrock. This has resulted in aquatic transport of phosphorus to the oceans, and also uptake by aquatic organisms, the remains of which lie embedded in the sediments. As such, the ocean floor is potentially a vast mine of phosphorus. However, the uncertainty regarding the concentrations, the difficult accessibility and the high extraction costs means that the reserves remain untapped.

The uncertainties in the estimates of the current phosphate rock reserves are due to several parameters, as is the timing of the peak and future depletion. Data regarding the global phosphate rock reserves have varied strongly, partly due to the inherent

Table 6.2 Phosphate rock production and estimated reserves 2010 (USGS 2011 Mineral Commodity Summary).

Country	2010 phosphate rock production mass/Mt	2010 phosphate rock reserves mass/Mt
China	65	3700
USA	26.1	1400
Morocco/Western Sahara	26	50 000
Russia	10	1300
Tunisia	7.6	100
Jordan	6	1500
Brazil	5.5	340
Egypt	5	100
World total	**176**	**65 000**

dynamics in the definition of reserves, the lack of transparency in the available data and insufficient research on this issue. Most studies are based on the US Geological Survey's Mineral Commodity Summaries of reserves for many countries. These potential reserves vary from year to year. Between 2010 and 2011, the total global reserves increased enormously from 16 Gt (16 billion tonnes) to 60 Gt. This changed followed a report[11] that included data from two publications from 1989 and 1998[12,13] and included their higher estimations for Morocco/Western Sahara reserves. The total global increase was thus mainly caused by a 10-fold higher estimate for the Morocco/Western Sahara data, which changed from 5 to 50 Gt of phosphate rock, whereas the reserves for most other countries increased only marginally or remained unchanged. These estimates for 2011 suggest that 77% of the reserves are situated in one geographic area, which unfortunately is in the internationally disputed territory of Morocco/Western Sahara.[14,15] This unfavourable geopolitical distribution of the economically and technically extractable reserves underlines the uncertainty of future phosphorus supply.

While this exemplifies the great contingency in reserve estimations, the unknown quality of the reserves with regard to the P_2O_5 concentration and also impurities and contamination further contributes to an uncertain timeline. A simple calculation of dividing the known reserves with the same production rate as today and assuming a consistent phosphorus concentration in the mined rock results in a timeline of 300–400 years. However, a growing global population with its increasing demand for food and energy, together with other factors that potentially influence the global consumption of phosphorus, makes this a very optimistic result. Peak phosphorus calculations, using Bayesian statistical methods to predict the increase in phosphorus consumption, arrive at the conclusion that phosphorus production will peak between 2050 and 2090, followed by a rapid decline in production.[16]

6.5 INDUSTRIAL MINING OF PHOSPHORUS

In 2010, the global production of phosphate rock was 176 Mt, with the largest producing countries being China, USA and Morocco/Western Sahara, followed by Russia, Tunisia, Jordan and Brazil.[15] Global production is expected to continue to increase, with new mines opening regularly.

Extractable reserves in varying quality and concentration can be found all over the world. The majority of global phosphate rock mining is from sedimentary rock, but in some countries, mainly Russia, Canada, South Africa and Brazil, phosphorus is extracted from igneous phosphate rock. Furthermore, some phosphorus is mined as a by-product to the mining of some metals, although some of this is not recovered owing to the low concentrations and economic viability.

One of the uncertainties in global phosphate rock mining relates to its geopolitical situation, in terms of the market dominance of only a few companies and few producing nations, but also in terms of the great volatility in price, as shown by the 2008 price spike, when the commodity price of phosphate rock increase by over 800% over an 18 month period.[10]

The mining and processing of phosphate rock are dependent on technical development and research. This could have an important effect on improving the mining efficiency and also on the amount of phosphorus that can potentially be extracted. At present the mining, processing and inter-continental transport of phosphates from the producers to farmers across the globe all use large amounts of energy. As a result, any global increase in energy costs will impact strongly on the feasibility and on the cost of phosphate rock extraction. Furthermore, as the cost of phosphate compounds is closely related to global food prices, the economics of global phosphorus reserves need to be linked to global food production and consumption.

6.5.1 Global Production and Consumption

Global consumption of P_2O_5 increased from 11 Mt in 1961 to 39 Mt in 2010, with a drop between 1988 and 1994 (Figure 6.1).[17] Since this drop, consumption in the developed countries has remained low and in some regions is continuing to decrease, which is partly the effect of decades of intensive addition to phosphorus supplements to soils.

6.5.2 Pollution

Phosphate minerals occur naturally with a number of elements that are potentially hazardous to the environment and to human health.

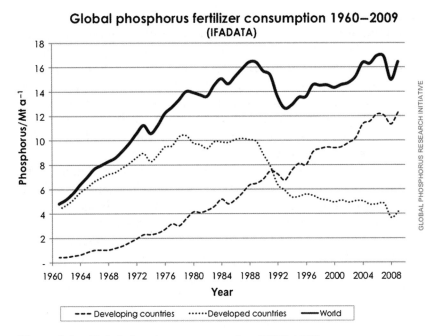

Figure 6.1 Global phosphorus consumption (IFADATA).

Among these are the heavy metals and radioactive substances, of which cadmium, arsenic, chromium, lead, selenium, mercury, uranium and vanadium are the most common. The transfer to soils via mineral fertilisers of such elements has been the focus of environmental concern. In particular, one of the by-products of phosphate rock mining is radioactive phosphogypsum. This involves relatively large quantities, with a current annual production rate of 40–47 Mt. These issues remain unresolved.[18]

6.6 ALTERNATIVE SOURCES AND SUSTAINABLE PATHWAYS

Until the mid-twentieth century, phosphate rock, as a fertiliser, had only a minor share of the total phosphorus use for global food production. In those days, the fertility of the soil was maintained largely by livestock manure and to some minor extent by human excreta and guano (bird droppings that have been deposited on predominantly coastal rock formations). Guano, which has a similar composition to the fossil sources of phosphorus, together with human excreta and manure and also other organic residues, are considered sustainable alternatives to fossil phosphorus in global food production. Livestock manure is the key organic source of phosphorus for agriculture and only during the last decade has been the object of debate, since infrastructural changes such as monoculture land use and centralised livestock breeding have separated these two sectors and as a result limited our ability to close the nutrient cycle.

The use of human excreta has in the past contributed to the fertility of agricultural fields. While the development of modern water and sanitation infrastructures in urban areas in many developed countries has inhibited this historical resource, new research and development are creating a knowledge base and novel opportunities for source separated systems and the recovery of struvite (magnesium ammonium phosphate) directly from waste-water streams.[19]

6.7   CONCLUSION

Given the dependence of food production on phosphorus to feed the global population, several sustainable measures are required to secure both an increased supply of alternative phosphorus sources, such as manure, excreta and organic residues, and also measures to decrease the total demand on phosphorus. The global supply of phosphorus is a sustainability challenge with many dimensions that needs to be approached from a number of sectors and disciplines. Sustainable pathways need to take into account a decrease in the use of phosphorus for food production through improved and efficient agricultural practices, which include recovered and recycled phosphorus from food production, a decrease in other uses of phosphorus, such as in detergents, reduced consumption of meat and dairy products (which have a higher demand on phosphorus input), ways of decreasing the losses from transport and storage and processing, and a reduction in the wastage of household phosphorus compounds.

Phosphorus is a natural resource that demands more focus. This is particularly true when one considers the number of positive synergies that would be generated from an increased efficiency in food production and from improving the global sustainability of phosphorus.[20] Again, the need for greater focus is vital when considering the amount of time required for changes in agriculture, industrial production, waste water treatment, nutrient recovery, policy and behavioural changes.

REFERENCES

1. K. Ashley, D. Cordell and D. Mavinic, *Chemosphere*, 2011, **84**, 737.
2. J. Emsley, *The Shocking History of Phosphorus*, Macmillan, London, 2000.
3. J. Schröder, D. Cordell, A. L. Smit and A. Rosemarin, *Sustainable Use of Phosphorus*, Report No. 357, Plant Research International, Wageningen University and Research Centre, Wageningen and Stockholm Environment Institute, Stockholm, 2010.
4. J. J Schröder, A. L. Smit, D. Cordell and A. Rosemarin, *Chemosphere*, 2011, **84**, 822.
5. International Fertilizer Industry Association, IFADATA, www.fertilizer.org (last accessed 20 November 2011).
6. International Food Additives Council, *Phosphates Used in Food*, http://www.foodadditives.org/index.html (last accessed 25 November 2011).
7. R. al Rawashdeh and P. Maxwell, *Miner. Econ.*, 2011, **24**, 15.
8. C. M. Park and H. J. Sohn, *Adv. Mater.*, 2007, **19**, 2465.
9. Encyclopaedia Britannica, *Phosphorus* http://www.britannica.com/EBchecked/topic/457568/phosphorus-P (last accessed 20 November 2011).
10. D. Cordell, J.-O. Drangert and S. White, *Global Environ. Change*, 2009, **19**, 292.
11. S. J. Van Kauwenbergh, *World Phosphate Rock Reserves and Resources*, IFDC Report, IFDC, Washington, DC, 2010.
12. OCT, The Phosphate Basins of Morrocco, in *Phosphate Deposits of the World, Vol. 2, Phosphate Rock Resources*, ed. A. J. G. Notholt, R. P. Sheldon and D. F. Davidsson, Cambridge University Press, Cambridge, 1989, pp. 301–311.
13. A. Gharbi, Les phosphates marocains, in *Chronique de la Recherche Miniére*, No. 531–532, Bureau de Recherches Géologiques et Minières, Orléans, 1998, pp. 127–138.

14. US Geological Survey, *Mineral Commodity Summaries 2010, Phosphate Rock*, US Geological Survey, Reston, VAm 2010.
15. US Geological Survey, *Mineral Commodity Summaries 2011, Phosphate Rock*, US Geological Survey, Reston, VA, 2011.
16. D. Cordell, S. White and T. Lindström, Peak phosphorus: the crunch time for humanity?, *The Sustainability Review*, 2011, **2**(2), 1–1.
17. International Fertilizer Industry Association (IFA), *Feeding the Earth: Global Phosphate Rock Production Trends from 1961 to 2010*, International Fertilizer Industry Association, Paris, 2011, www.fertilizer.org (last accessed 20 November 2011).
18. US Environmental Protection Agency, *About Phosphogypsum*. http://www.epa.gov/radiation/neshaps/subpartr/about.html (last accessed 25 November 2011).
19. D. Mavinic, K. Ashley and F. Koch (eds), *International Conference on Nutrient Recovery from Wastewater Streams, Vancouver, 2009*, IWA Publishing, London, 2009.
20. T.-S. S. Neset and D. Cordell, *J. Sci. Food Agric*, 2012, **92**, 2–6.

CHAPTER 7

Uranium

GAVIN M. MUDD

Environmental Engineering, Monash University, Clayton, VIC 3800,
Australia
Email: gavin.mudd@monash.edu

7.1 INTRODUCTION

Despite the utopian promise of electricity 'too cheap to meter'*,
nuclear power remains a minor source of electricity worldwide. In
2009, nuclear power accounted for 5.79% of the total primary
energy supply and was responsible for 13.46% of global electricity
supply – and both contributions have been declining through the
2000s.[1] Concerns about hazards and unfavourable economics have
stopped the growth of nuclear energy in all but two Western
countries, Finland and France. In the USA, no orders for nuclear
power stations placed after 1978 have been completed and all
plants ordered after 1973 have been cancelled. However, there is
still growth in nuclear energy in some countries, mostly in China,
Russia and India. Over the next 15–20 years, many more nuclear

---

*Although this statement was made by Lewis L. Strauss on 16 September 1954 mainly in
reference to the future potential of hydrogen fusion power, it was made during the early
years of the Cold War when nuclear fission power was being actively promoted through the
'Atoms for Peace' programme and remains a useful metaphor generally.

Materials for a Sustainable Future
Edited by Trevor M. Letcher and Janet L. Scott
© The Royal Society of Chemistry 2012
Published by the Royal Society of Chemistry, www.rsc.org

power stations will reach retirement age than those contracted for actual construction.[2]

World primary energy production and electricity generation are summarised in Table 7.1, including the International Energy Agency (IEA)'s projection of world energy and electricity demands in 2035 under the scenario of current policies continuing (this will be further discussed later). It can be seen that nuclear energy's share of electricity continues to decline by 2035 – yet we are led to believe that it is the saviour of the world's power-hungry demands. Also of note is the massive proportional increase in renewable energy-derived electricity. By 2035, all renewable electricity sources (wind, solar, hydro, geothermal and biomass) will double, especially led by hydro and wind and also with rapid growth in solar. There are more radical scenarios for 100% renewable energy on a global scale,[3,4] 100% renewable energy for the European Union[5] and 100% renewable electricity for New Zealand[6] and Australia,[7] among others.

There are numerous critical issues facing the energy and electricity sectors globally, such as:

- greenhouse gas emissions released by fossil fuel sources contributing to anthropogenic climate change;
- peak oil, the end of the era of cheap conventional oil, together with concerns about energy security, especially ongoing supply issues and maintaining resources.

It is argued by some that nuclear power can effectively address these problems on the basis that it has a lower carbon intensity than fossil fuels, is largely independent of oil or coal supplies and, if plutonium breeder reactors can be made commercial (while ignoring or, at best, downplaying severe proliferation risks), could provide a substantial energy resource which could last a considerable period of time (centuries or more).

In this context, what is the basis upon which nuclear power could be argued to increase beyond the IEA's current projections? In other words, can the nuclear fuel chain be considered a sustainable option for future electricity generation? These questions are of more than minor consequence to address. The contribution of nuclear power to nuclear weapons proliferation remains paramount and urgent (*e.g.* North Korea, India, Pakistan, Iran) and

Table 7.1 World primary energy demand and electricity generation by source for 2009 and projected 2035 demand under current energy policies.[1]

Primary energy demand[a]			Electricity generation[a]		
Energy source	2009 (EJ, %)	2035 (EJ, %)[b]	Energy source	2009 (TWh, %)	2035 (TWh, %)[b]
Coal	137.9 (27.15%)	171.7 (24.18%)	Coal	8118 (40.50%)	12 035 (33.20%)
Natural gas	106.3 (20.93%)	164.5 (23.16%)	Gas	4299 (21.45%)	7923 (21.86%)
Crude oil	166.9 (32.86%)	194.5 (27.38%)	Oil	1027 (5.12%)	533 (1.47%)
Biomass and other	51.5 (10.14%)	80.0 (11.27%)	Biomass	288 (1.44%)	1497 (4.13%)
Nuclear	29.4 (5.79%)	50.7 (7.15%)	Nuclear	2697 (13.46%)	4658 (12.85%)
Hydro	11.7 (2.31%)	19.9 (2.80%)	Hydro	3252 (16.23%)	5518 (15.22%)
Other renewables	4.1 (0.82%)	28.9 (4.07%)	Wind	273 (1.36%)	2703 (7.46%)
			Geothermal	67 (0.33%)	271 (0.75%)
			Solar photovoltaics	20 (0.10%)	741 (2.04%)
			Solar thermal	1 (<0.01%)	307 (0.85%)
			Tidal and wave	1 (<0.01%)	63 (0.17%)

[a]EJ, exajoules (10^{18} J); TWh, terawatt-hours (10^{12} Wh).
[b]IEA current policies scenario.

nuclear reactor safety and long-term stewardship of nuclear waste (*i.e.* for longer than 100 000 years) still represent fundamental concerns, especially when comparing these risks with energy efficiency and various renewable energy technologies such as baseload solar thermal technologies, solar photovoltaics and wind.

This chapter presents a detailed study of the ability for uranium mining and resources to meet possible future scenarios for expanded nuclear power. An extensive array of data is compiled and analysed, focusing on known economic resources and mine production, with these aspects then briefly discussed in the light of other sustainability issues such as greenhouse gas emissions (GGEs) intensity, water resources impacts and uranium mine rehabilitation. The chapter thus provides a detailed case study of the uranium sector of the global mining industry, which, albeit a somewhat small player in value terms, attracts significant political and corporate support while remaining deeply controversial in public debate.

7.2   URANIUM AND ITS USES

There are two primary uses for uranium – as a fuel in a nuclear power plant to produce electricity and as the ingredient for nuclear weapons. In both applications, the fissile uranium-235 atom (^{235}U) is forced to undergo a chain reaction and split (*i.e.* fission) and release energy in the process. In a weapon, the release of energy is uncontrolled and leads to a severe explosion. In a nuclear power plant, the fission rate of ^{235}U is controlled to limit energy release, which is used to produce high-temperature steam, which in turn spins a turbine to generate electricity.

Under the Nuclear Non-Proliferation Treaty (NPT), which all but five countries worldwide (India, Pakistan, Israel, Iran and North Korea) have signed, the traditional nuclear powers of the USA, Russia, UK, France and China are all obliged to work towards disarmament (although progress is arguably going backwards on this front internationally). Based on the NPT and a host of related nuclear safeguards, treaties and conventions, it is effectively illegal to sell U for nuclear weapons (although if a country mines uranium, there is virtually nothing to stop it using this uranium for its own internal purposes such as weapons; which has been the case for all nuclear weapons powers to date). As such, this chapter will exclude consideration of uranium in weapons, although in reality this remains a pivotal stumbling block in any consideration of uranium mining and nuclear energy's contribution to sustainable energy scenarios for the world.

Uranium is predominantly found in mineral deposits on its own, although it can be found in conjunction with certain types of vanadium deposits or, more rarely, in certain gold fields and a few copper deposits. There are a wide variety of deposit types (see later) and mining extracts and chemically refines the U to relatively pure oxide, mainly as triuranium octaoxide concentrate (*i.e.* >95% U_3O_8), although some projects may produce uranyl peroxide, UO_4.

Natural uranium consists primarily of two major isotopes, uranium-238 (^{238}U) and uranium-235 (^{235}U), comprising 99.28% and 0.715%, respectively.[8] The ^{235}U is fissile, meaning that it can readily be induced to undergo fission when bombarded with neutrons and split, whereas ^{238}U is fertile and tends to capture neutrons and undergoes radioactive decay to form plutonium-239 (^{239}Pu), which is highly fissile and can be used in nuclear weapons.

Most nuclear power plants require the ^{235}U fraction to be increased to ~ 3–5% to improve the fission process in a reactor. Some reactors, such as the CANDU type, can use natural uranium, but these are only a small proportion of the global nuclear reactor fleet. The process of increasing the fraction of ^{235}U is called enrichment and is technically very challenging and extremely energy intensive. To facilitate enrichment, the uranium is commonly converted to hexafluoride gas (UF_6) and the enriched uranium is then converted back to an oxide (UO_2) in order to make the fuel rods for the reactors.

In a reactor, the fission process leads to the formation of a wide range of highly radioactive elements and isotopes (fission products), with some of the ^{238}U also being converted to ^{239}Pu. The radiation intensity of spent nuclear fuel is several orders of magnitude higher than the that of the original uranium ore. Owing to the extreme radioactivity, the spent fuel is generally called high-level nuclear waste.

Overall, the use of uranium to produce nuclear-derived electricity leads to a chain of processes from mining to conversion, enrichment, reconversion, fuel manufacture, reactor use and finally high-level waste management – often called the nuclear fuel chain. In this chapter, we focus only on the mining stage, as this is a crucial aspect in underpinning possible long-term projections of nuclear power in this century (if it is to have a future).

7.3 BRIEF HISTORY OF URANIUM

Since the discovery of radioactivity in 1896, there have been arguably four major phases of U mining – the 'radium' phase from ~1900 to 1940, the Cold War (or military phase) from 1941 to the 1960s, the civilian phase from the 1970s to the mid-1990s and followed recently (since about 2003) by a resurgence in interest in U exploration and mining. At the start of each major phase, substantial concerns were raised about the extent of U resources – that is, the ability to meet rapidly growing demands. Very quickly, however, for each of the first three phases, new deposits and major fields were discovered that quickly led to large new supplies.

The radium mining phase saw the primary interest in the radioactive decay product of U – and radium reached around £300 000 per ounce (gold at that time was a mere £4.25 per ounce), mainly for medical uses in cancer treatment.[9] Initially, radium was procured from known deposits in the Erzgebirge ('Ore Mountains') in the then German Democratic Republic (GDR, part of the Soviet Union) and Joachimstal in Bohemia (today's Czech Republic), the region from where Martin Klaproth had first isolated U in 1789, and also the central Colorado Plateau in the USA. In 1922, the Belgians announced a rich discovery at Katanga in the Belgian Congo and went on effectively to monopolise the world radium market for a decade. Another rich discovery was made at Great Bear Lake in northern Canada in 1930, giving Canada a strong position in radium, although this was also the time when serious concerns began to arise with respect to cancer rates in exposed workers (especially the famous radium dial painters in the USA, who suffered horrendous bone cancers and other impacts from excessive occupational radium exposure).

The modern nuclear era was devastatingly ushered in with the bombings of Hiroshima and Nagasaki in Japan by the USA, courtesy of the secret military Manhattan Project which delivered the technology and weapons – half of the U was secured from the Belgians, one-quarter from Great Bear Lake and the Canadians and the remainder from the Colorado Plateau (all previously radium–vanadium mines). Following the end of World War II and the rapid escalation of the Cold War arms race between the USA and the Soviets, U exploration and mining became a priority all over the world. Exploration discovered a vast array of new fields and numerous new mines were developed to supply the respective military programmes. Although the USA and UK governments had incentive programmes to find and mine U, within a decade these had to be scaled back owing to the overwhelming success of finding new U sources. The major regions which saw a boom in U mining included the Elliot Lake district of northern Ontario in Canada, the Four Corners (mid-west) region of the USA, the Erzgebirge of the then GDR, Kazakhstan, the Witwatersrand gold fields in South Africa (where U was a by-product of gold), and minor production in other countries (*e.g.* Australia, France, Eastern European countries). Typically, most ores were between 0.05 and 0.5% U_3O_8 in grade and deposits typically contained from 1000 to 20 000 t of U_3O_8. With the scaling back of military purchasing in the early- to mid-1960s, most of these regions began to wind down production or it collapsed altogether, although the low-grade Elliott Lake district ($\sim 0.1\%$ U_3O_8) continued with major subsidies from the Canadian Government.

By the late 1960s, nuclear power for civilian electricity was becoming an economic prospect and the optimistic growth scenarios quickly encouraged miners to begin exploring again – with success even beyond that experienced during the height of the Cold War. In northern Saskatchewan, several high-grade deposits (0.5–5% U_3O_8) were discovered in the late 1960s to mid-1970s (*e.g.* Key Lake, Rabbit Lake, Cluff Lake), with super-rich deposits discovered at McArthur River and Cigar Lake in the 1980s (both averaging $\sim 15\%$ U_3O_8). These new deposits were also very large, each containing from 20 000 to 200 000 t of U_3O_8. In Australia, similar success was enjoyed in finding major new deposits, albeit at more typical grades of 0.05–0.5% U_3O_8 (*e.g.* Ranger, Jabiluka, Olympic Dam, Yeelirrie). Conversely, across Africa, the giant but very low-grade Rössing U deposit ($\sim 0.04\%$ U_3O_8) in Namibia was recognised as a major project, along with major discoveries in Niger by the French. Globally, U resources surged and production shifted from older Cold War centres to new regions such as northern Saskatchewan, northern and central Australia, Namibia and Niger.

By the late 1980s, nuclear power had not expanded as rapidly as the optimists had expected and this led to the U market being strongly over-supplied by mines – crashing the U price and numerous mines and fields in the process. A considerable proportion of demand was then supplied by large inventories, held either by governments or power companies, leading to new mine production being lower than demand from the late 1980s to the present. In the mid-1990s, some of the nuclear weapons stockpile of highly enriched uranium (*i.e.* $>20\%$ ^{235}U) was downblended to produce normal enrichment reactor fuel (*i.e.* 3–5% ^{235}U). The programme, known as 'Megatons to Megawatts', has supplied about half of the nuclear reactors in the USA since that time, effectively providing about 20% of global U supply. Another aspect of the 1980s–1990s was the considerable effort directed at remediating abandoned mines from the Cold War era, with programmes in the USA and the reunified Germany each costing billions of dollars.

At the start of the 2000s, U prices were at historic lows (in real terms) of US$16.42 per kilogram of U_3O_8 (US$7.45 per pound of U_3O_8), but in early 2003 began to rise in concert with hopes that nuclear power could provide a solution to replacing coal-fired electricity and address climate change risks from GGEs associated with burning coal. Remarkably, by mid-2007 U spot prices had reached the staggering level of about US$300 per kilogram (US$136 per pound) of U_3O_8 but quickly crashed to around US$97 per kilogram (US$44 per pound) of U_3O_8 by early 2009 – and have stayed in this vicinity ever since.

The multiple reactor accident at Fukushima in Japan following the severe earthquake and tsunami in March 2011 appears not only to have dampened U prices, but also to have given further impetus for some countries to formalise their exit from nuclear power (*e.g.* Germany). Although the IEA still project global growth in nuclear power capacity by 2035 (Table 7.1), this is mostly from planned economies and certainly not western-style democracies. The future demand for U remains very difficult to predict indeed.

Overall, U has had a roller-coaster history over the past century, arguably being one of the mining industries' most widely varying sectors – from major boom times to severe downturn several times. It is important to keep this history in mind when examining

U resources and mining, as it can inform trends and suggest useful assumptions for the future of the sector.

7.4 ECONOMIC RESOURCES OF URANIUM

7.4.1 Basic Geology of Uranium and Deposit Types

Uranium can be found in a wide variety of mineral deposit types, mainly related to its geochemical versatility.[10,11] According to the International Atomic Energy Agency (IAEA), there are 11 primary types of U deposits: (1) unconformity-related; (2) sandstone; (3) haematite breccia complex; (4) quartz–pebble conglomerate; (5) vein; (6) intrusive; (7) volcanic and caldera-related; (8) metasomatite; (9) surficial; (10) collapse breccia pipe; and (11) phosphorite.[12] In addition, minor deposit types include metamorphic, limestone-paleokarst and U-enriched coal deposits.[12] Historically, most U production has been derived from sandstone, unconformity and quartz–pebble conglomerate deposits. It is common for a country to be dominated by a single deposit type (*e.g.* South Africa by quartz–pebble conglomerates; Kazakhstan by sandstones), but not always (*e.g.* numerous types in Canada, Australia and the USA).

7.4.2 Economic Uranium Resources

Given the political prominence of U as a strategic national resource, especially during the Cold War years, there is a reasonable amount of data on U resources since 1950, including:

- *Uranium: Resources, Production and Demand* (aka the 'Red Book', published every 2 years since 1965);[13]
- *Canadian Minerals Yearbook* (1944 to present);[14]
- US Bureau of Mines' *Minerals Yearbook* (1933 to 1993);[15]
- US Department of Energy's *Uranium Industry Annual* (1992 to 2005);[16] and
- South African Chamber of Mines' *Facts and Figures.*[17]

In addition, other reports also provide data (*e.g.* Australia[18–21]). Numerous U companies also publish their ore reserves and mineral resources in annual corporate reports, as required for publicly listed companies in developed countries (*e.g.* Cameco, BHP Billiton, Rio Tinto, Uranium One, Paladin Energy, Denison Mines, Areva). This chapter compiles an extensive and global data set on reported U reserves and resources for 2010 by project (455 in total), including details such as ore type, ore tonnage, grade and contained U. Although the list is extensive, it is by no means complete, as many countries do not allow full reporting.

Based on methodology adopted by the 2009 Edition of the OECD/IAEA Red Book, U resources are classified slightly different to the normal 'ore reserves and mineral resources' (*e.g.* JORC Code[22]), instead using 'identified resources', which consists of 'reasonably assured resources' (RAR) and 'inferred resources' with deposits categorised into predicted cost ranges (*e.g.* for 1 kg of U, < US$40, < US$80, < US$130). Although there are differences in the various codes or standards for reporting U resources, they are broadly similar and provide a realistic basis for comparison.[23] It must also be stated that there are legitimate technical concerns regarding the reliability and accuracy of Red Book U resource data,[24] although we will assume that it is sufficiently accurate for this chapter.

The trends in remaining economic resources and country average ore grades are shown in Figure 7.1, with Red Book and national resource data for 2009 in Table 7.2, and with other tables including:

- Table 7.3 – country totals by ore, grade, contained uranium and number of deposits;
- Table 7.4 – deposit type totals by ore, grade, contained uranium and number of deposits;
- Table 7.5 – largest 30 deposits by ore, grade, contained uranium and deposit-type;
- Table 7.6 – top 20 deposits by highest ore grade.

Two main trends are evident in Figure 7.1: (i) global economic U resources have continued to grow over time; and (ii) most countries, except Canada, show declining average ore grades with time. The rise in the average ore grade in Canada in the 1990s is mainly due to the large, high-grade deposits discovered at McArthur River and Cigar Lake (see Table 7.5), although recent years have seen a declining trend emerge as lower grade deposits such as

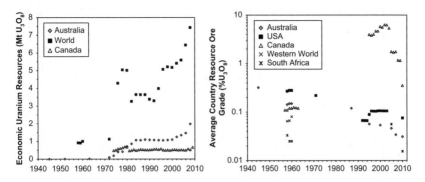

Figure 7.1 Reported economic uranium resources by country over time: contained uranium (left) and average country ore grade (right).

Table 7.2 Uranium resources by country: Red Book (2009) and selected national data (kt U3O8).

Country	Red Book	National	Country	Red Book	Country	Red Book
Australia	1979.5	1739[25]	Tanzania	33.5	Romania	7.9
Kazakhstan	981.0		Algeria	23.0	Japan	7.8
Russia	667.7		Argentina	22.5	Vietnam	7.5
Canada	642.1	663.8[26]	Malawi	17.8	Italy	7.2
USA	556.6	556.6[27,a]	Central African Republic	14.1	Indonesia	7.1
South Africa	348.5	435	Spain	13.3	Gabon	6.8
Namibia	335.1		Slovakia	12.1	Democratic Republic of Congo	3.2
Brazil	328.6	364.7[28]	Sweden	11.8	Peru	3.2
Niger	324.8		Slovenia	10.8	Iran	2.5
Ukraine	263.6		France	10.7	Egypt	2.2
China	202.1		Hungary	10.1	Mexico	2.1
Uzbekistan	135.1		Somalia	9.0	Chile	1.8
Jordan	131.8		Turkey	8.6	Zimbabwe	1.7
Denmark	100.9		Germany	8.3	Finland	1.3
India	94.4		Greece	8.3	Czech Republic	0.6
Mongolia	58.1		Portugal	8.3	**Total**	**7435.1**

[a]Excludes copper and phosphate resources.

Kiggavik-Sissons Schultz and others are now considered as potential economic prospects. In reality, however, almost all of these lower grade deposits were discovered in the 1970s–1980s and so if they were included in the resource data set over time, the

Table 7.3 Country totals by ore, grade, contained uranium and number of deposits.

Country	Ore/Mt	% U_3O_8	U_3O_8/kt	No.	Country	Ore/Mt	% U_3O_8	U_3O_8/kt	No.
Australia	10795.3	0.031	3344.7	95	Mali	10.8	0.090	9.8	1
Canada	559.5	0.165	922.4	44	Hungary	58.2	0.070	40.4	2
USA	725.3	0.082	593.3	109	Cameroon	13.1	0.100	13.1	1
Kazakhstan	801.5	0.063	506.2	10	Gabon	1.4	0.367	5.0	1
Namibia	2658.1	0.025	661.5	16	Guinea	17.7	0.030	5.2	1
Malawi	100.7	0.029	29.4	3	Senegal	0.7	0.236	1.7	1
Mongolia	164.3	0.052	84.8	8	Zimbabwe	0.3	0.707	2.1	1
Niger	660.8	0.093	613.5	10	Bulgaria	4.1	0.017	0.7	1
France	6.7	0.205	13.7	1	Finland	1559.8	0.002	38.7	13
South Africa	3618.0	0.019	685.2	40	Italy	4.5	0.115	5.2	2
Zambia	414.3	0.009	39.1	6	Portugal	1.7	0.130	2.3	1
Brazil	163.1	0.144	234.3	3	Spain	105.3	0.046	48.3	6
Argentina	93.7	0.015	14.4	5	Ukraine	102.5	0.118	120.8	3
Central African Republic	19.3	0.342	65.9	1	Paraguay	10.0	0.051	5.1	1
Russia	488.7	0.123	602.8	8	Peru	309.0	0.013	39.2	4
Botswana	780.1	0.015	118.6	1	India	229.4	0.037	85.7	5
Somalia	13.2	0.085	11.3	2	Iran	1.0	0.072	0.7	2
Slovakia	17.3	0.185	32.0	7	South Korea	102.7	0.032	32.7	7
Greenland	619.0	0.026	159.1	1	Pakistan	1.4	0.047	0.7	1
Tanzania	274.1	0.025	67.5	2	Saudi Arabia	385.0	0.014	53.9	1
Sweden	4,773.8	0.018	836.5	12	Turkey	22.8	0.047	18.7	6
Mauritania	96.5	0.031	30.4	3	Vietnam	1.5	0.040	0.6	1
Guyana	3.3	0.098	3.2	1	Jordan	153.3	0.050	76.6	4
Columbia	12.9	0.130	16.8	1	**Total**	**30 956**	**0.033**	**10 294**	**455**

Table 7.4 Deposit-type totals by grade, contained uranium and number of deposits.

Deposit type	No.	U_3O_8/kt	Minimum U_3O_8/kt	Maximum U_3O_8/kt	Minimum % U_3O_8	Maximum % U_3O_8
Alaskite Intrusive	6	457.8	11.37	165.9	0.011	0.045
Breccia pipes	2	3.02	0.52	2.51	0.61	0.79
Coal	2	97.1	0.52	96.62	0.017	0.050
Intrusive	11	108.5	0.40	53.90	0.0075	0.16
IOCG Breccia complex	5	2552.7	3.5	2443.0	0.010	0.062
Metamorphic	3	2.96	0.10	1.66	0.060	0.100
Metasomatite	21	809.8	0.5	376.3	0.041	0.354
Phosphorite	9	197.1	0.87	76.64	0.0023	0.342
Quartz–pebble conglomerate – tailings	10	130.4	0.40	34.29	0.003	0.0125
Quartz–pebble conglomerate	30	544.6	0.32	93.46	0.005	0.702
Sandstone	155	2087.5	0.02	341.3	0.014	0.707
Shale-hosted	10	972.6	2.38	485.7	0.002	0.230
Surficial/calcrete	57	411.4	0.03	76.26	0.0063	0.340
Unconformity	40	1071.3	0.005	227.9	0.03	15.74
Vein	20	230.6	0.005	159.1	0.01	3.537
Volcanic	34	234.5	0.28	46.94	0.01	0.25
Unclassified	40	370.8	0.1	136.0	0.0045	0.578

extent of Canada's increase in country average ore grade would not be as pronounced.

The 1950s–1960s U resources in Canada were dominated by the Elliot Lake district of northern Ontario, which produced ∼165 kt of U_3O_8 from ore averaging ∼0.11% U_3O_8 by field closure in 1996.[29] Substantial mineralised ore remains at Elliot Lake, given that only about half of the identified ore has been mined (a 1957 estimate of ore resources was 342 Mt whereas ore production totalled ∼156 Mt). Whether the remaining Elliot Lake ore could be classified as economic is speculative at best.[30]

Canadian U resources are very much dominated by two classes – the super-rich deposits of northern Saskatchewan's Athabasca Basin, especially McArthur River, Cigar Lake, Wheeler River-Phoenix and Roughrider East, but also the numerous low-grade deposits in Elliot Lake in northern Ontario, the Baker Lake district of Nunavut containing the undeveloped Kiggavik-Sissons/Schultz deposits and the unusually low-grade Hidden Bay deposits of

Table 7.5 Largest 30 deposits by contained uranium.

Project, country	Ore/Mt	% U_3O_8	U_3O_8/ kt	Others[a]	Deposit type
Olympic Dam, Australia	9075	0.027	2443.0	Cu–Au– Ag	IOCG Breccia complex
Viken, Sweden	2854	0.017	485.7		Shale-hosted
Elkon, Russia	217.1	0.173	376.3	Au	Metasomatite
Imouraren, Niger	435.2	0.078	341.4		Sandstone
Häggån, Sweden	1790	0.016	286.4		Shale-hosted
McArthur River, Canada	1.613	14.13	227.9		Unconformity
Inkai, Kazakhstan	353.7	0.056	199.0		Sandstone
Husab-Rössing South, Namibia	366.5	0.045	165.9		Alaskite Intrusive
Kvanefjeld, Greenland	619	0.0257	159.1	REEs	Vein
Cigar Lake, Canada	1.062	14.70	156.1		Unconformity
Jabiluka 2, Australia	29.24	0.486	142.0	Au	Unconformity
Ranger, Australia	156.1	0.089	139.1		Unconformity
Priargunsky, Russia	113.4	0.12	136.0		Not classified
Itataia-Santa Quitéria, Brazil	129.3	0.0998	129.0	PO_4	Metasomatite
Rössing, Namibia	417.3	0.030	126.1		Alaskite Intrusive
Letlhakane-Moko-baesi, Botswana	780.1	0.0152	118.6		Shale-hosted
Lagoa Real/Caetité, Brazil	28.5	0.35	100.8		Metasomatite
Springbok Flats, South Africa	193.2	0.050	96.6	Coal	Shale-hosted
Etango-Goanikontes, Namibia	500.8	0.0193	96.5		Alaskite Intrusive
Elliott Lake (Denison), Canada	154.0	0.061	93.5	Th	Quartz–pebble conglomerate
Ezulwini, South Africa	120.8	0.076	91.5	Au	Quartz–pebble conglomerate
Arlit-Somair, Niger	52.62	0.172	90.3		Sandstone
Jordan Phosphates, Jordan	153.3	0.05	76.6	PO_4	Phosphorite
Langer Heinrich, Namibia	127.1	0.06	76.3		Surficial/calcrete
Potchefstroom, South Africa	250	0.03	75.0	Au	Quartz–pebble conglomerate
Bakouma-Patricia. Cameroon	19.29	0.342	65.9	PO_4	Phosphorite
Akouta-Cominak, Niger	18.61	0.352	65.6		Sandstone
Kiggavik, Canada	11.15	0.54	60.2		Unconformity
Severinskoye, Russia	50	0.12	59.0		Metasomatite
Munkuduk, Kazakhstan	140	0.041	57.8		Sandstone

[a]REEs, rare earth elements; PO_4, phosphate.

Table 7.6 Top 20 deposits by highest ore grade.

Project, country	Ore/Mt	% U_3O_8	U_3O_8/kt	Deposit type
Wheeler River-Phoenix, Canada	0.114	15.74	17.9	Unconformity
Cigar Lake, Canada	1.062	14.70	156.1	Unconformity
McArthur River, Canada	1.613	14.13	227.9	Unconformity
Roughrider East, Canada	0.118	11.58	13.7	Unconformity
Tamarack, Canada	0.229	3.74	8.59	Unconformity
Millenium, Canada	0.780	3.70	28.9	Unconformity
Kunnansuonjuoni, Finland	0.0001	3.54	0.005	Vein
Paul Bay, Canada	0.213	3.53	7.52	Unconformity
Roughrider, Canada	0.438	2.89	12.6	Unconformity
Midwest, Canada	0.852	2.85	24.3	Unconformity
Dawn Lake, Canada	0.347	1.69	5.86	Unconformity
Shea Creek, Canada	2.942	1.36	40.0	Unconformity
McLean Lake Group, Canada	1.132	1.22	13.8	Unconformity
Koongarra 1, Australia	0.624	1.055	6.58	Unconformity
Rabbit Lake, Canada	2.235	0.81	18.1	Unconformity
Wate, USA	0.065	0.79	0.52	Breccia Pipes
Kanyemba, Zimbabwe	0.30	0.71	2.12	Sandstone
Otish Mountain, Canada	0.385	0.70	2.70	Quartz-Pebble Conglomerate
Angilak-Lac Cinquante, Canada	1.119	0.609	6.81	Vein
Arizona Strip, USA	0.414	0.61	2.51	Breccia pipes

Saskatchewan. As can be seen in Figure 7.1 and comparing data in Tables 7.2–7.6, when all known U deposits are included the total significantly exceeds the national or Red Book estimate – largely since Natural Resources Canada (NRC) only includes operating mines and those committed to development in their national resource estimates – but also that the average Canadian ore grade declines significantly.

The extent of U resources in Australia is increasingly dominated by the giant Olympic Dam deposit in South Australia. The Olympic Dam (Cu–U–Au–Ag) project is based on an underground mine, concentrator, U–Cu hydrometallurgical complex, Cu smelter and Cu refinery. The most recent (2011) reported mineral resource is 9129 Mt ore grading 0.86% Cu, 0.027% U_3O_8, 0.32 g t^{-1} Au and 1.50 g t^{-1} Ag – a contained U of $\sim$2486 kt U_3O_8. In October 2011, Olympic Dam was given approval to convert to a large open cut project, whereby some U would be produced locally but a significant portion would be contained in U-rich Cu concentrates proposed to be exported to a new, specially constructed Cu smelter in China[31] – for total production of 19 kt y^{-1} of U_3O_8 (although it will take close to a decade for the expansion to reach this scale).

In general, based on deposit data in Tables 7.3–7.6, there remain extensive U resources identified in existing producer countries, but almost all of these resources are of low grade (0.02–0.05% U_3O_8). In the longer term, continuing U demand will have to mean mining these lower grade U ore resources with the resulting large increases in CO_2 emissions (see below). Furthermore, while the total of 10 294 kt of U_3O_8 significantly exceeds the 2009 Red Book estimate of 7435 kt, major deposits and/or countries are still missing, and also numerous small deposits of <5 kt U_3O_8. According to Xun,[32] China's U resources as of about 1998 consisted of more than 200 U deposits containing at least 70 kt of U_3O_8, with hope for between 1 and 2 Mt of U_3O_8 in the long-term. The deposit types are 37.05% granite-related, 23.53% sandstone, 18.97% volcanic, 15.94% mudstone-related and 4.51% other.[32] For Brazil, the Itataia region alone is reported to contain ∼3.5 Gt of phosphate ore containing 634 kt of U_3O_8[33](*i.e.* ore grade of ∼0.02% U_3O_8) – compared with the current resource of 129 kt of U_3O_8[28] – demonstrating that Brazil has considerably more U resources than the data in this chapter (and even the Red Book) would indicate.

Plots of ore grade versus contained U and ore tonnage versus ore grade for all reported U deposits are given in Figure 7.2, both

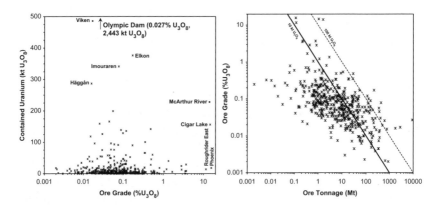

Figure 7.2   Ore grade versus contained U (left) and ore tonnage versus ore grade (right) for all reported U deposits.

showing a broad inverse correlation – as grade declines, ore tonnage and contained U tend to increase. The Olympic Dam, McArthur River, Häggån, Elkon, Imouraren, Viken and Cigar Lake deposits are clearly unusual in a global sense, given their relationship to all other U resources (*i.e.* they appear to be outliers in Figure 7.2).

In summary, history is indeed repeating itself yet again – with a new rush based on higher prices and a perceived hope for nuclear power, global U exploration has delivered outstanding results within the past 5 years. The major success continues to be expansion at known projects (*e.g.* Ranger and Olympic Dam in Australia, McArthur River in Canada) and major new discoveries around the world. These range from super-low-grade ore (*e.g.* Etango-Goanikontes, Namibia, at 0.0193% U_3O_8) to super-high-grade ore (*e.g.* Wheeler River-Phoenix, Canada, at 15.74% U_3O_8.). The latter is the highest grade deposit currently known in the world. Although there can be a small time lag with reporting through the Red Book, it is clear that the trajectory for known global U resources is still upwards, although the global average ore grade will continue to decline based on the dominance of lower grade ores and deposit types.

7.5 URANIUM MINES AROUND THE WORLD

7.5.1 Overview

The uranium mining sector is a relatively minor area of the global mining industry, with a global value of about US$6.56 billion (*i.e.* 63 269 t of U_3O_8 in 2010 valued at about US$103.6 per kilogram or about US$47 per pound of U_3O_8) compared with coal and iron ore, which are of the order of hundreds of billions of dollars. The U sector, however, is widely regarded as a key strategic asset and is intimately linked to considerations by countries of national security and energy independence – despite the reality of the ongoing menace of nuclear weapons and the broader energy debate (especially in the context of GGEs and climate change risks). Overall, there are only a small number of U mines, perhaps of the order of 50 or so globally, shared between Kazakhstan, Canada, Australia, Russia, Namibia, Niger, USA, South Africa and smaller countries. Most of the largest mines are operated by publicly listed mining companies, such as Cameco, Rio Tinto, Areva, BHP Billiton, Uranium One and ARMZ. In Kazakhstan, every project has a significant shareholding by the government through its state-owned enterprise KazAtomProm, although most U mines are now operated by western companies in joint ventures.

7.5.2   Brief Review of Uranium Ore Mining and Milling

The U industry uses conventional mining and milling methods and has also been a pioneer in developing new technologies in the mining industry, such as solvent extraction and *in situ* leaching. Mining is typically through open cut or underground methods, depending on depth, size and other factors (*e.g.* rivers or lakes). Given that it is common for U deposits to occur in clusters, as in the Athabasca Basin of Canada and Colorado Plateau of the USA, a central mill can often process ores from a variety of mines. Milling begins with fine grinding, followed by either acid or alkaline leaching, solvent extraction, chemical precipitation and finally calcination to produce triuranium octaoxide (U_3O_8). Acid-based leaching is most common, as it is faster and often achieves a more complete extraction, in addition to being a relatively cheaper reagent. Alkaline leaching is suited for particular ore types which contain significant calcite (or limestone), the most common of which is surficial (or carnotite/calcrete) type deposits. Further discussion of U ore mining and milling can be found elsewhere.[34–37]

In situ leaching (ISL) is a very specialised form of U production and is commonly only suitable for sandstone-type deposits. The process of ISL involves drilling hundreds of groundwater bores into the sandstone ore, using some as injection bores and most as extraction bores. The reagents are added to the recirculating solutions, including an acid (e.g. sulfuric acid) or alkali (sodium bicarbonate) plus a strong oxidant (e.g. oxygen, hydrogen peroxide or hypochlorite), thereby dissolving the U 'in situ' in the ore formation and bringing it to the surface in the extracted solutions. Although ISL used to be a relatively minor U source, restricted to a handful of mines throughout the world, the rapid growth of ISL mines in Kazakhstan in the past several years has seen ISL now dominate conventional sources – reaching about 41% of global U production in 2010.[38] A major issue with ISL is the challenge of

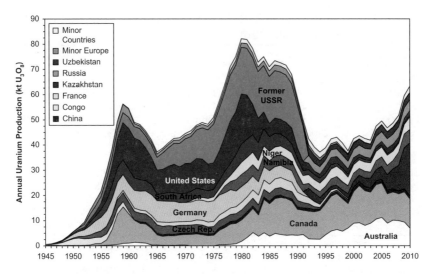

Figure 7.3 Historical world U production by country (data compiled from the literature[13,38,43]).

remediating impacted groundwater resources after mining, with the extent of groundwater contamination often severe at Cold War-era acid ISL sites[39,40] – although civilian-era alkaline sites have also proved much more difficult to remediate than anticipated.[41,42]

Historical U production by country is given in Figure 7.3, showing the major dominance of a select handful of countries such as Canada, the USA, Germany (dominantly East), the former Soviet Union and its now component states since 1992 (Kazakhstan, Uzbekistan, etc.), Niger and Australia. By the collapse of the Soviet Union in 1991, it is interesting that the USSR, USA, Canada and East Germany had produced 445.2, 397.5, 292.2 and 256.5 kt of U_3O_8, respectively. Cumulative world production by 2010 was ~2966.3 kt of U_3O_8 (which compares with the 7435 kt of U_3O_8 reported by the 2009 Red Book), given by country in Table 7.7 with 2010 production.

7.5.3 Statistics of Current Uranium Mines

A compilation of production statistics for hard rock U projects around the world is given in Table 7.8 and for ISL mines in

Table 7.7 Cumulative and 2010 uranium production $(t\,U^3O^8)$ by country (1945–2010).

Country	Cumulative	2010	Country	Cumulative	2010
Canada	525 913	11 534	Kazakhstan[b]	99 924	20 990
USSR/Russia[a]	511 184[a]	4200	France	89 600	8
USA	432 405	1957	Uzbekistan[b]	46 886	2830
Germany	258 845	0	China	38 879	975
Australia	200 729	6956	Democratic Republic of Congo	30 182	0
South Africa	190 789	687	Gabon	29 950	0
Niger	135 352	4949	Hungary	24 864	0
Namibia	123 097	5301	Romania	21 895	91
Czechoslovakia	120 546	299	Rest of the world	85 278	2491
			Total	**2 966 319**	**63 239**

[a]Much of USSR production (445 206 t U_3O_8 up to 1991) was from Kazakhstan, Uzbekistan and related countries.
[b]Production from 1992 to 2010 only (production previously part of the USSR).

Table 7.9. On comparing the tables, the dominance of a small number of mines (the top five mines, McArthur River, Ranger, Rössing, Priargunsky and Olympic Dam, represented 37.5% of 2010 global mine production) and also the popularity of ISL for sandstone ores is clear. The more than two orders of magnitude difference in ore grades, from McArthur River at $\sim 15\%$ U_3O_8 (before blending at Key Lake with very low-grade material to facilitate milling) to $\sim 0.04\%$ U_3O_8 at Rössing, is extremely unusual in global mining (compared with, say, maximum and minimum Cu and Au ore grades).This highlights the decidedly variable nature of individual U projects (scale, technical challenges, economics, ore processing, related environmental issues, etc.).

Another important aspect of the data in Tables 7.8 and 7.9 is that not all companies report complete production statistics. For example, some mines do not report waste rock (even open cut mines), despite this often being a major portion of mine waste, and some do not report complete mill data (e.g. ore milled, ore grades, U extraction or solution volumes and concentrations for ISL).

The available data for average country ore grades over time are compiled in Figure 7.4, showing the relative magnitude of different countries such as Namibia *versus* Australia, and also the increasing ore grades in Canada as the dominance of

Table 7.8 Hard rock uranium mine production around the world (2010 data).[a]

Mine, country	Ore/Mt	% U_3O_8	U_3O_8/t	Type[b]	Mt waste rock	Company
McArthur River, Canada	0.196	4.68	9027	UG	–	Cameco[70%], Areva[30%]
Ranger, Australia	2.400	0.23	4493	OC	~9.2	Rio Tinto[68.4%], Public[31.6%]
Rössing, Namibia	11.598	~0.04	3628	OC	41.955	Rio Tinto[68.6%], Iran[15%], IDC[10%], Namibia[3%]
Priargunsky, Russia[c]	~1.7[c]	~0.21[a]	3443	UG	nd	ARMZ[100%]
Somaïr, Niger[c]	~2[c]	~0.172	~3125	UG	–	Areva[63.4%], Sopamin[36.6%]
Olympic Dam, Australia[d]	7.046	0.054	2747	UG	nd	BHP Billiton[100%]
Langer Heinrich, Namibia	2.202	0.096	2108	OC	11.595	Paladin Energy[100%]
Cominak, Niger[c]	~0.7[c]	~0.352	~1825	OC	–	Areva[34%], Sopamin[31%], OURD[25%], Enusa[10%]
Rabbit Lake, Canada	0.199	0.78	1769	UG	–	Cameco[100%]
Kayalekera, Malawi	0.884	0.138	1222	OC	2.023	Paladin Energy[85%], Malawi Government[15%]
McLean Lake, Canada	0.097	0.80	785	UG	–	Areva[70%], Denison Mines[22.5%], OURD[7.5%]
Hérault, France	nd	nd	8	nd	nd	Areva[100%]
Approximate totals	**~29.0**	**~0.13**	**34 180**		**>55.6**	

[a]Data preceded by ' ~ ' are approximate only and are commonly estimated based on other reported data (*e.g.* resource ore grades); nd, no data.
[b]OC, open cut; UG, underground.
[c]Ore tonnage based on Red Book. All data sourced from the respective company annual (and/or quarterly) reporting (unless noted).
[d]Olympic Dam also extracts copper, gold and silver (but not rare earths).

Table 7.9   ISL uranium mine production around the world (2010 data).[a]

Mine, country	GL process solutions	mg U L^{-1}	Ore[b] % U$_3$O$_8$	t U$_3$O$_8$	Company
Karatau, Kazakhstan	~9.5	~182	~0.114	2016	Uranium One[50%], KazAtomProm[50%]
South Inkai, Kazakhstan	~20.1	~86	~0.044	2005	Uranium One[70%], KazAtomProm[30%]
Inkai, Kazakhstan	nd	nd	~0.056	1970	Cameco[60%], KazAtomProm[40%]
Akdala, Kazakhstan	~15.2	~70	~0.069	1218	Uranium One[70%], KazAtomProm[30%]
Zarechnoye, Kazakhstan[c]	nd	~48[a]	~0.067	917	Uranium One[49.7%], KazAtomProm[49.7%]
Akbastau, Kazakhstan[c]	nd	~288[a]	~0.108	872	Uranium One[50%], KazAtomProm[50%]
Smith Ranch, USA	nd	nd	~0.09	820	Cameco[100%]
Dalur, Russia	nd	nd	~0.05[d]	599	ARMZ[100%]
Beverley, Australia[e]	~8[e]	~100[e]	~0.2[e]	~417[e]	General Atomics[100%]
Crow Butte, USA	nd	nd	~0.12	320	Cameco[100%]
Kharasan, Kazakhstan	~4.4	~58	~0.122	303	Uranium One[30%], KazAtomProm[30%], Energy Asia[40%]
Khiagda, Russia	nd	nd	~0.06[b]	159	ARMZ[100%]
Approximate totals			**~0.083**	**11 616**	

[a]Data preceded by '~' are approximate only and are commonly estimated based on other reported data (*e.g.* resource ore grades); nd, no data.
[b]Based on resource grade only since ore grades of zones being leached are not reported.
[c]Uranium One acquired their interest in Akbastau and Zarechnoye on 27 December 2010, hence concentration data are indicative only.
[d]Data based on Red Book.
[e]Data estimated from available historic production data[21]. All data sourced from respective company annual (and/or quarterly) reporting (unless noted).

Saskatchewan's rich deposits grows (especially the start of the high-grade McArthur River project in 2000). Given the increasing proportion of low-grade projects under development in Canada, Australia and Namibia, and especially ISL production in Kazakhstan, the overall global average ore grade will continue to decline over time.

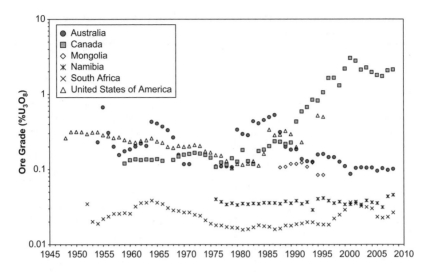

Figure 7.4 Average country U ore grades over time.

7.6 SUSTAINABILITY ISSUES, CONSTRAINTS AND THE FUTURE OF URANIUM

The extent to which U will continue to be mined in the future is extremely difficult to predict – it is certainly considerably more complex to assess than merely applying a regression curve to historical production or a Hubbert-style peak model. The primary factors which affect uranium will be the future of nuclear power, public concern and/or opposition, the energy and GGEs intensity of lower grade mines, effective mine rehabilitation, markets and economic conditions and government policies, amongst others. In this section, potential growth scenarios of nuclear power, energy and GGEs intensity of mining and finally mine rehabilitation are discussed (albeit somewhat briefly), although other issues may prove to be just as important.

7.6.1 Projections of Nuclear Power

As noted in the Introduction, there is only a modest growth in nuclear power predicted by the IEA by 2030 (Table 7.1). The IEA also projects other future energy scenarios, depending on various assumptions about policy issues such as GGEs and climate change,

Table 7.10 World primary energy demand and electricity generation by source projected in 2035 by IEA scenario (Current, New, 450).[1]

Primary energy source	Current (EJ, %)	New (EJ, %)	450 (EJ, %)
Coal	171.7 (24.18%)	226.9 (29.61%)	97.0 (16.23%)
Natural gas	164.5 (23.16%)	209.0 (27.28%)	163.5 (27.37%)
Crude oil	194.5 (27.38%)	176.1 (22.98%)	135.5 (22.68%)
Biomass and other	80.0 (11.27%)	71.5 (9.33%)	86.0 (14.40%)
Nuclear	50.7 (7.15%)	44.1 (5.76%)	60.8 (10.18%)
Hydro	19.9 (2.80%)	18.5 (2.42%)	20.4 (3.42%)
Other renewables	28.9 (4.07%)	20.1 (2.63%)	34.2 (5.73%)

Electricity source	Current (TWh, %)	New (TWh, %)	450 (TWh, %)
Coal	12 035 (33.20%)	16 932 (43.01%)	4797 (14.89%)
Gas	7923 (21.86%)	8653 (21.98%)	5608 (17.40%)
Oil	533 (1.47%)	591 (1.50%)	360 (1.12%)
Biomass	1497 (4.13%)	1150 (2.92%)	2025 (6.28%)
Nuclear	4658 (12.85%)	4053 (10.30%)	6396 (19.85%)
Hydro	5518 (15.22%)	5144 (13.07%)	6052 (18.78%)
Wind	2703 (7.46%)	2005 (5.09%)	4320 (13.41%)
Geothermal	271 (0.75%)	200 (0.51%)	407 (1.26%)
Solar photovoltaic	741 (2.04%)	435 (1.10%)	1332 (4.13%)
Solar thermal	307 (0.85%)	166 (0.42%)	845 (2.62%)
Tidal and wave	63 (0.17%)	39 (0.10%)	82 (0.25%)

energy security and markets, amongst others. The three scenarios of energy and electricity demand in 2035 are shown in Table 7.10 and are defined briefly as follows:[1]

- *'Current Policies'* includes all policies in place and supported through enacted measures as of mid-2011.
- *'New Policies'* is based on broad policy commitments and plans that have been announced by countries around the world to address energy security, climate change and local pollution and other pressing energy-related challenges, even where the specific measures to implement these commitments have yet to be announced.
- *'450 Policies'* set out a scenario for an energy pathway that is consistent with a 50% chance of meeting the goal of limiting the increase in average global temperature to 2 °C compared with pre-industrial levels.

Based on IEA data for nuclear power growth and U requirements, the Current, New and 450 growth scenarios were projected

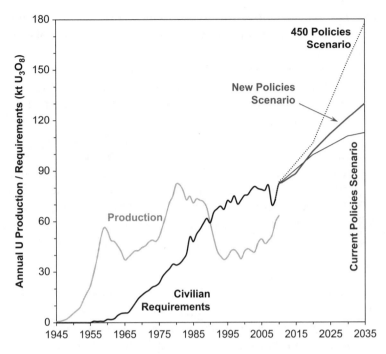

Figure 7.5 Historical uranium production, civilian reactor requirements and U demand under the three IEA future energy policy scenarios (Current, New and 450).

for U requirements, shown with historical U production and civilian reactor requirements in Figure 7.5. Assuming that new mine production supplies all U requirements from 2015 (*i.e.* no further downblending of nuclear weapons material into civilian fuel or stockpile drawdowns), this means that by 2035, the Current, New and 450 scenarios project cumulative U requirements of 2493, 2630 and 3095 kt of U_3O_8, respectively.

All estimates are well within known reasonable cost U resources. According to the Red Book, of the 7435 kt of U_3O_8 globally, about 59% (or 4412 kt of U_3O_8) is within the cost category of <US$80 kg^{-1} U – compared with the average 2010 spot price of about US$103.6 kg^{-1} U_3O_8 (US$47 lb^{-1} U_3O_8). This means that future production is still within economic reach; however, given the sluggish growth in western U production, this places more pressure on countries such as Kazakhstan, which have seen considerable growth in ISL-based U production over the past several years (see

Figure 7.3). In addition, this requires new production from the numerous low-grade projects in South Africa and Namibia, with grades ranging as low as 0.015% U_3O_8 at Trekkopje – which is being developed as a heap leach project especially with a seawater desalination plant to provide its water supply. It is possible, based on operating sites, U resource data and assumptions regarding the sequence of developing mines, to project the future ore grade out to 2035, but the high uncertainty in making such estimates precludes any meaningful insight. For example, new deposits could be discovered (or increases to existing ones), the development sequence could vary, economics could radically alter prices and supply–demand balances, decommissioned nuclear weapons material could be delivered into the civilian market or even nuclear power could decline and not grow at all. On the other hand, some countries could use imported U to free up local sources of U to increase their nuclear weapons production.

Overall, it can be concluded that sufficient low-cost U resources are already known to meet the IEA's projected growth in nuclear power to 2035 – the question is not a matter of 'how much', but the actual production rates and from which deposits or mines that nuclear reactor requirements will be met.

7.6.2   Energy and Greenhouse Gas Emissions Intensity of Mining

A welcome trend across all sectors of global industry is the strong emergence of sustainability reporting over the past decade. Numerous mining companies have certainly been at the forefront of this change in corporate accountability by publishing annual sustainability reports alongside statutory financial reports.[44] The most popular protocol is the Global Reporting Initiative (GRI)[45] – a coalition of the United Nations, industry, government and civil society groups. The use of the GRI for reporting is (still) voluntary and it includes core and voluntary indicators covering economic, social, environmental, human rights and labour aspects of an organisation's activities, with some being qualitative whereas others are quantitative. A specific sector supplement was recently finalised to facilitate improved and more relevant sustainability reporting for mining.[46] Some mining companies continue to rely on internally developed systems for sustainability reporting, with variable comparison with the GRI. The extent to which a report meets GRI requirements can also be assessed, giving a company's report an 'application level', essentially a measure of thoroughness or quality assurance. The issue of external auditing is emerging as a key test regarding the credibility of reports[47] – that is, the old 'spin *versus* substance' debate.

Given the growing extent of reporting, it is possible to use the data to analyse uranium mining and derive 'sustainability metrics' such as energy and GGEs intensity. Although some U mines and companies do publish sustainability reports and data, many (still) do not – often despite their rhetoric about the safety credentials of U mining and nuclear power. This chapter summarises the available data only; more detailed discussions on this area can be found elsewhere.[29,44] The energy and GGEs intensity of some uranium mines are given in Table 7.11 and the relationships between energy and GGEs intensity with ore grade are shown in Figure 7.6.

The first key observation from Table 7.11 and Figure 7.6 is the inverse ore grade–energy/GGEs intensity relationship. Comparing the Ranger and Rössing mines, they have average ore grades of $\sim 0.3\%$ and $\sim 0.04\%$ U_3O_8, respectively, with Ranger having both a lower energy and GGEs intensity for U production. The high grade Saskatchewan U mines, however, do not necessarily have a lower intensity – such as McLean Lake with a higher energy/GGEs intensity than Rössing. For McArthur River, the energy cost of mining each tonne of ore is clearly considerable (~ 6000 MJ t^{-1} ore), much higher than for other mines (~ 120–600 MJ t^{-1} ore), showing that the energy-intensive ground freezing and remote mining methods used more than offset the higher grade since energy intensity is only marginally lower than Ranger (*i.e.* 153 *versus* 266 GJ t^{-1} U_3O_8). Finally, the Beverley ISL mine has a similar energy intensity to conventional mines, showing that although there is no ore excavation and processing, the energy costs are offset by the lower grade ores, intensive solution pumping and lower overall extraction efficiencies.

The second key observation from Table 7.11 and Figure 7.6 is the gradually increasing energy/GGEs intensity over time. At the Ranger mine, ore grades have declined from 0.28% to 0.23% U_3O_8 from 2005 to 2010 – leading to a strong increase in energy/GGEs intensity over this time. Ore grade declines are also apparent for the McLean Lake and Olympic Dam mines. For the Rössing mine, the increasing intensity is mainly related to an expansion of open pit

Table 7.11 Energy and greenhouse gas emissions intensity of some uranium mines (further details in and data updated from ref. 29).[a]

Project	Ore grade % U_3O_8	Electricity Ore/kWh t^{-1}	Energy		CO_2 emissions	
			Ore/MJ t^{-1}	U_3O_8/GJ t^{-1}	kg(CO_2) t^{-1} (ore)	t(CO_2) t^{-1} (U_3O_8)
Olympic Dam[18.4%,b]	0.053–0.114	17.1 ± 1.4 (6)	127 ± 26 (20)	266 ± 58 (20)	22.8 ± 3.3 (18)	47.0 ± 11 (18)
Ranger	0.231–0.423	nd	594 ± 152 (20)	224 ± 64 (20)	44 ± 13 (20)	16.7 ± 5.5 (20)
Rössing	~0.03–0.04	19.5 ± 0.8 (5)	120 ± 23 (16)	393 ± 76 (16)	15 ± 2.0 (15)	49.1 ± 7.5 (16)
Beverley (ISL)	~0.18	–	–	216 ± 63 (5)	–	11.2 ± 3.3 (5)
McArthur River	3.91	1083 (1)	6027 (2)	153 (2)	782 (2)	19.9 (2)
Rabbit Lake	0.78	438 (1)	2513 (2)	355 (2)	337 (2)	47.5 (2)
McLean Lake	0.53–2.29	249 ± 17 (7)	4848 ± 868 (7)	415 ± 293 (7)	198 ± 43 (7)	20.7 ± 16 (7)
Cluff Lake	2.71	nd	5187 (1)	194 (1)	325 (1)	12.1 (1)
Production weighted average		**~92 (20)**	**1055 (68)**	**272 (73)**	**106 (66)**	**31.2 (71)**

[a]nd, no data.
[b]Olympic Dam is presented on the basis of attributing 18.4% of inputs and outputs to U production, since this is the long-term average proportional revenue from U (Cu is 75.7%, Au–Ag 5.9%).[21]

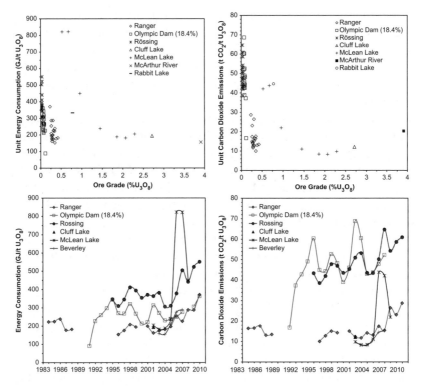

Figure 7.6 Energy intensity versus ore grade (top left); greenhouse gas emissions intensity versus ore grade (top right); energy intensity over time (bottom left); greenhouse gas emissions intensity over time (bottom right).

mining scale, whereby waste rock mined has increased from 7.5 to 42 Mt y^{-1} over the period 2005 to 2010 whereas ore milled has been stable at ~ 12 Mt y^{-1}.

Taken together, these two key findings are crucial in understanding the future of U mining. As shown with U resources, although some high-grade mines will continue for some time, the future of U mining will be increasingly dominated by lower grade ores. In conjunction with the energy/GGEs intensity issues, this means that there will be significant upward pressure on energy/GGEs intensity as ore grades decline. The growing intensity will be very gradual and each mine will have its own particular circumstances to address, such as old infrastructure, larger scales (deeper pits, longer haul routes for waste rock), declining ore grades and possible constraints such as carbon pricing and water resources.

Another major aspect which is often overlooked in understanding energy/GGEs intensity is U mineralogy. Some U minerals, such as uraninite, carnotite and coffinite, are readily soluble in acidic or alkaline solutions (especially with an added oxidant such as pyrolusite, oxygen or hydrogen peroxide), whereas others, such as brannerite and davidite, have very low solubility without extremely aggressive conditions (*i.e.* extremely strong acids, high temperatures and/or high pressures). Some of the world's major U deposits have a significant proportion of brannerite and hence are not readily processed using conventional milling methods. For example, Olympic Dam ore has 54% coffinite, 34% brannerite and 12% uraninite[48] – with a historical extraction efficiency of 67.1%[21] showing that most uranium is dissolved from the coffinite and uraninite with only a very small fraction derived from the brannerite. The Elkon deposit in Russia was first discovered in 1960 and remains one of the largest U deposits in the world (third after Olympic Dam and Viken) – but has yet to be developed since the U mineralogy is dominantly brannerite.[49] If the mineralogy is refractory, this means low recoveries and high unit production costs – often making a project uneconomic. It is possible to treat refractory ores, using a combination of high-strength acids or alkalis with high temperature and/or pressure (*e.g.* the former Elliot Lake district, Canada, and former Radium Hill mine, Australia), but this invariably increases chemical and energy inputs and places even greater pressure on the energy/GGEs intensity of U production.

7.6.3 Uranium Mine Rehabilitation – Some Key Issues

A major problem with the early era U mines was poor mine waste management and invariably a lack of mine site rehabilitation. Along with the rise of environmental regulation from the 1970s in most countries, U mining has also been required to undertake mine rehabilitation after operations cease. The community typically expects that this will lead to a stable site, with no ongoing pollution or radioactivity issues. The experience internationally, however, is varied and this sub-section only briefly touches on this area. Further details, including numerous papers on various mines, are found in references 50–56 among a plethora of others.

As noted above, countries such as Germany and the USA have spent billions of dollars on remediation of U mines from the Cold War era.

In the former GDR, the Soviets operated large-scale, low-grade mines in the Saxony–Thuringia region. At the time of reunification in 1991, West Germany had to accept the cost of all remediation of these mines – around €6.6 billion (or 13 billion Deutschmarks).[57] The Wismut legacy included five large underground mines, one large open cut mine (84×10^6 m^3), waste rock dumps (311×10^6 m^3) and tailings (160×10^6 m^3). The major problems included excessive radiation exposure risks (especially due to radon build-up), leaching of heavy metals and radionuclides to surface water ecosystems due to growing acid mine drainage, groundwater contamination, physical safety risks and aesthetic issues, amongst others. All of these issues were made more pressing by the high population density of the Saxony–Thuringia region. The programme was planned to last for more than 20 years and by 2011 most of the work was completed with efforts moving to ongoing maintenance and monitoring. Many of the challenges had never before been attempted – such as remediating extremely soft tailings with a density half that of water, backfilling a large open cut, groundwater remediation from underground (or in-place) leaching, and large scale recontouring of waste rock dumps. Although early indications are that the programme has been relatively successful, there remains a need for caution as ongoing monitoring indicates the true status of the former sites.

In the USA, many of the Cold War era U mines supplied the military programme of the day and the US government has therefore taken on the liability of rehabilitation of these old sites. The large effort was known as the Uranium Mill Tailings Remediation Action (UMTRA) programme and included numerous sites in midwest states (Colorado, Utah, New Mexico, Arizona) and others. At some sites, tailings dams and waste rock dumps were covered and revegetated, and at some the tailings were excavated and moved to a new, more suitably engineered site. The programme has cost billions of dollars, although a thorough review of the entire programme does not appear to have been published (yet).

In Australia, there is often a widely held belief that rehabilitating old U projects has been successful – but invariably this view is held by those who have never visited the sites. In brief, the major Cold

War-era U mines in Australia were the Mary Kathleen, Rum Jungle, Radium Hill–Port Pirie and the Upper South Alligator Valley, with the latter rehabilitated only in the 2000s (after the Coronation Hill saga), while all others were rehabilitated in the mid-1980s. Further small U projects were also developed at Pandanus Creek-Cobar 2, Fleur de Lys, George Creek, Brock's Creek and Adelaide River in the Northern Territory and Myponga in South Australia,[21] although no substantive rehabilitation work is known for each site. The Nabarlek project, which operated from 1979 to 1988, was a 'modern U mine' and was approved and operated under strict regulations and supervision, being rehabilitated in the mid-1990s. Other 'modern U mines' are still in operation at Ranger, Olympic Dam and more recently Beverley. At present, surprisingly, there is no former U project in Australia which can be claimed as a successful, long-term rehabilitation case study – all still require ongoing monitoring and maintenance and some remain mildly to extremely polluting. Further details and photographs are available elsewhere.[56,58,59]

In summary, successful rehabilitation of former U mines still remains challenging, with issues varying widely and being dependent on a range of site-specific factors (climate, site history, scale, geology, rehabilitation approach, etc.). In many parts of the world, and especially in developing countries (*e.g.* Niger, Gabon, Argentina, Brazil), community concern is often related to the negative legacy of past U projects – and this can lead to significant opposition. There is no room for complacency or over-confidence in U mine rehabilitation, with the need for high standards and sustainable rehabilitation outcomes clear throughout the world.

7.7 CONCLUSION

This chapter has provided an extensive review of the mining and milling of U ore in the context of perceived future increase in global nuclear power. Despite its early promise, nuclear power remains a minor source of primary world energy supplies and electricity – and its share is projected as most likely to continue to decline, especially as renewable energy sources enter substantial growth in the coming decades. Although the detailed evaluation of known U resources shows that there is sufficient low-cost U available to meet expected nuclear power demands by 2035, this will increasingly have to be from lower grade deposits. A detailed compilation and analysis of the sustainability metrics of U production, such as energy inputs and GGE outputs, shows that they are inversely related to ore grade – meaning that as global average ore grades decline, the unit intensity of U production will increase. This means that GGEs from U mining will begin to increase significantly over this time frame, effectively reducing any perceived benefit of nuclear power as a low carbon intensity electricity source. Overall, on moving beyond simple rhetoric and considering the factual implications of the sustainability of U mining and nuclear power, reality clearly demonstrates that nuclear power is not a viable strategy to address burgeoning global GGEs and the serious risks of climate change.

REFERENCES

1. IEA, *World Energy Outlook 2011*, International Energy Agency, Paris, 2011.
2. M. Schneider, A. Froggatt and S. Thomas, *Nuclear Power in a Post-Fukushima World: 25 Years After the Chernobyl Accident – The World Nuclear Industry Status Report 2010–2011*, Worldwatch Institute, Washington, DC, 2011.
3. M. Jacobson and M. Delucchi, *Sci. Am.*, 2009, **301**, 58.
4. B. Sørensen and P. Meibom, *Int. J. Global Energy Issues*, 2000, **13**, 4.
5. A. Zervos, C. Lins and J. Muth, *Rethinking 2050: a 100% Renewable Energy Vision for the European Union*, European Renewable Energy Council, Brussels, 2010.
6. I. G. Mason, S. C. Page and A. G. Williamson, *Energy Policy*, 2010, **38**(8), 3973.
7. M. Wright and P. Hearps (eds), *Australian Sustainable Energy: Zero Carbon Australia – Stationary Energy Plan*, Beyond Zero Emissions and University of Melbourne Energy Research Institute, Melbourne, 2010.
8. B. Barré, in *Handbook of Nuclear Engineering, Volume 1: Nuclear Engineering Fundamentals*, ed. D. G. Cacuci, Springer, Berlin, 2010, Ch. 25, pp. 2896–2933.
9. G. M. Mudd, *Hist. Rec. Aust. Sci.*, 2005, **16**, 169.
10. F. J. Dahlkamp, *Uranium Deposits of the World – Asia*, Springer, Berlin, 2009.

11. F. J. Dahlkamp, *Uranium Deposits of the World – USA and Latin America*, Springer, Berlin, 2010.
12. IAEA, *World Distribution of Uranium Deposits (UDEPO) with Uranium Deposit Classification – 2009 Edition*, TECDOC-1629, International Atomic Energy Agency, Vienna, 2009.
13. OECD–NEA and IAEA, *Uranium: Resources, Production and Demand*, Nuclear Energy Agency, Organisation for Economic Co-operation and Development and International Atomic Energy Agency, Paris, Years 1965–2009.
14. NRC, *Canadian Minerals Yearbook*, Mining Sector, Natural Resources Canada, Ottawa, Years 1944–2009.
15. US Bureau of Mines, *Minerals Yearbook*, US Bureau of Mines, Washington, DC, Years 1933–1993.
16. EIA, *Uranium Industry Annual*, Energy Information Administration, US Department of Energy, Washington DC, Years 1992–2005.
17. CMSA, *Facts and Figures 2010*, Chamber of Mines of South Africa, Johannesburg, 2010.
18. G. C. Battey, Y. Miezitis and A. D. Mackay, *Australian Uranium Resources*, Bureau of Mineral Resources, Geology and Geophysics, Canberra, 1987.
19. S. B. Dickinson, in *Report on Investigation of Uranium Deposits at Mt Painter, South Australia, June 1944 to September 1945 – Undertaken at the Request of the British Government by the Government of the Commonwealth of Australia in Conjunction with South Australian Government, November 8, 1945*, ed. S. B. Dickinson, South Australia Department of Mines, Adelaide, 1945.
20. A. D. McKay and Y. Miezitis, *Australia's Uranium Resources, Geology and Development of Deposits*, Geoscience Australia, Canberra, 2001.
21. G. M. Mudd, *Compilation of Uranium Production History and Uranium Deposit Data Across Australia*, SEA-US, Melbourne, last updated December 2011.
22. AusIMM, MCA, AIG, *Australasian Code for Reporting of Exploration Results, Mineral Resources and Ore Reserves: the JORC Code*, Joint Ore Reserves Committee of the Australasian Institute of Mining and Metallurgy (AusIMM), Minerals Council of Australia and Australian Institute of Geoscientists, Parkville, VIC, 2004.

23. I. Lambert, Y. Meizitis, A. D. McKay, *AusIMM Bull.*, 2009, December, 52.
24. M. Dittmar, *Physics and Society*, 2009, arXiv:0909.1421v1 [physics.soc-ph].
25. Geoscience Australia, *Australia's Identified Mineral Resources – Preliminary Resources Table (online)*, Geoscience Australia, Canberra, 2011, www.ga.gov.au (last accessed 31 March 2012).
26. NRC, *Canadian Minerals Yearbook 2009*, Mining Sector, Natural Resources Canada, Ottawa, 2010.
27. EIA, *U.S. Uranium Reserves Estimates (2008)*, Energy Information Administration US Department of Energy, Washington, DC, 2010.
28. INB, *Uranium Reserves – Brazil and World*, Indústrias Nucleares do Brasil, Rio de Janeiro, Brazil, 2001, www.inb.gov.br/ (last accessed 23 December 2011).
29. G. M. Mudd and M. Diesendorf, *Environ. Sci. Technol.*, 2008, **42**, 2624.
30. L. B. Cochrane and L. R. Hwozdyk, *Technical Report on the Elliot Lake District, Ontario, Canada*, prepared for Denison Mines Corp. by Scott Wilson Roscoe Postle Associates Inc., Toronto, 2007.
31. BHPB, *Olympic Dam Expansion – Draft Environmental Impact Statement*, BHP Billiton, Adelaide, 2009.
32. Z. Xun (ed.) *Mineral Facts of China*, Science Press, Beijing, 2006.
33. S. Saad, in *The Uranium Production Cycle and the Environment: an International Symposium*, ed. J.-P. Nicolet and D. H. Underhill, International Atomic Energy Agency, OECD Nuclear Energy Agency, Nuclear Energy Institute, World Nuclear Association and the Office of the Supervising Scientist (OSS), Vienna, 2002, p. IAEA-SM-362/4P.
34. IAEA, *Uranium Extraction Technology*, International Atomic Energy Agency, Vienna, 1993.
35. IAEA, *Establishment of Uranium Mining and Processing Operations in the Context of Sustainable Development*, International Atomic Energy Agency, Vienna, 2009.
36. OECD–NEA and IAEA, *Environmental Activities in Uranium Mining and Milling*, Nuclear Energy Agency, Organisation for Economic Co-operation and Development and International Atomic Energy Agency, Paris, 1999.

37. E. Özberk and A. J. Oliver (eds), *Uranium 2000 – International Symposium on the Process Metallurgy of Uranium*, Metallurgical Society, Canadian Institute of Mining, Metallurgy and Petroleum, Saskatoon, 2000.
38. WNA, *World Uranium Mining*, World Nuclear Association, London, last updated December 2011, www.world-nuclear.org/info/inf23.html (last accessed 31 March 2012).
39. G. M. Mudd, *Environ. Geol.*, 2001, **41**, 390.
40. G. M. Mudd, *Environ. Geol.*, 2001, **41**, 404.
41. S. Hall, *Groundwater Restoration at Uranium In-Situ Recovery Mines, South Texas Coastal Plain*, Open-File Report 2009-1143, US Geological Survey, Denver, CO, 2009.
42. J. K. Otton and S. Hall, in *International Symposium on Uranium Raw Material for the Nuclear Fuel Cycle: Exploration, Mining, Production, Supply and Demand, Economics and Environmental Issues (URAM-2009)*, International Atomic Energy Agency, Vienna, 2009, p. IAEA-CN-175/87.
43. OECD–NEA and IAEA, *Forty Years of Uranium Resources, Production and Demand in Perspective*, Nuclear Energy Agency, Organisation for Economic Co-operation and Development and International Atomic Energy Agency, Paris, 2006.
44. G. M. Mudd, in *SDIMI 2009 – Sustainable Development Indicators in the Minerals Industry Conference*, Australasian Institute of Mining and Metallurgy, Gold Coast, 2009, pp. 377–391.
45. GRI, *Sustainability Reporting Guidelines*, Global Reporting Initiative, Amsterdam, 2006.
46. GRI, *Sustainability Reporting Guidelines and Mining and Metals Sector Supplement*, Global Reporting Initiative, Amsterdam, 2010.
47. A. Fonseca, *Corp. Soc. Respons. Environ. Manage.*, 2010, **17**, 355–370.
48. S. Hayward and S. O'Connell, in *AusIMM Uranium Reporting Workshop*, Australasian Institute of Mining and Metallurgy, Adelaide, 2007.
49. A. Boytsov, in *International Symposium on Uranium Raw Material for the Nuclear Fuel Cycle: Exploration, Mining, Production, Supply and Demand, Economics and Environmental Issues (URAM-2009)*, International Atomic Energy Agency, Vienna, 2009.

50. B. J. Merkel and A. Hasche-Berger (eds), *Uranium Mining and Hydrogeology III (UMH-3) – 3rd International Conference*, Springer, Freiberg, 2002.

51. B. J. Merkel and A. Hasche-Berger (eds), *Uranium in the Environment: Uranium Mining and Hydrogeology IV (UMH-4) – 4th International Conference*, Springer, Freiberg, 2005.

52. B. J. Merkel and A. Hasche-Berger (eds),. *Uranium Mining and Hydrogeology V (UMH-5) – 5th International Conference*, Springer, Freiberg, Germany, 2008.

53. B. J. Merkel, M. Schipek (eds), *Uranium Mining and Hydrogeology V (UMH-6) – 6th International Conference*, Springer, Freiberg, 2011.

54. IAEA, *Uranium Mining Remediation Exchange Group (UMREG): Selected Papers 1995–2007*, International Atomic Energy Agency, Vienna, 2011.

55. IAEA, *Environmental Contamination from Uranium Production Facilities and Their Remediation: an International Workshop, Lisbon, Portugal, 11–13 February 2004*, International Atomic Energy Agency, Vienna, 2005.

56. G. M. Mudd and M. Diesendorf, in *Sustainable Mining Conference 2010*, Australasian Institute of Mining and Metallurgy, Kalgoorlie, 2010, pp. 315–340.

57. M. Hagen and A. T. Jakubick, in *Uranium in the Environment: Uranium Mining and Hydrogeology IV (UMH-4) – 4th International Conference*, ed. B. J. Merkel and A. Hasche-Berger, Springer, Freiberg, 2005, pp. 11–26.

58. B. G. Lottermoser and P. M. Ashley, in *Uranium in the Environment: Uranium Mining and Hydrogeology IV (UMH-4) – 4th International Conference*, ed. B. J. Merkel and A. Hasche-Berger, Springer, Freiberg, 2005, pp 357–362.

59. G. M. Mudd and J. Patterson, *Environ. Pollut.*, 2010, **158**, 1252.

II
Sustainability Related to Biomass

CHAPTER 8

Aquatic Biomass for the Production of Fuels and Chemicals

ANGELA DIBENEDETTO*[1,2] AND ANTONELLA COLUCCI[2]

[1]Department of Chemistry, Campus Universitario, University of Bari, 70126 Bari, Italy; [2]CIRCC, Consorzio Interuniversitario Reattività Chimica e Catalisi, Via Celso Ulpiani 27, 70126 Bari, Italy
*Email: a.dibenedetto@chimica.unica.it

8.1 INTRODUCTION

It is now clear that burning coal, natural gas and oil is the main source of the CO_2 in the atmosphere that traps heat from radiating from the Earth. The need to reduce atmospheric CO_2 is documented by several global and national agreements.[1] Consequently, techniques for CO_2 capture and sequestration in geological formations or chemical, biological and technological utilisation have been identified. The utilisation of CO_2 is of great interest as it is possible to use it to produce valuable products, thus contributing to recycling of carbon and reducing the extraction of fossil fuel.

The enhanced fixation of CO_2 into biomass is a challenging option under intensive investigation. A high concentration of CO_2 can be used to stimulate the growth of seaweeds and micro-algae

Materials for a Sustainable Future
Edited by Trevor M. Letcher and Janet L. Scott
© The Royal Society of Chemistry 2012
Published by the Royal Society of Chemistry, www.rsc.org

off-shore or in on-shore culture and also to promote their growth in fresh water and wastewater (municipal effluents or selected process waters).

In this chapter, an analysis is made of the products that can be obtained from aquatic biomass. Biomass, in general, is considered an alternative to fossil fuels as a 'quasi-zero emission' option in the production of fuels for the transport sector. It has been predicted that biomass may contribute to the global energy balance with a share of more than 10% by 2050.[2] This requires that new biomass for energy be specifically grown and used in addition to terrestrial or residual biomass used today. Aquatic biomass may represent a convenient solution, because it has a higher (6–8%) solar energy utilisation efficiency with respect to terrestrial plants (1.5–2.2%); this being due to a higher photosynthetic efficiency. Micro-algae have been extensively studied,[3,4] but more recently marine macro-algae have attracted increasing attention.[5] CO_2 recovered from power plants or industrial flue gases could be distributed into algae ponds under controlled conditions, implementing an enhanced recycling of carbon.

8.2   THE AQUATIC BIOMASS: MACRO-ALGAE, MICRO-ALGAE AND PLANTS

Aquatic biomass includes macro-algae, micro-algae and emergents (plants), growing in either salty or fresh water (Figure 8.1).

Macro-algae, also known as 'seaweeds', are multicellular organisms that can reach sizes of up to 60 m in length. They are now utilised for the production of human food and animal feed and the extraction of hydrocolloids. Micro-algae are microscopic organisms, the most used being Cyanophyceae (blue–green algae), Chlorophyceae (green algae), Bacillariophyceae (including the diatoms) and Chrysophyceae (including golden algae). Natural macro-algae have been used since the beginning of civilisation as food, feed and fertilisers. Currently, the main producers are China, the Philippines, North and South Korea, Japan and Indonesia, with a world annual production of seaweeds of several million tonnes. They have been cultivated and used at an industrial level as a source of agar, alginate, carrageenans and fucerellans. The use of algae for energy production is very recent (after 1970) and the technical and economic feasibility of producing energy from

Chlorella sp   Spirulina   Nannochloropsis   Dunaliella salina   Botryococcus braunii

Ulva lactuca   Caulerpa taxifolia   Codium vermilaria   Pterocladiella capillace   Chaetomorpha linum

Egeria densa   Lemna minor   Poseidonia

Figure 8.1   Aquatic biomass (micro-algae, macro-algae, plants).

marine biomass, i.e. macro-algae,[6] is even more recent.[7] This recent enthusiasm is justified by the high productivity per hectare of aquatic compared with terrestrial biomass; the productivity of macro-algae ranges from 150 to 600 t_{fw} ha^{-1} y^{-1} (where t_{fw} refers to tonnes fresh weight), this being much higher than the productivity obtained from sugarcane (70–170 t_{fw} ha^{-1} y^{-1}).

To capitalise on the commercialisation of biofuels from aquatic biomass, several barriers and technical challenges must be overcome; in particular, large ponds (several hectares) need to be built, which requires large investments in terms of capital and operational (cultivation, harvesting, work-up) costs. A key issue is to maintain the quality of the biomass and the productivity for an extended period of time.

8.2.1   Micro-algae

Micro-algae can be grown in open ponds or in photobioreactors (PBRs); open ponds are overall economically favourable,[8] whereas

Table 8.1 Comparison between open ponds and photobioreactors (PBRs).[10]

Parameter	Open ponds and raceways	PBRs
Capital costs	High, ~ US$100 000 per hectare	Very high, ~ US$1 000 000 per hectare (PBR plus supporting systems)
Operating costs	Low (paddle wheel, CO_2 addition)	Very high (CO_2 addition, pH control, oxygen removal, cooling, cleaning, maintenance)
Contamination risk	High (limiting the number of species that can be grown)	Low
Harvesting costs	High, species dependent	Lower due to high biomass concentration and better control over species and conditions

PBRs are more expensive to build (Table 8.1). The former option raises the issues of land cost and water availability, appropriate climatic conditions, nutrient cost and production. In open ponds, the maintenance of long-term cultures of a desired algal strain is an issue and subject to interference from competitor strains, grazers or pathogens that may affect the productivity.[9]

Several options are available to grow micro-algae. Figure 8.2 shows the most common cultivation techniques.[11–13] Bioreactors have the advantage that the micro-algae can grow under light irradiation (either through walls or via fibres or tubes) with temperature control and with an enhanced fixation of carbon dioxide that can be bubbled through the culture medium.

The production of micro-algae in open ponds depends on solar irradiation and temperature, and these strongly influence the farming process, the productivity and the economics of the process. The availability of land and water is a key factor: marginal lands unsuitable for tourism, industry, agriculture or municipal development are good candidates. Water can be saline groundwater or local industrial water, drained from agricultural areas and recycled after harvesting algae. Nutrients (N- and P-containing compounds, micronutrients) represent one of the major costs of algal growth. The use of wastewater (sewage, fisheries, some industrial waters) rich in N- and P-containing nutrients is essential; the recovery of

Figure 8.2 (a) Raceway pond; (b) circular pond; (c) horizontal glass tube; (d) plastic bag; (e) vertical algae reactor; (f) water-supported flexible films.

useful inorganic compounds will produce clean water that, finally, can be reused or discharged into natural basins.

The direct bubbling of industrial flue gases containing CO_2 requires that algae are resistant to nitrogen oxides and sulfur oxides; 150 ppm of NO_2 and 200 ppm of SO_2 seem not to affect the growth of only a few algal species.[14] This therefore limits the direct use of flue gases.

In order to prevent CO_2 loss (13–20% of CO_2 is used), a different method to supply CO_2 is a gas exchanger, consisting of a plastic frame, covered with transparent sheeting and immersed in the pond culture. A drawback is the need to use very concentrated and pure CO_2. As 1 g of algal biomass requires 1.8–2 g of CO_2, for a ~ 6000 m^2 pond (single algal pond) a total amount of 300 kg of algal biomass will be produced per day, which uses 540–600 kg of CO_2 per day, only 0.006% of the CO_2 produced by a 500 MW power station. These values depend on the growth rate of the micro-algae and on the pond system used.[15,16]

Micro-algae will easily adapt to the culture conditions[17,18] if the parameters that influence the rate of growth and cell composition of microorganisms are kept under strict control. Light availability is of paramount importance; shadow or short light-cycles may cause a slowing of growth, whereas intense light (as may occur in desert areas or bioreactors) may modify the cell functions.[19,20] Tropical or semi-tropical areas are the most practical locations for algal culture systems.[21] Humidity must also be considered as low relative humidity results in high rates of evaporation, which can have a cooling effect on the medium.[11,17]

8.2.2 Macro-algae

Macro-algae (or seaweeds) have different properties to micro-algae. The productivity of macro-algae in natural basins is in the range 1–$20 \, kg \, m^{-2} \, y^{-1}$ dry weight (10–$150 \, t_{dw} \, ha^{-1} \, y^{-1}$; $t_{dw} =$ tonnes dry weight) for a 7–8 month culture. Owing to their effectiveness in nutrient (N, P) uptake from sewage and industrial wastewater (estimated at $16 \, kg \, ha^{-1} \, d^{-1}$),[22] they have been used in Europe and Japan for cleaning municipal wastewater,[23,24] the treatment of fishery effluents[24,25] and the recycling of nutrients. Macro-algae, grown on nets or lines or seeded on to thin, light-weight lines suspended over a larger horizontal rope,[26] have been tested in the north-western Mediterranean Sea, along the French coast[27] using *Ulva lactuca* or *Enteromorpha intestinalis*.

Climatic factors may affect the productivity by changing either the rate of growth or the growing season, or both. The Mediterranean Sea has ideal climatic conditions for a long growing season, owing to good solar irradiation intensity and duration and to acceptable temperature conditions. Moreover, along the coasts of Italy, Spain, France and Greece there are fish ponds that may prove to be the ideal locality for algae ponds. The photosynthesis of macro-algae responds to concentrations of CO_2 at levels ranging from 500 to 2000 ppm.[28,29] This means that under optimum conditions, macro-algae may grow even better than they do in their natural environments.[30] *Gracilaria bursa-pastoris*, *Chaetomorpha linum* and *Pterocladiella capillacea* have been used at concentrations of 150 times their natural concentrations.[31]

Today, brown seaweeds represent 63.8% of macro-algae production, red seaweeds 36.0% and green seaweeds 0.2%. China is the main producer[32] and $\sim 1 \times 10^6$ t (1 million tonnes) of wet seaweed is harvested and extracted to produce about 55 000 t of hydrocolloids, valued at almost US$ 600 million.[33] A knowledge of the physiological conditions is essential for the definition of the best operating parameters (optimised growing conditions).[34]

8.2.3 Aquatic Plants

Aquatic plants, also known as hydrophytes, grow in ponds, shallow lakes, marshes, ditches, reservoirs, swamps, canals and sewage lagoons; they rarely grow in flowing water, streams or rivers. They include the aquatic angiosperms (flowering plants), pteridophytes (ferns) and bryophytes. They grow in many forms: floating unattached; floating attached; submersed; or emergent. Macrophytes play a key role in nutrient cycling to and from the sediments and are often used for water phytodepuration, as some species can also concentrate heavy metals.[35]

8.3 HARVESTING OF AQUATIC BIOMASS

The different kinds of aquatic biomass require harvesting techniques that differ in cost and energy. While macro-algae and plants require simple operations, micro-algae, owing to their size and, sometimes, fragility, demand more sophisticated equipment and handling operations.

The choice of harvesting methods depends on factors such as the type of algae that have to be harvested (filamentous, unicellular), whether harvesting occurs continuously or discontinuously, what the energy demand is per cubic metre of algal suspension and what the investment costs are.[17,20,36] The main techniques used for harvesting micro-algae include centrifugation; sedimentation, filtration, screening and straining and flocculation.[37–39] Problems that can occur include biological aggregation of algae[40,41] and the rate of separation, which can be an issue when centrifugation[42] is used. The harvesting of macro-algae and plants requires less sophisticated techniques, but this depends on whether the algae are floating unattached, or attached to and growing on a hard substrate.[43]

8.4 AQUATIC BIOMASS COMPOSITION

Aquatic biomass contains a number of different chemicals, which depend on the strain and species of biomass and can be used for a specific purpose as shown in Scheme 8.1. The chemical concentration depends on the growing environment.[44] Chemicals can be extracted from the aquatic biomass by using a variety of non-destructive techniques. Soft techniques will not affect the complex

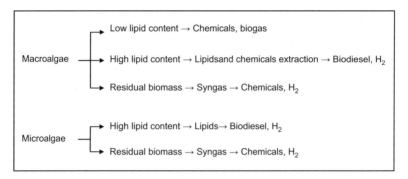

Scheme 8.1 Relation between composition of biomass and final uses.

molecular structures that can be recovered unchanged. In general, micro- and macro-algae can be used for the production of:

- energy products (hydrocarbons, hydrogen, methane, methanol, ethanol, *etc.*);
- foods and chemicals (proteins, oils and fats, sterols, carbohydrates, sugars, alcohols, *etc.*);
- other chemicals (dyes, perfumes, vitamins/supplements, enzymes, *etc.*).

Aquatic biomass is rich in carotenoids, chlorophyll, phycocyanin, amino acids, minerals and bioactive compounds; it can be used as food, as feed and in pharmaceutical preparations related to immune-stimulating, metabolism-increasing, cholesterol-reducing, anti-inflammatory and antioxidant agents.[45] Compounds such as alginate, carrageenan and agar have commercial value and represent potential co-products for existing markets. These compounds are unique to macro-algae and some, such as alginates, which occur in high concentrations in brown seaweeds, are considered recalcitrant to fermentation since the redox balance favours the formation of pyruvate as the end product.[46] Polysaccharides and sugar alcohols in brown algae, e.g., laminarin and mannitol, are candidate feedstocks for conversion to liquid fuels.

Micro-algae are rich in essential vitamins (A, B_1, B_2, B_6, B_{12}, C, E, nicotinate, biotin, folic acid and pantothenic acid).[47] Algae are also rich in omega-3 fatty acids, which have significant therapeutic importance inherent in the ability to act as an anti-inflammatory agents and to treat heart disease. Furthermore, some species can be

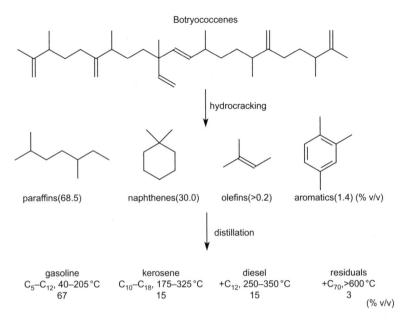

Scheme 8.2   Structure of botryococcene and its hydrocracking products.

engineered to produce a specific product. As an example, *Botryo-coccus braunii* has been engineered to produce the terpenoid C_{30} botryococcene, a hydrocarbon similar to squalene in structure.[48,49] Botryococcene is a triterpene and can be used as a feedstock for hydrocracking in an oil refinery to produce octane, kerosene and diesel (Scheme 8.2)

Table 8.2 shows a selection of microalgal species and their products and application areas.[36]

The ability of algal organisms to concentrate proteins, starch or lipids upon stress may help to reduce the product entropy. Because of the cost of cultivation, harvesting and processing, the use of algae for energy purposes does not meet economic criteria today. These criteria can be overlooked if the algal biomass is a by-product of a wastewater treatment system or if high-value compounds (astaxanthin or β-carotene) can be extracted.[57]

8.4.1   Extraction Techniques

The extraction of chemicals from different micro- and macro-algae requires different techniques, depending on the algae species.

Table 8.2 Products from micro-algae and application areas.

Micro-algae	Product	Application areas	Ref.
Spirulina	Phycocyanin, Biomass	Health foods, cosmetics	50
Dunaliella salina	Carotenoids, β-carotene	Health foods, food supplement, feed	51, 52
Haematococcus pluvialis	Carotenoids, astaxanthin	Health foods, pharmaceuticals, feed additives	52
Odontella aurita	Fatty acids	Pharmaceuticals, cosmetics, baby food	36
Porphyridium cruentum	Polysaccharides	Pharmaceuticals, cosmetics, nutrition	53
Isochrysis galbana	Fatty acids	Animal nutrition	36, 54
Phaeodactylum tricornutum	Lipids, fatty acids	Nutrition, fuel production	55, 56

For example, the cell membrane of an alga can be very hard or very elastic. If the cellular walls have to be broken to increase the extraction yield, liquid nitrogen (183 K) can be used. This will obviously increase the processing cost.

The techniques available to produce chemicals from biomass include solvent extraction. Organic solvents (which may be toxic to animals and humans) are very often used, with the drawback of the possible retention of the solvent. Supercritical carbon dioxide (scCO$_2$) is not toxic and can be very suitable to replace organic solvents.[58] The production costs are of the same order of magnitude as those of classical processes based on organic solvents. For such techniques, a relatively anhydrous material is recommended (water content below 5%). All the bench-scale scCO$_2$ has so far been used for the extraction of paraffinic and natural waxes from *Botryococcus* and *Chlorella*, strong antioxidants (astaxanthin, β-carotene) from *Chlorella* and *Dunaliella* and linolenic acid from *Arthrospira*. Owing to the moderate operating temperatures and the inert nature of CO$_2$, degradation of the product extracted is prevented. The possibility of significantly varying the CO$_2$ solvation power by changes in pressure adds operational flexibility to the scCO$_2$ extraction process that no other extraction method can claim.[59]

Other methods[60] include the use of methanol, which can be added as a co-solvent in order to increase the extraction yield, mechanical extraction (ultrasound, disintegration, mechanical shear, *etc.*), which can be used to extract starch/cellulose from cells, and enzymes.

8.5 ENERGY PRODUCTS

Biodiesel, biogas, bioethanol and biohydrogen can be produced from aquatic biomass. Macro-algae, in general, are produced at lower cost than micro-algae. The energetic value of an alga cannot be stated only on the basis of its oil content but must take into account all the forms of energy that can be produced (e.g. ethanol, hydrogen gas or biogas).

The quality and composition of the biomass will suggest the best option for the biofuel to be produced. A biomass rich in lipids will be suitable for the production of bio-oil and biodiesel, whereas a biomass rich in sugars will be better suited for the production of bioethanol. Anaerobic fermentation of the biomass (sugars, proteins, organic acids) will produce biogas.

8.5.1 Bio-oil Production

Species of micro-algae rich in lipids (30–70% dry weight) are suitable for bio-oil production. In a commercial culture, what is of interest is the productivity of a pond, which means the production per unit time. Often a fast-growing alga is not the most specialised in lipid production.

Table 8.3 compares the volume of oil per hectare per year for different type of biomass.[61,62] Macro-algae, in general, produce a lower content of lipids than do micro-algae, with greater variability[63] depending on the cultivation technique and on the time of year when they are havested.[64,65]

Table 8.3 Amounts of oil in various types of biomass.

Biomass	Yield/$L\ ha^{-1}\ y^{-1}$
Corn	170
Safflower	785
Sunflower	965
Coconut	2840
Palm	6000
Micro-algae	47 250–142 000

Table 8.4 Companies currently engaged in macro-algae to fuels production.

Company	Activity
Seaweed Energy Solutions	Offshore cultivation and conversion to biogas and bioethanol
Green Gold Algae and Seaweed Sciences	Land-based ponds and conversion to ethanol
Butamax Advance Fuels, DuPont, Bio Architecture Lab, Statoil	Offshore kelp cultivation and conversion to ethanol and butanol
Seambiotic	Land-based ponds and CO_2 absorption of power plant flue gases
Blue Sun Energy	Jet fuel production
Holmfjord	Biofuel production from seaweed
Oil Fox	Biodiesel from seaweed

Several companies are engaged in producing fuels from macro-algae (Table 8.4), but as they have low lipid levels, production of biofuels from macro-algae is expected to depend on conversion of carbohydrate feedstocks, rather than extraction of energy-rich oils that can be processed to biodiesel or hydrocarbons.

8.5.1.1 The Quality of Bio-oil. Algal biomass thermally processed to produce an oily product also produces highly acidic compounds not suited for use as fuels. Lipids contain more than a single type of fatty acid (FA), with different numbers of unsaturated bonds that determine the usability of the oil as a fuel. A biodiesel with good combustion properties must have only one unsaturated bond in the molecule.[66] Hydrogenation treatment can be used to reduce the unsaturation number[67] and produce a better quality fuel. It has been found that the degree of unsaturation may increase with the concentration of CO_2 bubbled in the pond/PBR.[68,69]

Bio-oil, as extracted, can be converted into biodiesel through a transesterification reaction (Scheme 8.3) in order to increase the low heating value (LHV) to 36 MJ kg^{-1}.

From the environmental point of view, biodiesel is better than fossil fuel because of the reduction of carbon monoxide (CO) (50%) and CO_2 (78%) emissions;[70] and furthermore SO_2 is not emitted as biodiesel does not contain sulfur. As a result, biodiesel

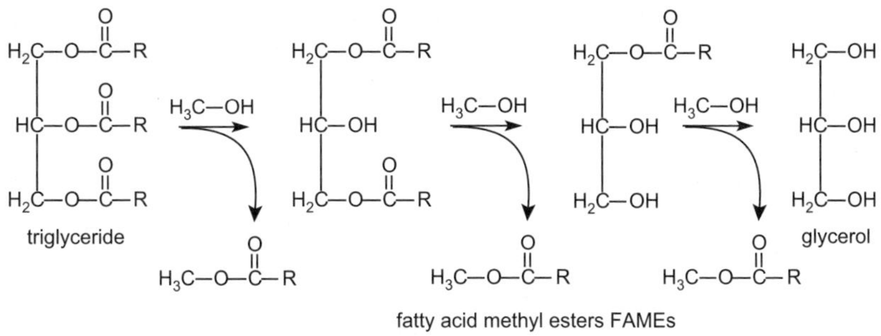

Scheme 8.3 Conversion of lipids (tryglycerides) into fatty acid methyl esters (FAMEs) (biodiesel).

Table 8.5 Fuel characteristics of different bio-oils.

Source	Density/ kg L^{-1}	Ash content/%	Flash point/°C	Pour point/°C	Cetane number	Calorific value/ MJ kg^{-1}	Ref.
Algae	0.801	0.21	98	−14	52	40	71
Peanuts	–	–	271	−6.7	41.8	–	72
Soya bean	0.885	–	178	−7	45	33.5	72
Sunflower	0.860	–	183	–	49	49	72
Diesel	0.855	–	76	−16	50	43.8	72
Biodiesel from marine fish oil	–	–	103	–	50.9	41.4	73

production is rapidly expanding, especially in Europe, the USA and Asia. A growing number of fuel stations make biodiesel available to consumers. Table 8.5 compares the fuel properties of different biodiesels.

Algal biomass has the potential to produce a number of secondary metabolites that have characteristics much closer to those of existing petroleum fuels. The most promising of these are the terpenes, which offer a potentially new fuel source as an alternative to fatty acids that is compatible with our existing fuel framework. Terpenes are polymers of isoprene (C_5H_8). If produced in high quantities, hemiterpenes (single isoprene units) and monoterpenes ($C_{10}H_{16}$) could function in current gasoline engines with only minor engine modifications.[74] Longer isoprene chains could be used for biodiesel or cracked into biogasoline or jet fuels. To date, terpene accumulation for fuels has not been a major focus of algal biology, since many terpenes and terpene derivatives have properties that make them more valuable for other uses, such as flavours, fragrances and antibiotics.[75,76]

8.5.1.2   Other Applications of Fatty Acids Present in Aquatic Biomass.   It well known that fish such as tuna, salmon, mackerel, herring and sardines are common sources of long-chain polyunsaturated fatty acids (PUFAs). Unfortunately, safety issues have been raised about using these chemicals because of the possible accumulation of toxins in fish[77] and also for other reasons such as a typical fishy smell, unpleasant taste and poor oxidative stability.[78–80] Therefore, it is logical to consider micro-algae as potential sources of PUFAs.[81] The PUFAs of particular interest are γ-linolenic acid (GLA), arachidonic acid (AA), eicosapentaenoic acid (EPA) and docosahexaenoic acid (DHA). The extraction of DHA from micro-algae has several advantages, which include the absence of contaminants, making it is suitable for pregnant women, infants and vegetarians, and the possibility of its extraction being less expensive that the extraction of DHA from fish. Micro-algae which are rich in DHA include autotrophic (*Isochrysis*, *Pavlova*) or heterotrophic algae (*Crypthecodinium*, *Schizochytrium*, *Thraustochytrium*, *Aurantiochytrium*) and several companies have begun to commercialise DHA oil extracted from micro-algae, as reported in Table 8.6.

8.5.2   Production of Bioalcohol

The production of bioethanol from biomass, using an alternative method to the corn route, is a leading research theme today. Aquatic biomass, poor in cellulose and lignin, contains variable

Table 8.6   Companies commercialising DHA oil extracted from micro-algae.

Company	Algal species	Applications
Martek (USA)	*Crypthecodinium cohnii*	Infant milk, feed
OmegaTech (USA)	*Schizochytrium* sp.	Health foods, feed
Bio-Marine (USA)	*Schizochytrium* sp.	Feed
Advanced BioNutrition (USA)	*Schizochytrium* sp.	Feed
Nutrinova (Germany)	*Ulkenia*	Health foods

Table 8.7 Starch contents of some micro-algae.

Algal strain	Starch (g/dry weight)/%	Ref.
Chlorella	12–37	86, 87, 88
Chlamydomonas reinhardtii	17–53	86, 87, 89
Scenedesmus	7–24	88, 90
Spirulina	37.3–56.1	91

quantities of simple sugars and starch. Therefore, ethanol can be obtained by fermentation of aquatic biomass with other alcohols. Micro-algae such as *Chlorella*, *Chlamydomonas*, *Scenedesmus* and *Spirulina*, rich in starch and glycogen,[82,83] can be used to produce bioethanol (Table 8.7).[84,85]

Similarly, macro-algae can be used for ethanol production by converting their storage material into fermentable sugars.[92] Bioethanol can be obtained after extraction of oil from the residue which contains fermentable sugar. Finally, purification processes are required to obtain ethanol at concentrations >95%.[93] Algae can accumulate cellulose as carbohydrate which is present in the cell wall. The cellulosic biomass can be hydrolysed enzymatically using cellulase enzyme and converted into simple sugars that can be easily fermented to ethanol.

8.5.3 Production of Biogas and Biomethane

Biogas, a mixture of methane and CO_2, can be produced from aquatic biomass. The CH_4:CO_2 ratio depends on the feed and the operating conditions.[94,95] Reactors of different types can be used with both mesophilic microorganisms and thermophilic bacteria.[96] Literature data indicate that some species of micro-algae are suited for biogas production.[97]

Biomethane can be produced by aquatic biomass with high potential compared with terrestrial biomass as the growth rates of marine macro-algae exceed those of land plants. It was found that marine algae, such as *Gracilaria* sp. and *Macrocystis* (kelp), produce methane with a good yield (0.28–0.4 m^3 kg^{-1})[98] considering that some waste sources produce 0.54 m^3 kg^{-1} of methane.[99]

The production costs are not low, but biomethane production could be made profitable if one considers the possibility of producing gas from the digestion of the residues after extraction.

8.5.4   Production of Materials

As already stated, micro-algae contain oil that can be used for biodiesel production whereas macro-algae contain sugar or cellulose that can be used for ethanol production. The remaining part (almost 50–60%) of the biomass remains useful for further purposes such as in pharmaceuticals and nutraceuticals, pigments, proteins and new materials.

8.5.4.1   Pharmaceuticals and Nutraceuticals.   Different species of micro-algae are used for pharmaceutical and nutraceutical purposes. *Chlorella*, for example, produced by several companies (the largest producer is Taiwan Chlorella Manufacturing, which produces $400 \, t \, y^{-1}$ of dry algal biomass),[86] is reported to have therapeutic effects such as healing of gastric ulcers and wounds and relieving constipation, together with preventive action against both atherosclerosis and hypercholesterol and anti-tumour activity. The most important active substance is β-1,3-glucan, which is believed to be an active immune stimulator, a free radical scavenger and a reducer of blood lipids.

Other micro-algae are used in human nutrition. in particular *Spirulina* (*Arthrospira*), because of it high protein content and excellent nutrient value;[86] it is considered to be a valuable source of essential fatty acids such as linolenic acid that cannot be synthesised in any other way.[93] Unfortunately, the validations of many of these claims is still awaited.[86,100–103] Of interest is the production from micro-algae of the so called 'nutraceuticals', food supplements with claimed nutritional and medicinal benefits. Many companies are producing 'nutraceuticals' from *Spirulina*. The largest enterprise is Earthrise Farms in California (owned by DIC, Japan), covering over 444 000 m^2 and producing algal tablets and powder sold in over 20 countries.[104] Cyanotech, in Kona, Hawaii, produces a powder named Spirulina Pacifica.[105] The market for dried *Spirulina* was estimated to be US$40 million in 2005.[96]

8.5.4.2   Pigments.   According to their chemical structure, algal pigments can be divided into different categories such as chlorophyll *a* and *b* (chlorins), chlorophyll *c* (porphyrins), phycobilipigments (open tetrapyrroles) and carotenoids (polyisoprenoids with terminal cyclohexane rings; carotenes and xanthophylls) (Figure 8.3).

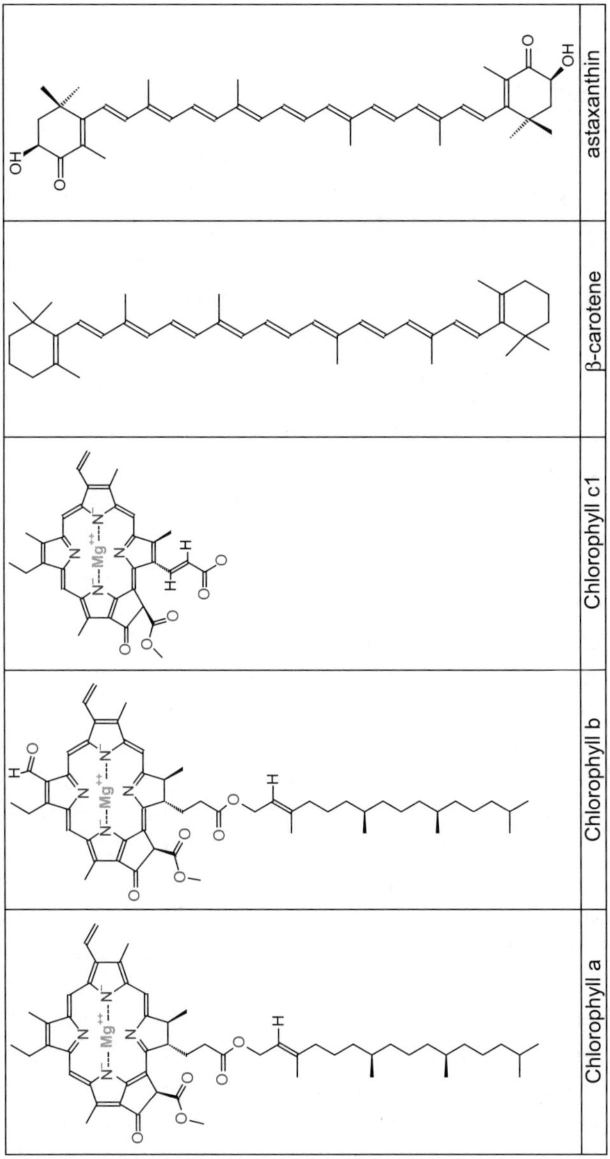

Figure 8.3 Structures of some pigments.

The pigment composition of an algal species is not entirely uniform; for example, within *Cyanobacteria* the most abundant are chlorophyll *a*, phycobilins and carotenoids, but the cyanobacterial carotenes include ε-carotene, lycopene, γ-carotene and β-carotene and xanthophylls include astaxanthin, canthaxanthin, β-crypto-xanthin, echinenone, myxoxanthophyll and oscillaxanthin.[106] The most commercially valuable groups of pigments are carotenoids (β-, ε- and α-carotenes, xanthophylls) and phycobilins with several available products derived mainly from green micro-algae (e.g. *Dunaliella*) and *Cyanobacteria* (e.g. *Spirulina*).[107,108]

Carotenoids are generally powerful antioxidants[109] and have reported anti-cancer activity.[110,111] Essentially they are used as food colorants and as supplements for human and animal feeds.[83,109] The percentage of carotenoids in most algae is only 0.1–2%, but it is possible to increase the amount to 14% by using the optimum conditions of high salinity and light intensity (as happens with *Dunaliella* for the production of β-carotene).[58,86,112,113] The major producer of β-carotene is Cognis Nutrition and Health, whose farms cover 800 ha in Western Australia.

Lutein, zeaxantin and canthaxantin are also used commercially, *e.g.* in poultry feed, and astaxanthin is used as a colorant in aquaculture[36] and also in human health and nutrition.[114,115] Astaxanthin can be used in fish farming and as a dietary supplement or antioxidant. It can be produced by *Haematococcus*, a freshwater alga that normally grows in puddles, birdbaths and other shallow fresh water depressions. *Haematococcus* can contain up to 3% astaxanthin, but it requires a two-stage culture process which is not suited to open pond cultivation. The first stage of the process is designed to optimise algal biomass (green, thin-walled flagellated stage with optimum growth at 22–25 °C) and the second stage (thick-walled resting stage) under intense light and nutrient-poor conditions during which astaxanthin is produced.[86] Although natural astaxanthin is expensive, its application is preferred in, for example, carp, chicken and red sea bream diets owing to enhanced natural pigment deposition, regulatory requirements and consumer demand for natural products.[86] Commercial production is being carried out in Hawaii, India and Israel, where Algatech sell a cru-shed *Haematococcus* biomass on the pharmaceutical market.[86,116,117]

Phycobiliproteins (phycoerythrin, different phycocyanins and allophycocyanins) from red algae and *Cyanobacteria* have a long tradition of use as dyes in food and cosmetics and as fluorescent markers in biomedical research.[106,118,119]

8.5.4.3 Proteins. Proteins are present in algae in a diverse range of forms and cellular locations, *e.g.* as a component of the cell wall, as enzymes and bound to pigments and carbohydrates. Green algal cell walls contain hydroxyproline-rich glycoproteins, one of the few protein groups to contain high levels of hydroxyproline.[120]

Their potential is demonstrated by the fact that in several cases extracts from a single species have been reported to have multiple bioactivities, *e.g.* pepsin digests from *Porphyra yezoensis* have been reported to have anti-mutagenic, blood sugar-reducing, calcium precipitation inhibition, cholesterol-lowering, antioxidant and improved hepatic function activities.[121]

Algal proteins are of importance as a source of dietary protein and essential amino acids, and also for the bioactive potential of specific lectins, enzymes and protein derivatives such as peptides. Macro-algae contain relatively high protein levels (10–47% dry weight), and in particular are rich in aspartic and glutamic acids,[122] but their presence is strictly dependent on the season, *e.g.* glutamic acid, serine and alanine were found to be present in high concentration in *Palmaria palmata* harvested in late winter/early spring but absent at other times of the year.

Algal biomass contains proteins, phycobiliproteins and lectins, with great bioactive potential related to antibiotic, mitogenic, cytotoxic, anti-inflammatory, anti-adhesion, anti-HIV, pain sensitivity reduction, human platelet aggregation inhibition and anti-cancer, anti-microbial, anti-viral and anti-hypertensive properties.[123]

8.5.4.4 New Materials. Of particular interest is the utilisation of synthetic biology and genomics to enhance the productivity and increase the utility of algae to produce advanced plastics and chemicals from biopolymers. The productivity of aquatic biomass can be improved through genetic engineering of the (micro)organism. An alternative is to generate new plants through the inclusion of bacterial genes in the plant so that the latter may produce new materials that are extracted and processed. In this way, it should be possible to produce plastic materials with new properties and meeting economic standards.

This development requires a multidisciplinary team. Such materials would have the advantage of being derived from renewable resources, fully biodegradable or compostable and produced through eco-friendly synthesis. They would be characterised by high processability with good mechanical properties and low cost. The engineering of such new materials would take profit from a possible functionalisation *in vivo*, introducing in the molecular structure functional groups that would be useful for interactions with other fibres or for making materials with defined properties such as thermal stability or brittleness. The production of copolymers using only biomass-derived monomers or a mix of fossil fuel-derived and biomass-derived monomers is also an interesting approach. Several polymers have been produced, including collagen, gelatine, alginates, casein, elastin and zein, proteins used in the production of medical textiles.

The market for bioplastics is currently a tiny niche in the global plastics market but is expected to double in the near future as rising oil prices and environmental regulations reduce the number of petroleum-based products. The various plastics that can be made from algal feedstock include the following: (i) hybrid plastics obtained by adding denatured algal biomass to petroleum-based plastics such as polyurethane and polyethylene as fillers (filamentous green algae are suited for this use); (ii) cellulose-based plastics derived from cellulose (algal strains that contain cellulose are suited as feedstocks for cellulose-based plastics); (iii) polylactic acid (PLA) where lactic acid, produced by bacterial fermentation of algal biomass, is polymerised to produce PLA; (iv) biopolyethylene, where ethanol is used to produce ethylene, can be derived from bacterial digestion of algal biomass or directly from algae. Although the above types of plastics from algae are technically feasible, their economic (cost) feasibility is still being analysed.

Alginate. As a naturally occurring biopolymer, alginate has been used successfully in the food and beverage industry as a gelling agent and colloidal stabiliser and holds strong potential in the area of drug delivery. Alginate polymers are extracted from brown algae including *Laminaria hyperborea*, *Ascophyllum nodosum* and *Macrocystis pyrifera*[124] and consist of linear, unbranched polysaccharides with acid residues of 1,4′-linked-β-D-mannuronic acid (M) and α-L-gluronic acid (G) residues (Figure 8.4).

Alginate matrices are very biodegradable and can be broken down under normal physiological conditions.[125]

Chitins and Chitosan. Chitin can be extracted from several bio-systems, including algae, and has anti-bacterial, anti-fungal and anti-viral properties. It is non-toxic and non-allergenic. The fibres are soft to handle and they have good breathability, smoothness and absorbency.

Chitosan is derived from chitin via deacetylation. Chitosan is fully biocompatible and non-toxic. It can be used as a support matrix, as a gel or in the form of beads and also to cover alginate and to adsorb antibiotics.

Hybrid scaffolds for tissue repair have been produced from collagen and chitin. The latter together with chitosan have great potential as materials for medical care. Figure 8.5 shows the structures of chitin, chitosan and cellulose.

Polyalkanoates. Another class of polymers extracted from algae are the polyalkanoates, synthesised from coenzyme-A thioesters of hydroxyalkanoic acids polymerised by a synthase. Such poly-mers can be made from a single monomer such as hydroxy-butyric acid, which forms a polyhydroxybutyrate (PHB). The PHB is a polyester with thermoplastic properties that is naturally occurring and produced by bacteria such as *Ralstonia eutropha* H16 and *Bacillus megaterium*. PHB is biodegradable and its pro-duction is not dependent on fossil resources, making this bioplas-tic interesting for various industrial applications. Today, the highest levels of PHB synthesis in plants with fertile offspring are obtained in the plastid of *Nicotiana tabacum*, resulting in up to 18% PHB cellular dry weight. Hempel *et al.*[126] recently demon-strated that the production of the bioplastic PHB is feasible in the diatom *Phaeodactylum tricornutum* by introducing the bacter-ial PHB pathway into the cytosolic compartment. PHB levels of up to 10.6% algal dry weight were obtained, revealing the great potential of this low-cost and environmentally friendly expression system. Of course, PHB production in *P. tricornutum* cannot at present compete with bioplastic production in *R. eutropha*, which was established commercially many years ago.

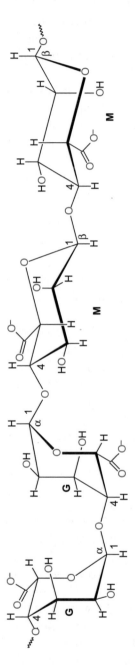

Figure 8.4   Structure of alginate.

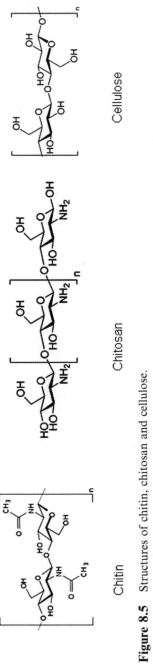

Figure 8.5 Structures of chitin, chitosan and cellulose.

Alternatively, polymers can be made from two different mono-mers such as Poly(3HB-co-3MP), a polymer with a thioester link-age made of 3-hydroxybutyrate and 3-mercaptopropionate. Tri-block copolymers can be made, for example, from caprolactam copolymerised with a hydrophilic natural oligoagarose and a polypeptide (polylysine). In this way, a large number of polymers having different properties can be built. The biogenic part can be used to modify some properties of oil-derived polymers.

Algaenan, a non-hydrolysable, insoluble biopolymer[127] isolated from a variety of unicellular algae, results from the reticulation of low molecular weight aldehydes and unsaturated hydrocarbons (long-chain, up to C_{40}). Reticulation occurs *via* acetal and ester links that build the primary structure of the polymer. Such material is produced by *Botryococcus braunii*, *Chlorella* and other micro-algae.

8.6 CONCLUSION

Aquatic biomass, *i.e.* micro-, macro-algae and plants, has a che-mical composition that depends on the species and on the growing conditions (within the same strains). To produce energy, the wild types are not always the most suitable as they can be sensitive to the culture conditions, so it is better to use a selected cultivated strain in order to have an optimum energetic yield.

In particular, aquatic biomass can be a source of several che-micals, some of which can be used to produce energy. The co-production of chemicals and fuels can be of great importance in order to make the economic balance of growing algae for com-mercial purposes a positive endeavour. In fact, the cost of fuels derived from aquatic biomass is not competitive with that from fossil fuels. The correct application of the biorefinery concept is to produce fuels at low cost together with high-value chemicals. Algae bioplastics can play a vital role as an environmentally friendly, biodegradable alternative to conventional plastics.

Algae-based biofuels are often quoted as the only plausible biofuel solution to the world's oil crisis. However, the economic viability of algae-based fuels is still in doubt. Exploring the pro-duction of non-fuel products such as bioplastics could play a major role in shaping the economics and viability of algal biofuel solutions.

REFERENCES

1. K. Nuortimo and M. Hajda, presented at the ICPQR Conference, Oulu, 2008.
2. M. Aresta, A. Dibenedetto and I. Tommasi, *Energy Fuels*, 2001, **15**, 269.
3. E. W. Wilde and J. R. Benemann, *Biotechnol. Adv.*, 1993, **11**, 781.
4. K. L. Kadam, *Energy Convers. Manage.*, 1997, **38**, 505.
5. K. Gao and K. R. McKinley, *J. Appl. Phycol.*, 1994, **6**, 45.
6. D. P. Chynoweth, *HortScience*, 2005, **40**, 283.
7. (a) Algae World 08 International Workshop and Conference, 17–18 November 2008, Singapore, http://www.cmtevents.com/eventdatas/081167/081167.pdf; (b) Green Chip Stoks, www.greenchipstocks.com (last accessed 31 March 2012).
8. D. A. Johnson and S. Sprague, presented at the Energy Conversion Engineering Conference, Philadelphia, PA, 10–14 August 1987.
9. A. Darzins, P. Pienkos and L. Edye, *Report to IEA Bioenergy Task 39*, Report T39-T2, Don O'Connor of (S&T) Consultants Inc., NREL, 6 August 2010.
10. O. Pulz, *Appl. Microbiol. Biotechnol.*, 2001, **57**, 287.
11. A. Richmond, *Handbook of Microalgal Mass Culture*, CRC Press, Boca Raton, FL, 1986.
12. M. A. Hitchings, *Fuel*, Fourth Quarter, Hart Energy Publishing, Houston, TX, 2007.
13. J. K. Bourne Jr, *Natl. Geogr. Mag.*, October 2007, p. 41.
14. K. G. Zeiler, D. A. Heacox, S. T. Toon, K. L. Kadam and L. M. Brown, *Energy Convers. Manage.*, 1995, **36**, 707.
15. S. Y. Chiu, C. Y. Kao, C. H. Chen, T. C. Kuan, S. C. Ong and C. S. Lin, *Bioresource Technol.*, 2008, **29**, 3389.
16. E. Jacob-Lopes, C. H. G. Scoparo, M. I. Queiroz and T. T. Franco, *Energy Convers. Manage.*, 2010, **51**, 894.
17. S. Collins, D. Sueltemeyer and G. Bell, *Plant Cell Environ.*, 2006, **29**, 1812.
18. E. Cecere, M. Aresta, G. Alabiso, M. Carone, A. Dibenedetto and A. Petrocelli, presented at the 10th International Conference on Applied Phycology, Kunming, 2006.
19. A. Dibenedetto and I. Tommasi, in *Carbon Dioxide Recovery and Utilisation*, ed. M. Aresta, Kluwer, Dordrecht, 2003, p. 315.

20. (a) E. Ono and J. L. Coello, in *Greenhouse Gas Control Technologies*, Special Issue dedicated to GHGT, 2002, pp. 1503–1508; (b) E. Ono and J. L. Coello, *Biosystems Eng.*, 2007, **96**, 129.

21. M. A. Borowitzka and L. J. Borowitzka, *Micro-algae Biotechnology*, Cambridge University Press, Cambridge, 1988.

22. J. H. Ryther, J. A. DeBoer and B. E. Lapointe, *Proceedings of the 9th International Seaweed Symposium*, 1979, p. 1.

23. W. Schramm, in *Seawed Resources in Europe: Uses and Potential*, ed. M. D. Guiry and G. Blunden. Wiley, Chichester, 1991, p. 149.

24. I. Cohen and A. Neori, *Bot. Mar.*, 1991, **34**, 977.

25. H. Hirata and B. Xu, *Suisan Zoshoku*, 1990, **38**, 177.

26. J. M. Adams, J. A. Gallagher and I. S. Donnison, *J. Appl. Phycol.*, 2009, **21**, 569.

27. F. Sauze, in *Energy from Biomass*, ed. A. Stub, A. Chartier, P. Schleser and G. Schleser, Elsevier Applied Science, London, 1983, p. 324.

28. D. L. Brown and E. B. Tregunna, *Can. J. Bot.*, 1967, **45**, 1135.

29. R. G. Smith and R. G. S. Bidwell, *Plant Physiol.*, 1987, **83**, 735.

30. K. Gao, Y. Aruga, K. Asada and M. Kiyohara, *J. Appl. Phycol.*, 1993, **5**, 563.

31. M. Aresta, G. Alabiso, E. Cecere, M. Carone, A. Dibenedetto and A. Petrocelli, *VIII Conference on Carbon Dioxide Utilization*, Oslo, June 2005, Book of Abstracts, **56**, 20–23.

32. FAO, *The State of World Aquaculture*, Food and Agriculture Organization, Rome, 2006.

33. D. J. McHugh, *A Guide to the Seaweed Industry*, FAO Fisheries Technical Paper No. 441, Food and Agriculture Organization, Rome, 2003.

34. M. Aresta, A. Dibenedetto, I. Tommasi, E. Cecere, M. Narracci, A. Petrocelli and C. Perrone, in *Greenhouse Gas Control Technologies*, Special Issue dedicated to GHGT-6, Kyoto, 2002.

35. S. Dhote and S. Dixit, *Environ. Monit. Assess.*, 2009, **152**, 149.

36. O. Pulz and W. Gross, *Appl. Microbiol. Biotechnol.*, 2004, **65**, 635.

37. W. F. R. Bare, N. B. Jones and A. J. Middlebrooks, *J. Water Pollut. Control Fed.*, 1975, **47**, 153.

38. M. W. Tenny, W. F. Echelberger, R. G. Scnessler and J. L. Pavoni, *Appl. Microbiol.*, 1969, **18**, 965.
39. B. P. Nigam, P. K. Ramanathan and L. V. Venkataram, *Arch. Hydrobiol.*, 1980, **88**, 378.
40. J. Benemann, B. C. Koopman, J. R. Weissman, D. M. Eisenberg and R. P. Goebel, in Algal Biomass, ed. G. Shelef and C. J. Soeder, Elsevier/North-Holland Biomedical Press, Amsterdam, 1980, p. 457.
41. K. Kogure, U. Simidu and N. Taga, *J. Exp. Marine Biol. Ecol.*, 1981, **5**, 197.
42. H. F. Mohn, in *Algae Biomass: Production and Use*, ed. G. Shelef and C. J. Soeder, Elsevier/North Holland Biomedical Press, Amsterdam, 1980, 547.
43. O. Morineau-Thomas, P. Legentilhomme, P. Jaouen, B. Lepine and Y. Rince, *Biotech. Lett.*, 2004, **23**, 1539.
44. Mathias Symposium, University of California, Bodega Marine Laboratory, 2006, http://nrs.ucop.edu/grants/mathias_symposium/images/06_symposium.pdf.
45. G. S. Kim, M. K. Shin, Y. J. Kim, K. K. Oh, J. S. Kim, H. J. Ryu and K. H. Kim, *US Patent Application* 2010/0124774 A1.
46. J. Forro, in *Seaweed Cultivation for Renewable Resources*, ed. K. T. Bird and P. H. Benson, Elsevier, Amsterdam, 1987, p. 305.
47. W. Becker, in *Handbook of Microalgal Culture*, ed. A. Richmond, Blackwell, Oxford, 2004, p. 312.
48. C. Arnaud, *Chem. Eng. News*, 2008, 45.
49. T. D. Niehaus, S. Okada, T. P. Devarenne, D. S. Watt, V. Sviripa and J. Chappell, *Proc. Natl. Acad. Sci. USA*, 2011, **108**, 12260.
50. J. A. V. Costa, L. M. Colla and P. Duarte, *Z. Naturforsch.*, 2003, **58**, 76.
51. E. S. Jin and A. Melis, *Biotechnol. Bioprocess Eng.*, 2003, **8**, 331.
52. J. A. Del Campo, M. Garcia-Gonzales and M. G. Guerrero, *Appl. Microbiol. Biotechnol.*, 2007, **74**, 1163.
53. M. M. F. Fuentes, J. L. G. Sanchez, J. M. F. Sevilla, F. G. A. Fernandez, J. A. S. Perez and E. M. Grima, *J. Biotechnol.*, 1999, **70**, 271.

54. E. M. Molina Grima, J. A. S. Perez, F. G. Camacho, J. M. F. Sevilla and F. G. A. Fernandez, *Appl. Microbiol. Biotechnol.*, 1994, **41**, 23.
55. W. Yongmanitchai and O. P. Ward, *Appl. Environ. Microbiol.*, 1991, **57**, 419.
56. F. G. A. Acien Fernandez, D. O. Hall, E. C. Guerrero, K. K. Rao and E. M. Grima, *J. Biotechnol.*, 2003, **103**, 137.
57. T. van Harmelen and H. Oonk, in *TNO Built Environment and Geosciences*, Apeldoorn, The Netherlands, May 2006.
58. S. Singh, B. N. Kate and U.C. Banerjee, *Crit. Rev. Biotechnol.*, 2005, **25**, 73.
59. R. L. Mendes, B. P. Nobre, M. T. Cardoso, A. P. Pereira and A. F. Palavra, *Inorg. Chim. Acta*, 2003, **356**, 328.
60. F. Gaspar and G. Leeke, *J. Essential Oil Res.*, 2004, **16**, 64.
61. M. Briggs, University of New Hampshire Biodiesel Group, 2004, http://www.energybulletin.net/node/2364.
62. T. F. Riesing, Cultivating algae for liquid fuel production, 2006, http://oakhavenpc.org/cultivating_algae.htm (last accessed 31 March 2012).
63. M. Aresta, A. Dibenedetto, M. Carone, T. Colonna and C. Fragale, *Environ. Chem. Lett.*, 2005, **3**, 136.
64. S. V. Khotimchenko, *Bot. Mar.*, 2003, **46**, 455.
65. R. H. Al-Hasan, F. M. Hantash and S. S. Radwan, *Appl. Microbiol. Biotechnol.*, 1991, **35**, 530.
66. S. M. Renaud and J. T. Luong-Van, *J. Appl. Phycol.*, 2006, **18**, 381.
67. *European Directive EN 14214:* "Automotive fuels - Fatty acid methyl esters (FAME) for diesel engines - Requirements and test methods", http://www.novaol.it/novaol/export/sites/default/allegati/EN14214.pdf.
68. F. X. Fu, M. E. Warner, Y. Zhang, Y. Feng and D. A. Hutchins, *J. Phycol.*, 2007, **43**, 485.
69. T. Andersen and F. Andersen, *Aquat. Bot.*, 2006, **84**, 267.
70. A. Ben-Amotz, J. E. W. Polle and D. V. Subba Rao (eds), *The Alga Dunaliella: Biodiversity, Physiology, Genomics and Biotechnology*. Science Publishers, Enfield, NH, 2008.
71. K. Vijayaraghavan and K. Hemanathan, *Energy Fuels*, 2009, **23**, 5448.
72. G. Knothe, R. O. Dunn and M. O. Bagby, in Fuels and Chemicals from Biomass, ACS Symposium Series,

Vol. 666, American Chemical Society, Washington, DC, 1997, p. 172.
73. C. Y. Lin and R. J. Li, *Fuel Proc. Technol.*, 2009, **90**, 130.
74. B. Sialve, N. Bernet and O. Bernard, *Biotechnol. Adv.*, 2009, **27**, 409.
75. J. D. Connolly and R. A. Hill, *Nat. Prod. Rep.*, 2008, **27**, 79.
76. J. Kirby and J. D. Keasling, *Annu. Rev. Plant Biol.*, 2009, **60**, 335.
77. K. E. Apt and P. W. Behrens, *J. Phycol.*, 1999, **35**, 215.
78. M. Certik and S. Shimizu, *J. Biosci. Bioeng.*, 1999, **87**, 1.
79. E. E. M. Luiten, I. Akkerman, A. Koulman, P. Kamermans, H. Reith, M. J. Barbosa, D. Sipkema and R. H. Wijffels, *Biomol. Eng.*, 2003, **20**, 429.
80. R. Abril, J. Garrett, S. G. Zeller, W. J. Sander and R. W. Mast, *Regul. Toxicol. Pharm.*, 2003, **37**, 73.
81. Y. Jiang, F. Chen and S. Z. Liang, *Process Biochem.*, 1999, **34**, 633.
82. M. Huntley and D. G. Redalje, *Mitig. Adapt. Strat. Global Change*, 2007, **12**, 573.
83. J. N. Rosenberg, G. A. Oyler, L. Wilkinson and M. J. Betenbaugh, *Curr. Opin. Biotechnol.*, 2008, **19**, 430.
84. J. Sheehan, T. Dunahay, J. Benemann and P. A. Roessler, NREL close out report, NREL/TP-580-24190, 1998.
85. R. P. John, G. S. Anisha, K. M. Nampoothiri and A. Pandey, *Biores. Technol*, 2011, **102**, 186.
86. P. Spolaore, C. Joannis-Cassan, E. Duran and A. Isambert, *J. Biosci. Bioeng.*, 2006, **101**, 87.
87. A. Hirano, R. Ueda, S. Hirayama and Y. Ogushi, *Energy*, 1997, **22**, 137.
88. S. Rodjaroen, N. Juntawong, A. Mahakhant and K. Miyamoto, *J. Nat. Sci.*, 2007, **41**, 570.
89. M. S. Kim, J. S. Baek, Y. S. Yun, S. J. Sim, S. Park and S. C. Kim, *Int. J. Hydrogen Energy*, 2006, **31**, 812.
90. R. Harun, M. K. Danquah and G. M. Forde, *J. Chem. Technol. Biotechnol.*, 2010, **85**, 199.
91. I. M. Rafiqul, A. Hassan, G. Sulebele, C. A. Orosco, P. Roustaian and K. C. A. Jalal, *J. Biol. Sci.*, 2003, **6**, 648.
92. R. Ueda, S. Hirayama, K. Sugata and H. Nakayama, *US Patent* 5 578 472, 1996.
93. J. M. Adams, J. A. Gallagher and I. S. Donnison, *J. Appl. Phycol.*, 2009, **21**, 569.

94. A. Demirbas, *Energy Convers., Manage.*, 2001, **42**, 1357.
95. P. S. Nigam and A. Singh, *Prog. Energy Combust. Sci.*, 2011, **37**, 52.
96. M. Aresta, M. Narracci and I. Tommasi, *Chem. Ecol.*, 2003, **19**, 451.
97. A. Vergara-Fernández, G. Vargas, N. Alarcón and A. Velasco, *Biomass Bioenergy*, 2008, **32**, 338.
98. K. T. Bird, D. P. Chynoweth and D. E. Jerger, *J. Appl. Phycol.*, 1990, **2**, 207.
99. D. P. Chynoweth, C. E. Turick, J. M. Owens, D. E. Jerger and M. W. Peck, *Biomass Bioenergy*, 1993, **5**, 95.
100. American Cancer Society, *Chlorella*, American Cancer Society, Atlanta, GA, 2007, http://www.cancer.org/docroot/ ETO/content/ETO_5_3X_Chlorella.asp (last accessed 31 March 2012).
101. US Department of Health and Human Services, *Dietary Fads and Frauds*, US Department of Health & Human Services, Washington, DC, 2001, http://profiles.nlm.nih.gov/NN/B/C/ R/R/_/nnbcrr.pdf (last accessed 31 March 2012).
102. S. Singh, K. N. Bhushan and U. C. Banerjee, *Crit. Rev. Biotechnol.*, 2005, **25**, 73.
103. R. Henrikson, "*Earth food spirulina: how this remarkable blue-green algae can transform your health and our planet*", Published 1989 by Ronore Enterprises in Laguna Beach, Calif, ISBN 10 0962311103, pp 174.
104. Earthrise Nutritionals LLC, 2004, http://www.earthrise.com/ company.asp?page=page5.html (last accessed 31 March 2012).
105. Cyanotech Corporation, *Hawaiian Spirulina Pacifica*, 2007, http://www.cyanotech.com/spirulina.html (last accessed 31 March 2012).
106. R. Prasanna, A. Sood, S. Jaiswal, S. Nayak, V. Gupta, V. Chaudhary, M. Joshi and C. Natarajan, *Appl. Biochem. Microbiol.*, 2010, **46**, 119.
107. G. Chaneva, S. Urnadzhieva, K. Minkova and J. Lukavsky, *J. Appl. Phycol.*, 2007, **19**, 537.
108. R. A. Prasanna, A. Sood, S. Suresh, S. Nayak and B. D. Kaushik, *Acta Bot. Hung.*, 2007, **49**, 131.
109. N. M. Sachindra, E. Sato, H. Maeda, M. Hosokawa, Y. Niwano, M. Kohno and K. Miyashita, *J. Agric. Food Chem.*, 2007, **55**, 8516.

110. M. Hosokawa, M. Kudo, H. Maeda, H. Kohno, T. Tanaka and K. Miyashita, *Biochim. Biophys. Acta*, 2004, **1675**, 113.
111. H. Nishino, M. Murakoshi, T. Ii, M. Takemura, M. Kuchide, M. Kanazawa, X. Y. Mou, S. Wada, M. Masuda, Y. Ohsaka, S. Yogosawa, Y. Satomi and K. Jinno, *Cancer Metastasis Rev.*, 2002, **21**, 257.
112. E. W. Becker, Micro-algae. *Biotechnology and Microbiology, Cambridge*: Cambridge University Press, Cambridge, 1994.
113. R. H. Wijffels, *Trends Biotechnol.*, 2007, **26**, 26.
114. G. Hussein, U. Sankawa, H. Goto, K. Matsumoto and H. Watanabe, *J. Nat. Prod.*, 2006, **69**, 443.
115. C. Vilchez, E. Forjan, M. Cuaresma, F. Bedmar, I. Garbayo and J. M. Vega, *Mar. Drugs*, 2011, **9**, 319.
116. Algatech, *Astaxanthin – The Algatech Story*, 2004, http://www.algatech.com (last accessed 31 March 2012).
117. M. A. Borowitzka, *Biotechnological and Environmental Applications of Micro-algae*, Murdoch University, 2006, http://www.bsb.murdoch.edu.au/groups/beam/BEAM-Appl0.html (last accessed 31 March 2012).
118. N. Eriksen, *Appl. Microbiol. Biotechnol.*, 2008, **80**, 1.
119. S. Sekar and M. Chandramohan, *J. Appl. Phycol.*, 2008, **20**, 113.
120. I. B. Gotelli and R. Cleland, *Am. J. Bot.*, 1968, **55**, 907.
121. P. Harnedy and R. J. FitzGerald, *J. Phycol.*, 2011, **47**, 218.
122. J. Fleurence, *Trends Food Sci. Technol.*, 1999, **10**, 25.
123. C. Fitzgerald, E. Gallagher, D. Tasdemir and M. Hayes *J. Agric. Food Chem.*, 2011, **59**(13), 6829–6836.
124. O. Smidsrod and G. Skjak-Braek, *TIBTECH*, 1990, **8**, 71.
125. W. R. Gombotz and S. F. Wee, *Adv. Drug Rev.*, 1998, **31**, 267.
126. F. Hempel, A. S. Bozarth, N. Lindenkamp, A. Klingl, S. Zauner, U. Linne, A. Steinbüchel and U. G. Maier, *Microbial Cell Factories*, 2011, **10**, 81.
127. E. W. Tegelaar, J. W. de Leeuw, S. Derenne and C. Largeau, *Geochim. Cosmochim. Acta*, 1989, **53**, 3103.

CHAPTER 9

Chemicals from Sugarcane

FERNANDO GALEMBECK,*[1,2] GABRIELA ALVES MACEDO[3] AND YARA CSORDAS[1,4]

[1] Institute of Chemistry, University of Campinas, Campinas, SP 13083-970, Brazil; [2] National Nanotechnology Laboratory, National Center for Energy and Materials Research, Campinas, SP 13083-970, Brazil; [3] Faculty of Food Engineering, University of Campinas, Campinas, SP 13083-970, Brazil; [4] Chem-Trend Southern Hemisphere, Valinhos, SP, Brazil
*Email: fernagal@iqm.unicamp.br

9.1 INTRODUCTION

The growing of sugarcane is an effective way of removing carbon dioxide from the atmosphere and using it to build complex molecules which can then be used by industry as raw materials for the synthesis of value-added and useful chemicals.

Over the past five centuries, sugar has been an important commodity in South America and in the Caribbean and has had a strong influence on both the economic and political history of the region. It was accompanied by liquor production – rum and cachaça – and later ethanol, obtained by fermentation of the mother liquor from sucrose crystallisation. A factory producing motor fuel based on ethanol from sugarcane – with 10% diethyl ether and 1% castor

Materials for a Sustainable Future
Edited by Trevor M. Letcher and Janet L. Scott
© The Royal Society of Chemistry 2012
Published by the Royal Society of Chemistry, www.rsc.org

oil – operated in north-eastern Brazil during the 1920s, until low oil prices forced it to close.

The first sugarcane plantation for the production of chemicals in Brazil was established by Rhodia in the Campinas area (São Paulo state) in 1942; the chemicals produced were diethyl ether, ethyl chloride, acetic acid, cellulose acetate and many other related products. Other companies produced solvents such as butyl and amyl chlorides and acetates and a thriving ethanol-based chemical industry existed in the 1960s–1970s in Brazil. The oil crisis of the 1970s led to a renewed interest in ethanol as a fuel,[1] resulting in a steep increase in production and, together with the concurrent decrease in price of producing the ethanol, it became clear by 2002 that ethanol from sugarcane was not only competitive but even advantageous compared with fuels derived from oil.[2] This was soon noticed by the chemical industry and by 2005 it was clear that ethanol was also a competitive raw material, provided that oil prices remained above US$50–60 per barrel.

Today, sugarcane provides inputs for the food, fuel, electricity, chemical and materials industries. Depending on market conditions, sugarcane producers can change the balance of their output; this is becoming more and more complex since many new products are being offered on the market or are under development.

A major advantage of growing sugarcane is the high productivity of the biomass. Moreover, sugarcane shows a positive response to the increased CO_2 atmospheric concentration and that can make a further positive contribution to its currently high productivity.[3]

9.2 CHEMICALS FROM SUCROSE AND ETHANOL

In 2010–2011, sugarcane crushing in Brazil was projected at 660 Mt (660 million tonnes) and sugar production was forecast to increase to 40.7 Mt, while ethanol production for 2009–2010 was in the 30 GL (30 billion litres) range. Both sucrose and ethanol are thus obtained in large amounts with attractive prices and they are both easily converted to many other chemicals. As a result, they are listed as major platform chemicals.[4] The development of sucrose as a raw material was already noted in the mid-twentieth century, benefiting from the technology and scale advantages derived from its use as a food. Research and development by consulting groups such as the Sugar Research Foundation[5] have resulted in an intensified industrial activity over the past 10 years; this has been driven by the perceived value of sucrose in Green Chemistry processes. Large and potentially interesting markets were identified (*e.g.* surfactants, plastics, polysaccharides) that could benefit from the low cost, high purity, availability, safety and easy storage/transportation properties of sucrose as compared with oil and first- and second-generation petrochemicals. This stimulated fundamental and applied studies on sucrose chemistry, leading to an array of products that are now found in global and regional markets. Sucrose is chemically interesting for its complex structure and for the diverse reactivity of its many hydroxyl groups, especially in enzymatic reactions. It is a versatile chemical building block and is also a source of the reducing sugars, glucose and fructose, which can be interconverted, fermented or chemically transformed into many other products.[6,7]

9.2.1 Esters from Sucrose and Molasses

Sucrose esters (or sucroesters)[8,9] are polyester polyols,[10] which are usually synthesised in a neutral or weakly basic medium by esterification[11] of sucrose with chlorides or anhydrides of short-chain carboxylic acids (C_4–C_8) or by transesterification of sucrose with fatty acids (C_8–C_{22}). These esters cannot be produced by direct synthesis or by esterification in acidic media because under these pH conditions sucrose tends to invert.

The stereochemical configuration of sucrose allows the formation of one to eight ester groups; the extent of the esterification reaction is controlled by the acid to sucrose mole ratio. Furthermore, the different hydroxyl groups in the sucrose molecule have different reactivities, resulting in the formation of stereoisomers during the esterification reaction. The yield of a preferred isomer[12] can be increased by judicious control of the sucrose particle size, reaction temperature, pH, catalyst selection and the concentration of reactants.

Animal fat and vegetable oils can be used directly as reagents in the transesterification reaction with significant economic and logistical advantages. Owing to the distribution of different fatty acids in each of the oils, the transesterification process can generate a mixture of sucroesters that requires further fractionation to reach any desired product composition. Corn, palm, safflower, canola, sunflower, coconut and soybean oils and also animal tallow or lard are used by the chemical industry before or after purification and refining of the oil by clarification, isomerisation or hydrogenation.[13,14]

Besides the advantages described above, the production of sucroesters brings other valuable benefits to replace successfully non-ionic surfactants and emulsifiers based on petrochemical-sourced hydrocolloids or polymers [especially poly(vinyl alcohol)] in traditional applications (food, textiles, chemicals and paper), since:

- their production costs can be competitive when using abundant vegetable oils as raw materials (*e.g.* soybean oil);
- they are biodegradable under aerobic and anaerobic conditions;
- unlike cationic or anionic surfactants, they do not form complexes or aggregates with counter-ions or alkali metals, which usually affect the performance of surfactants;
- they are solids or viscous liquids that crystallise at temperatures slightly below room temperature and can thus be processed in different particle sizes for incorporation into powders, polymers and composites;
- the formation reactions can be made in aqueous media;
- current production routes allow the adjustment of the hydrophilic/lipophilic balance (HLB),[15] which is not significantly altered by variations in temperature (as in ethoxylated surfactants) and concentration but is regulated by the degree of substitution and the fatty acids selected as raw material; the alkyl chain length and the number of double bonds define specific surfactant characteristics and especially its emulsifying properties;[16]
- the lauric, myristic, palmitic and stearic sucrose monoesters have a critical micellar concentration (CMC) much lower than those of the current main commercial non-ionic surfactants.

Sucroesters can also be prepared by the direct esterification route following acid prehydrolysis of sucrose or its modification by enzymes, including oxidation of hydroxyl bonds and isomerisation. These sucroesters are used as raw materials for the production of polymers and other complex molecules via condensation,

Grignard, sulfonation, amination and Raney nickel hydrogenation reactions. Sulfonated sucroesters are non-toxic, biodegradable, non-ionic detergents with antibacterial properties and are specifically designed for cosmetics and toiletries for babies and for washing delicate cloth.

The solubility of sucroesters in water is inversely proportional to the length of the alkyl chains in the corresponding carboxylic acids. Sucrose esters of fatty acids with chain lengths above C_{16} are practically insoluble in water. Sucrose esters with alkyl chain lengths between C_8 and C_{16} are non-toxic, odourless and have no after-taste. They are completely digestible by humans when their degree of esterification is ≤ 3. Their core application lies in the food industry as stabilisers for emulsions such as margarine, mayonnaise and dairy desserts, plasticisers for candies and chewing gum, emulsifiers for bakery products, surfactants for soft drinks, fruits juice and milk drinks, film-forming agents for fresh fruit covering and antibacterial agents (especially against Gram-positive bacteria) in fat and slightly acidic media (*e.g.* chocolate, dairy products, vegetables and canned fruit). Low molecular weight sucroesters have excellent tenso-active properties and they are replacing petrochemical non-ionic surfactants in their traditional applications, while gaining new markets as agents for chemical spill treatment, to minimise environmental impacts.

Short alkyl chain sucrose esters are also being studied as replacements for glycols and other polyols in antifreeze fluids, helping to reduce their toxicity when in contact with the skin, and also the biochemical oxygen demand required for their degradation. The use of sucrose esters also minimises (or eliminates) the need to add inorganic salts to these fluids at room temperatures around or slightly less than $0\,°C$, contributing to reductions in corrosion and environmental impact.

Some short-chain esters of the family, especially butyric, valeric and caproic, possess insecticidal properties,[17] thus being interesting additives for agrochemical formulations, especially those applied by spraying. Di-and trisucroesters of carboxylic acids with low HLB have been particularly valued by the flavour industry owing to their ability to act as microreactors for flavour synthesis or as flavour carriers in food and cosmetics. Mono- and disucroesters of carboxylic acids can be used as lubricants in the extrusion of PVC compounds, thus improving PVC flow properties and reducing the product losses which currently occur with traditional waxes and plasticisers used by the plastics industry.

Sucrose octaacetate (SOA), sucrose octaisobutyrate (SOIB) and sucrose diacetate hexaisobutyrate (SAIB) are high added-value products within the group of sucroesters because of their widespread consumption and application. They are produced by controlled esterification reactions of sucrose with a mixture of anhydrides (*e.g.* acetic, isobutyric) in aqueous media.

SOIB is primarily used as a weighting agent for soft drinks and emulsion beverages. It dissolves in the emulsion oil phase which contains the flavours and dyes, thus increasing the emulsion density without destabilising the emulsion. SOA is an excellent fixative of essences and flavours and is used in perfumes, nail polishes, insect repellents, sunscreens, deodorant sprays and hair products such as sprays, shampoos, dyes and smoothing and curl formulations. It unfortunately has an intensely bitter taste and as a result is not used in food applications. For this reason, SOA is often added to alcohol and other hazardous liquids to prevent poisoning by accidental ingestion.

SAIB[18,19] it is a food additive approved by the European Community in 1968 under the code E444 and it has been extensively used worldwide as a weighting agent and flavour carrier for juices, energy drinks and carbonated soda. In 1999, after their approval by the US Food and Drug Administration (FDA) as an additive for soft drinks under the classification 21CFR 172.833b, its global consumption grew rapidly at a rate of 7% by volume per year, reaching 100 000 t per year in 2007. Consumption trends for these sucroesters are strongly connected to the global supply and the market price of glycerine, another important weighting agent for emulsion beverages.

The hexa-, hepta- and octasucroesters of myristic (C_{14}), palmitic (C_{16}) and stearic (C_{18}) acids are completely insoluble in water and they present sensorial characteristics very close to those of fats in food applications. Since they cannot be processed by pancreatic lipases responsible for breaking triglycerides into smaller lipids in the human digestive system, they are expelled as such from the body. Hence pharmaceutical and food companies replaced lipids by these sucroesters in sweets, pasta sauces, bakery products, snacks and fried foods in general, especially potato chips and popcorn, in order to reduce body weight and the food level of low-density lipoprotein (LDL) cholesterol. The most widely used product in the market is olestra, an octaester also marketed under the brand Olean, developed by Procter and Gamble in the early 1970s and, after years of testing, approved by the FDA as a food additive in January 1996.[20] Olestra resembles vegetable oils in their sensorial and physicochemical properties, but with a high resistance to degradation when exposed to high temperatures. Although the Procter and Gamble patent has now expired, new applications may still be found; recent information on its possible role as an antidote to dioxins was recently disclosed.[21]

When highly polar functional groups are inserted in hexa-, hepta- and octapolysucroesters during transesterification, the esters develop lubricant properties[22] and can be applied as working fluids for metal and plastic moulded parts subjected to extreme friction and environmental variations. These modified polysucroesters are also monomers for the production of high-performance alkyd resins in the paint industry.[23]

9.2.2   Sucrose-based Food Ingredients

Although sucrose is a disaccharide whereas starch is a polysaccharide, hydrolysis of the former generates two distinct molecules without any intermediates, which can make sucrose more interesting, from a chemistry point of view, than starch. As a fermentable feedstock, sucrose is equivalent to glucose, hence there is no restriction on producing the same products as obtained from starch fermentation. Fructose can be isomerised to glucose, if required, and this is done without the simultaneous formation of dextrin. Isomerisation is also used to produce fructose syrups from starch glucose.

A comparison of starch and sucrose as sources for fructose syrups highlights the fact that sucrose can be commercialised as such or as inverted syrup containing sucrose, glucose and fructose. Sucrose syrup with nearly 100% inversion is equivalent to high-fructose corn syrup (HFCS). However, HFCS cannot substitute for sucrose (crystal sugar) in some important applications, for instance, when a dry sweetener is required or when water addition must be avoided.[24]

9.2.2.1   Sweeteners.   The global sweetener market has changed considerably in the last 10 years. In the 1970s, the market was dominated by sucrose, glucose, sorbitol, mannitol, saccharin and cyclamates. Today, fructose and fructose syrups and also new polyols hold a market share. Many non-cariogenic molecules with higher sweetening power and low caloric value have been synthesised since then.[25] The global artificial sweetener market is forecast to reach US\$1.5 billion in 2015 and is largely used to make soft drinks.[26] In 1999, aspartame was the most important sweetener, with 90% of the market, but it later suffered strong competition from acesulfame K and sucralose. Sweeteners derived from sucrose are fructooligosaccharides, isomaltulose, isomalt (not obtained by fermentation), leucrose and fructose syrups. A summary of some of their main characteristics is presented below.

The fructooligosaccharides[27] (FOSs) (1-kestose, nystose, 1-fructosilnistose) are sugars that can promote the growth of bifid bacteria that lower the intestinal pH, reduce pathogenic bacteria and are less metabolised than sucrose. They thus act as soluble fibres. The sweetening power is equivalent to 20–40% of that of sucrose but the enzymes in the digestive tract, such as invertases, β-fructofuranosidases and α-amylase, do not hydrolyse FOSs. FOSs can be found in Nature in some vegetables. Commercial production today is based on enzymatic hydrolysis of inulin (a fructose polymer) extracted from vegetables and FOSs can also be produced from sucrose by fermentation using *Aspergillus niger* or *Kluyveromyces bulgaricus*.

Isomaltulose is a sugar found in bee honey and in sugarcane molasses. It is a disaccharide from fructose and glucose with a glycoside α-1,6 link (in sucrose the link is α-1,2). It is less cariogenic than sucrose, since the bacteria in the mouth do not synthesise dextran (to form plaque) from isomaltulose. It is only slowly degraded by the human metabolism. Conversion of sucrose to isomaltulose is effected by glycosyltransferase produced by bacteria such as *Protaminobacter rubrum*, *Erwinia* sp. and *Klebsiella* sp.

Leucrose[28] is another sucrose isomer: glucose and fructose linked by an α-1,5 glycoside bond. It is a by-product of dextran production, with a sweetening power 50% lower than that of sucrose. It is also non-cariogenic. Enzymes in the human body slowly metabolise it and it is therefore recommended for diabetic patients.

Sucralose (1,6-dichloro-1,6-dideoxy-β-D-fructofuranosyl-4-chloro-4-deoxy-α-D-galactopyranoside) is a natural sweetener developed by Tate and Lyle in the 1970s[29] and is produced by adding three chlorine atoms to sucrose. It is almost 500 times sweeter than sucrose and very stable; it is commercialised in many countries today, with the brand name Splenda, and under strong industrial property protection.[30]

Fructose syrups are important sweeteners and the most commercialised are 42 HFCS and 55 HFCS, which contain 42 and 55% fructose, respectively. They are produced from starch, but they can also be derived from sucrose. The production process used by Archer Daniels Midland (ADM) (enzymatic, from corn starch) is as follows: 55 HFCS is refined from 42 HFCS, the dextrin and other saccharides are returned to the production of HFCS 42, then syrup with 80–95% fructose is produced and diluted to 55%. The 42 HFCS syrup may have crystallisation problems due to the presence of dextrin and higher molecular weight saccharides.

It is expected that the number of new products will grow in the coming years, since a large number of research projects are under way. Products from sucrose fermentation show better prospects than those obtained by extraction from various plants.

Polyols are currently used in different industrial applications. Sorbitol, maltitol, mannitol, xylitol, lactitol, erythritol and isomalt are used as sweeteners and their physicochemical properties are similar to those of common sugars. They are non-cariogenic and present low caloric values, owing to their limited transformation by the human metabolism. Sorbitol is the most commonly used polyol, typically in toothpaste and shampoos. It is also an intermediate for ascorbic acid production. The market has recently grown by 2–3% per year and this increase has come from the USA, Europe, Japan and other Asian countries. Production costs are dominated by feedstock and capital costs; with low-cost substrates, a multifunctional plant to produce different polyols can be attractive. Sorbitol and mannitol are largely available in Nature: sorbitol is found in some fruits and marine vegetation and mannitol is found in exudations of 'grey' tree, fungi and seeds. However, commercial production by extraction is not economically viable today. The usual processes are based on the hydrogenation of glucose and mannose, when sucrose or corn glucose is used as raw material. Hydrogenation of inverted sugar leads to 75% sorbitol and 25% mannitol simultaneously, yielding $\sim$0.272 Mt

$(600 \times 10^6$ lb) in 2004. Xylitol is a C_5 polyalcohol with a sweetening power equivalent to that of sucrose, found in small amounts in fruits and vegetables. It is the most often used sweetener for 'diet' chewing gums.

9.2.2.2 Amino Acids and Vitamins. Amino acid fabrication is a multi-billion dollars per year enterprise. All 20 amino acids are sold, albeit in greatly different quantities. Lysine, methionine and threonine are used as animal feed additives, monosodium glutamate, aspartic acid and serine are flavour enhancers and all of these are human nutrition supplements. Glutamic acid, lysine and methionine account for the majority of amino acids sold, by weight. Glutamic acid and lysine are made by fermentation whereas methionine is made by chemical synthesis. The major amino acid producers are based in Japan, the USA, South Korea, China and Europe.[31]

Seaweeds have been used as flavouring ingredients for centuries in Asia. In 1908, Kikunae Ikeda of Tokyo Imperial University isolated the flavour-enhancing principle from the seaweed *konbu* (also spelled *kombu*; *Laminaria japonica*, related to kelp), as crystals of monosodium glutamate (MSG). Adding MSG to meat, vegetables and just about any other type of prepared food makes it savoury, a property referred to as umami. Soon after Ikeda's discovery and recognising the market potential of MSG, Ajinomoto in Japan began to extract MSG from acid-hydrolysed wheat gluten or defatted soybeans and to market it as a flavour enhancer. Following the isolation of microorganisms able to produce L-glutamate from sugar fermentation, MSG is currently made from sucrose. The same applies to other amino acids but methionine and glycine are produced only by chemical synthesis. However, fermentation has led to more economical large-scale production for other amino acids. Biochemical pathways for the industrial production of L-glutamic acid, L-lysine and threonine are similar and the substrates are glucose or sugarcane molasses, using *Corynebacterium glutamicum*, *C. lactofermentum* or *C. flavum*. In addition to the processes based on sucrose or glucose, there are also techniques to produce L-lysine from ethanol or acetic acid, using selected genetically modified organisms (GMOs). Yields reach 45–50 g of lysine per gram of sugar and the final concentration may reach 170 g L^{-1} fermented broth. There are a large number of world producers and Ajinomoto has three sites in Brazil, using sugarcane molasses.

Some microorganisms are able to synthesise a great number of vitamins but the yields are usually low. Vitamins B_{12} and riboflavin are obtained commercially from fermentation with *Propionibacterium* and *Ashbya gossypii*, respectively, but all the other commercial vitamins are obtained by chemical synthesis, although the development of fermentative processes is being pursued in many cases. For instance, vitamin E can be produced by chemical synthesis; extraction from vegetables or fermentation and it has important uses as an antioxidant in human health and also in the plastics industry. Ascorbic acid production is an example of a hybrid process using chemical and enzymatic synthesis where the substrate can be glucose or sorbitol.

9.2.2.3 Polysaccharides. Many microorganisms produce polysaccharides from excess carbon in the growing medium. Some polysaccharides, such as glycogen, are used for energy storage and are kept within the cell; others are released to the medium and some of these have commercial importance. These macromolecules modify the rheological characteristics of a medium, increasing the viscosity; they are therefore used as gelling agents and also as emulsifiers. They can present acidic, neutral or basic characteristics and their polyelectrolyte behaviour influences their interactions with the medium. The conformation in solution depends on the ionic strength, pH and the concentration of some specific ions such as Ca^{2+}. Production of exo-polysaccharides (released by the cell to the medium) is usually done by batch fermentation under stirring and aeration. Polysaccharide biosynthesis is concurrent with the growth of the microorganism population and ceases when growth ends; the viscosity increases as the material is released, limiting production due to lower agitation and aeration efficiencies.

The carbon:nitrogen ratio in the culture medium is a fundamental parameter and the nitrogen source is a limiting factor for the growth and release of polysaccharides by the microorganism. Two well-known industrial products are xanthan and dextran. Recent research and development have resulted in new products in recent years:[32] curdlan (Pureglucan) and gellan. Pureglucan is a polysaccharide (β-glucan) produced by fermentation, with very interesting gelling and emulsifying properties. It is extensively used in Japan and throughout Asia in meat emulsions and its use is spreading to the USA and other countries, expanding its market to the US$1 billion per year range.

Xanthan[33] is produced by Gram-negative bacteria, *Xanthomonas campestris*. It is the best-known and well-characterised exo-polysaccharide, with a molar weight in the 10^6 Da range. It is a glucose polymer with glucan β-1,4 links. Acetate and pyruvic acid residues found in the product depend on the process and fermentation conditions and also on the *Xanthomonas* species employed. It is widely used in industry as a food suspension stabilising agent and in the oil industry as a lubricating additive.

Dextran[34] is an α-glucan with many branches in the main chain, depending on the type of microorganism used in the fermentation. Dextran is produced from sucrose with an extracellular enzyme from *Leuconostoc mesenteroides* and *Streptococcus* sp. Over the past 50 years, Pharmacia has produced it as a human blood plasma substitute. Today it has many other clinical applications, including thrombosis prevention. Various dextran grades are used in the food and chemical industry and also as a support for resins in chromatographic separation systems, such as the well-known Sephadex products. Prices have been strongly influenced by inexpensive Chinese products, but producers expect that by controlling the production volumes it will be possible to stabilise (and increase) prices. Market growth is now expected mainly in the food sector.

9.2.2.4 Producing Enzymes Used as Food Additives. The market for food additives is estimated at US$ 30 billion and is growing at 5% per year. The enzyme share is US$ 1.5 billion and it has been growing at 12% per year over the past 10 years. Until recently, it was largely concentrated in Europe, the USA and Japan.[35] Approximately 400 companies are involved in enzyme production worldwide, targeting the following markets: food 45%, detergents 34%, textiles 11%, leather 3%, paper and pulp 1% and other 6%. Enzymes can be produced from plants, animals and microorganisms but currently the largest proportion of enzymes used in food processing is produced by microorganisms, in controlled fermentations. The number of commercial enzymes produced by GMOs is growing rapidly, since these processes have many advantages: GMO-produced enzymes can be purified more easily than those derived from natural organisms and the productivity is higher than using extraction methods. Enzyme production processes are already well known, but intense development is continuously under way for some of the key enzymes. The substrate for enzyme production depends on the type of process adopted. For solid-state fermentation, it is possible to use agricultural by-products or wastes such as sugarcane bagasse, straw and other plant residues; for liquid fermentation, glucose, sucrose and molasses are the most often employed substrates. Research and development are aimed at new microorganisms able to produce different enzymes with new properties of industrial interest or at new applications for commercial enzymes. Large cost reductions can be expected over the next few years, depending on the efforts to bring down the costs of liquid fuel (ethanol, butanol, hydrocarbon) production from lignocelluloses for the energy market. It is possible that this development will lead to small local enzyme production facilities near the main factory, using one or more process streams from lignocellulose transformation processes.

9.3 CARBOXYLIC ACIDS

Sugarcane is a valuable source of carboxylic acid production since it offers the highest yield of sucrose per hectare planted. The existence of primary and secondary hydroxyl sites with different reactivities in the molecular structure of carbohydrates is a major advantage for their use as precursors of carboxylic acids and their derivatives *via* chemical or biochemical routes with high process conversion and low generation of by-products. Hexoses are most commonly used as feedstock for the production of carboxylic acids because they provide higher process yields than pentoses. Most processes are based on glucose but, owing to competition with fuel production, many routes may become impractical. For this reason, Research and development efforts are being channelled into exploring renewable and abundant raw materials for glucose replacement. Low-cost mono- and disaccharides such as sucrose or inverted sugar, starch and cellulose are potential glucose substitutes.

Table 9.1 presents a summary of the short-chain carboxylic acids that can be produced by means of chemical or biotransformation processes using carbohydrates and biomass as raw materials, including those derived from sugar cane.

Formic acid can be used to remove adjacent hydroxyl groups from carbohydrates (didehydroxylation) that was successfully tested using sucrose. The process is still not commercial but it offers an enormous potential for the production of olefins in sugarcane biorefineries.[36]

3-Hydroxypropionic acid (3-HP, also called hydracrylic acid) is an isomer of lactic acid and an important precursor of acrylic acids, anhydrides and esters, acrylonitrile, acrylamides and β-lactones. 3-HP was historically obtained from the acrylic acid process based on the hydrolysis of nitriles, but since this process was replaced by the oxidation of propylene, which does not produce 3-HP as an intermediate, it became commercially unavailable. Since 3-HP is a highly polar, water-soluble and highly reactive liquid, it is regaining interest as a building block for the development of new acrylic polymers and the production of 1,3-propanediol; a competitive commercial process is expected by 2013.[37,38]

The first patent on the production of succinic acid from carbohydrates *via* aerobic fermentation was published in November 2010.[39] In this process, the microorganism is grown under aerobic conditions while the production of succinic acid is carried out under aerobic or anaerobic conditions (N_2 or CO_2 atmosphere). The acid is purified using well-known procedures and the FDA approves it as 'generally recognised as safe' (GRAS) as an additive for food, cosmetics and pharmaceuticals. Owing to its higher purity compared with the petrochemical-sourced succinic acid, it is used as a comonomer in the polymerisation of special polyurethanes, polyamides and biodegradable alkyd resins and polyesters destined for the paint industry. It is also a precursor for many chemical products: tetrahydrofuran (THF), 1,4-butanediol and γ-lactones can be produced from the catalytic hydrogenation of the succinic acid. Polybutylsuccinate is a thermoplastic suitable for packaging, which replaces polypropylene (PP) and low-density polyethylene (LDPE) with many advantages.[40] Diethyl succinate is a solvent that replaces methylene chloride in many traditional applications and has been added to diesel fuel in order to reduce the amount of particulates emitted during its combustion in engines.[41] Succinic

Table 9.1 Carboxylic acids obtained from sugarcane products and other biomass.

Carboxylic acid	Process	Raw material	By-products	Status
Formic acid[42] and levulinic acid	Heavy acid hydrolysis of carbohydrates (mainly hexoses) under catalysis[43–45]	Sucrose	5-HMF	Commercial process
		Sugarcane bagasse or cellulose pulp	5-HMF, high-purity charcoal	Commercial process
Acetic acid	Aerobic fermentation of carbohydrates or biomass by *Acetobacter* or *Gluconobacter*	Sucrose solutions and sugarcane molasses	Water	Commercial process
	Aerobic fermentation of carbohydrates by a biocatalytic enzymatic process[46]	Sugarcane biomass	Water	R&D screening
	Anaerobic fermentation of carbohydrates[47–49]	Sugarcane molasses	Water	R&D screening
Hydracrylic acid[50] (3-HP)	Aerobic fermentation of glycols or carbohydrates by genetically modified organisms	Glycerol, sucrose and hexoses	3-HPA (anhydride), pyruvate, acrylic salts	Pilot scale, very low commercial volumes
Succinic acid[51]	Anaerobic[52] or aerobic fermentation of carbohydrates by natural or genetic modified organisms	Sucrose, dextrose and other hexoses, lactose, polyols, sugarcane and corn molasses, glutamate and lysine intermediates	Succinic anhydride, succinate salts, succinate esters (lactates, acetates, formates), isoamyl acid	Commercial process
Glutaric acid[53]	Aerobic fermentation of carbohydrates by *Pseudomonas*	Glucose (no routes at present from sucrose or sugarcane derivatives)	α- and β-ketoglutaric acids	R&D screening
Adipic acid[54–57]	Microbial oxidation of 1,6-hexanediol produced via 5-HMF obtained by hydrolysis of biomass	Pulp and paper, wood, corn, sugarcane, oilseeds and other fibrous crop residues	5-Hydroxymethylfurfural, 1,6-hexanediol	R&D screening

	Process	Feedstock	By-products	Status
	Chemical conversion of biomass to carbohydrates + catalytic oxidation + selective hydrogenation	Maize and sugarcane biomass, lignocellulose	Water	Commercial process
	Catalytic hydrogenation of cis,cis-muconic acid produced by aerobic fermentation of carbohydrates	Glucose, sucrose, lignocellulose, maize and sugarcane biomass, vegetable oils, short-chain alkanes	Catechols, complex pyruvate acids	Commercial process
Glucaric acid[58]	Catalytic oxidation of carbohydrates	Sucrose and other hexoses	Water and aldaric acids	Commercial process
	Aerobic fermentation of hexoses by GMO	Sucrose and other hexoses	Glucuronic acid	R&D screening
Kojic acid[59,60]	Aerobic fermentation of hexoses by *Aspergillus* or GMO	Sucrose, dextrose, lactose, polyalcohols, sugarcane and corn molasses, glutamate and lysine intermediates, biomass containing hexoses	Different by-products depending on the fungi selected as fermentation agent	Commercial process
L-Aspartic acid	Chemical synthesis catalysed by lyases	Fumaric acid + ammonia	Short-chain amino acids, D-aspartic acid	Commercial process
	Aerobic fermentation of hexoses by GMO	Sucrose	Malic acid, D-aspartic acid	R&D screening
Itaconic acid[61]	Aerobic fermentation of hexoses in acid aqueous media by *Aspergillus*	Glucose (Sucrose and other hexoses, pentoses, glycerol, sugarcane and corn biomasses also being studied)	Unconverted salts and sucrose/glucose esters	Commercial process

acid reacts with urea, ammonia or its salts, forming succinates that can be converted to succinimides which are the active base for the formulation of anticonvulsant medicines, precursors of hormones, steroids and complex proteins and are also used as cross-linking agents in the polymerisation of thermosetting plastics. Aromatic succinates are precursors for pharmaceuticals and perfumes.

Itaconic, glutaric, adipic, glucaric, L-aspartic, kojic and levulinic acid are all important chemicals, but a detailed discussion of these and other carboxylic acids sourced from hydrocarbons goes beyond the scope of this chapter.

9.4   ETHERS AND SUCROGELS

Sucrose has been used for more than a century as a raw material for the production of ethers with the general structure [nR–O–$C_{12}H_{18}O_{10}$–O–nR]. Sucrose ethers are water soluble and are stable amphiphiles, especially under mild alkaline conditions. They are good solvents and emulsion stabilisers that do not pose significant toxicity problems to humans and animals. Synergy among these properties makes them valuable surfactants and emulsifiers for pharmaceutical and cosmetic applications.[62]

The reactivity patterns of sucrose ethers also allow them to perform as very important precursors of a myriad of chemical compounds, *e.g.* ethyl cyanide ethers for aerospace and automotive parts (made by reacting acrylamide and sucrose in an aqueous alkaline medium) and formaldehyde resins (by reaction of β-starch ethyl sucrose ethers with formaldehyde).[63]

Short alkyl chain water-soluble sucroethers such as hydroxypropylsucrose are obtained from the reaction of sucrose with propylene oxide. Owing to their low viscosity, they are largely used as tasteless thickening agents for foods or as prepolymers for the production of semi-rigid polyurethanes for furniture, insulation and roofing. Long alkyl chain sucroethers (octyl and octadienyl ethers) are additives for personal care products, soaps and cosmetics.[64] Branched sucrose ethers used in detergents and in washing powders improve the removal of dirt and stains from fabrics in room-temperature washing.[65]

Polyether esters of fatty acids and sucrose can be made from sucrose ethers with degrees of esterification between 4 and 8 units.[66] They are sucrose polyether esters of C_6–C_{14} carbon chains similar to olestra and are currently used in medicines to reduce the bioavailability of unwanted lipids contributing to LDL cholesterol. They form a gel layer over the gut walls, decreasing the adsorption of lipids but without the side effects presented by olestra.[67]

The market for sucrose ethers has been historically low in volume and limited to small niche applications because they are more expensive than other substances that compete for the same applications, such as polyester polyols. Production routes for highly lipophilic and hydrophobic ethers sourced from cellulose or cheaper polysaccharides (especially dextran) are technically feasible, but the yields of these processes are much lower than the processes using polyols and disaccharides: sucrose, sorbitol, mannitol, raffinose and trehalose. Research groups are nowadays focusing on the use of sugarcane bagasse for sucroether production owing to its low cost, abundance and easy handling while not competing with food and liquid fuel production from sugarcane. These routes are still not economically feasible owing to the cost of removal of the by-products derived from lignin and hemicellulose. Moreover, they depend on highly selective catalysts.

Industrial processes for carbohydrate ethers involves the reaction of disaccharides (especially sucrose) with a bifunctional organic halide (such as a dichloride or epichlorohydrin) in the presence of a Lewis catalyst in a highly alkaline biphasic medium composed of water and a solvent (usually hexane or pentane). The specificity of the ether produced – regarding its solvation and emulsion stabilisation properties – depends on the precise control of the reaction, which occurs at the water/solvent interface. With the choice of different raw materials and catalysts and by thorough control of the concentration of reactants and the temperature and pressure of the reaction mass, it is possible to produce different functional ethers, such as totally water-soluble or insoluble polyether sucrose esters, dimers of sucrose ester ethers, cyclic sucrose esters and mainly alkyl, vinyl and alkyl benzyl ethers.[68] These ethers find applications as defoamers, viscosity control agents for agrochemical formulations, for carbonated beverages and for fruit juices, for the partial replacement or enhancement of the sweetening power of sweeteners, fat substitutes in functional foods, stabilisers for paints, lotions and different types of creams and emulsions, as a precursor of biofibres, bioplastics, bioadhesives and biofilms for food/feed packaging and medical applications, surface agents for paper and fabrics before dye processes, precursors for pharmaceutical intermediates, and plasticisers for resins and coatings.

Other manufacturing routes involve the reaction of disaccharides with epoxides. Initially, the epoxy reacts with the disaccharide in an aqueous alkaline solution and the ether intermediate reacts with the same (or a different) epoxy in a solvent-based medium in the presence of a Lewis catalyst and/or metal ions such as molybdenum, tungsten, vanadium and titanium.[69] The ether produced needs to be soluble in the solvent-based medium for the halide salts to be removed from the reaction medium by water washing. The resulting ethers from this reaction are strongly lipophilic substrates, which are mainly used as column substrates for liquid chromatography gels and other similar separation techniques.[70] This route can also be used to produce sucrose–epoxy monomers that are precursors of high-performance epoxy resins mostly known in the market as eco-epoxies. These epoxies are intended for very specific niches, such as adhesives, inks and composites for home flooring and for the production of boats and art objects.[71] This reaction is historically based on dextrose and vegetable oils (mainly from soybeans and peanuts), but recent efforts to produce economically viable eco-epoxy from commercial sucrose by using other oils and by developing new catalysts to improve the selectivity of the reaction have been reported.[72]

Sucrose ethers are precursors of sucrogels, which are polyester or polyether ester hydrocolloids produced by the reaction of sucrose with acrylates.[73] Sucrogels are capable of absorbing water at a rapid rate and at volumes up to several times their original volume. Furthermore, they can maintain a steady gel state over a wide range of pH and temperature. These gels are reversible, releasing the water when subjected to controlled drying conditions and they can be regenerated many times, over a long period time.

Sucrogels are known commercially as *carbomers* and they comprise high molecular weight branched polysaccharide acrylic acid ethers, such as the allyl ethers of sucrose, glucose and pentaerythritol. They are mainly used as thickening, dispersing, suspending and emulsifying agents used to adjust the rheology of pharmaceutical and cosmetic formulations such as creams, shampoos and liquid soaps. These ethers are largely used as a component of teeth whitening gels because they prolong the oxygen release time when added to carbamide peroxides. High molecular weight sucrogels are also replacing poly(vinyl alcohol) as an encapsulation agent for bacterial/fungicidal agrochemicals, for hormones and plant control growth formulae and for other agrochemical purposes. Since the complete removal of acrylic acid from a reaction medium is a highly costly process, the FDA approves few commercial gels as food additives. Therefore, in the food market, sucrogels suffer strong competition from other sucrose-based hydrogels such as dextran and gellan gum.

Their production process involves an enzymatic synthesis of sucrose by lipases or proteases immobilised in a solvent reaction medium that contains the acrylic acid or its derivative. Different carbohydrates can be used as starting material but there is a noted advantage with sucrose because it is a non-reducing disaccharide, which prevents the formation of by-products from side reactions.[74] The gel produced is very stable and insoluble in the reaction medium, being easily separated from the reaction medium by filtration. The main obstacle for a high conversion of the process in gels lies in the enantioselectivity of the biocatalyst, which must have the ability to select which functional groups of the sucrose should be esterified or not.

Since the sucrogel market is expanding rapidly, research efforts are being focused on the development of high molar yield processes based on sucrose fermentation. These routes are technically feasible but depend on the development of specific GMO recombinant microorganisms to improve the conversion of sucrose into the gel. A cost-effective process will also depend on the market price of acrylic acid, which is the limiting raw material since its price is historically higher than that of sucrose. An economic fermentative route to sucrogels would represent a large strategic advantage as soon as acrylic acid and its derivatives can be produced in large volumes and at a lower price by biotechnological processes.

9.5 CHEMICALS FROM BAGASSE

Sugarcane bagasse is one of the main residues from agricultural production in Brazil, constituting 25% of the overall harvested cane weight[75] and amounting to 155 Mt in 2009.[76] Bagasse is currently largely burned to produce heat that is used in the distillation of ethanol or in thermal electricity generation. Up to 30%

of the bagasse in Brazil goes to producing power for the national grid or is sold as a solid fuel.

Bagasse is also used to feed cattle, grow mushrooms[77] and make paper and cardboard. The last two applications are still on a small scale of less than 1 Mt y^{-1} (1 million tonnes per annum).[78] Its lignin content is 21.5%, which is significantly less than that of the *Eucalyptus* species which is largely used to make paper and pulp in Brazil. Moreover, the available amounts of bagasse are much higher than the overall amount of *Eucalyptus* used for making paper and cellulose in Brazil.

Bagasse is a candidate for conversion into 'second-generation' ethanol but there is some evidence that its cellulose may be less susceptible to hydrolysis than some wood celluloses.[79] On the other hand, hydrolysis of bagasse hemicelluloses produces xylose and xylitol.[80]

Bagasse cellulose can be used to produce a large number of chemicals, from polyesters to ascorbic acid and butanol,[81] and it is also a prospective source of regenerated cellulose, ethers, esters and carboxymethylcellulose. Various cellulose fractions[82,83] are used in polymer composites and in special concrete for use in areas susceptible to earthquakes. Regenerated cellulose holds a large market in the textiles industry[84] and films made using environmentally friendly processes can be important in the packaging industry.

Current production of lysine in Brazil is largely based on bagasse fermentation.[85] This process can also yield alkaloids, enriched cattle fodder, enzymes and penicillin. Lignin is also a prospective source of many chemicals and it is an excellent fuel, with a heating power equivalent to that of high-quality coals.

Any biomass residues can be used to make water gas that is then used in hydrocarbon production, benefiting from developments resulting from the production of liquid fuels from coal, dating back to the first half of the twentieth century. Relevant results in this direction continue to appear in the literature, with the route for production of C_8–C_{15} hydrocarbons based on xylose oligomers.[86] This work is still in the early stages of development and the anticipated production cost is between US$0.54 and US$ 1.16 L^{-1} (2.06–4.39 US$ per US gallon).

Activated charcoal from bagasse pyrolysis shows singular adsorbent properties, especially a high capacity for retention of chlorine[87] and microcystine toxins.[88] It also performs very well in the removal of colour during sugar processing.[89,90] A recent application for carbon fine particles is in agricultural soil conditioning and decontamination.[91,92] Another rapidly growing application for activated charcoal is the fabrication of carbon electrodes for use in super-capacitors. Some specific requirements are described in the literature,[93] reporting specific energy up to 36 J g^{-1} (10 Wh kg^{-1}) and specific capacitance close to 300 F g^{-1}. To achieve the desired pore structure, bagasse must first be impregnated with $ZnCl_2$ solution, which unfortunately makes this product unfit for some applications.

9.6 CO_2: FROM EXHAUST TO THE MARKET

CO_2 is a by-product of sucrose fermentation to ethanol. As such, it does not contain carbon monoxide, nitrogen oxides, particulate matter or any other contaminants which are formed during the combustion of fuels. This pure CO_2 from sucrose fermentation can therefore very easily be used in a number of applications, as a gas, a solid or a raw material for the chemical industry.

CO_2 can be used for cooling, as an intensifier of plant growth, in the gasification of food and beverages, as a pressurising agent in oil production and as a supercritical solvent, and solid CO_2 is a unique abrasive leaving no solid or liquid residues. It is used as a cooling fluid in supermarkets[94] and as dry-ice. Its use as a cooling fluid has been growing, especially in the USA thanks to the increased supply from corn-based ethanol production. Its use as a cooling agent in the form of dry-ice is important especially when one considers that its cooling capacity is twice that of water ice and it cools to lower temperatures.[95]

The effect of CO_2 on plant growth is spectacular even at a concentration of only twice that of the current atmospheric concentration. Encouraging results have been obtained from growing flowers and vegetables in glasshouses,[96] but it is possible to conceive of CO_2 being released in controlled ways in open fields, perhaps as carbonated water or aqueous calcium bicarbonate solution.

Increased oil recovery from CO_2 injection in oil wells has been reported.[97] It contributes to lowering oil viscosity and it also reacts with rock in alkaline formations, contributing to the release of oil.

CO_2 is the most widely used supercritical solvent in industry in spite of the actual results still lagging behind some initial expectations. Perrut has made a valuable critical assessment of its possibilities.[98]

Finely powdered dry-ice is a unique abrasive for blast cleaning in open environments, especially in building construction and repair, with obvious advantages over other solid or liquid abrasives. However, the complexity of the equipment and personnel training has restricted its application to professional service providers.[95]

CO_2 also holds great potential in fuel production. It is a limiting raw material in microalgae cultivation that has been attracting great attention as a strategy for fuel oil production[99] and its availability can provide an important contribution to make this viable. Algenol, Solix, Sapphire Energy, Seambiotic and Solazyme were among the companies operating development plants for this purpose in 2010.

The use of CO_2 in liquid fuel production has been defended by many authors, using different approaches, often coupled with the storage of photovoltaic or wind electricity, to buffer against the dependence on sunshine and wind strengths. The formation of hydrocarbons has been examined in great detail[100,101] and this also fits within the 'methanol economy' strategy presented by Olah.[102] The use of hydrogen obtained from natural gas is often considered as a way of reducing CO_2 to methanol.[103] In a sugarcane processing plant, methane can also be obtained from the anaerobic fermentation of biomass residues, including straw and bagasse.

Another hitherto little explored possibility is the thermal oxidation of residual biomass in a CO_2 atmosphere, producing syngas. This has been described in recent publications from Columbia University on the gasification of municipal waste.[104,105] Not only was mass conversion enhanced but also tar formation was considerably reduced, with energy and environmental benefits.

The use of CO_2 as a reagent for chemical synthesis has attracted great attention and it is now acquiring an increasing role as a raw material for the chemical industry. An important focus is the replacement of phosgene ($COCl_2$), a by-product from many industrial processes, which is extremely toxic and was once used as a chemical weapon. Unfortunately, phosgene reactions often produce large amounts of chlorinated residues. Some authors have focused on specific objectives: a group from Monsanto showed the possibility of making alkyl carbonates, carbamates, polyurethanes and other chemicals normally made using phosgene, giving special attention to the production of substances used as agrochemicals.[106]

The use of CO_2 as a chemical reagent and possibly as an industrial raw material was reviewed a few years ago.[107] It covers dimethyl carbonate synthesis from methanol and CO_2, synthesis of cyclic carbonates, alternating polymerisation of oxiranes and CO_2, synthesis of urea and urethane derivatives, synthesis of carboxylic acids, esters and lactones, isocyanates, hydrogenation and hydroformylation, alcohol homologation and electrochemical reactions.

Large-scale reactions with a potential for CO_2 mitigation were also recently reviewed.[108] The reactions included chemical transformations, photochemical reductions, chemical and electrochemical reductions, biological conversions, reforming and inorganic transformations.

New fundamental results are appearing constantly, such as the possibility of making methyl acrylate from CO_2[109] and its photochemical conversion to hydrocarbons.[110]

9.7 FROM ASHES TO NEW RAW MATERIALS

Sugarcane bagasse is the fibrous material that remains after sugarcane stalks have been crushed to extract the sugar juice. It is usually burned to generate heat for ethanol distillation, sucrose processing and power generation. This is essential to achieve a highly positive carbon balance and for the economic competitiveness of ethanol production from sugarcane. The current annual amount of ash produced by burning bagasse is ~ 4 Mt. This ash remains largely underused and discarded, but there are possible uses for it.

Ash contains silica and varying amounts of Al, Fe, Ca, Na, K, Mn, Mg and P together with fine particles. Depending on the burning conditions, ash can be used as a pozzolanic material, partly replacing cement in the fabrication of high-performance concrete.[111,112] If the burning temperature exceeds $600\,°C$, the amount of crystalline silica increases and the pozzolanic activity decreases. Concrete made with addition of ash may show lower setting times, reduced porosity and improved mechanical resistance compared with Portland cement.

Ash can also be used advantageously to make glass, glazed ceramics,[113] charcoal briquettes,[114] zeolites and silicon carbide thanks to the high reactivity of the finely divided amorphous silica.

Ash is also an abundant and inexpensive filter aid and a possible adsorbent for effluent treatment[115] involving the removal of phenol,[116] heavy metals,[117] dyes and insecticides,[118] thus offering the possibility of easy recycling and recovery. In these applications, an intentionally high content of carbon particles may be advantageous.

A relatively widespread application is as a complement to fertilisers used in rice plantations in Korea and Japan.[119] It has been estimated that rice uses up to 300 kg of silicon per hectare. Wheat, sugarcane, oats and barley plants also absorb silicic acid from the soil and incorporate it as amorphous silica[120,121] in the leaf epidermis, tricomas and spines, protecting the plant from drought and saline stress, and also from attack by fungi, insects and bacteria.[112] It has been estimated that sugarcane uses between 300 and 700 kg of silicon per hectare.

Many questions are still open concerning bagasse ash properties and its possible uses. For instance, detailed speciation of ash and information on how the conditions of ash formation affect its composition and properties are still largely unknown. There are also only limited data on the availability for plants of K, P, C, Fe and Mg found in the ash and also on the solubility of Si species relevant for use with different crops. This information is essential for optimising the use of ash, especially concerning the fertilisation and conditioning of soil that in turn is essential for the sustainability of sugarcane plantations.

Ash can probably also play a role in animal nutrition, considering the chemical composition and the industrial production of silicates and silica for that purpose.[123]

9.8   CONCLUSION

Sugarcane is an important source of chemicals and its role is rapidly increasing, thanks to its availability, productivity gains and also the properties and rapid diversification of its first-generation products: sucrose, ethanol, bagasse, ash and CO_2.

Acknowledgements

This work is a summary of a study prepared for the company ETH in São Paulo, Brazil. The authors thank Dr Carlos Calmanovici for authorising this publication. This is a contribution from INCT Inomat (Institute for Complex Functional Materials), supported by CNPq and FAPESP. The authors thank Maria do Carmo V. M. da Silva for her help in organising the text.

REFERENCES

1. J. R. Moreira and J. Goldemberg, *Energy Policy*, 1999, **27**, 229.
2. J. Goldemberg, *Science*, 2007, **315**, 808.
3. A. P. de Souza, M. Gaspar, E. A. da Silva, E. C. Ulian, A. J. Waclawovsky, M. Y. Nishiyama Jr., R. V. dos Santos, M. M. Teixeira, G. M. Souza and M. S. Buckeridge, *Plant Cell Environ.*, 2008, **31**, 1116.
4. *Medium and Long-term Opportunities and Risks of the Bio-technological Production of Bulk Chemicals from Renewable Resources – The Potential of White Biotechnology – The BREW Report*, European Commission GROWTH Programme (DG Research), Utrecht, September 2005, isi.fraunhofer.de/isi-de/publ/download/isi06b47/brew.pdf (last accessed 15 December 2011).
5. Sugar Research Foundation, *Science*, 1945, **101**, 426.
6. J. K. Hwang, C. J. Kim and C. T. Kim, *Starch–Stärke*, 1998, **50**, 463.
7. M. McCoy, *Chem. Eng. News*, 1998, **76**, 13.
8. C. J. Drummond, C. Fong, I. Krodkiewska, B. J. Boyd and I. J. Baker, Sugar fatty acid esters, in *Novel Surfactants*, 2nd edn, ed. K. Holmberg, Marcel Dekker, New York, 2003, p. 35.
9. B. A. P. Nelen and J. M. Cooper, Sucrose esters, in *Emulsifiers in Food Technology*, ed. R. J. Whitehurst, Blackwell, Oxford, 2004.
10. V. F. Ferreira, D. R. da Rocha and F. de C. da Silva, *Quím. Nova*, 2009, **32**(3), 623.
11. A. M. Schwartz and C. A. Rader, *J. Am. Oil Chem. Soc*, 1965, **42**, 800.
12. Y. Queneau, S. Chambert, C. Besset and R. Cheaib, *Carbohydr. Res.*, 2008, **343**, 1999.
13. H. Schnell, K. Uerdingen, H. Dohm, Frefeld and J. Nentwig, Krefeld-Bockum, process for the production of sugar esters, *US Patent*, 3 251 827, 1966.
14. S. S. Narine and X.-H. Kong, in *Bailey's Industrial Oil and Fat Products*, ed. F. Shahidi, Wiley-Interscience, New York, 2005, p. 279.

15. C. R. de Schaefer, M. E. F. de Ruiz Holgado and E. L. Arancibia, *J. Argent. Chem. Soc*, 2002, **90**, 55.
16. B. C. Youan, A. Hussain and N. T. Nguyen, *AAPS PharmSci*, 2003, **5**, 1.
17. R. F. Severson, R. F. Arrendale, O. T. Chortyk, C. R. Green, F. A. Thome, J. L. Stewart and A. W. Johnson, *J. Agric. Food Chem.*, 1985, **33**(5), 870.
18. S. J. Byrd, *SAIB – The Oldest New Ingredient for the Beverage Market*, Eastman Chemical Company Technical Report, Eastman Chemical, Kingsport, TN, 2006.
19. R. C. Reynolds and C. I. Chappel, *Food Chem. Toxicol.*, 1998, **36**, 81.
20. Olean (olestra), http://www.pgfoodingredients.com/ or http://www.olean.com/ (last accessed 27 September 2011).
21. University of Cincinnati Health News, *Olestra Could Be Antidote to Toxins*, 2005, http://healthnews.uc.edu/publications/findings/?/466/1576/ (last accessed 27 September 2011).
22. P&G Chemicals, http://www.pgchemicals.com (last accessed 27 September 2011).
23. *Chempol*, http://en.wikipedia.org/wiki/chempol, (last accessed 27 September 2011).
24. F. W. Schenk, *Int. Sugar J.*, 2000, **102**, 285.
25. M. G. Lindley, *Int. Sugar J.*, 2002, **104**, 346.
26. PRWeb, *Global Artificial Sweeteners Market to Reach US$1.5 Billion by 2015, According to a New Report by Global Industry Analysts, Inc.*, 2010, www.prweb.com/releases/artificial_sweeteners/aspartame_sucralose/prweb4563584.htm (last accessed 31 March 2012).
27. W. J. Yun, *Enzyme Microb. Technol.*, 1996, **19**, 107.
28. K. Buchholz, M. Borchers and D. Schwengers, *Starch–Stärke*, 1998, **50**, 164.
29. Tate & Lyle, www.tateandlyle.com/Pages/default.aspx (last accessed 10 January 2011).
30. M. R. Jenner and D. Waite, Crystalline 4,1′,6′-trichloro–4,1′,6′-trideoxy-galactosucrose, *US Patent*, 4 343 934, 1982; L. Hough, S. P. Phadnis, R. A. Khan and M. R. Jenner, Sweeteners, *US Patent*, 4 435 440, 1984; P. K. Beyts and Z. Latymer, Sweetening agents containing chlorodeoxysugar, *US Patent*, 4 495 170, 1965.
31. M. Ikeda, *Adv. Biochem. Eng. Biotechnol.*, 2003, **79**, 1.

32. M. A. Godshall, *Int. Sugar J.*, 2001, **103**, 1233.
33. J. A. Casas, V. A. Santos and F. Garcia-Ochoa, *Enzyme Microb. Technol.*, 2000, **26**, 282.
34. R. W. Bailey and E. A. Oxford, *J. Gen. Microbiol.*, 1958, **19**, 130.
35. R. Nieves, Enzyme based biomass-to-ethanol technology: au update, presented at the NREL-INT Development Seminar on Fuel Ethanol, Washington, DC, 2001.
36. Alkenes made simple from biomass polyols, *Chem. Eng.g News*, 2009, **87**(21), 35, http://pubs.acs.org/isubscribe/journals/cen/87/i21/html/8721scic5.html (last accessed 31 March 2012).
37. Cargill, *Cargill and Novozymes to Enable Production of Acrylic Acid via 3HPA from Renewable Raw Materials*, http://www.cargill.com/news-center/news-releases/2008/NA3007665.jsp (last accessed September 2011).
38. Cargill and Novozymes to enable production of acrylic acid via 3HPA from renewable raw materials, http://novozymes.com/en/news/news-archive/Pages/44469.aspx (last accessed 4 May 2012).
39. C. Koseki, K. Fukui, J. Nakamura and H. Kojima, Process for production of succinic acid, *US Patent*, 7 829 316, 2010.
40. Big plans for succinic acid, *Chem. Eng. News*, 2009, **87**(50), 23, http://pubs.acs.org/doi/pdf/10.1021/cen–087n050.p023.
41. *Shell Oil Company Magazine*, 1st Quarter 2011.
42. P. J. Fagan and L. E. Manzer, *Preparation of Levulinic Acid Esters and Formic Acid Esters from Biomass and Olefins*, U.S. Patent 7,153,996, Dec. 26th, 2006.
43. J. Lewkowski, Synthesis, chemistry and applications of 5-hydroxy methylfurfural and its derivatives. *ARKIVOC General Papers,* 2001 (i), pp. 17–54.
44. *Chem. Eng. News*, 2006, **84**(34), 7.
45. J. C. Serrano-Ruiz, R. M. West and J. A. Dumesic, *Annu. Rev. Chem. Biomol. Eng.*, 2010, (1), 79.
46. D. Weiner, *Verenium Technology Presentation*, http://www.verenium.com/webinars/technology/index.htm (last accessed 7 May 2012).
47. J. L. Gaddy, E. C. Clausen, C. W. Ko, L. E. Wade and C. V. Wikstrom, *Microbial Process for the Preparation of Acetic Acid, as well as Solvent for its Extraction from the Fermentation Broth*, U.S. Patent 7 196 218, Mar. 27th 2007.

48. P. J. Brumm and R. Datta, *Production of Acetic Acid by an Improved Fermentation Process*, U.S. Patent 4 830 963, May 16th 1989.
49. R. D. Schwartz and F. A. Keller Jr., *Acetic Acid by Fermentation*, U.S. Patent 4,371,619, Feb. 1st, 1983.
50. Cargill and Novozymes to Enable production of acrylic acid via 3HPA from Renewable Raw Materials, http://www.cargill.com/news/releases/2008/NA3007665.jsp (last accessed 7 May 2012).
51. Biobased Chemicals: Myriant Technologies to build the largest biobased Succinic Acid plant. *Chem. Eng. News*, 2011, **89**(2), 24.
52. C. Koseki, K. Fukui, J. Nakamura and H. Kojima, *Process for production of succinic acid*, U.S. Patent 7 829 316, Nov. 11th 2010.
53. http://www.carboshale.ee/glutaric_acid.htm
54. K. M. Draths and J. W. Frost, *J. Am. Chem. Soc.*, 1994, **116**, 399.
55. W. Niu, K. M. Draths and J. W. Frost, *Biotechnolol. Prog.*, 2002, **18**, 201.
56. J. W. Frost and K. M. Draths, *Synthesis of Adipic Acid from Biomass-Derived Carbon Sources*, U.S. Patent 5 487 987, Jan 30th 1996.
57. Adipic Acid, http://www.verdezyne.com/Verdezyne/Products/adipic.cfm (last accessed 7 May 2012).
58. D. E. Kielv and L. Chen, *Process for preparing poly(glucaramides)*, U.S. Patent 5 473 035, Dec 5th 1995.
59. M. Kitada, H. Ueyama, E. Suzuki and T. Fukumbara, *Journal of Fermentation Technology*, 1967, **45**, 1101.
60. M. Rosfarizan, S. M. Mohd, S. Nurashikin, M. S. Madihah and B. A. Arbakariya, *Biotechnology and Molecular Biology Reviews*, 2010, **5**(2), 24.
61. Itaconic acid produced by fermentation of carbohydrates with Aspergillus terreus, www.itaconix.com/technology.html (last accessed 7 May 2012).
62. V. F. Ferreira, D. R. da Rocha and F. de C. da Silva, *Quím. Nova*, 2009, **32**(3), 623.
63. H. Zoebelein, *Dictionary of Renewable Resources*, 2nd ed, Wiley-WCH, Weinheim, 2001, p. 282.

64. B. Gruber, K. J. Weese, H. P. Mueller, K. Hill, A. Behr and J. R. Tucker, *Octyl ethers and octadienyl ethers*, US Patent 5 236 909, 1993.
65. S. K. Rogers, B. J. Royles and M. S. White, *Polymers and their synthesis*, US Patent 7 041 730, 2006.
66. R. B. Hutchison, *Sucrose polyester useful as fat substitute and preparation process*, US Patent, 5 504 202, 1996.
67. A. S. Hofman and N. S. Choi, *Method for reducing absorption of unde-sired lipids in the gastrointestinal tract*, US Patent 6 180 617, 1995.
68. J. F. Robyt and R. Mukerjea, *Linear and cyclic sucrose reaction products, their preparation and their use*, US Patent, 6281351, 2001.
69. J. Ellingboe, E. H. Nystrom and J. B. Sjovall, *Polysaccharide polyols*, US Patent, 4 076 930, 1978.
70. Y. Queneau, S. Jarosz, B. Lewandowski and J. Fitremann, *Adv. Carbohydr. Chem. Biochem.*, 2007, **61**, 217.
71. Ecopoxy - The greener choice, http://www.ecopoxysystems canada.com/ (last accessed 4 May 2012).
72. N. D. Sachinala and M. H. Litt, *Epoxy monomers from sucrose*, US Patent 5 646 226, 1997.
73. K. Park and J. Chen, *Carbohydr. Polym.*, 2000, **41**, 259.
74. L. Ferreira, M. M. Vidal, C. F. G. C. Geraldes and M. H. Gil, *Carbohydr. Polym.*, 2000, **41**, 15.
75. L. Bustani, *Surgane Bagasse: a Profitable Residue*, http:// archive.psfk.com/2009/03/sugarcane-bagasse-a-profitable-residue.html (last accessed 2 May 2012).
76. Empresa de pesquisa energetica, *Análise de Conjuntura dos Biocombustíveis Jan. 2009-Mar 2010*, http://www.advivo. com.br/blog/roberto-sao-paulo-sp-2010/analise-de-conjuntura-dos-biocombustiveis-jan-2009-%E2%80%93-mar-2010pdf (last accessed 2 May 2012).
77. Champignon de Paris (Agaricus bisporus), http:// www.brasmicel.com.br/cchamp.htm (last acessed 2 May 2012).
78. Papel de bagaço de cana de açúcar, *GCE Papeis*, http:// www.gcepapeis.com.br/ (last acessed 2 May 2012).
79. D. Bhattacharya, L. T. Germinario and W. T. Winter, *Carbohydr. Polym.*, 2008, **73**, 371.

80. J. E. de Paiva, I. R. Maldonade and A. R. P. Scamparini, *Rev. Bras. Eng. Agríc. Ambient.* [online], vol. 13, n. 1 [cited 2010-07-19], pp. 75–80, 2009. Available from: http://www.scielo.br/scielo.php?script=sci_abstract&pid=S1415-43662009000100011&lng=pt&nrm=iso&tlng=pt (last accessed 3 May 2012).

81. J. O. B. Carioca, S. Macambira, R. G. Corrêa and P. Alcântra, *Quím. Verde Brasil*, 2010, 219.

82. M. N. Belgacem and A. Gandini, *Composite Interfaces*, 2005, **12**, 41.

83. R. Silva, S. K. Haraguchi, E. C. Muniz and A. F. Rubira, *Quím. Nova*, 2009, **32**, 661.

84. É. Borbély, *Acta Polytechnica Hungarica*, 2008, **5**, 3. In http://old.bmf.hu/journal/Borbelyne_15.pdf (last accessed 3 May 2012).

85. F. Galembeck, *Energy Environ. Sci.*, 2010, **3**, 393.

86. R. Xing, A. V. Subrahmanyam, H. Olcay, W. Qi, G. P. van Walsum, H. P. Pendse and G. W. Huber, *Green Chem.*, 2010, **12**, 1933.

87. E. F. Jaguaribe, L. L. Medeiros, M. C. S. Barreto and L. P. Araujo, *Braz. J. Chem. Eng.* [online], vol. 22, n. 1, 2005. [cited 2010-07-19], pp. 41–47. Available from: http://www.scielo.br/scielo.php?script=sci_arttext&pid=S0104-66322005000100005 (last accessed 3 May 2012).

88. UNICAMP, *Biblioteca Digital da UNICAMP*, http://libdigi.unicamp.br/document/?code=000391015

89. K. Qureshi, I. Bhatti, R. Kazi and A. K. Ansari, *Int. J. Chem. Biomol. Eng.*, 2008, **1**, 3.

90. R. J. Kuhn, *Designing and Assessing the Feasibility of an Active Learning Approach to the Teaching of Legal Research* http://researchspace.ukzn.ac.za/xmlui/handle/10413/187 (last accessed 3 May 2012).

91. B. Liang, J. Lehmann, D. Solomon, J. Kinyangi, J. Grossman, B. O'Neill, J. O. Skjemstad, J. Thies, F. J. Luizão, J. Petersen and E. G. Neves, *SOILSCI*, 2006, **70**, 1719.

92. Department of Applied Environmental Science, Stockholm University, Gerard Cornelissen, www.itm.su.se/page.php?pid=536&id=275 (last accessed 31 March 2012).

93. T. E. Ruford, D. Hulicova-Jurcakova, K. Khosla, Z. Zhu and G. Q. Lu, *J. Power Sources*, 2010, **195**, 912.

94. S. Sawalha, Emergency Applications for CO_2, http://www.energy.kth.se/index.asp?pnr=10&ID=285&lang=1 (last accessed 3 May 2012).

95. Universal Industrial Gases, *Carbon Dioxide (CO_2) Properties, Uses, Applications: CO_2 Gas and Liquid Carbon Dioxide*, http://www.uigi.com/carbondioxide.html (last accessed 3 May 2012).

96. R. A. Furlan, F. C. Rezende, D. R. B. Alves and M. V. Folegatti, *Hortic. Bras.*, 2002, **20**, 547.

97. M. Perrut, *Ind. Eng. Chem. Res.*, 2000, **39**, 4531.

98. GreeneServices, *CO_2 Blast Cleaning Applications*, http://www.greeneservices.com/co2-blasting.html (last accessed 3 May 2012).

99. Energias Renováveis, *Algafuel Avança com Primeiro Projecto de Microalgas para Sequestro de CO_2*, http://energiasrenovaveis.wordpress.com/2007/09/11/algafuel-avanca-com-primeiro-projecto-de-microalgas-para-sequestro-de-co2/ (last accessed 3 May 2012).

100. A. Symposium, Carbon Dioxide a Raw Material for Sustainable Development, http://www.emrsstrasbourg.com/index.php?option=com_abstract&task=view&i=01&year=2010&Itemid=&id_season=3 (last accessed 3 May 2012).

101. M. Yamasaki, H. Habazaki, T. Yoshida, E. Akiyama, A. Kawashima, K. Asami, K. Hashimoto, M. Komori and K. Shimamura, *Appl. Catal. A: Gen*, 1997, **163**, 187.

102. K. Bullis, *The Methanol Economy*, http://www.technologyreview.com/biztech/wtr_16466,296,p1.html (last accessed 3 May 2012).

103. M. Steinberg, *Int. J. Hydrogen Energy*, 1998, **23**(6), 419.

104. E. Kwon, K. J. Westby and M. J. Castaldi, *Transforming Municipal Solid Waste (MSW) into Fuel via the Gasification/Pyrolysis Process*, www.seas.columbia.edu/earth/wtert/sofos/nawtec/nawtec18/nawtec18-3559.pdf (last accessed 2 April 2012).

105. J. Dooher and M. J. Castaldi, *Enhanced product gas and power evolution from carbonaceous materials via gasification*, US Patent Application, 20100129691, 2010.

106. W. McGhee and D. Riley, *J. Org. Chem.*, 1995, **60**, 6205; D. C. Dean, M. A. Wallace, T. M. Marks and D. G. Melillo, *Tetrahedron Lett.*, 1997, **38**, 919; W. D. McGhee, Y. Pan and D. P. Riley, *J. Chem. Soc., Chem. Commun.*, 1994, 699.

107. T. Sakakura, J.-C. Choi and H. Yasuda, *Chem. Rev.*, 2007, **107**, 2365.
108. M. Mikkelsen, M. Jørgensen and F. C. Krebs, *Energy Environ. Sci.*, 2010, **3**, 43.
109. C. Bruckmeier, W. L. Maximilian, R. Reichardt, S. Vagin and B. Rieger, *Organometallics*, 2010, **29**, 2199.
110. C. R. Somnath, K. V. Oomman, P. Maggie and A. G. Craig, *ACS Nano*, 2010, **4**, 1259.
111. G. C. Cordeiro, R. D. Toledo-Filho and E. de M. R. Fairbairn, *Quim. Nova*, 2009, **32**, 82.
112. M. O. Oliveira, I. de, F. F. Tinôco, C. de, S. Rodrigues, E. N. da Silva, C. de and F. Souza, *Engenharia Agric*, 2009, **17**, 15.
113. S. R. Teixeira, A. E. de Souza, G. T. de, A. Santos, A. F. V. Peña and A. G. Miguel, *J. Am. Ceram. Soc.*, 2008, **91**, 1883.
114. S. R. Teixeira, A. F. V. Peña and A. G. Miguel, *Waste Manage*, 2010, **30**, 804.
115. V. K. Gupta, P. J. M. Carrott, M. M. L. Ribeiro Carrott and Suhas, *Crit. Rev. Environ. Sci. Technol.*, 2009, **39**, 783.
116. V. C. Srivastava, M. M. Swamy, I. D. Mall, B. Prasad and I. M. Mishra, *Colloids Surf. A: Physicochem. Eng.*, 2006, **272**, 89; V. K. Gupta, S. Sharmas, I. S. Yadav and D. Mohan, *J. Chem. Technol. Biotechnol.*, 1998, **71**, 180.
117. G. M. Taha, *Monit. Remediat*, 2006, **26**, 137.
118. V. K. Gupta, C. K. Jain, I. Ali, S. Chandra and S. Agarwal, *Water Res*, 2002, **36**, 2483.
119. O. F. de Lima Filho, *O Silício é um fortificante e Anti-estressante Natural para as plantas*, http://www.silifertil.com.br/artigos/silicio02.pdf (last accessed 3 May 2012).
120. D. J. Belton, S. V. Patwardhan, V. V. Annenkov, E. N. Danilovtseva and C. C. Perry, *Proc. Natl. Acad. Sci. USA*, 2008, **105**, 5963.
121. C. Aguirre, T. Chávez, P. Garcia and J. C. Raya, *Interciencia*, 2007, **32**, 503.
122. G. H. Korndörfer, H. S. Pereira and M. S. Camargo, *Rev. STAB*, 2002, **21**, 6.
123. Animal Nutrition, *Tixosil®: High Performance Silica,* http://www.rhodia.com/en/binaries/Tixosil%20range%20for%20feed%20market.pdf (last accessed 3 May 2012).

CHAPTER 10

Chemicals from Biomass

JANET L. SCOTT*[1] AND GIANFRANCO UNALI[2]

[1] Centre for Sustainable Chemical Technologies, University of Bath,
Bath BA2 7AY, UK; [2] Unilever R&D Port Sunlight, Quarry Road
East, Bebington CH63 3JW, UK
*Email: j.l.scott@bath.ac.uk

10.1 INTRODUCTION

Obtaining chemicals from biomass is not a new concept; indeed, the term 'chemurgy' was coined in the early 1900s for chemicals derived from agricultural biomass.[1,2] However, the non-renewable, or fossil, carbon-based chemical industry and specifically that based on oil (with lesser contributions from coal and natural gas), has been so successful that the industry is currently almost exclusively dependent on non-renewable carbon sources. For example, at the turn of the twenty-first century in the USA, only 2% of chemical feedstock came from biomass,[3] although, as by far the greatest majority of 'chemical' production is for fuels, this figure is higher in certain speciality areas, such as surfactants. By 2008 in Germany the proportion of chemical raw materials drawn from renewable (bio) sources was 11%, reflecting both a gradual shift in raw materials sources due to economics[4] and also a regional predilection. Awareness that reserves of fossil carbon are limited and

Materials for a Sustainable Future
Edited by Trevor M. Letcher and Janet L. Scott
© The Royal Society of Chemistry 2012
Published by the Royal Society of Chemistry, www.rsc.org

that supply is likely to become constrained in future (resulting in higher prices, if not actual shortages), alongside a desire for a more sustainable chemical industry, has led to a resurgence in efforts to produce chemicals from bio-based renewable resources.[5] Over 30 years ago, Lipinsky argued that biomass feedstocks were well suited for the production of chemicals owing to the flexibility resulting from the ability to plant appropriate crops selectively (with all the usual caveats about economies of scale).[6] He used a series of biological and ecological analogies to discuss direct substitution *versus* functional substitution, *i.e.* producing exactly the same chemical from a bio-sourced feedstock *versus* producing a new chemical that performs that same function from a bio-based feedstock. The example of Brazil as an early adopter of the strategy of using renewable resources strategically to reduce long-term reliance on imported oil or petrochemicals, even where this was temporarily expensive in the early stages of development of the new industry, has been vindicated by the development of a significant bio-based ethanol capacity, which in turn has led to a developing bio-based industry with some of the biggest chemical producers beginning to commercialise bio-based bulk and fine chemicals, particularly monomers.

10.1.1 Sources of Biomass and the Biorefinery Concept

Analogously to the petrochemical refinery, the commercial viability of a biorefinery will generally be dependent on the production of large-volume, low-value products such as transportation fuels, while the smaller-volume, higher-value co-products or secondary products (possibly produced by further chemical or biochemical upgrading) will account for a significant part of the profitability of the refinery. Thus, integrated production of chemicals and fuels is mooted.[7] The concept of the biorefinery has been described by a plethora of authors over the last 30 years[8] and is illustrated schematically in Figure 10.1. While oil feedstocks are (reasonably) homogeneous from batch to batch, at least with regard to chemical composition, refinery feedstocks derived from biomass may vary considerably from season to season and crop to crop or where wastes derived from agro-industry are used. Many analysts believe that to be economically viable a biorefinery must be able to accept a range of feedstocks that may change over time, although most of

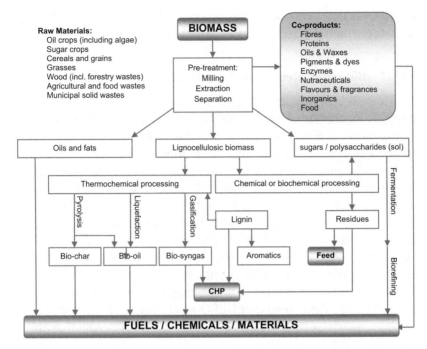

Figure 10.1 Schematic representation of the biorefinery concept illustrating the flexibility that could be required with respect to feedstocks and conversion technologies. In reality it is likely that oil crop-based and lignocellulosic biorefineries will operate separately, although, following the 'industrial metabolism' concept, it is conceivable that wastes from the former might be additional raw materials for the latter. Lignin processing is vastly oversimplified in this schematic.

the commercial biorefineries currently in operation rely on single feedstocks, such as wood.[9]

Many so-called 'first generation' biorefineries used feedstocks that competed directly with food production; indeed, in many cases these *were* foodstuffs, such as sugars, starch and vegetable oils, but developments towards use of whole crops and even wastes (so-called third-generation or 'phase III' biorefineries[8b]) proceed apace. Different conversion technologies are required, depending on the feedstock and product mix, which must be optimised to achieve an economically viable biorefinery.[10,11] This concept is taken further as the development of an 'industrial metabolism'[12] where the integrated biorefinery 'feeds' off the wastes of other plants. Where this approach is not possible, the relatively low density of the feedstocks makes optimal location an important factor, particularly for highly distributed facilities.[13]

10.1.2 Bio-based/Renewable/Sustainable

It is worth emphasising that 'bio-based' does not automatically imply 'sustainable'. Few would argue that the whaling industry, as practiced in the early part of the nineteenth century and which resulted in a catastrophic decline in some populations of whales to feed the lamp-oil industry, was sustainable.[14] A more recent example is the discovery of the anti-cancer drug paclitaxel, originally isolated by extraction from the bark of the Pacific yew tree, *Taxus brevifola*, in a process that resulted in the death of the tree. Extensive harvesting (~ 30 t of bark are required to produce 100 g of taxol) resulted in drastic losses in the *T. brevifola* population and also highlighted the inadequate nature of the supply chain. Ultimately, a semi-synthetic route to the drug from 10-deacetylbaccatin (10-DAB) was developed and 10-DAB is extracted from the needles of the English yew, *T. baccata*, which can be harvested without damage to the tree.[15]

Thus, the first requirement for a sustainable feedstock is that it be renewable on a time frame that is consistent with that of consumption of the resource. Depleting the resource generating the useful chemical or starting material may be likened to 'bio-mining', while harvesting a resource at a rate that allows continual replacement is sustainable use of a renewable resource. This, however, raises the spectre of the use of land for growing feedstocks for production of chemicals and fuels instead of food for a growing population.

Even if the supply chain for biomass for chemical production is carefully constructed to avoid conflicts with food production, increased demand for feedstocks, driven by all users (including biofuels producers), may result in an increased rate of expansion of land used for agriculture at the expense of pristine environments.[16] From a global perspective, the loss or downgrading of forests may also be implicated in the loss of previously sequestered carbon (from sources both above and below ground) and forfeited carbon sequestration opportunities.[17] This is a complex topic, which goes beyond the 'food *versus* fuel (or chemicals)' debate, to include political and social issues, such as security of supply and the rights of developing nations to exploit their resources or to expect recompense for not doing so, *i.e.* for protecting a *global* resource. Although a full discussion requires a more extensive treatment than is possible here, it serves to emphasise the importance of remaining cognisant of the *three* pillars of sustainability: people, profit and planet. Socioeconomic aspects of sustainability cannot be ignored and it should be emphatically stated that moving to bio-based chemicals is not a goal unto itself, but is a possible route to a more sustainable supply of raw materials for the chemical industry. In many of the examples below, the use of a waste product or by-product as the raw material source will be considered.

10.1.3 Scope

In this chapter, we do not focus on large-volume fuels, which include bioethanol, biobutanol and biodiesel, but we will refer to these where they fulfil specific non-fuel functions or are (or may be) useful as raw materials for the production of other chemical entities. Similarly, bio-oil,[18] *i.e.* fractionated products of biomass pyrolysis, is not a focus, even though this may well prove to be a large scale source of industrial chemicals in the future. Processing of bio-oils is not hugely dissimilar to that of fossil oils, at least at a conceptual level – details of processes are surely far more complex.

After fuels, the second largest sector of the chemical industry is that of polymers. Biomass and composite materials are discussed in Chapter 22 and this topic will not be duplicated here, although, as many monomers are also useful reactive intermediates for the production of other chemicals, the development of viable routes to bio-based monomers will also yield bio-based fine chemical raw materials.

Niche 'chemicals', such as drugs, or drug precursors, derived either by direct extraction (*e.g.* artemesinin) or by semi-synthetic routes following extraction (*e.g.* paclitaxel), are not considered, nor are drug products or nutraceuticals produced by genetically engineered cells or higher organisms – so-called 'pharming'.[19] Finally, we do not discuss chemicals from aquatic biomass in detail as this topic is dealt with in Chapter 8.

10.2 BIO-BASED PLATFORM CHEMICALS

So-called 'platform chemicals' are those that can be, or are expected to be, produced from renewable bio-resources in large quantities and so to form a 'platform' for the development of derivatives. Thus, a platform chemical is one which is accessible (or expected to be so) and may change as technology, either to produce the molecular entity or to convert it into more useful and usually higher value compounds, advances. This point was eloquently made by Bozell and Peterson.[20]

10.2.1 Platform Chemicals from Carbohydrates

In a very detailed report prepared for the US Department of Energy (DoE) published in 2004, the authors outlined a shortlist of 12 'building block' chemicals (from a longlist of 30) that might be produced largely from carbohydrates.[21] This list was the result of a very extensive screening exercise that included market and volume considerations (largely based on the then current petrochemical market and projections), and also the technical complexity associated with the best available processing pathway *at the time*. In addition, 'strategic fit criteria' defined in the report included 'direct replacement' and 'novel properties', criteria that are often at odds with each other. Six years on, two of the original authors revisited the original report and prepared an updated list of target structures (Table 10.1).[20]

Table 10.1 The top chemical candidates (largely) from carbohydrates in 2004 and 2009.

2004	2009
Succinic, fumaric and malic acids	Ethanol
2,5-Furandicarboxylic acid (FDCA)	Furans (furfural, hydroxymethylfurfural, FDCA)
3-Hydroxypropionic acid	Glycerol and derivatives[a,b]
Aspartic acid	Biohydrocarbons[b]
Glucaric acid	Lactic acid
Glutamic acid	Succinic acid
Itaconic acid	Hydroxypropionic acid/aldehyde
Levulinic acid	Levulinic acid
3-Hydroxybutyrolactone	Sorbitol
Glycerol[a]	Xylitol
Sorbitol	
Xylitol/arabinitol	

[a]Glycerol is discussed in detail in a following section.
[b]Not from carbohydrates.

Many compounds appear on both lists, but one of the most notable changes is the reduction in emphasis of the importance of a range of organic acids accessible by fermentation processes. This is partly due to the added initial screen of 'research activity reported in the literature', included to reflect the search for applicable conversion technology for a specific building block. Ethanol was omitted from the first list because it was expected that its main use would be as a fuel, hence it was designated a 'supercommodity' and not included in the list of building blocks. It is now clear that dehydration of bioethanol to ethylene will become a large-scale process: recent commercial developments, using reinvigorated old technology in the form of alcohol dehydration, include new ethanol to ethylene production facilities built by Dow, Braskem and Solvay.[20] The ethylene produced is a direct drop-in replacement for petrochemically derived material in polymerisation processes such as the production of polyethylene and (post further reaction) poly(vinyl chloride).

The upgrading and use of many of these platform chemicals are discussed in the sections below, particularly where these are currently commercially available, or promising to become so in the near future.

10.2.2 Chemicals from Lignin

The use of lignin as a feedstock for the production of chemicals from biomass is predicated on the expectation that very large quantities of lignin will become available as co-products of bio-refineries and the fact that lignin is one of the few sources of aromatic molecules readily available from biomass (Scheme 10.1). However, the variability of the complex structure of lignin, its stability (particularly resistance to depolymerisation) and high energy content have led many knowledgeable authors to conclude that combustion for energy recovery, or conversion to synthesis gas or bio-oil, is the best use for this biomass fraction. Currently, much of the lignin isolated, *e.g.* from the production of pulp, is burned for energy recovery, but it would seem that this is a less than optimal use of the material if we are likely to be in need of chemicals derived from sources other than fossil carbon. A second, less often cited, US DoE report published in 2007 covers 'potential biorefinery candidates from lignin'.[22] The production of aromatics from lignin (a focus of investigation in many academic

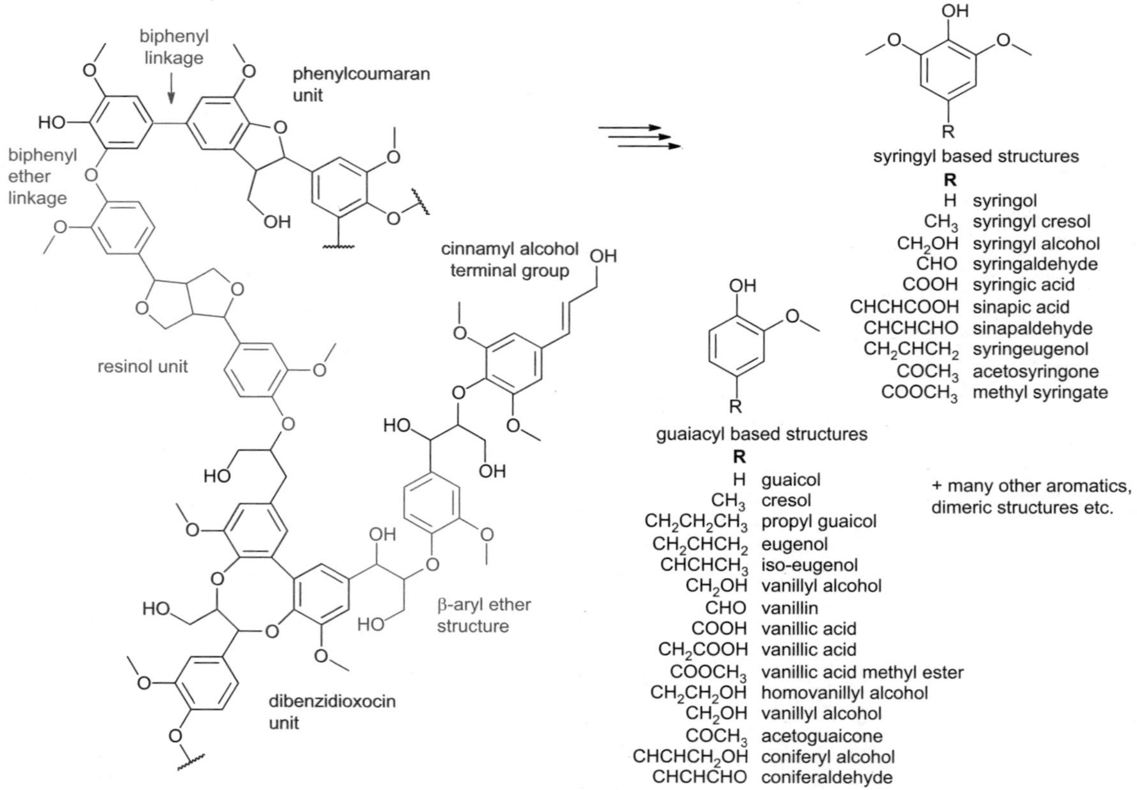

syringyl based structures

R	
H	syringol
CH_3	syringyl cresol
CH_2OH	syringyl alcohol
CHO	syringaldehyde
COOH	syringic acid
CHCHCOOH	sinapic acid
CHCHCHO	sinapaldehyde
CH_2CHCH_2	syringeugenol
$COCH_3$	acetosyringone
$COOCH_3$	methyl syringate

+ many other aromatics, dimeric structures etc.

guaiacyl based structures

R	
H	guaiacol
CH_3	cresol
$CH_2CH_2CH_3$	propyl guaiacol
CH_2CHCH_2	eugenol
$CHCHCH_3$	iso-eugenol
CH_2OH	vanillyl alcohol
CHO	vanillin
COOH	vanillic acid
CH_2COOH	vanillic acid
$COOCH_3$	vanillic acid methyl ester
CH_2CH_2OH	homovanillyl alcohol
CH_2OH	vanillyl alcohol
$COCH_3$	acetoguaicone
$CHCHCH_2OH$	coniferyl alcohol
CHCHCHO	coniferaldehyde

phenylcoumaran unit

biphenyl linkage

biphenyl ether linkage

resinol unit

cinnamyl alcohol terminal group

β-aryl ether structure

dibenzidioxocin unit

Scheme 10.1   A model lignin structure showing some of the possible structural motifs in red or blue, based on motifs published by Ralph et al.[23] and Rencoret et al.[24] Illustrative examples of aromatic chemicals that might be obtained from lignin, based on only two of the simpler structural units; fuller details are given in ref. 22. Many other simple and dimeric aromatic structures are also possible, but the complexity of the mixtures obtained can render separation difficult.

laboratories) is designated as providing 'long-term opportunities', while macromolecule production is considered more likely in the medium term. Lignin products are divided into the categories listed in Table 10.2 and some examples of the plethora of products that might be accessible are illustrated in Scheme 10.1.

Lignin, as a renewable raw material, has recently been reviewed,[33,34] as has catalytic valorisation of lignin.[35] Previously commercialised industrial products from lignin include vanillin, a range of phenols, cresols and guaiacols, aromatic acids (and acetic and formic acids) and deoxygenated aromatic hydrocarbons[36,37] (Scheme 10.2); vanillin and vanillin derivatives are currently produced commercially by Borregaard[38] in an advanced wood-based biorefinery.

Although the range of products that can be imagined as derived from lignin is huge, this very complexity creates significant separation challenges post-depolymerisation, whether this is achieved catalytically or thermochemically.[39,40]

Lignin is transformed during the process of separation from the polysaccharide fractions of biomass, and sulfur-free lignins from biomass conversion technologies and solvent pulping are likely to be of utility as a feedstock for high-value applications,[41] although lignosulfonates have long been used in relatively low-value applications, such as concrete additives,[42] where the lignin acts as a plasticizer.[43] Lignosulfonates are also used in small quantities in Portland cements as retarders to provide extended slurry pumping times (preventing premature setting of the concrete in oil wells) and may be applied in mixtures with non-sulfonated lignins where cements are foamed.[44] Other applications include grinding aids and dispersants, sequestrants to reduce scale formation and to deliver micronutrients, such as complexed metal ions, in agriculture and in drilling muds.[42]

Using the polymeric structure of lignin in blends with other polymers may provide the most logical high-value use of the material, particularly as larger volumes of non-sulfonated lignins become available from biorefineries. Indeed, the authors of the 2007 DoE report suggested that 'Complete lignin depolymerisation is an energy-negative process aimed at undoing what Nature has done in biosynthesis. In the chemical and commercial lignin industry, one recurrent theme is that research is needed to enhance the uses and add value to the polymer that nature has already provided.'[22] Choice of the lignin source or modification *via* reaction with hydroxyl or carboxyl functional groups allows tuning of

Table 10.2 Potential products derived from lignin, drawn from ref. 22 with added notes.

Type	Specific products	Notes
Process heat and power		Lignin has a high energy content and is frequently burned for energy recovery in pulp plants
Syngas and syngas products	Methanol and dimethyl ether Ethanol and mixed alcohols Fischer–Tropsch liquids Mixed liquid fuels By-product C_1–C_7 gases, hydrocarbons or oxygenates	Using syngas as a feed for upgrading to higher value products is effectively a means of transferring bio-based feedstocks to the current Fischer–Tropsch-type plants already in operation, by first degrading these to CO and H_2.[25] Fermentation processes using syngas,[26,27] or producer gas (CO, CO_2, CH_4, H_2 and N_2)[28] as feedstocks are known
Hydrocarbons	Benzene, toluene, xylenes (BTX) and higher alkylates Cyclohexane Styrenes Biphenyls	Require advanced hydroliquefaction technology From further reaction of deoxygenated BTX aromatics Biphenyl 'monomers' occur in most lignin structures, but quantities may be enhanced by oxidative coupling during/post-depolymerisation
Phenols	Phenol Substituted phenols Catechols, cresols, resorcinols Eugenol Syringols Coniferols Guaiacols	Lignin 'monomer' structures. While the structures are already present or result from hydrolysis of the lignin polymer, depolymerisation is energy intensive and must be followed by purification/separation without repolymerisation

Category	Product	Description
Oxidised products	Vanillin	Carried out on an industrial scale,[29] by mild oxidation under basic conditions
	Vanillic acid	Previously produced industrially by reaction with sulfur followed by strong oxidation
	DMSO	Previously produced industrially under conditions of vigorous oxidation. Also possible from biotechnological conversion of lignin[30]
	Aromatic acids	
	Aliphatic acids	
	Syringaldehydes and aldehydes	Oxidised 'lignin monomers'
	Quinines	From further reaction of benzene (above)
	Cyclohexanol/one	
	β-Ketoadipate	Produced by fermentation
Macromolecules	Carbon fibre	Replacement for polyacrylonitrile-based fibres.[31] Currently largely in the form of blends with other conventional polymers. *e.g.* polypropylene or poly(ethylene terephthalate)[32]
	Polyelectrolytes	Provide sequestrant and dispersant activity for metal ions and solids in water, respectively (*e.g.* in inks or dyes); also 'descaling' in industrial cleaning
	Polymer alloys	In some applications the brown colour imparted to polymer blends by lignin addition is unacceptable. Lignin tends to phase separate into lignin-rich domains.[45]
	Fillers or polymer extenders	
	Substituted lignins: carbonylated, ethoxylated, carboxylated, epoxidised, esterified	There are a range of products already on the market that reflect these catch-all categories and significant numbers of papers appear annually extending the possible applications of derivatised lignins
	Thermosets	
	Composites	
	Formaldehyde-free adhesives and binders	
	Wood preservatives	
	Nutraceuticals/drugs	
	Mixed aromatic polyols	

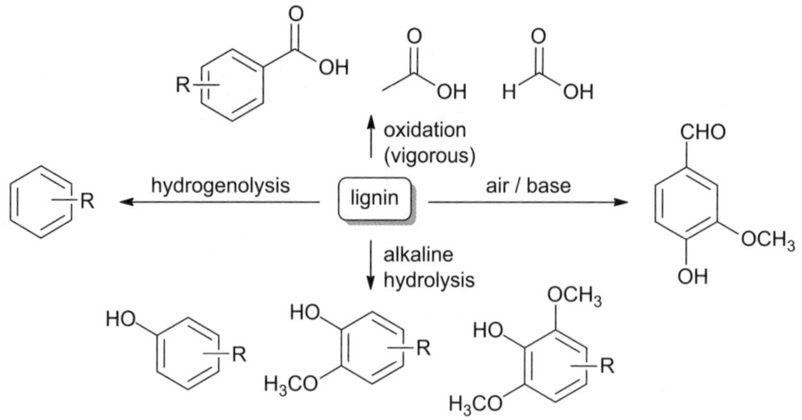

Scheme 10.2   Products previously produced commercially from lignin. Adapted from ref. 22.

hydrophobicity and compatibility with other polymers. Blends of lignin with a range of polymers and also in epoxides and phenol–formaldehyde resins (where the lignin forms all or part of the phenolic component) and polyolefin or vinyl polymers (lignin acts as a UV and thermal stabiliser), in addition to lignin graft copolymers, have recently been reviewed by Doherty *et al.*[45] and lignin and synthetic polymer blends by Feldman[46] and Stewart.[47]

10.2.3   Chemicals from Fats and Oils

Oleochemicals are some of the oldest chemicals intentionally produced by humans (the oldest soap recipe is said to be from ancient Babylon).[48] Naturally occurring fats and oils consist mainly of (tri)glycerides with smaller percentages of waxes, free fatty acids, phospholipids, sterols, hydrocarbons, pigments and vitamins.[49] Triglycerides may be used without transformation (*e.g.* in polymer coatings), but are more commonly sources of fatty acid moieties (frequently converted to alkyl esters, amines or alcohols) and glycerol. Aside from biofuel production and polymer applications, the largest proportion of fatty acid and fatty alcohol moieties is used in surfactant production for detergent applications, as discussed in detail in a later section.

   Waxes are esters of fatty acids and fatty alcohols, while free fatty alcohols and ethers of glycerol and fatty alcohols may also be isolated from animal tissues.[49] Although most plants produce oils consisting mainly of triglycerides, a few accumulate seed oil in the form of wax esters. For example, the jojoba plant (*Simmondsia chinensis*) produces esters of predominantly long-chain, monounsaturated, fatty acids and fatty alcohols, at levels as high as 50% liquid wax,[50] but, although these are excellent lubricants, the plant is low yielding.

Most chemical transformations of the glyceride portion of fats and oils involve the fatty acid carboxylate group,[51] but Biermann *et al.*, also summarised the opportunities arising from synthetic transformations of various parts of both saturated and unsaturated fatty acid moieties, as the free fatty acids or triglycerides.[52] The point is made that selective breeding of plants provides access to a plethora of fatty acids, but that a relatively small range are currently in bulk use.[53] As growth in vegetable oil supply is driven, at least in part, by biodiesel production[54] (and also the growth in consumer markets for other fatty acid derivatives such as surfactants), this is unlikely to change in the near future. The use of edible oils for production of fuels is an emotive issue[55] and highlights a potential consequence of an uncontrolled shift from fossil resources to renewable (particularly plant-based) resources: that of competition between the use of agricultural land for the production of fuel or chemical feedstocks and for the production of food (although clearly the loss of fossil resources for fuel will also frustrate food production). Gunstone highlighted the increase in production of vegetable oils using figures from the US Department of Agriculture (USDA) and interesting trends may be discerned in a plot of vegetable oil use over a recent 14 year period (Figure 10.2).[54] Demand for vegetable oils is growing, even on a per person basis (*i.e.* taking population growth into account), and the rise in the proportion used for non-food purposes has increased by roughly 10% over the period examined. This is ascribed, in part, to the growth in the oleochemical industry in south-east Asia.[54] Carlsson *et al.*, in their rather aptly titled paper 'Replacing fossil oil with fresh oil – with what and for what?', suggest that significant enhancements in oil crop productivity will be required to meet demand, which they estimate will require a trebling of plant oil production to provide 40% of the input oil raw materials to the chemical industry in 20 years, without compromising food production.[56]

Interestingly, recent attempts to 'green' the energy and fuels sector may be creating a threat to the established and growing

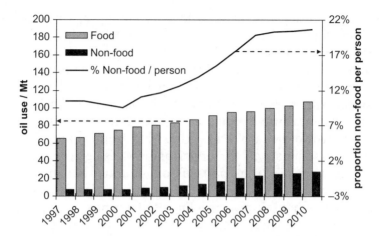

Figure 10.2   Comparison of vegetable oil use over a 14 year period. Data are drawn from Gunstone[54] and replotted here. Bars represent the relative absolute quantities of vegetable oils for food and non-food applications and the solid line is the percentage of non-food use of vegetable oil on a per person basis (*i.e.* accounting for increases due to population.

oleochemical sector: APAG [European Oleochemicals and Allied Products Group, a Cefic (European Chemical Industry Council) sector group], in a position paper published in 2005, suggested that the use of Emissions Trading Schemes 'credits' and other incentives designed to promote the use of biomass (including animal fats and vegetable oils) as fuels or renewable energy sources creates competition for resources between (subsidised) biofuel or bioenergy sectors and (non-subsidised) value-added oleochemicals.[57] Hence the 'fuel *versus* chemicals' tension in the petrochemical industry may well prove to be reflected in the non-petrochemical (bio-based) industry. Non-carbon-based energy vectors may ultimately provide the only answer to this conundrum.

10.2.4   Chemicals from Glycerol

Strictly, bio-based glycerol is derived from fats and oils and therefore belongs in the section above, but there has been such an explosion in potential and realised use of glycerol as a raw material for the production of chemicals, that a separate section is warranted for this useful molecular building block.

Although purified glycerol has myriad uses (*e.g.* food additive, pharmaceutical excipient, cosmetic or personal care product additive and in the industrial production of a range of polymers and other products[58]), most new developments focus on the use of low-quality glycerol arising as a by-product of biodiesel production.

The rapid growth of the biodiesel industry is projected to result in huge (over)production of glycerol from the transesterification of oils and fats to produce fatty acid methyl esters (FAMEs). The by-product, unpurified glycerol, is produced at a level of ~ 10 wt%, *i.e.* 100 kg of crude glycerol (containing 60–75% glycerol[59] for every tonne of biodiesel produced.[60] This huge resource is driving a shift from glycerol production (largely from petrochemically derived propylene) to the utilisation of glycerol as a feedstock.[61] Biodiesel feedstock and production methods have an impact on the impurities in the glycerol produced and significant purification is required to produce glycerol for conventional uses, although there is the added benefit of being able to market this product as 'natural' glycerol. Here we focus on the further transformation of glycerol, *i.e.* on glycerol as a 'platform' molecule derived from the integrated biorefinery.

Conversion of glycerol to value-added products, such as those depicted in Scheme 10.3, has been extensively reviewed,[62,63,64] as have the conversions required.[65,66] Although glycerol itself has been produced from petrochemical feedstocks (from propylene *via* epichlorohydrin), this trend is now being reversed in a pleasing symmetry.

10.2.4.1 Epichlorohydrin. Epichlorohydrin is produced from glycerol via a process commercialised by Dow Chemical as the GTE process[68] and by Solvay as the EPICEROL process.[67] This two step process (Scheme 10.4) not only uses a renewable raw material, but also has higher atom efficiency with respect to chlorine.[68] Solvay claims that their refined process, soon to be rolled out in China, also cuts energy use, reduces water consumption by an order of magnitude and hugely reduces chlorinated by-products.[69] Epichlorohydrin is widely used in the manufacture of epoxy resins and epichlorohydrin prepared from glycerol represents a 'drop-in' replacement for a petrochemically derived monomer.

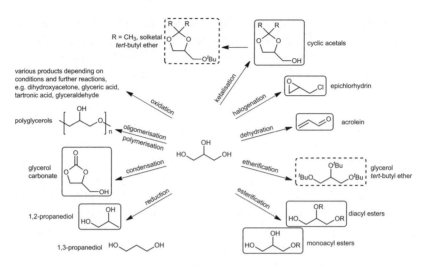

Scheme 10.3   Glycerol as a platform molecule: some possible conversions to value-added products. Already commercialised products are framed and those probably close to commercialisation are framed with dashed lines. Scheme extensively modified from Katryniok *et al.*[67]

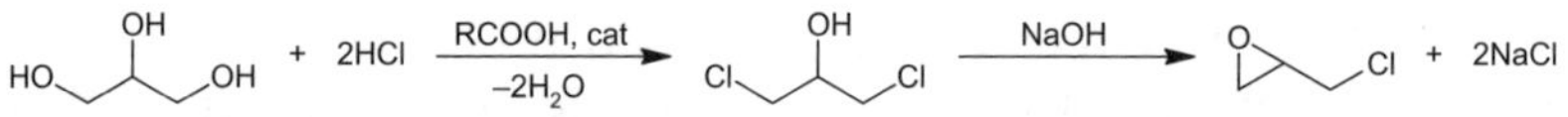

$$HO\underset{OH}{\overset{OH}{\diagdown\!\diagup}}OH \; + \; 2HCl \xrightarrow[-2H_2O]{RCOOH,\, cat} Cl\underset{OH}{\overset{OH}{\diagdown\!\diagup}}Cl \xrightarrow{NaOH} \overset{O}{\triangle}\diagdown Cl \; + \; 2NaCl$$

Scheme 10.4   Both Solvay's and Dow Chemical's route to prepare epichlorohydrin from glycerol comprises reaction of glycerol with HCl, giving primarily 1,3-dichloropropan-2-ol (2,3-dichloropropan-1-ol is minor contaminant), followed by reaction with base.

10.2.4.2   Acrolein and Acrylic Acid.   Production of acrolein is achieved by dehydration of glycerol[70] over a solid acid catalyst in the presence of gaseous SO_2, SO_3 or NO_2[71] (a process commercialised by Arkema). Acrolein is highly toxic and is immediately converted to polymers,[72] or to acrylic acid and thence to polymers.[73] Rising acrylic acid prices make bio-based acrylic acid (and acrolein) highly desirable and a number of large manufacturers are pursuing different routes, such as Dow Chemical from 3-hydroxypropionic acid (derived from sugar, not glycerol).[74] Dehydration of glycerol to acrolein has been reviewed.[75]

10.2.4.3 Butyl Ethers and Cyclic Acetals. Both glycerol *tert*-butyl ether (GTBE) and solketal *tert*-butyl ether (STBE) (Scheme 10.3) have been mooted as fuel additives for the replacement of the anti-knock compound methyl *tert*-butyl ether (MTBE).[76] Cyclic acetals, such as that formed by reaction of glycerol with acetone to produce 2,2-dimethyl-4-hydroxymethyl-1,3-dioxolane, are beginning to be marketed as solvents. For example, Rhodia (now a member of the Solvay group) offers Rhodia Augeo SL 191, developed in Brazil, for use as a thinner in paint and coating formulations and in the leather industry. Interestingly, it is likely that the process used was first patented as a means of purifying crude glycerol by reacting the crude glycerol fraction (from biodiesel production) with a ketone in the presence of an acid catalyst, *e.g.* sulfuric acid, recovering the pure dioxolane by distillation and then converting it back to glycerol by hydrolysis in an aqueous acidic medium (Scheme 10.5).[77] As acetone can be derived from ABE fermentation of biomass, such cyclic ketals of glycerol may be entirely derived from renewable bio-sources. The company's promotional literature suggests that there are further Augeo products in the pipeline.

10.2.4.4 Glycerol Mono- and Diacyl Esters. Glycerol mono- and diacyl esters are well known and find broad application as emulsifiers, plasticizers and solvents.[67] These esters are excellent emulsifiers and their application as non-ionic surfactants is described in Section 10.4.2.

10.2.4.5 Glycerol Carbonate. Glycerol carbonate may be prepared directly from glycerol and carbon dioxide in the presence of appropriate catalysts,[78] or by reaction of glycerol

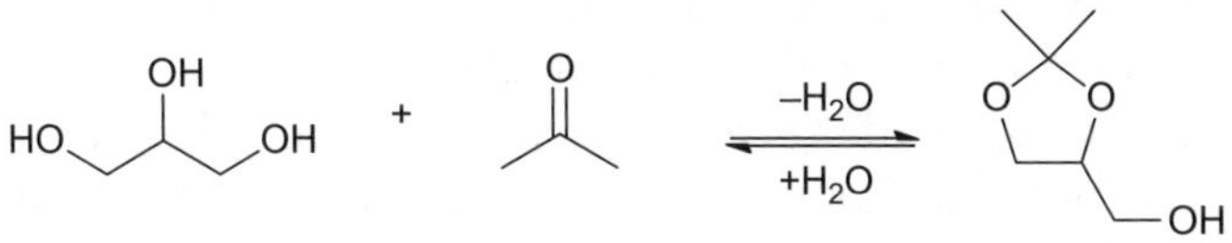

Scheme 10.5 Preparation of the 2,2-dimethyl-4-hydroxymethyl-1,3-dioxolane by reaction of glycerol with acetone. Distillation of the product and then hydrolysis are described as a purification method for crude glycerol, but the dioxolane is also a useful chemical product.

with urea in the presence of a dehydrating agent.[79] It is now offered by Huntsman Chemicals under the tradename JEFFSOL as a solvent or reactive intermediate of particular application in polymer synthesis.[80]

10.2.4.6 Propylene Glycol. A plant producing bio-based 1, 2-propanediol, also known as propylene glycol, was opened by Archer Daniels Midlands Company in 2011 to add to their 'Evolution Chemicals' line of bio-based chemicals, largely based on raw materials from soy.[81] Conversion of glycerol to propylene glycol is achieved by hydrogenolysis of glycerol (or sorbitol) over heterogeneous catalysts at elevated temperature, followed by purification of the product.[82] The catalysts used are rhenium, or rhenium–nickel, supported on carbon and may be used under different conditions to yield either propylene or ethylene glycol from sugars or sugar alcohols in addition to glycerol.[83,84]

10.2.4.7 Smaller Volume Chemicals from Glycerol. A range of chemicals with a three-carbon backbone and with established markets might be prepared by oxidation or reduction of glycerol. These include very small-volume, potentially high-value products such as dihydroxyacetone, glyceric acid, tartronic acid and glyceraldehydes, small-volume chemicals such as propionic acid, *n*-propanol, propionaldehyde and allyl alcohol and large-volume products such as acrylic acid, 2-propanol, acetone and propylene oxide.[60]

It is possible to envisage a range of dehydration steps followed by further reaction to yield a range of three-carbon compounds and even, ultimately, aromatics,[84] but one should also keep in mind the direct use of glycerol as, for example, a reaction solvent offering some of the benefits of water (low toxicity, low price) and ionic liquids (low vapour pressure and high boiling point, low solubility in extraction solvents and supercritical CO_2).[85]

Finally, there are the chemicals accessible from glycerol that are currently prepared from other bio-based feedstocks such as 1, 3-propanediol. Produced *via* a fermentation route from corn syrup, it is commercially manufactured by a joint venture between DuPont and Tate & Lyle and marketed as Bio-PDO, and is used in the production of polymers such as poly(trimethylene tere-phthalate) marketed as Sorona.[86] Although this route is not from glycerol, it is clear that there is a market for the product and that both bio-catalytic[87–89] and chemo-catalytic[90] routes from glycerol exist.

10.2.5 Chemicals from Proteins and Other High-value Products

Although most plant biomass is comprised predominantly of carbohydrates, polysaccharides and lignin, there are also myriad lesser components, many of which are of high value. Adequate separation and purification of these fractions to provide speciality chemicals may contribute significantly to the profitability of an integrated biorefinery and could constitute an important part of pretreatment or upstream processing.

10.2.5.1 Protein. Some biorefinery residues, such as seed cake remaining after vegetable oil extraction, have a high protein content and are of value as animal feed.[91] If extraction techniques are consistent with this application, this use is preferred over further processing to extract and use the proteins. Amino acids may appear to be the most obvious chemicals, likely to be derived from this biomass fraction, but many of these are prepared synthetically, at least in part due to customer resistance to animal-derived products.[92] In addition, some of the protein-rich starting materials, such as keratin from feathers, are useful raw materials in the production of composite materials, as discussed in Chapter 22. Bio-based polymer materials derived from protein-rich biomass fractions, such as soy protein isolate,[93] may provide biodegradable packaging materials.

10.2.5.2 Waxes. Cuticular waxes constitute a very small proportion of the overall wet mass of most plants (exceptions, such as carnauba and candelilla waxes, are specialist products from tropical plants) yet can be converted to high-value products after removal from the bulk by scaleable extraction processes, including supercritical CO_2 extraction.[94] Careful fractionation into polycosanols, long-chain alkanes and sterols provides potentially very high-value materials of use in the cosmetic, nutraceutical and crop protection industries.[95]

10.2.5.3 Essential Oils. Extraction of crops specifically grown for their valuable essential oils is an ancient industry, but, as the biorefinery concept enters mainstream thought, essential oils from food-industry wastes, such as citrus peel, are attracting interest.[96] D-Limonene, a cyclic terpene, is the main product from this source (~90% of the oil, which in turn constitutes 0.5% of the fruit crop based on whole fruit).[96] In addition to fragrance and flavour applications, this terpene is an excellent solvent (see below).

10.3 THE REDUCTIONIST APPROACH *VERSUS* USING NATURE'S COMPLEXITY

The chemical industry has become more and more reliant on purified single feedstocks or fractions with relatively low complexity derived from the petrochemical industry by refining, cracking and fractionation. A similar reductionist approach to the utilisation of bio-based feedstocks is a good fit with current plants and processes, but this often means 'retroengineering' the feedstock, *i.e.* taking multifunctional, and usually highly oxygenated compounds derived from biofeedstocks, and reducing them to the simpler deoxygenated compounds familiar from petrochemical (or other fossil) feedstocks. Perhaps the most extreme example is the reduction of complex bio-based raw materials to syngas followed by reconstruction of familiar molecules by known Fischer–Tropsch processes.[97]

A crucial difference, then, between a complex primary petrochemical source of molecules and biomass is one of functionality. The collection of petrochemical molecules has little or no intrinsic functionality. The individual molecules must be, after isolation, further modified and functionality introduced *via* chemical derivatisation into structures that confer specific physicochemical properties or allow for further elaboration into structures that serve as surfactants, polymers, solvents, *etc.* Furthermore, the modified molecules are then reassembled into complex functional mixtures in commercial products. This functional complexity is already present in biological systems: amphiphilic molecules, polymers and functional organic molecules are all present and fulfil specific roles. A worthwhile ambition would be to maintain the functional complexity present in biological systems *via* selective (minimal) purification or derivatisation of biomass and transfer the functionality directly into the finished product. A diagrammatic illustration of the concept is presented in Figure 10.3.

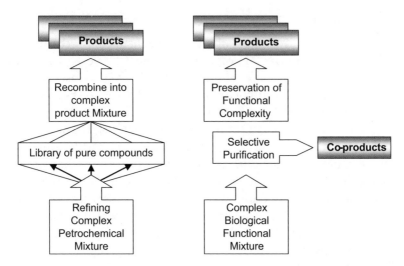

Figure 10.3 Schematic diagram illustrating the conceptual difference between petrochemical feedstock processing to provide a library of pure (or at least fractionated) chemicals for further functionalisation *versus* exploiting the complexity of bio-based feedstocks.

10.4 BIO-BASED CHEMICALS BASED ON CHEMICAL TYPE AND APPLICATION

Chemical use is not driven by the availability of a resource, but by the need for a product, or, more correctly, the effect or 'function' of that product. Thus, the consumer selecting a washing liquid for clothes purchases a formulation designed to clean cloth, not a 25% aqueous solution of a particular surfactant mixed with salt, perfume and a plethora of additives designed to enhance the effect of the main ingredient and to make the formulation easy and pleasant to handle. The formulator, in turn, selects an active chemical ingredient based on efficacy, cost, compatibility with other ingredients used, security of supply chain and (although perhaps not overtly) familiarity. In the sections below, both drop-in replacement and functional replacement bio-based chemicals are considered, on the basis of their chemical class or function. This is a very large topic and we have therefore chosen as exemplars two large classes of chemicals: solvents and surfactants.

10.4.1 Solvents

Although solvents are almost ubiquitous in chemical processing and synthesis (particularly of organic chemicals), very significant volumes are used in coatings, inks and adhesives, consumer products and industrial cleaning applications.[98] Extractions of valuable natural products such as flavours and fragrances and drug substances or precursors are also processes that are 'solvent heavy', but in many cases these have been, or are being, replaced by processes such as super- or near-critical CO_2 extraction (see Chapter 15). Remarkably, a growing area of solvent use is in the extraction of oils for biodiesel production, where hexane is widely used. Chronic exposure to hexane causes peripheral nervous system damage, but the cost of replacing this with, for example, heptane renders many processes uneconomic due to the high energy costs incurred in subsequent evaporation of the heptane!

Solvents derived from renewable resources come in one of two types: drop-in replacements that are identical to the petrochemically derived materials, and different molecular entities with similar properties that 'do the same job'. The latter may often be improvements over the solvents that they replace as greater attention is now paid to issues of health and safety, *e.g.* emissions due to evaporation of volatile organic compounds (VOCs) and to end-of-life considerations, *e.g.* biodegradability.

10.4.1.1 Like-for-Like 'Drop-in Replacement' Solvents

Alcohols and Other Fuels. Bulk chemicals produced for their fuel value, such as ethanol, butanol and dimethyl ether, are often useful solvents and, as bio-based versions of these are developed as fuels, large volumes will become available for solvent use. Fermentation technologies based on the acetone, butanol, ethanol (ABE) process for the production of alcohols from sugars (which may, in turn, be derived from non-food polysaccharides such as cellulose) are advancing rapidly and the price of these fuels is likely to make such solvents accessible from bio-sources.

Ethyl Acetate. Ethyl acetate may be prepared from bio-based acetic acid and ethanol available by fermentation of sugars. So-called 'acetogens' may be used to convert both C_5 and C_6

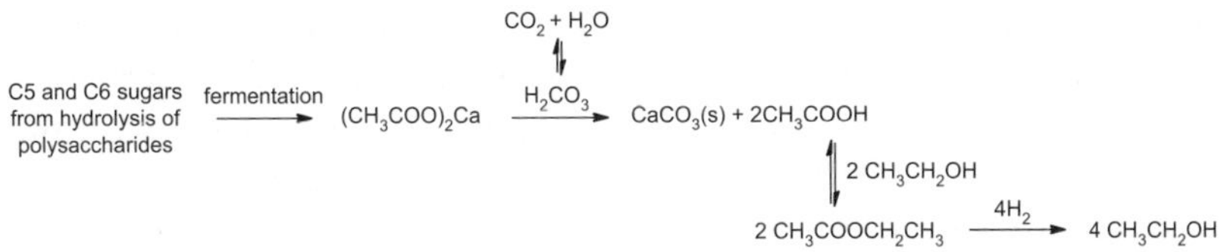

$$CO_2 + H_2O$$

C5 and C6 sugars — fermentation → $(CH_3COO)_2Ca$ — H_2CO_3 → $CaCO_3(s) + 2CH_3COOH$
from hydrolysis of
polysaccharides

$2\ CH_3CH_2OH$

$2\ CH_3COOCH_2CH_3$ — $4H_2$ → $4\ CH_3CH_2OH$

Scheme 10.6   Zeachem's process for ethyl acetate and ethanol production.

sugars to acetic acid, producing no carbon dioxide in the process. For example, in the Zeachem process, calcium acetate is produced directly from sugars at near neutral pH, acetic acid is generated from the acetate in solution by addition of CO_2 and the ethyl ester is formed by direct reaction with ethanol (Scheme 10.6) in a process that includes reactive distillation (of the ethyl acetate–water–ethanol ternary azeotrope, composition 82.6:9.0:8.4) to shift the equilibrium and achieve high conversion.[99] Ethanol is recycled in the process and, in the original patent application, Zeachem described further hydrogenation of ethyl acetate to produce ethanol.

Hydrocarbons.   Even ignoring the huge demand for hydrocarbons as fuels, the current chemical industry is set up to use hydrocarbons as feedstocks for a vast range of processes (which was also Bozell and Petersen's rationale for including biohydrocarbons in the new 'top ten' platform chemicals[20]), so offering chemically identical materials from bio-sources serves to facilitate the transition from petrochemicals to bio-based chemicals.[100]

Hydrocarbons are produced directly by a range of microorganisms, including cyanobacteria, bacteria, yeasts and fungi.[101] Long chain ($C_{25}–C_{35}$) hydrocarbons are found intracellularly, where they are thought to fulfil physiological functions. This hydrocarbon content, based on dry mass, is usually $<1\%$, although some higher values have been reported.[42] Many organisms produce extracellular (usually lighter) hydrocarbons on a much larger scale. For optimised processes with selected organisms, such as the hydrocarbon-producing bacterium *Vibrio furnissii* M1, values of 60% based on cell mass dry weight have been reported.[102] Algae such as *Botryococcus braunii*, which are both rich in oil and grow at relatively high density in 'blooms' in the wild, are also possible sources of bio-produced hydrocarbons.[103] The hydrocarbons produced tend to be both long chain and branched, with significant levels of unsaturation; the type and distribution of hydrocarbons are dependent on organism type, species and carbon source (and other nutrients), and also conditions; some representative hydrocarbon structures are presented in Chart 10.1. For long-chain alkenes and alkanes, it is suggested that the most likely biosynthetic route is by decarbonylation of fatty acid aldehydes.[104]

Inspection of recent patents filed suggests that, in most cases, hydrocarbons produced from bio-sources such as algae would be subjected to cracking or hydrodeoxygenation, possibly following hydrogenation, or followed by isomerisation.[105] Although the chemical processing following bio-production of hydrocarbons might appear onerous (and extra cell lysing and oil extraction steps are also required), it is worth noting that these, or very similar, processes are already widely applied in upgrading of fossil oil in a petrochemical refinery. Indeed, Solazyme, who produce Solajet, a prototype aviation fuel derived from cracked algal oils, have supplied jet fuel tested in commercial airliners, albeit as a 60:40 blend with conventional aviation fuel.[106]

Although fuels were ruled out of scope for this chapter, it is clear that the development of large-scale hydrocarbon manufacture for fuels could also provide a chemical feedstock and this is particularly evident if the oils produced are subjected to processing in a conventional or minimally modified petrochemical refinery, thus producing a range of products that are already familiar to the industry. Although the processes might be similar to those used in

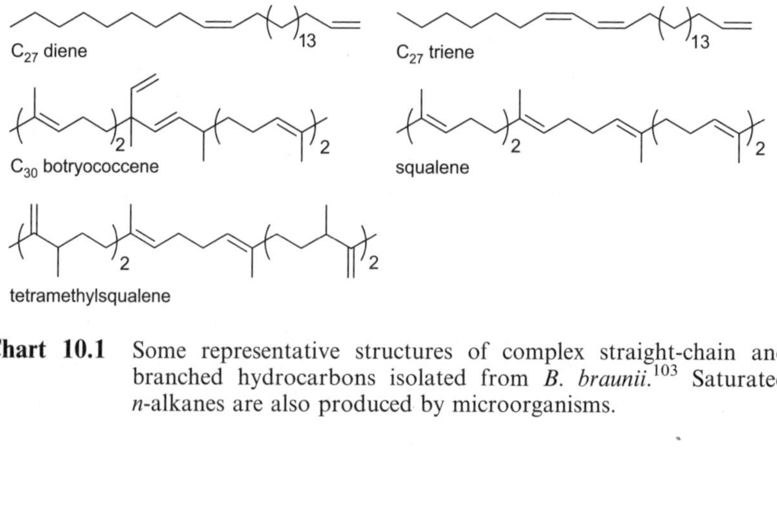

C$_{27}$ diene

C$_{27}$ triene

C$_{30}$ botryococcene

squalene

tetramethylsqualene

Chart 10.1 Some representative structures of complex straight-chain and branched hydrocarbons isolated from *B. braunii*.[103] Saturated *n*-alkanes are also produced by microorganisms.

conventional refineries, it is likely that new catalysts will be required.[107]

A less widely publicised approach to bio-based linear hydrocarbons entails metathesis of long-chain fatty acids (or triglycerides) with ethylene in the presence of catalysts such as the well-known Grubbs or Grubbs–Hoyveyda catalysts. (The ethylene used may be derived from ethanol, obtained by fermentation.) The process may follow isomerisation of polyunsaturated fatty acid esters, as in the Cargill process for the preparation of a range of abbreviated fatty acid esters bearing terminal alkene groups and 1-octene (Scheme 10.7).

Terminal alkenes, such as isoprene and isobutylene, may be produced by over-expression of specific enzymes in robust hosts such as *Escherichia coli*, with the added advantage that they are easily stripped from the broths as gases.[100]

The use of terpenes as solvents is hardly new; turpentine (derived from wood or trees, *e.g.* by distillation of resin) having been in wide use for many years. The displacement of such bio-derived solvents by petroleum-based products may be beginning to undergo a partial reverse; increased products from wood biorefineries and terpenes may also be converted to fine chemicals.[108]

10.4.1.2 Functional Replacement Solvents. A significant number of solvents based on esters of various organic acids and alcohols produced from bio-sources are appearing on the market. In some cases these have been available for some time, whereas in others they are offered as alternatives to less desirable solvents which result in significant VOC emissions in use. Thus, many of these

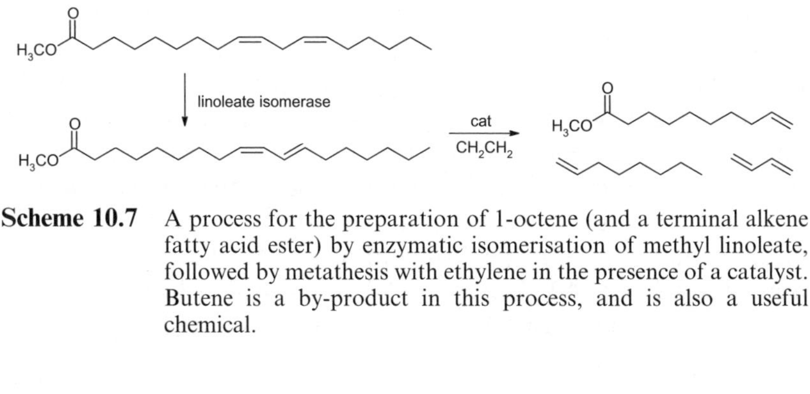

Scheme 10.7 A process for the preparation of 1-octene (and a terminal alkene fatty acid ester) by enzymatic isomerisation of methyl linoleate, followed by metathesis with ethylene in the presence of a catalyst. Butene is a by-product in this process, and is also a useful chemical.

solvents are offered to the market as formulated mixtures for specific purposes, such as paint strippers or degreasers. Below we attempt to focus on the molecular form of the solvents, but it should be recognised that in most applications, except perhaps synthetic chemistry, single-component solvents are not required or even desirable. The function of dissolution of specific substrates may be achieved by careful blending of molecular components.

Glycerol Carbonate. Glycerol carbonate may be prepared from glycerol by reaction with dimethyl carbonate in the presence of a biocatalyst: a lipase such as *Candida arctica* lipase B (CalB).[109] The process may be carried out in tandem with biodiesel production so that both FAMEs and glycerol carbonate are produced in one pot.[110] As many enzymes remain active in the presence of glycerol carbonate, this also provides a bio-based solvent for biotransformations,[111] including as a co-solvent with ionic liquids.[112] Currently, glycerol carbonate (along with solketal, see above) is widely used as a solvent in cosmetic and personal care formulated products and opportunities exist for further functionalisation via reaction of the free alcohol functional group.[113]

Ethyl Lactate and Other Lactate Esters. Ethyl lactate, the ester of ethanol and lactic acid (usually L-lactic acid produced by fermentation of sugars), may be prepared entirely from carbohydrate feedstocks, is biodegradable and is approved as a food additive. Ethyl lactate is usually prepared by the esterification reaction between ethanol and lactic acid using acid catalysis and most proprietary technologies pertain to energy- (and thus cost-) effective separation processes and purification.[114] Cargill Dow and Ashland have reportedly teamed up to prepare 'ultra-pure' ethyl lactate for use in the electronics industry, from ethanol and lactide.

The solvent characteristics of ethyl lactate are often described as similar to those of *N*-methylpyrrolidone, although both the flashpoint and boiling point are lower (91 and 204 °C versus 46 and 151–155 °C, respectively). Ethyl lactate solvent is marketed by a large number of companies for use in the formulation of inks and coatings and other formulated products such as resin cleaners and paint strippers (including graffiti removers). As it is approved as a list 4A 'inert ingredient' by the US Environmental Protection Agency (EPA),[115] it is 'exempt from the requirement of tolerance' and may be used, at any level, in the formulation of agricultural chemicals such as pesticides and herbicides. It and other lactate esters are included in the Significant New Alternatives Programme (SNAP) lists for electronics and precision cleaning substitutes.[116]

A characteristic odour, that is considered by many to be unpleasant, has been a barrier to the acceptance of ethyl lactate solvent in many applications, but at least one manufacturer offers blends with ethanol or other lower alcohols, which serves both to reduce odour and an associated 'bite' sensation and to modulate solvency and drying rate.[117] Addition of blends of terpenes and terpenoids (primarily limonene and linalool) derived from citrus, or even pure D-limonene, to lactate esters yields solvents that may be more effective at dissolving polymer and paint residues and thus for use as line flush solvents.[118] D-Limonene alone is not water miscible, restricting its use to cleaning applications that either allow a non-aqueous final rinse or are tolerant of surfactant-containing rinse water, but some combinations of D-limonene and ethyl lactate (marketed by Vertec BioSolvents as VertecBio Citrus I20) may be rinsed from a surface using high-pressure water and reportedly outperform toluene and xylene in paint removal applications.[119]

An interesting application of ethyl lactate solvent is in recycling of domestic food packaging plastics. These are shredded and washed with ethyl lactate, followed by liquid CO_2, which provides plastic clean enough for recycling back into the packaging stream.[120] Both solvents are recycled by distillation and the relatively low-volume residues are disposed of as solid waste. In contrast to water-based cleaning, no liquid wastes emanate from the process.[40]

A range of lactate esters are produced by Purac as part of their Purasolv range, which includes methyl, ethyl and *n*-butyl esters of L-lactic acid. As the market for biodegradable lactide-based polymers grows, prices of other lactic acid derivatives should also become more favourable and their use as chemical intermediates may also grow. Lactic acid or derivatives, such as lactamides,[121] provide potentially useful chiral building blocks in organic synthesis.

Other esters of acids produced by fermentation, such as ethyl valerate or ethyl pentanoate, may follow as 'functional replacement' fuels are developed.[122]

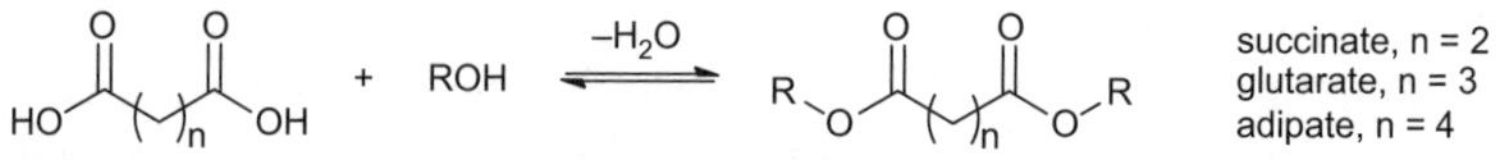

Scheme 10.8 Dibasic esters used as solvents. The alcohol used to prepare the ester and thus R may vary, but the most common are methyl esters, R = CH$_3$, ethyl esters, R = CH$_3$CH$_2$, and isobutyl esters, R = (CH$_3$)$_2$CHCH$_2$.

Dibasic Esters. Diesters of dicarboxylic acids, dibasic esters, have been promoted for some time as replacements for methylene chloride in paint stripping applications. These are commonly mixtures of glutarates, succinates and adipates (Scheme 10.8), which are produced from the by-products of adipic acid production[123] (probably the adipic acid crystallisation step). Although ratios vary from producer to producer, the distribution is approximately succinate 15–28%, glutarate 55–67%, adipate 9–25%, reflecting the separation of the major glutaric acid contaminant by exploiting its greater solubility in water (at 20 °C the solubility of both adipic and succinic acid is <60 g L^{-1}, whereas that of glutaric acid is >1200 g L^{-1}).[124]

Adipic acid is overwhelmingly produced from petrochemical feedstocks, so, at present, these are not bio-based solvents (although many of the alcohols used in ester synthesis may be obtained from fermentation processes). However, routes to adipic acid from renewable raw materials[125] are beginning to be commercialised[126] and succinic acid is one of the platform chemicals, which may be produced by fermentation[127] thus, as bioadipic and biosuccinic acid production grows, it is expected that dibasic esters derived from renewable resources will become available. Industrial biotechnology required for the large-scale production of these and other diacids will be driven by the demand for monomers for polymer production, but esterification is already mooted as a means of separating the acids from the fermentation broths.[128] The dibasic esters are already extensively marketed, as summarised in Table 10.3.

FAMEs. Fatty acid methyl esters (FAMEs), or biodiesel, are excellent solvents in their own right (a characteristic that can

Table 10.3 Summary of dibasic esters marketed for different solvent applications; many of these or related compounds could become available from fully bio-based sources as bio-synthesis of adipic and other diacids develops.

Manufacturer	Product	Acid from which derived	R	Applications and benefits
DuPont Rhodia	DBE Rhodiasolv RPDE	Glutaric, adipic, succinic mixture	Methyl	Coating formulations; application and stripping. Coalescing agent in water-borne formulations. Other formulation applications including pesticides and insecticides, liquid detergents, corrosion inhibitors and printing inks. Low toxicity and vapour pressure, high flashpoint, biodegradable
Rhodia	Rhodiasolv DEE	Glutaric, adipic, succinic mixture	Ethyl	As above, but particularly as a paint coalescence agent
	Rhodiasolv DIB	Glutaric, adipic, succinic mixture	Isobutyl	Metal degreasing (dissolves most oils), coalescent agent in paint formulations. Low odour/low VOCs, good ecotoxicity profile, hydrolytically stable
	Rhodiasolv IRIS	Various isomers of adipic acid: 1–5% adipic, 80–94% 2-methylglutaric, 7–14% 2-ethylsuccinic acid	Methyl	Paint stripping, graffiti removal, industrial cleaning of paint and resin vessels and lines

bedevil the production of biodiesel by the unwary, sometimes resulting in swelling or dissolution of seals and fuel lines). Methyl soyate is widely marketed as a solvent and a study of the solvent properties characterised by inverse gas chromatography, and solvent parameters calculated, has recently been published.[129] Perhaps not surprisingly, the extrapolated total solubility parameter of methyl soyate is found to be similar to that reported for soybean oil. As the triglycerides of soybean oil contain a range of fatty acid moieties, methyl soyate and other FAMEs produced from vegetable (or indeed animal) fats and oils will have variable composition. For example, soybean oil typically contains the following fatty acid distribution: α-linolenic 7–10%, linoleic 51%, oleic 23%, stearic 4%, palmitic 10%. As linolenic acid is prone to oxidation (indeed, soybean oil is a 'drying oil', *i.e.* undergoes cross-linking by autoxidation), this could limit its use in certain applications, although strains producing low linolenic oils are available.[130]

γ-Valerolactone and 2-Methyltetrahydrofuran. Both γ-valerolactone (GVL) and 2-methyltetrahydrofuran (2-Me-THF) may be prepared from levulinic acid, which is, in turn, derived from sucrose under acid treatment (Scheme 10.9) (a range of carbohydrate starting materials, including cellulose and acid catalysts, may be used). Dumesic and co-workers have described an improved process of reactive distillation of butene esters of levulinic and formic acids, which spontaneously phase separate from

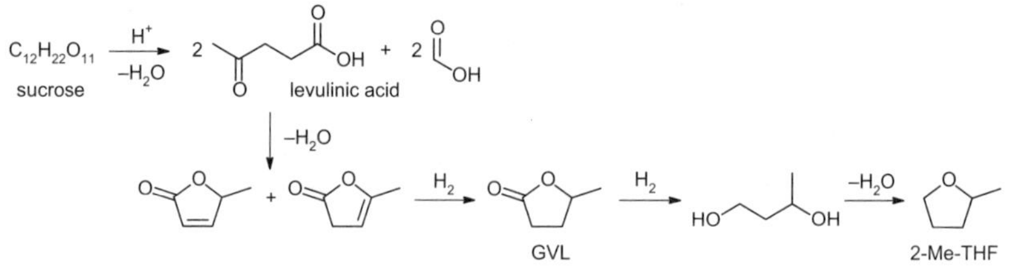

Scheme 10.9   Levulinic acid (and formic acid) are derived from sucrose. Dehydration of levulinic acid followed by hydrogenation yields GVL and further dehydration/hydrogenation steps yield 2-Me-THF.

aqueous sulfuric acid (the acid of choice in levulinic acid production by the Biofine process[131]), so simplifying purification.[132] (In this process, the products are further reacted to yield C_8 alkenes suitable as fuels, but also providing a new route to unsaturated hydrocarbons.)

Horváth *et al.* proposed that GVL forms an ideal energy storage liquid,[133] *i.e.* a fuel or fuel additive, but its favourable flash point (96 °C), low toxicity (LD_{50} rat, oral 8800 mg kg^{-1}) and remarkably low vapour pressure, even at elevated temperatures, would also render it useful as a solvent under conditions that do not lead to hydrolysis.

2-Me-THF, prepared directly from GVL or from furfural (from biomass), may be used as a higher boiling replacement for THF (2-Me-THF b.p. = 80.3 °C), but it is prone to formation of peroxides.[134]

10.4.2 Surfactants

If one includes soaps, surfactants are some of the oldest man-made chemicals in use and constitute the price and performance benchmark against which subsequent synthetic surfactants have been developed. Over decades of process, price and performance optimisation, synthetic surfactants have surpassed soaps in a number of important ways: greater availability of hydrophobic chain sources (over tallow and palm oil), easier production and processing, greater tolerance to hard-water ions, excellent detergency and superior foaming, so that industrially synthetic surfactants became the benchmark. Although not all surfactants are destined for detergent applications, they constitute about 50% of surfactant use. Global surfactant production is currently estimated to be in the region 12–13 million tonnes per annum, 65% of which is destined for use in consumer products.

Surfactants are extensively used in aqueous detergent and personal care formulations. They are multifunctional molecules acting simultaneously as stabilisers, emulsifiers, solubilisers, viscosity modifiers, degreasers and foamers. Classically, formulations have exploited this multifunctionality. However, often more surfactant is used than is strictly necessary to perform the primary function of the formulation (for example, detergency), which incurs consequences, *e.g.* increase in carbon footprint. Substituting petrochemically derived surfactants with judiciously selected renewable surfactants can, in principle, go some way towards reducing the carbon footprint.

Renewable surfactants have been reviewed.[134] The renewable hydrophobic moiety of the amphiphile comes predominantly from the transformation of triglycerides, chiefly from palm kernel oil and coconut oil.[135–137] Alternative sources of hydrophobes can be found in by-products of biomass conversion, most notably tall oil, which can constitute up to 0.5% of Kraft pulp production.[138,139] The recent drive towards biofuel production, however, has shifted the focus to algae as the alternative source of triglycerides. Advantages include potentially high yields (with respect to algal mass) of selected oils from a non-food competitive source.

Triglycerides then undergo a number of standard chemical processes to produce (in addition to soaps) the fatty acids (FAs), FA esters, FA chlorides, FA amines, FA anhydrides and fatty alcohol intermediates which are the precursors to a large number of different surfactants classes.[136]

Chemical coupling of hydrophobic precursors with suitable hydrophiles yields a variety of surfactants and determines the type of surfactant: non-ionic, anionic, zwitterionic or cationic.

10.4.2.1 Non-ionic Surfactants

Mono- and Diacyl Glycerides. Conceptually, monoglycerides (MGs) and diglycerides (DGs) of fatty acids represent the simplest form of non-ionic surfactant, as they are structurally very close to the originating triglycerides with no extraneous chemical moiety being added. Although synthesis via transesterification[140] of FAMEs in enzyme-mediated reactions,[141] or in supercritical CO_2,[142] have been proposed, MGs and DGs are generally obtained via glycerolysis (transesterification of triglycerides with excess glycerol), which yields predominantly MGs (50–70%) and a significant fraction of DGs. Principally used in the food industry[143,144] in dairy emulsions,[145] baking and nutraceuticals, they also find wide application in the formulation of personal care products, where they are employed for their structuring ability and good skin feel,[146] as well as in pharmaceutical formulations as binders and in trans-dermal patches.[147]

Saccharide-based Surfactants. Saccharide surfactants represent another important family of bio-based non-ionic surfactants. Given the ubiquitous nature of polyols in Nature, they are an obvious choice of hydrophile for the preparation of surfactants such as alkyl glucosides, glucamides and glucose and sucrose esters (Chart 10.2). Of these classes of saccharide surfactants, alkyl glucosides (if the sugar is glucose, otherwise 'glycosides') are among the most important in personal care because they are industrially available in appropriate quantities, are relatively mild and act as foam boosters. Von Rybinski and co-workers have written extensively about these surfactants.[148,149,150] Briefly, they are obtained either by transacetylation of a butyl or ethyl glycoside with the appropriate hydrophobic alcohol required for the target surfactant or *via* direct reaction of glucose (from starch) with the long-chain fatty alcohol of choice,[135] both under acid-catalysed conditions. There are efforts to use agricultural waste to overcome the use of a food competitive raw material in glycoside production.[151,152]

10.4.2.2 Anionic Surfactants. Other important ingredients in the manufacture of surfactants from natural biomass include molecular species carrying a charge. Those that carry a negative

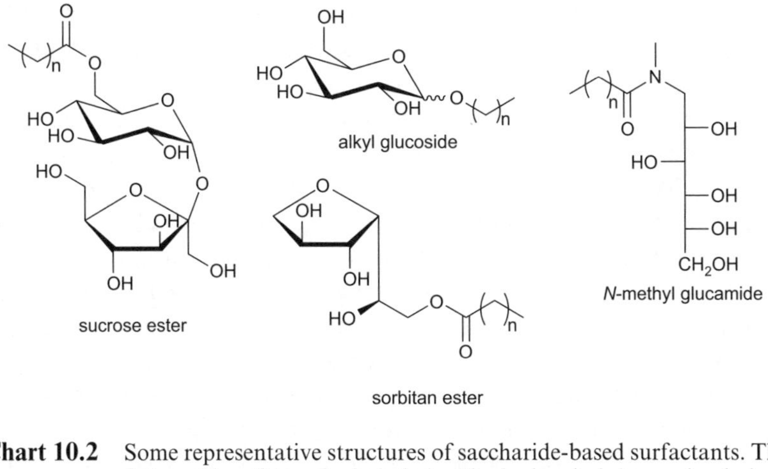

sucrose ester

alkyl glucoside

sorbitan ester

N-methyl glucamide

Chart 10.2 Some representative structures of saccharide-based surfactants. The fatty acid or fatty alcohol chain (the hydrophobe) may be derived from either plant or animal sources, but the former are more common.

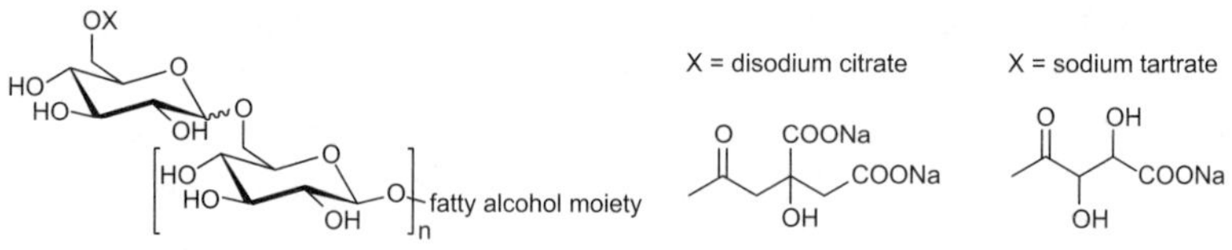

Chart 10.3 Structures of citrate and tartrate polyglycosides – bio-based anionic surfactants.

charge are particularly important. Carboxylates are the most obvious and widely used species of hydrophile. A large number of surfactant examples have been produced,[153] but only a few are of commercial importance. Surfactants with anionic headgroups from renewable hydrophiles have been reviewed.[136,155] Typically citric, tartaric and lactic acid are the carboxylic acids from biomass which have commercial significance for the manufacture of surfactants, such as in the citrate or tartrate esters of alkyl polyglucosides (marketed as AGEs by Lamberti) (Chart 10.3) or directly as the sodium salts of alkyl lactates. These surfactants are marketed as efficient and mild.

The surfactants mentioned above, including those obtained from amino acids,[154] although either wholly or in part based on renewable raw materials, must still undergo some chemical transformation.

10.4.2.3 Biosurfactants. Biosurfactants are obtained directly, often in their final form, from biomass and are therefore considered 100% natural. Other advantages include biodegradability (often rapid), low toxicity, consumer appeal and high efficacy, and depending on the raw material feedstock they can be obtained from waste.[155] Although some biosurfactants are starting to become the focus of commercial interest, they remain high-value materials when compared with commodities. Their adoption on a large scale is hampered by the problems typically encountered in microbial bioprocessing: dilute culture media, generally low overall yields and expensive product recovery.[156] Biosurfactants are a class of amphiphiles produced by a variety of microorganisms either as part of their cell membrane or excreted as metabolites. A number of biosurfactants have been identified and studied to date. Mulligan[157,158] and Rahman and Gakpe[159] have provided excellent overviews of the production and applications of biosurfactants.

Because they are the result of microbial activity, the type of biosurfactant produced then depends crucially on the micro-organism, the culture conditions and the feedstock source. The most important among these biosurfactants are glycolipids, where the hydrophilic portion of the molecule consists of mono- and oligo-saccharides of glucose, mannose, galactose, rhamnose, *etc.* Although there are a large number of biosurfactant classes, the two most important classes of glycolipids are rhamnolipids and sophorolipids.

Rhamnolipids. Rhamnolipids,[160,161] produced by *Pseudomonas* bacterial strains, are reported to be excellent surfactants, exhibiting all the usual properties of surface-active molecules, such as detergency, foaming, emulsifying, solubilising and wetting action, but appear to have a particular potential in bioremediation for the removal of hydrophobic contaminants. 'As-prepared' rhamnolipids tend to be mixtures of mono- and di-rhamnolipids (Chart 10.4), with the former predominating in commercial preparations from *P. aeruginosa*.[162] Rhamnolipid surfactant preparations find application in the cosmetics and personal care industry,[163] but also show promise in bioremediation of heavy metal-contaminated sites[164] and enhanced oil recovery.[165]

Sophorolipids. Sophorolipids[166,167] (Chart 10.4) represent a promising class of biosurfactants, which can be produced in high yields (400 g L^{-1}) by strains of the non-pathogenic microorganisms *Candida* yeast. They are low-foaming surfactants with a

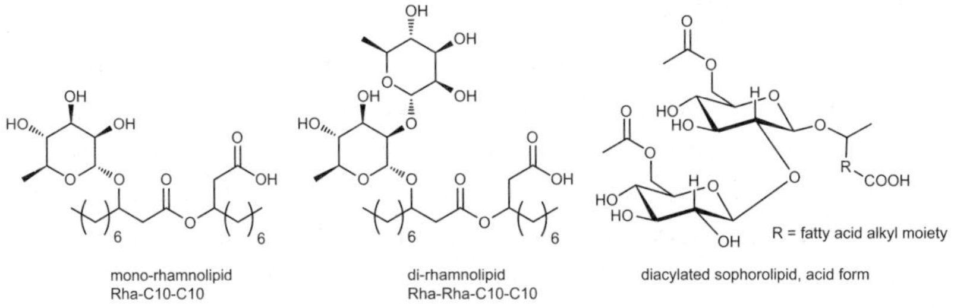

mono-rhamnolipid
Rha-C10-C10

di-rhamnolipid
Rha-Rha-C10-C10

diacylated sophorolipid, acid form

R = fatty acid alkyl moiety

Chart 10.4 Representative structures of biosurfactants. Rhamnolipids bearing a range of fatty acid moieties (C$_8$–C$_{14}$) are known. Sophorolipids in commercial preparations are present in both non-acylated and mono- and diacylated forms and may be present as a lactone form post-cyclisation.

surprising tolerance for water hardness and with detergency comparable to that of sodium soap and synthetic surfactants. Additionally, they present low toxicity and are rapidly biodegradable.

10.4.2.4   Opportunities and Challenges.   The increasing interest in renewable surfactants provides the industry with both opportunities and challenges. The opportunities lie in being able to break free of a resource which is finite, with a largely negative CO_2 impact and increasingly poor public relations image and to move towards the use of renewable resources that have a more benign environmental profile. Additional opportunities lie in the intellectual property associated with surfactant development and their use in formulations, and also in the marketing value (for now) of positive environmental credentials.

The challenges have typically been in developing renewable alternatives which are cost competitive and at least as efficacious as the synthetics they aim to replace, without incurring additional CO_2 penalties. Doing so on a scale which is relevant to the major global users will be an additional challenge, physicochemical considerations notwithstanding. A further challenge perhaps lies in the marketing approach to perception of product performance. Consumers associate high foaming with high detergency, but it is well known among product scientists that this does not always follow. In this regard, some formulations may be considered as technically over-specified, with subsequent environmental implications for rinse water use.

10.5   CONCLUSION

The development of both drop-in and functional replacement chemicals from biomass is proceeding apace, but generally it is direct drop-in replacement chemicals that are being readily and quickly adopted. Functional replacements appear to be more commonly employed when there is a distinct hazard associated with a chemical, *e.g.* replacement of methylene chloride in paint strippers.

From a formulator's point of view, the exploitation of materials that are complex mixtures of functional molecules appears exciting, but requires a 'blank canvass' approach, *i.e.* reformulation in a manner which takes full advantage of the properties of renewable raw material, including (and in particular) material inhomogeneity.

A survey of the patent literature yields a vast number of recently filed patents pertaining to the preparation and use of bio-based chemicals, in all areas, from fuels (which may also provide bulk chemical building blocks) through bulk chemicals and monomers to speciality chemicals, suggesting a potential explosion of activity in the area in the future, particularly if truly integrated biorefineries become widespread.

REFERENCES

1. W. J. Hale and W. Jay, *The Farm Chemurgic: Farmward the Star of Destiny Lights Our Way*, Stratford, Boston, MA, 1934.
2. P. J. Clark, Chemurgy, in *Kirk–Othmer Encyclopedia of Chemical Technology*, Wiley, New York, 2000, p. 553.
3. J. S. McLaren, *J. Chem. Tech. Biotechnol.*, 2000, **75**, 927.
4. R. Diercks, J.-D. Arndt, S. Freyer, R. Geier, O. Machhammer, J. Schwartze and M. Volland, *Chem. Eng. Technol.*, 2008, **31**, 631.
5. M. Eissen, J. O. Metzger, E. Schmidt and U. Schneidewind, *Angew. Chem. Int. Ed.*, 2002, **41**, 414.
6. E. S. Lipinsky, *Science*, 1981, **212**, 1465.
7. J. Bozell, *Clean*, 2008, **346**, 641.
8. Useful reviews include: (a) P. F. Levy, J. E. Sanderson and R. G. Kispert, *Enzyme Microb. Technol.*, 1981, **3**, 207; (b) B. Kamm and M. Kamm, *Appl. Microbiol. Biotechnol.*, 2004, **64**, 137; (c) A. J. Ragauskas; C. K. Williams, B. H. Davison, G. Britovsek, J. Cairney, C. A. Eckert, W. J. Frederick Jr., J. P. Hallet, D. J. Leak, C. L. Liotta, J. R. Mielenz, R. Murphy, R. Templer and T. Tschaplinski, *Science*, 2006, **311**, 484; (d) A. Dermirbas, *Energy Convers. Manage.*, 2009, **50**, 2782.
9. H.-L. Kangas, J. Lintunen, J. Pohjola, L. Hetemaki and L. J. Uusivuori, *Energy Econ.*, 2011, **33**, 1165.
10. J. E. Santibañez-Aguilar, J. B. González-Campos, J. M. Ponce-Ortega, M. Serna-González and M. M. El-Halwagi, *Ind. Eng. Chem. Res.*, 2011, **50**, 8558.
11. P. Sharma, B. R. Sarker and J. A. Romagnoli, *Comput. Chem. Eng.*, 2011, **35**, 1767.
12. S. Octave and D. Thomas, *Biochimie*, 2009, **91**, 659.
13. I. M. Bowling, J. M. Ponce-Ortega and M. M. El-Halwagi, *Ind. Eng. Chem. Res.*, 2011, **50**, 6276.

14. R. R. Reeves, *J. Cetacean Res. Manage.*, 2001, **2**, 187.
15. F. Stephenson, *Florida State Univ. Res. Rev.* 2002, **12**(3), http://www.rinr.fsu.edu/fall2002/taxol.html (last accessed 1 April 2012).
16. R. E. Green, *Science*, 2005, **307**, 550.
17. L. P. Koh, J. Miettinen, S. C. Liew and J. Ghazoul, *Proc. Natl. Acad. Sci. USA*, 2011, **108**, 5127.
18. D. Mohan, C. U. Pittman Jr and P. H. Steele, *Energy Fuel*, 2006, **20**, 848.
19. A. Kind and A. Schnieke, *Trangenic Res.*, 2008, **17**, 1025.
20. J. J. Bozell and G. R. Petersen, *Green Chem.*, 2010, **12**, 539.
21. T. Werpy and G. Petersen (eds), *Top Value Added Chemicals from Biomass. Volume I – Results of Screening for Potential Candidates from Sugars and Synthesis Gas*, US Department of Energy, Washington, DC, 2004.
22. J. E. Holladay, J. J. Bozell, J. F. White and D. Johnson (eds), *Top Value Added Chemicals from Biomass. Volume II – Results of Screening for Potential Biorefinery Candidates Lignin*, US Department of Energy, Washington, DC, 2007.
23. J. Ralph, K. Lundquist, G. Brunow, F. Lu, H. Kim, P. F. Schatz, J. M. Marita, R. D. Hatfield, S. A. Ralph, J. H. Christensen and W. Boerjan, *Phytochem. Rev.*, 2004, **3**, 29.
24. J. Rencoret, A. Gutierrez, L. Nieto, J. Jimenez-Barbero, C.B. Faulds, H. Kim, J. Ralph, A. T. Martinez and J. C. del Rio, *Plant Physiol.*, 2011, **155**, 667.
25. M. Siedlecki, W. de Jong and A. H. M. Verkooijen, *Energies*, 2011, **4**, 389.
26. H. Younesi, G. Najafpour and A. R. Mohamed, *Biochem. Eng. J.*, 2005, **27**, 110.
27. A. M. Henstra, J. Sipma, A. Rinzema and A. J. M. Stams, *Curr. Opin. Biotechnol.*, 2007, **18**, 200.
28. R. P. Datar, R. M. Shenkman, B. G. Cateni, R. L. Huhnke and R. S. Lewis, *Biotechnol. Bioeng.*, 2004, **86**, 587.
29. M. B. Hocking, *J. Chem. Educ.*, 1997, **74**, 1055.
30. H. Priefert, J. Rabenhorst and A. Steinbüchel, *Appl. Microbiol. Biotechnol.*, 2001, **56**, 296.
31. (a) J. F. Kadla, S. Kubo, R. D. Gilbert and R. A.Venditti, Lignin based carbon fibres, in *Chemical Modification, Properties and Usage of Lignin*, ed. T. Q. Hu, Kluwer Academic/Plenum Publishers, New York, 2002, p. 121; (b) J. F. Kadla,

S. Kubo, R. A. Venditti, R. D. Gilbert, A. L. Compere and W. Griffith, *Carbon*, 2002, **40**, 2913.

32. S. Kubo and J. F. Kadla, *J. Polym. Environ.*, 2005, **13**, 97.
33. F. G. Calvo-Flores and J. A. Dobado, *ChemSusChem*, 2010, **3**, 1227.
34. X. Zhang, M. Tu and M. G. Paice, *Bioenerg. Res.*, 2011, **4**, 246.
35. J. Zakzeski, P. C. A. Bruijnincx, A. L. Jongerius and B. M. Weckhuysen, *Chem. Rev.*, 2010, **110**, 3552.
36. J. L. McCarthy and A. Islam, Lignin chemistry, technology and utilization: a brief history, in *Lignin: Historical, Biological and Materials Perspectives*, ed. W. G. Glasser, R. A. Northey and T. P. Schultz, ACS Symposium Series, Vol. 742, American Chemical Society, Wsahington, DC, 1999, p. 2.
37. J. D. Gargulak and S. E. Lebo, Commercial use of lignin-based materials, in *Lignin: Historical, Biological and Materials Perspectives*, ed. W. G. Glasser, R. A. Northey and T. P. Schultz, ACS Symposium Series, Vol. 742, American Chemical Society, Wsahington, DC, 1999, p. 304.
38. H. Evju, *US Patent*, 4 151 207, 1979.
39. C. Amen-Chen, H. Pakdel and C. Roy, *Bioresource Technol.*, 2001, **79**, 277.
40. M. P. Pandey and C. S. Kim, *Chem. Eng. Technol.*, 2010, **34**, 29.
41. J. H. Lora and W. G. Glasser, *J. Polym. Environ.*, 2002, **10**, 39.
42. S. E. Lebo Jr, J. D. Gargulak and T. J. McNally, Lignin, in *Kirk–Othmer Encyclopedia of Chemical Technology*, Wiley, New York, 2001, published online.
43. N. B. Singh and S. P. Singh, *J. Sci. Ind. Res.*, 1993, **52**, 661.
44. S. E. Lebo Jr and S. L. Resch, US Patent, 6 372 037, 2002.
45. W. O. S. Doherty, P. Mousavioun and C. M. Fellows, *Ind. Crops Prod.*, 2011, **33**, 259.
46. D. Feldman, Lignin and its polyblends – a review, in *Chemical Modification, Properties and Usage of Lignin*, ed. T. Q. Hu, Kluwer Academic/Plenum Publishers, New York, 2002, p. 81.
47. D. Stewart, *Ind. Crops Prod*, 2008, **27**, 202.
48. M. Willcox, Soap, in *Poucher's Perfumes, Cosmetics and Soaps*, 10th edn, ed. H. Butler, Kluwer Academic, Dordercht, 2000, p. 453.
49. A. Thomas, Fats and fatty oils. in *Ullmann's Encyclopedia of Industrial Chemistry*, Wiley-VCH, Weinheim, 2012, published online.

50. A. Benzioni, M. Van Boven, S. Ramamoorthy and D. Mills, *Ind. Crops Prod.*, 2007, **26**, 337.
51. H. Baumann, M. Bühler, H. Fochem, F. Hirsinger, H. Zoebelein and J. Falbe, *Angew. Chem. Int Ed. Engl.*, 1988, **27**, 41.
52. U. Biermann, W. Friedt, S. Lang, W. Lühs, G. Machmüller, J. O. Metzger, M. R. Klaas, H. J. Schäffer and M. P. Schneider, *Angew. Chem. Int Ed.*, 2000, **39**, 2206.
53. U. Biermann, U. Bornscheuer, M. A. R. Meier, J. O. Metzger and H. J. Schäfer, *Angew. Chem. Int. Ed.*, 2011, **50**, 3854.
54. F. D. Gunstone, *Eur. J. Lipid Sci. Technol.*, 2011, **113**, 3.
55. D. Graham-Rowe, *Nature*, 2011, **474**, S6.
56. A. S. Carlsson, J. L. Yilmaz, A. G. Green, S. Stymne and P. Hofvander, *Eur. J. Lipid Sci. Technol.*, 2011, **113**, 812.
57. APAG, *EU Renewable Energy Policy – Position Paper*, European Oleochemicals and Allied Products Group, 2005, http://www.apag.org/issues/index.htm (last accessed 2 April 2012).
58. R. Christoph, B. Schmidt, U. Steinberner, W. Dilla and R. Karinen, Glycerol, in *Ullmann's Encyclopedia of Industrial Chemistry*, Wiley-VCH, Weinheim, 2006.
59. J. C. Thompson and B. B. He, *Appl. Eng. Agric.*, 2006, **22**, 261.
60. D. T. Johnson and K. A. Taconi, *Environ. Prog.*, 2007, **26**, 338.
61. Many opportunities are presented in recently published books, such as: M. Pagliaro and M. Rossi, *Future of Glycerol: New Usages for a Versatile Raw Material*, RSC Green Chemistry Series, RSC Publishing, Cambridge, 2008.
62. M. Pagliaro, R. Ciriminna, H. Kimura, M. Rossi and C. D. Pina, *Angew. Chem. Int. Ed.*, 2007, **46**, 4434.
63. A. Behr, J. Eilting, K. Irawadi, J. Leschinski and F. Lindner, *Green Chem.*, 2008, **10**, 13.
64. A. Behr, J. Eilting, K. Irawadi, J. Leschinski and F. Lindner, *Chim. Oggi*, 2008, **26**, 32.
65. D. T. Johnson and K. A. Taconi, *Environ. Prog.*, 2007, **26**, 338.
66. C. Santibáñez, M. T. Varnero and M. Bustamante, *Chil. J. Agric. Res.*, 2011, **71**, 469.
67. B. Katryniok, H. Kimura, E. Skrzyńska, J.-S. Girardon, P. Fongarland, M. Capron, R. Ducoulombier, N. Mimura, S. Paul and F. Dumeignil, *Green Chem.*, 2011, 13, 1960.
68. B. M. Bell, J. R. Briggs, R. M. Campbell, S. M. Chambers, P. D. Gaarenstroom, J. G. Hippler, B. D. Hook, K. Kearns,

J. M. Kenney, W. J. Kruper, D. J. Schreck, C. N. Theriault and C. P. Wolfe, *Clean*, 2008, **36**, 657.

69. Epichlorohydrin: Solvay to build 100 kt/y plant in China, *Process Worldwide*, 2011, http://www.process-worldwide.com/management/markets_industries/articles/305447/ (last accessed 2 April 2012).

70. J. Deleplanque, J.-L. Dubois, J.-F. Devaux and W. Ueda, *Catal. Today*, 2010, **157**, 351.

71. J. L. DuBois, *US Patent Application*, 2011082319, 2011.

72. J. L. DuBois, *US Patent Application*, 20110136954, 2011.

73. M. S. Reisch, *Chem. Eng. News*, 2010, **88**, 20.

74. M. M. Bomgardner, *Chem. Eng. News*, 2011, **89**, 9.

75. B. Katryniok, S. Paul, V. Bellière-Baca, P. Rey and F. Dumeignil, *Green Chem.*, 2010, **12**, 2079.

76. J.-C. M. R. Monbaliu, M. Winter, B. Chevalier, F. Schmidt, Y. Jiang, R. Hoogendoorn, M. Kousemaker and C. V. Stevens, *Chim. Oggi*, 2010, **28**, 8.

77. R. Macre and C. F. W. Lourenco, *US Patent Application*, 2011112336, 2011.

78. M. Aresta, A. Dibenedetto, F. Nocito and C. Pastore, *J. Mol. Catal. A: Chem*, 2006, **257**, 149.

79. M. Okutsu and T. Kitsuki, *US Patent*, 6 496 703, 2002.

80. JEFFSOL, JEFFSOL glycerine carbonate, *JEFFSOL Technical Bulletin*, http://www.huntsman.com/performance_products/Media/JEFFSOL%20Glycerine%20Carbonate.pdf (last accessed 20 January 2012).

81. ADM press release, *ADM Announces Industry-First Biobased Propylene Glycol USP*, 2 November 2011, http://origin.adm.com/en-US/news/_layouts/PressReleaseDetail.aspx?ID=368 (last accesed 2 April 2012).

82. J. R. Beggin, T. P. Binder, A. K. Hilaly, L. P. Karcher and B. Zenthoefer, *PCT Patent*, WO2008133939, 2008.

83. T. A. Werpy, J. G. Frye Jr, A. H. Zacher and D. J. Miller, *US Patent*, 6 479 713, 2002.

84. J. ten Dam and U. Hanefeld, *ChemSusChem*, 2011, **4**, 1017.

85. Y. Gu and F. Jerome, *Green Chem.*, 2010, **12**, 1127.

86. J. V. Kurian, *J. Polym. Environ.*, 2005, **13**, 159.

87. M. Chateau, J. Y. Dubois and P. Soucaille, *PCT Patent*, WO2010128070, 2010.

88. A. P. Zeng and W. Sabra, *Curr. Opin. Biotechnol.*, 2011, **22**, 749.

89. J. Kretschmann, F.-J. Carduck, W.-D. Deckwer, C. Tag and H. Biebl, *US Patent*, 5 254 467, 1993.
90. Y. Nakagawa and K. Tomishige, *Catal. Sci. Technol.*, 2011, **1**, 179.
91. N. Saoharit and K. S. Kumar, *Crit. Rev. Environ. Sci. Technol.*, 2012, **42**, 1.
92. K. Drauz, I. Grayson, A. Kleemann, H.-P. Krimmer, W. Leuchtenberger and C. Weckbecker, Amino acids, in *Ullmann's Encyclopedia of Industrial Chemistry*, Wiley-VCH, Weinheim, 2012, published online.
93. F. Song, D.-L. Tang, X.-L. Wang and Y.-Z. Wang, *Biomacromolecules*, 2011, **12**, 3369.
94. F. E. I. Deswarte, J. H. Clark, A. J. Wilson, J. Hardy, R. Marriott, S. P. Chahal, C. Jackson, G. Heslop, M. Birkett, T. J. Bruce and G. Whiteley, *Biofuels Bioproducts Biorefining*, 2007, **1**, 245.
95. E. I. Deswarte, J. H. Clark and J. Hardy, *PCT Patent*, WO2006082437, 2006.
96. J. A. S. López, Q. Li and I. P. Thompson, *Crit. Rev. Biotechnol.*, 2010, **30**, 63.
97. M. Siedlecki, W. de Jong and A. H. M. Verkooijen, *Energies*, 2011, **4**, 389.
98. E. M. Kirschner, *Chem. Eng. News*, 1994, **72**, 13.
99. D. W. Verser, *European Patent*, EP1813590, 2007.
100. P. Dominguez de Maria, *ChemSusChem*, 2011, **4**, 327.
101. N. Ladygina, E. G. Dedyukhina and M. B. Vainshtein, *Process Biochem.*, 2006, **41**, 1001.
102. M.-O. Park, K. Heguri, K. Hirata and K. Miyamoto, *J. Appl. Microbiol.*, 2005, **98**, 324.
103. P. Metzger and C. Largeau, *Appl. Microbiol. Biotechnol.*, 2005, **66**, 486.
104. A. Schirmer, M. A. Rude, X. Li, E. Popova and S. B. Del Cardayre, *Science*, 2010, **329**, 559.
105. D. E. Trimnur, C.-S. Im, H. F. Dillon, A. G. Day, S. Franklin and A. Coragliotti, *US Patent Application*, 20110190522, 2011.
106. Solazyme press release, *Solazyme Announces First U.S. Commercial Passenger Flight on Advanced Biofuel*, 7 November 2011, http://solazyme.com/media/2011-11-07 (last accessed 2 April 2012).

107. N. H. Tran, J. R. Bartlett, G. S. K. Kannangara, A. S. Milev, H. Volk and M. A. Wilson, *Fuel*, 2010, **89**, 265.
108. J. L. F. Monteiro and C. O. Veloso, *Top. Catal.*, 2004, **27**, 169.
109. K. S. Jung, J. H. Kim, J. H. Cho and D. K. Kim, *US Patent Application*, 20100209979, 2010.
110. P. J. Seong, B. W. Jeon, M. Lee, D. H. Cho, D. K. Kim, K. S. Jung, S. W. Kim, S. O. Han, Y. H. Kim and C. Park, *Enzyme Microb. Technol.*, 2011, **48**, 505.
111. G. N. Ou, B. Y. He and Y. Z. Yuan, *Enzyme Microb. Technol.*, 2011, **49**, 167.
112. M. Benoit, Y. Brissonnet, E. Guelou, K. D. Vigier, J. Barrault and F. Jerome, *ChemSusChem*, 2010, **3**, 1304.
113. M. Selva, V. Benedet and M. Fabris, *Green Chem.*, 2012, **14**, 188.
114. C. S. M. Pereira, V. M. T. M. Silva and A. E. Rodrigues, *Green Chem.*, 2011, **13**, 2658.
115. US Environmental Protection Agency, *Inert Ingredients Eligible for FIFRA 25(b) Pesticide Products*, updated 20 December 2010, http://www.epa.gov/opprd001/inerts/section25b_inerts.pdf (last accessed 2 April 2012).
116. US Environmental Protection Agency, *Vendor List: Alternatives for CFC-113 and Methyl Chloroform in Electronics Cleaning*, Significant New Alternatives Policy Program, Stratospheric Protection Division, 1998, http://www.epa.gov/ozone/snap/solvents/sol_elec.pdf (last accessed 2 April 2012).
117. R. Datta and J. E. Opre, *PCT Patent*, WO2008016805, 2008.
118. M. Henneberry, J. A. A. Snively, G. J. Vasek and R. Datta, *PCT Patent*, WO03077849, 2003.
119. Vertec BioSolvents, *VertecBio Citrus I20 Technical Information Sheet*, http://www.vertecbiosolvents.com/ (last accessed 25 January 2012).
120. M. Reilly, *New Sci.*, 2007, **2603**, 28.
121. J. Van Krieken, *US Patent Application*, 20110178339, 2011.
122. J. Bozell, *Science*, 2010, **329**, 522.
123. N. E. Kob, Clean solvents, in Alternative Media for Chemical Reactions and Processing, ed. M. A. Abraham and L. Moens, *ACS Symposium Series*, Vol. 819, American Chemical Society, Washington, DC, 2002, p. 238.
124. E. C. Attané and T. F. Doumani, Ind. Eng. Chem., 1949, 41, 2015.

125. W. Niu, K. M. Draths and J. W. Frost, *Biotechnol. Prog.*, 2002, **18**, 201.

126. S. Picataggio and T. Beardslee, *US Patent Application*, 20120021474, 2012.

127. J. G. Zeikus, M. K. Jain and P. Elankovan, *Appl. Microbiol. Biotechnol.*, 1999, **51**, 545.

128. D. Dunuwila, *Canadian Patent*, 2 657 666, 2009.

129. K. Srinivas, T. M. Potts and J. W. King, *Green Chem.*, 2009, **11**, 1581.

130. K. Wu, P. McLaird, J. Byrum, R. Reiter and M. Erickson, US Patent Application, 2009193547, 2009.

131. S. W. Fitzpatrick, *US Patent*, 5 608 105, 1997.

132. E. I. Gürbüz, D. M. Alonso, J. Q. Bond and J. A. Dumesic, *ChemSusChem*, 2011, **4**, 357.

133. I Horváth, H. Mehdi, V. Fábos, L. Boda and L. T. Mika, *Green Chem.*, 2008, **10**, 238.

134. K. Holmberg, *Curr. Opin. Colloid Interface Sci.*, 2001, **6**, 148.

135. P. Foley, A. Kermanshah, E. S. Beach and J. B. Zimmerman, *Chem. Soc. Rev.*, 2012, **41**, 1499.

136. T. Benvegnu, D. Plusquellec and L. Lemiègre, Surfactants from renewable sources: synthesis and applications, in *Monomers, Polymers and Composites from Renewable Resources*, ed. M. N. Belgacem and A. Gandini, Elsevier, Amsterdam, 2008, p. 153.

137. A. Lavergne, Y. Zhu, A. Pizzino, V. Molinier and J.-M. Aubry, *J. Colloid Interface Sci.*, 2011, **360**, 645.

138. J. M. F. Nogueira and M. A. R. B. Castanho, *Colloids Surf. A*, 2001, **191**, 263.

139. K. White, N. Lorenz, T. Potts, W. R. Penney, R. Babcock, A. Hardison, E. A. Canuel and J. A. Hestekin, *Fuel*, 2001, **90**, 3193.

140. A. Keskin, M. Guerue and D. Altiparmak, *Bioresource Technol.*, 2008, **99**, 6434.

141. W. Kaewthong, S. Sirisansaneeyakul, P. Prasertsan and A. H-Kittikun, *Process Biochem.*, 2005, **40**, 1525.

142. P. H. L. Moquin, *J. Supercritic. Fluids*, 2006, **37**, 417.

143. F. R. Lupi, D. Gabriele, B. de Cindio, M. C. Sanchez and C. Gallegos, *J. Food Eng.*, 2011, **107**, 296.

144. J. J. Kokelaar, J. A. Garritsen and A. Prins, *Colloids. Surf. A*, 1995, **95**, 69.

145. R. G. Jensen, J. Sampugna and G. W. Gander, *J. Dairy Sci.*, 1961, **44**, 1057.
146. Cognis, *HAPPI, Household Personal Prod. Ind.*, 2003, **40**, 92.
147. P. P. Constantinides and J.-P. Scalart, *Int. J. Pharm.*, 1997, **158**, 57.
148. W. von Rybinski and K. Hill, *Angew. Chem. Int. Ed.*, 1998, **37**, 1328.
149. D. Nickel, T. Förster and W. von Rybinski, Physicochemical properties of alkyl polyglycosides, in *Alkyl Polyglycosides: Technology, Properties and Applications*, ed. K. HiII, W.von Rybinski and G.Stoll, VCH, Weinheim, 1997, p. 39.
150. W. von Rybinski, *Curr. Opin. Colloid Interface Sci*, 1996, **1**, 587.
151. B. Estrine, S. Bouquillon, F. Henin and J. Muzart, *Green Chem.*, 2005, **7**, 219.
152. F. Bouxin, S. Marinkovic, J. Le Bras and B. Estrine, *Carbohydr. Res.*, 2010, **345**, 2469.
153. (a) R. O. Brito, S. G. Silva, R. M. F. Fernandes, E. F. Marques, J. Enrique-Borges and M. L. C. do Vale, *Colloids Surf. B*, 2011, **86**, 65; (b) S. G. Silva, J. E. Rodríguez-Borges, E. F. Marques and M. L. C. do Vale, *Tetrahedron*, 2009, **65**, 4156; (c) I. Johansson and M. Svensson, *Curr. Opinion Colloid Interface Sci.*, 2001, **6**, 178; (d) R. Bordes and K. Holmberg, *Colloids Surf. A*, 2011, **391**, 32.
154. M. R. Infante, *C. R. Chim.*, 2004, 7, 583.
155. M. Deleu and M. Paquot, *C.R. Chim.*, 2004, 7, 641.
156. M. H. M. Isa, R. A. Frazier and P. Jauregi, *Sep. Purif. Technol.*, 2008, **64**, 176.
157. C. N. Mulligan, *Proc. Indian Natl. Sci Acad*, 2004, **B70**, 31.
158. C. N. Mulligan, *Environ. Pollut.*, 2005, **133**, 183.
159. K. S. M. Rahman and E. Gakpe, *Biotechnology*, 2008, **7**, 360.
160. M. Nitschke, S. G. V. A. O. Costa and J. Contiero, *Process Biochem.*, 2011, **46**, 621.
161. C. N. Mulligan, *Curr. Opin. Colloid Interface Sci.*, 2009, **14**, 372.
162. O. Kaeppeli and L. Guerra-Santos, *US Patent*, 4 628 030, 1986.
163. J. D. Desai and I. M. Banat, *Microbiol. Mol. Biol. Rev.*, 1997, **61**, 47.
164. C. N. Mulligan, *Environ. Pollut.*, 2005, **133**, 183.

165. L. Torres, A. Moctezuma, J. R. Avendaño, A. Muñoz and J. Gracida, *J. Petrol. Sci. Eng.*, 2011, **76**, 6.
166. I. N. A. Van Bogaert, J. X. Zhang and W. Soetaert, *Process Biochem.*, 2011, **46**, 821.
167. Y. Hirata, M. Ryu, Y. Oda, K. Igarashi and A. Nagatsuka, *J. Biosci. Bioeng.*, 2009, **108**, 142.

III
Sustainability Related to Feedstocks – CH_4 and CO_2

CHAPTER 11

Methane for Transportation Fuel and Chemical Production

ARNO DE KLERK* AND VINAY PRASAD

Chemical and Materials Engineering, University of Alberta, Edmonton, AB, Canada
*Email: deklerk@ualberta.ca

11.1 INTRODUCTION

Methane (CH_4) is the richest carbon-based hydrogen source. So, why is this significant and why is it a material that is needed for a sustainable future?

First, we consider the simplest of applications, namely the direct combustion of methane to provide heat or power. The lower heating value (LHV) of methane combustion is $-802\,kJ\,mol^{-1}$, of which only half is derived from carbon combustion. Its CO_2 intensity is $1.25\,mol\,MJ^{-1}$ of CO_2 whereas that of coal combustion, as approximated by anthracene, is $2.02\,mol\,MJ^{-1}$ of CO_2. By employing methane instead of coal for direct heating applications, the CO_2 footprint is reduced by almost 40%, not to mention the other heteroatom-containing combustion products. The difference in CO_2 footprint is mainly due to the difference in hydrogen content of the carbon sources. It is the combustion of methane to produce

Materials for a Sustainable Future
Edited by Trevor M. Letcher and Janet L. Scott
© The Royal Society of Chemistry 2012
Published by the Royal Society of Chemistry, www.rsc.org

two-thirds water on a molar basis that gives methane its 'green' appeal.

The real advantage of methane for combustion applications is not in large-scale facilities where combustion gas cleaning can easily be incorporated in the design, but for distributed heating applications. In distributed heating applications, such as central heating of buildings in cold climates, there are two important benefits inherent in the use of methane (natural gas): ease of distribution and low heteroatom content. Being gaseous, methane can be distributed by pipeline infrastructure. Natural gas is clean burning, producing mainly CO_2 and H_2O as combustion products on account of its low heteroatom content. To emphasise the point: the properties of methane are so valued that in times of natural gas shortage, industrial facilities for the production of substitute natural gas (SNG) from coal were constructed at high cost.[1] Although we do not consider direct heating applications further in this chapter, the analogy with transportation fuels for mobile combustion applications is clear.

Second, we consider methane as a source of synthesis gas (H_2 and CO). Methane reforming (Section 11.5) can ideally produce synthesis gas with an H_2:CO ratio of 2:1 by partial oxidation of methane:

$$CH_4 + 1/2O_2 \rightarrow 2H_2 + CO, \quad \Delta H_{25°C} = -36\,kJ\,mol^{-1} \quad (11.1)$$

This is close to the correct usage ratio required by many of the industrially important syngas conversion technologies (Section 11.6). Furthermore, methane reforming is also the most efficient fossil fuel for the production of H_2. It is possible to produce 3 mol of H_2 per mole of CO_2 from methane:

$$CH_4 + 1/2O_2 + H_2O \rightarrow 3H_2 + CO_2, \quad \Delta H_{25°C} = -77\,kJ\,mol^{-1} \quad (11.2)$$

Lastly, we consider transportation fuels. Transportation fuels are the energy carriers employed for mobile applications. In most such applications, the transportation fuel is combusted with air to provide motive power. It is impractical to consider complex post-combustion treatment in mobile applications. Over time, transportation fuel specifications evolved to restrict the composition of transportation fuels to compounds containing only C, H and O, which results in CO_2 and H_2O as combustion products.

The advantage of directly using methane as a transportation fuel is immediately apparent. However, there are two complicating factors that restrict the direct use of methane as a transportation fuel: the engine types employed for mobile applications have different fuel requirements (Section 11.3) and methane is not a suitable fuel for all engine types. Methane is normally gaseous and direct liquefaction and storage of methane require cryogenic conditions (Section 11.4.1). Although methane can also be stored as a compressed gas, it then has a lower volumetric energy density, detracting from mobile applications where the distance that can be travelled before refuelling is important. Both of these disadvantages related to the direct use of methane as a transportation fuel can be overcome by indirect liquefaction technology (Sections 11.4.2 and 11.4.3) and appropriate refining (Section 11.7).

11.1.1 Sustainability

It is a misconception that sustainability is derived from the use of natural resources that are replenished on a non-geological time scale, such as wind, solar energy and biomass. Although biomass is the most obvious 'renewable' carbon source, it is only renewable if it is replenished and strictly its use is only sustainable if it is replenished at the same rate as it is consumed. Unfortunately, the history of energy exploitation is riddled with examples of the unsustainable use of biomass; burning wood and using whale oil have led to deforestation and the near extinction of whales.[2]

One should not rely only on Nature to replenish itself. Replenishment can take place by recycling and renewable carbon resources do not necessarily have to be of natural origin.

When we consider the sustainable use of methane, at the present time we cannot apply the rigorous requirement of balancing the rates of methane consumption and replenishment. Even though methane is a stable molecule and microorganisms readily produce methane from carbon-based materials, the current production rate by such and other means is much lower than the consumption rate. For this chapter, we adopt the sustainability definition of the Brundtlant Commission of the United Nations (1987):[3] 'Sustainable development is development that meets the needs of the present, without compromising the ability of future generations to meet their own needs.'

What can we do to use natural gas in a sustainable way? The value of natural gas lies in its high H:C ratio, low heteroatom content and fluid nature. These properties suggest uses such as distributed and mobile energy applications, where the clean-up of combustion gases is impractical or infeasible. It also suggests uses that require products with a high H:C ratio. If we restrict natural gas use to such applications, we can hopefully preserve our natural gas resources sufficiently long to enable future generations to devise technology that does not require natural gas to meet their needs. Hence the focus on transportation fuels, which are products with high H:C ratio and low heteroatom content.

11.2   NATURAL SOURCES OF METHANE

Natural gas is the primary source of methane and is found in crude oil reservoirs, natural gas reservoirs and condensate reservoirs. The natural gas produced from crude oil reservoirs is termed associated gas and the gas from other sources is non-associated gas.[4] The gas from condensate reservoirs is also non-associated gas and is termed wet gas.

With the advent of new technologies such as hydraulic fracturing ('fracking'), subterranean reservoirs such as coal beds and shale reservoirs can be exploited to a greater extent for natural gas production, and this can potentially change the outlook on proven resources for natural gas. There is also a large amount of natural gas in hydrate form under the sea and in arctic regions; however, there is no commercially viable method yet to extract natural gas from these hydrates.

In addition to natural gas of thermogenic origin, natural gas may also be created by biogenic mechanisms. Methanogenic organisms produce natural gas by decomposing organic waste. Landfill gas and the gas from sewage treatment plants also contain a substantial amount of methane.

11.2.1   Production of Natural Gas

The proven reserves of natural gas and production and consumption data are provided in Table 11.1.[5] The regions with the largest proven reserves are the Middle East and central and eastern Europe. The countries that comprised the former Soviet Union have

Table 11.1 Worldwide and regional proven natural gas reserves and annual production and consumption in 2010 in billions of cubic metres.

Region	Reserves/ $10^9 m^3$	Production/ $10^9 m^3 y^{-1}$	Consumption/ $10^9 m^3 y^{-1}$
North America	9000	840	760
Central and South America	7600	740	210
Western Europe	4400	460	410
Central and Eastern Europe	61 600	460	630
Africa	14 000	240	100
Middle East	75 200	220	400
Asia and Oceania	15 200	160	530
Worldwide	187 000	3120	3040

the largest single collection of reserves. North America is the largest producer and consumer of natural gas, with the USA being the major contributor. The other major producer is the Russian Federation, followed by Canada, the UK, Algeria, Indonesia, Iran, The Netherlands, Norway and Uzbekistan. In natural gas consumption, the USA and the Russian Federation are the countries with the largest consumption. The worldwide consumption is projected to grow at a rate of 1.6% annually up to 2035. Most of the reserves, and the production and consumption, relate to natural gas reservoirs, and the amount from other sources is not significant.

11.2.2 Composition of Natural Gas

Natural gas of thermogenic origin is composed primarily of methane, but does have other components in smaller quantities. The composition of natural gas from a typical (dry) reservoir is provided in Tables 11.2,[5] bearing in mind that relatively large variations are possible depending on the type of reservoir source. Methane is, of course, the component with the largest mole fraction; other components include nitrogen, carbon dioxide, hydrogen sulfide, ethane, propane and smaller quantities of butane and pentane and heavier components. Trace amounts of benzene, toluene, xylenes and mercury may also be present.

Note that the composition provided in Table 11.2 is of natural gas at the source and not after processing. The production of natural gas consists of many steps, which can be classified into water removal, nitrogen removal, enrichment and fractionation.

Table 11.2 Representative composition of natural gas at its source.

Component	Volume/%
Methane (CH_4)	>85
Ethane (C_2H_6)	3–8
Propane (C_3H_8)	1–2
Butanes (C_4H_{10})	<1
Heavier hydrocarbons	<1
Nitrogen (N_2)	1–5
Carbon dioxide (CO_2)	1–2
Hydrogen sulfide (H_2S)	<1
Helium (He)	<0.5

The enrichment process consists of acid gas removal and sweetening of the natural gas, using methods such as carbonate washing, purification using molecular sieves and metal oxide processes. Fractionation, which involves de-ethanisation, de-propanisation, de-butanisation and butane splitting, increases the concentration of methane in the natural gas stream. The methane content in the gas would naturally be higher after such processing.

For other sources, such as coal bed methane, the methane composition can vary significantly, but sulfur levels are typically low, with carbon dioxide, nitrogen and ethane sometimes being found in significant quantities. The situation is similar with shale gas.

For landfill gas and gas from sewage treatment, the composition is 50% and more of methane and 25–50% of CO_2, with small amounts of nitrogen, oxygen, hydrogen, hydrogen sulfide and siloxanes. The presence of siloxanes is a relatively recent phenomenon as a result of the changed nature of the material that is landfilled. The siloxanes are present in trace quantities,[6] but can cause downstream processing problems if not taken into consideration during process design.

Based on the sulfur content (usually as H_2S), natural gas is typically classified as sweet or sour. Sweet gas natural gas is classified as sweet contains less than 4 ppm on a volume basis (6 mg m^{-3}) of H_2S; above this value, the gas is considered to be sour. The liquid content leads to classification as lean or rich. Liquid content is quantified in gallons of liquid per 1000 standard cubic feet (scf) of gas measured, *i.e.* at 0 °C and 101.235 kPa (1 US gallon per 1000 scf = 0.134 L of liquid per cubic metre of gas) and the greater the liquid content, the 'richer' the gas is considered to be.

11.3 TRANSPORTATION FUELS

Generic attributes of all transportation fuels are that they should have a high volumetric and appropriate gravimetric energy density and that they should be compatible with the vehicle engine technology employed. Gaseous fuels are therefore not useful as transportation fuels unless compressed and the mass of the storage vessel is not too high. Although the concept of a single fuel is appealing, especially to the military,[7] it requires standardised engine technology for all vehicles and aeroplanes. This is currently not the case.

Over time, fuel specifications were developed for each engine type. The specifications were devised to ensure compatibility between the fuel and engine technology, which in turn permits reliable long-term engine performance. In addition to engine requirements, fuel specifications address emission performance. Although vehicles are equipped with some post-combustion treatment, it does not extend to the removal of heteroatom oxides (SO_x and NO_x). The fuels must be cleaned beforehand in a fuels refinery, where it is possible to eliminate and capture sulfur and nitrogen. The fuel specifications set limits on the heteroatom content of fuels, and also other fuel properties that lower fuel-derived emissions.

Almost all transportation fuels are currently refined from crude oil and engine systems are designed accordingly. It is important to distinguish 'standard' applications, where the fuels meet current fuel specifications and standard engine types, from 'special' applications, where the fuels do not conform to fuel specifications and/or require modifications to the engine or vehicle to be used reliably. There are also special engine types that have special associated fuel requirements, for example, fuel cells.

The use of 'special' fuels, such as compressed natural gas (CNG), liquefied natural gas (LNG), methanol and dimethyl ether (DME), requires changes to the fuel distribution network and vehicles to use and store the fuel safely. The use of compressed natural gas as fuel for auto-rickshaws in India (Figure 11.1) demonstrates that methane can be directly employed as a 'special' transportation fuel. However, in order to do so, the fuel storage and distribution infrastructure, and also the vehicles, had to be modified. Unless a 'special' fuel becomes widely adopted (*i.e.* it becomes a 'standard' fuel), the use of vehicles that are modified for 'special' fuel use is restricted to locations where the infrastructure for the distribution of that fuel exists.

Figure 11.1 An example of an auto-rickshaw powered by a spark ignition engine using compressed natural gas (CNG) and an example of a typical 200 cc CNG water-cooled engine.

If methane is to become a future source of transportation fuel, there are two options: either the methane must be converted into fuels compatible with present infrastructure or the infrastructure must be modified to allow the use of methane (as in the case of the auto-rickshaws). Crude oil is likely to remain the dominant source of transportation fuel and most consumers are unlikely to embrace the costs associated with changes in vehicles and infrastructure. At present, the most viable global approach to methane use for mobile applications is to convert the methane into transportation fuels meeting 'standard' fuel specifications so that it is compatible with the existing infrastructure.

A short overview of the main 'standard' fuel types is given below, and also typical specifications of each. This is necessary background for the discussion of refining requirements (Section 11.7) to produce such fuel types from methane.

11.3.1 Spark Ignition Engine Fuel

The fuel type used in spark ignition engines is motor gasoline (petrol). It usually contains hydrocarbons in the C_4–C_{10} range (naphtha, boiling range typically $< 180\,°C$).

Spark ignition engines are reciprocating engines. In a spark ignition engine, the fuel is compressed with air and, at the appropriate time, the fuel–air mixture is ignited. It is important that the fuel–air mixture does not ignite spontaneously, because it would degrade the engine performance, delivering less power and resulting in higher emissions. The ability of a fuel to resist autoignition is one of the key fuel properties.

The autoignition resistance of a fuel is expressed in terms of octane number. The higher the octane number, the better is the autoignition resistance of the fuel. Two octane numbers are distinguished: research octane number (RON), which describes ignition behaviour under mild conditions, and motor octane number (MON), which describes ignition under severe conditions. The relationship between octane number and the compound classes found in motor gasoline is given in Table 11.3. Methane has a very high octane number and it therefore works well in spark ignition engines.

In addition to the octane number, there are limitations on all compound classes in motor gasoline, except the alkanes. For

Table 11.3 Relationship between octane number and main compound classes found in spark ignition engine fuels (motor gasoline).

| Compound class | Octane number range | | Description |
	RON	MON	
Linear alkanes	<0 to >100	<0 to >100	Octane number increases with decreasing carbon number
Branched alkanes	<0 to >100	<0 to >100	Octane number usually increases with increasing degree of branching
Cycloalkanes	<0 to >100	<0 to >100	Mimics the properties of the alkane after ring opening
Linear alkenes	~0 to >100	~0 to 85	As with *n*-alkanes, but 1-alkenes have lower octane numbers than internal alkenes
Branched alkenes	~70 to >100	~65 to ~90	Fairly insensitive to carbon number and degree of branching
Aromatics	>100	~95 to >100	High octane numbers, degraded only by long *n*-alkyl groups
Alcohols	>100		Related to the structure of the alkyl groups, as with alkanes/alkenes.
Ethers	~70 to >100		Related to the structure of the alkyl groups, as with alkanes/alkenes.

environmental reasons, the sulfur content is limited (in many parts of the world, the maximum sulfur content is currently $10–15\,\mu g\ g^{-1}$). Other important specifications include vapour pressure, boiling range and oxidation stability during storage.

Although methane easily meets most motor gasoline specifications, it cannot meet the vapour pressure requirement. This implies that it cannot be stored and dispensed as a liquid fuel and that vehicles must also be adapted to carry such a high vapour pressure fuel. Methane is therefore considered a 'special' fuel, because it requires modifications to the infrastructure in order to be used. Some additional points with respect to the use of methane in spark ignition engines are as follows:[8]

- Methane is a good spark ignition engine fuel and it does not require a change in engine technology in order to be used.
- The flame speed of methane is different from that of motor gasoline and ignition timing for optimum engine performance is consequently also different. Engines operating on methane must therefore be tuned accordingly.
- The volumetric energy density of methane is lower than that of motor gasoline, which results in a higher volumetric fuel consumption per unit distance travelled. The lower flame speed of methane also leads to a larger timing loss (power loss due to non-instantaneous combustion) than for motor gasoline.
- The engine emissions from methane combustion (CO, NO_x and hydrocarbons) are consistently lower than those for motor gasoline over stoichiometric air-to-fuel ratios (λ) from 0.8 to 1.2.

11.3.2 Compression Ignition Engine Fuel

The fuel type used in compression ignition engines is diesel fuel. It contains hydrocarbons mainly in the $C_{11}–C_{22}$ range (distillate, boiling range typically $180–360\,°C$).

Like spark ignition engines, compression ignition engines are also reciprocating engines. As the name suggests, the engine relies on autoignition of the fuel–air mixture when it is compressed. This is exactly the opposite requirement of spark ignition engines and generally one can say that a good compression ignition fuel is a poor spark ignition fuel. The ability of the fuel to autoignite readily is important for good compression ignition performance.

The ease of autoignition of a fuel is expressed in terms of cetane number. The higher the cetane number, the easier it is for the compressed fuel–air mixture to autoignite. There is consequently an inverse relationship between cetane number and octane number.[9] Heavier linear alkanes have high cetane numbers, whereas aromatic compounds have low cetane numbers.

Methane, owing to its high autoignition resistance, makes an extremely poor diesel fuel. On its own, a methane–air mixture cannot be made to autoignite in an acceptable manner, and when methane is applied as fuel for compression ignition, it is always used in combination with a fuel that has the required autoignition properties.[10]

Cetane number is not the only fuel property of importance. Among others, lubricity, storage stability and ensuring fluidity at low temperature are also important. Distillate range compounds usually have higher freezing points than naphtha range compounds and good cold flow properties are required to ensure that diesel fuels can be used in winter and in colder climates. The sulfur content in diesel fuel is limited for environmental reasons. In many countries the maximum allowable sulfur content is $10–15\,\mu g\,g^{-1}$ at present.

11.3.3 Turbine Fuel

The fuel used in turbine engines contains hydrocarbons mainly in the C_{10}–C_{13} range (kerosene, boiling range typically 160–230 °C), but a wider range is possible. Since most turbine fuel is employed as aviation turbine fuel, it is often referred to as jet fuel.

Unlike spark ignition and compression ignition engines, turbines are continuous combustion engines. The fuel–air mixture must be able to maintain stable and continuous combustion. There are no equivalent concepts to octane number or cetane number. Heavy naphtha can be used as motor gasoline and light distillate (which can be used as diesel fuel); both are acceptable fractions for use as turbine fuel.

Cold flow properties are very important for turbine fuel, especially jet fuel. Aircraft operating at high altitude experience very low temperatures and Jet A-1 (commercial aviation turbine fuel) has a maximum freezing point specification of −47 °C. This requires the paraffinic molecules to be branched, since branched molecules have lower freezing points than their linear analogues. It also places a restriction on the composition of the aromatics; high freezing point aromatics must be avoided.

In addition to the cold flow properties, thermal stability, storage stability, flash point, smoke point and distillation profile are all important for stable and reliable engine operation. It can be appreciated that engine failure during flight has more severe consequences than engine failure on the road. Jet fuel is internationally regulated, with only minor variations in fuel specifications, and even synthetic jet fuels with similar properties as Jet A-1 are treated as 'special' fuels with additional specification requirements.[11]

11.3.4 Production of Transportation Fuels

Transportation fuels are produced mainly from crude oil. Global crude oil consumption is at present around 90 million barrels per day ($14.3 \times 10^6 \, m^3 \, d^{-1}$ or $5 \times 10^9 \, m^3 \, y^{-1}$). When the crude oil is refined to transportation fuels there is a net increase in volume, due to the decrease in density of the refined products. Crude oils typically have densities around $825–1000 \, kg \, m^{-3}$, although there are also lighter and heavier crudes. Transportation fuels have densities in the range $720–845 \, kg \, m^{-3}$.

Most of the global refining capacity is in the northern hemisphere. Table 11.4 gives production quantities of the major transportation fuels by region.[5] The values are provided for gasoline, jet fuel and diesel fuel (distillate fuel oil), which represent spark ignition engine fuel, turbine fuel and compression ignition engine fuel, respectively. At least in the main markets of the USA and Western

Table 11.4 Transportation fuel production in the northern hemisphere by region.

| Region | Production in 2009/$bbl \, d^{-1a}$ | | | |
	Gasoline	Jet fuel	Diesel fuel	Total
North America	11.2	1.7	6.0	18.9
Western Europe	8.8	2.7	15.1	26.6
Middle East	3.3	1.4	6.1	10.8
Asia	11.0	4.9	20.9	36.8
Total:	34.3	10.7	48.1	93.1

[a]Unit conversion: 1 barrel (bbl) = $0.159 \, m^3$; $1 \, m^3 = 6.29 \, bbl$.

Europe, the consumption is slightly more than the production, with the shortfall being met by imports.

The USA is the major producer (and consumer) of gasoline in the world. Trends in gasoline production in the USA have indicated a steady increase in production from the early 1980s onwards (~ 2300 million barrels annually or $370 \times 10^6 \, m^3 \, y^{-1}$) to 2009 (3200 million barrels annually or $509 \times 10^6 \, m^3 \, y^{-1}$), in response to increased demand. Similar trends can be observed for diesel (~ 900 million barrels or $143 \times 10^6 \, m^3$ in the USA in 1983 to almost 1500 million barrels or $239 \times 10^6 \, m^3$ in 2009). In the case of diesel, Western Europe and Asia are major producers, with China producing 2900 million barrels ($460 \times 10^6 \, m^3$) in 2009. Jet fuel production in the USA showed a sharp increase in the early 1980s from around 300 million barrels ($48 \times 10^6 \, m^3$) in 1970 to about 510 million barrels ($81 \times 10^6 \, m^3$) in 1983, and has stayed steady since then (510 million barrels in 2009). However, Asia is the major producer and China in particular produced 1017 million barrels ($162 \times 10^6 \, m^3$) in 2009.

The prices of transportation fuels are location dependent. To give an indication of value, only the pricing in the USA will be given. Of the transportation fuels, gasoline commanded the best price of 273.6 cents per gallon or 72.2 cents per litre in 2010 in the USA, with the price, rising from 116.4 cents per gallon (30.7 cents per litre) in 1990 to a high of 326.6 cents per gallon (86.1 cents per litre) in 2008. Diesel has had significant increases in price in recent years, peaking at 274.5 cents per gallon (72.4 cents per litre) in 2008 and was at 165.6 cents per gallon (43.7 cents per litre) in 2009. Jet fuel showed a similar trend, peaking at 302.0 cents per gallon (79.2 cents per litre) in 2008 and reducing to 171.2 cents per gallon (45.2 cents per litre) in 2009.

11.4   INDUSTRIAL GAS-TO-LIQUIDS (GTL) TECHNOLOGIES

Natural gas is useful in its own right as a clean-burning heating fuel. With adequate distribution infrastructure, there is no environmental incentive to practice GTL conversion. So, why do we see a gradual increase in industrial GTL facilities around the world?

There are a number of reasons that can be cited for the increase in GTL facilities, but the main driving force is economic. Crude oil is the benchmark liquid fossil fuel source and, in locations where natural gas is much cheaper on an energy basis than crude oil, it becomes economic to turn the natural gas into a liquid product. The price differential between natural gas and crude oil must be significant, because GTL conversion is expensive.

In regions where natural gas is available, but far from natural gas-consuming markets, the natural gas must be transported. When an adequate pipeline infrastructure does not exist, the natural gas is stranded and has a low market value. It is in these situations where there is an economic and a technical incentive to consider GTL conversion. Turning the natural gas into a liquid product facilitates transportation and adds value to the product. From a sustainability point of view, GTL conversion of stranded natural gas is preferable to wasteful practices, such as flaring.

Industrially there are three key technology groups for turning natural gas into liquid products: direct liquefaction to produce liquefied natural gas (Section 11.4.1), indirect liquefaction to produce methanol (Section 11.4.2) and indirect liquefaction to produce Fischer–Tropsch syncrude (Section 11.4.3).

11.4.1 Liquefied Natural Gas Production

Liquefied natural gas (LNG), as the name suggests, is produced by direct liquefaction with the aim of making it easier to store and transport. The gas is condensed into liquid form at close to atmospheric pressure at a temperature of about $-162\,°C$. A number of refrigeration loops in series are employed to lower the temperature of the natural gas. Each refrigeration loop lowers the temperature by $\sim 60\,°C$ and a typical LNG facility requires a cascade of three refrigeration loops in series to liquefy methane. The liquid density of LNG is 456 kg m^{-3} and it has a heating value of 24.8 GJ m^{-3}.[12]

The worldwide LNG demand has been growing steadily over the past few years and the same trend is projected to continue.[13] LNG trade volumes have increased from 140 million tonnes (Mt) in 2005 to about 200 Mt in 2009, with the LNG trade accounting for 28% of all natural gas trade. Japan (65 Mt in 2009), Korea (25 Mt) and Spain (22 Mt) are the major importers, while Qatar (37 Mt in 2009), Malaysia (22 Mt), Indonesia (19 Mt), Australia (18 Mt), Algeria (15 Mt) and Trinidad and Tobago (14 Mt) are the major exporters.

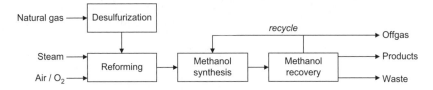

Figure 11.2 Flow diagram of a typical GTL facility for methanol synthesis.

The typical size of an LNG facility may vary from 0.9 to 10 Mt produced per year. However, there are facilities which are larger than 20 Mt y^{-1}, for example, the Incheon and Pyeong Taek (Korea Gas Corporation), Sodegaura (Tokyo Gas) and Sabine Pass (Cheniere Energy) facilities.

11.4.2 Industrial GTL Methanol Production

A block flow diagram of a typical GTL process for methanol synthesis is shown in Figure 11.2. The first step in a GTL methanol facility is to convert the natural gas into synthesis gas (Section 11.5), which is then converted into methanol. Methanol is predominantly produced using syngas-to-methanol technology based on the ICI 'low-pressure synthesis' process (Section 11.6.1). The methanol is recovered by condensation from the unconverted synthesis gas. The unconverted synthesis gas is recycled and the methanol is purified by distillation.

There are numerous methanol-producing facilities in the world, with the largest single-train facilities traditionally being of the order of 2500–3000 t d^{-1}. However, with new technology, a key aspect of which is combined reforming (steam reforming followed by autothermal reforming), facilities that can produce 5000–10 000 t d^{-1} are being built.

Table 11.5 lists the annual worldwide methanol capacity, production and consumption.[5] Not all of this methanol is produced from natural gas, with part of the installed capacity based on coal-to-liquids conversion, especially in Asia. Asia is the region with the highest capacity, production and consumption. China has more than 90% of Asia's capacity, 85% of its production and 70% of its consumption.

The worldwide methanol capacity is projected to increase by about 63% to around 125 000 kt y^{-1} by 2015.[5] The significant

Table 11.5 Worldwide and regional installed methanol production capacity and annual methanol production and consumption in 2010.

	Methanol/kt y^{-1a}		
Region	*Capacity*	*Production*	*Consumption*
North America	1200	800	6400
Central and South America	10 200	8100	1500
Western Europe	3000	1800	6400
Central and Eastern Europe	5400	3600	2900
Africa	1900	1600	400
Middle East	15 400	14 200	3100
Asia	39 200	18 100	27 400
Oceania	1000	900	100
Total (worldwide)	77 300	49 100	48 200

[a]Unit conversion: $1 \, kt \, y^{-1} = 3.46 \, m^3 \, d^{-1} = 21.8 \, bbl \, d^{-1}$.

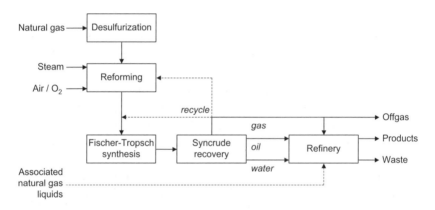

Figure 11.3   Flow diagram of a typical Fischer–Tropsch-based GTL facility.

surplus in methanol production capacity creates an opportunity for converting methanol into hydrocarbons,[14,15] which in turn will permit access to traditional transportation fuel markets. This conversion also forms the basis for the Mobil methanol-to-gasoline process.[16]

11.4.3   Industrial GTL Fischer–Tropsch Production

A block flow diagram of a typical Fischer–Tropsch-based GTL process is presented in Figure 11.3. Like methanol synthesis,

Table 11.6 List of current industrial Fischer–Tropsch-based GTL facilities.

| Name | Location | Year | Nameplate capacity/bbl d^{-1c} | | Fischer–Tropsch technology[d] |
			Syncrude	NGL	
Sasol 1	Sasolburg, South Africa	1955[b]	6750	–	Fe-LTFT fixed bed and Fe-LTFT slurry bed
Bintulu GTL[a]	Bintulu, Malaysia	1993	12 500	–	Co-LTFT fixed bed
PetroSA ('Mossgas')	Mossel Bay, South Africa	1993	22 500	10 500	Fe-HTFT fluidised bed and Co-LTFT slurry bed
Oryx GTL	Ras Laffan, Qatar	2007	34 000	–	Co-LTFT slurry bed
Pearl GTL	Ras Laffan, Qatar	2011	140 000	120 000	Co-LTFT fixed bed
Escravos GTL	Escravos, Nigeria	2012?	34 000	–	Co-LTFT slurry bed

[a]Shell Middle Distillate Synthesis (SMDS) process.
[b]Commissioned as coal-to-liquids facility, changed to gas-to-liquids in 2004.
[c]The original nameplate Fischer–Tropsch capacity and excludes later modifications, improvements or operational difficulties.
[d]LTFT = low-temperature Fischer–Tropsch; HTFT = high-temperature Fischer–Tropsch.

Fischer–Tropsch synthesis requires natural gas to be desulfurised and converted into synthesis gas first (Section 11.5). The synthesis gas is then used as the feed for Fischer–Tropsch synthesis to produce a multiphase mixture called syncrude (Section 11.6.2). The composition of the syncrude depends on the Fischer–Tropsch technology and it consists mainly of hydrocarbons, oxygenates and water. The last step in the process is refining (Section 11.7), which turns the syncrude into useful products, such as transportation fuels, petrochemicals and intermediates. In some GTL facilities, the natural gas feed contains associated natural gas liquids (NGLs) that are co-refined with the Fischer–Tropsch syncrude.

The industrial Fischer–Tropsch-based GTL facilities are listed in Table 11.6. Historically, the first of these facilities was the Hydrocol GTL facility in the USA. This facility was decommissioned in the 1950s mainly for economic reasons [the crude oil price was US$2 per barrel (1.26 cents per litre), which made GTL conversion unprofitable].[11] Renewed interest in GTL conversion in the past decade led to the construction of several industrial GTL facilities, together with various demonstration-scale facilities to evaluate new Fischer–Tropsch technologies. Despite the increase in global GTL Fischer–Tropsch production capacity, its contribution is small ($\sim 0.25\%$) compared with global crude oil production.

11.5   SYNTHESIS GAS PRODUCTION FROM NATURAL GAS

In the conversion of methane to transportation fuels, most processes involve the generation of synthesis gas ($CO + H_2$) by natural gas reforming, which is followed by the conversion of the synthesis gas into liquid products. Synthesis gas can likewise be generated by the gasification of solid fuels, such as coal and biomass, but that is not the focus here. Synthesis gas is also an intermediate in chemical production, *e.g.* ammonia (Section 11.8).

As described in Section 11.2.2, natural gas contains components other than methane, notably sulfur and higher hydrocarbons, that substantially affect the performance of natural gas reforming processes. Sulfur is removed by treatment with a small stream of hydrogen or hydrogen-rich gas to convert all of the sulfur-containing species into H_2S. The H_2S can then be removed by absorption over ZnO. Higher hydrocarbons are converted into C_1 species by using a pre-reforming step.

The conversion of the clean natural gas into synthesis gas is achieved using steam methane reforming (SMR), partial oxidation (POX) or autothermal reforming (ATR).[17–21] The respective reformer types are illustrated in Figure 11.4. The use of O_2 *versus* air as an oxidant and the required syngas composition are two important considerations in selecting the reforming technology.

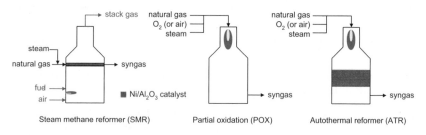

Figure 11.4   Main reformer types.

Since the downstream conversion of synthesis gas usually requires a specific H_2:CO ratio, water gas shift conditioning may additionally be performed in order to manipulate the H_2:CO ratio. The H_2:CO ratio can also be adjusted through the gas loop design (Section 11.6). Other types of conditioning include CO_2 removal, which is performed by standard gas cleaning techniques.[22] These techniques will not be discussed here.

11.5.1 Natural Gas Pretreatment and Pre-reforming

Pretreatment of natural gas essentially consists of sulfur removal, along with removal of liquid components using a flash drum. It should be noted that the pretreatment described here is in relation to the natural gas reforming process. Natural gas from the wellhead is typically pretreated for mercury, sour gas (CO_2/H_2S) and water removal and extraction of higher hydrocarbons; however, the natural gas typically requires further pretreatment before the main reforming steps.

Sulfur removal involves the addition of hydrogen, which results in the replacement of the S–C bond in organosulfur compounds to form hydrogen sulfide. Since the amount of sulfur present in natural gas at this stage is fairly small (of the order of 20 ppm by volume), the amount of hydrogen required is small. The reaction is typically carried out at 350–500 °C over a hydrogenation catalyst, which is based on cobalt or nickel and molybdenum. Once the H_2S is formed, it is separated from the natural gas stream by absorbing it in a bed of ZnO. The removal of sulfur to a very low level is essential, since sulfur is a poison for the catalysts in the downstream processes, including gas reforming.

Pre-reforming of natural gas is typically carried out adiabatically over a nickel catalyst and is used to convert the higher hydrocarbons (C_2+) by reaction with steam to a mixture of methane, hydrogen, carbon monoxide and carbon dioxide, as shown in the following equations:[23,24]

$$C_nH_m + nH_2O \rightarrow nCO + (n+m/2)H_2,$$
$$\Delta H_{25°C} = +1108 \text{ kJ mol}^{-1} \text{ for } n = 7 \tag{11.3}$$

$$3H_2 + CO \rightleftarrows CH_4 + H_2O, \quad \Delta H_{25°C} = -206 \text{ kJ mol}^{-1} \tag{11.4}$$

$$H_2O + CO \rightleftarrows H_2 + CO_2, \quad \Delta H_{25°C} = -41 \text{ kJ mol}^{-1} \tag{11.5}$$

The steam reforming reaction, Equation (11.3), is endothermic and essentially irreversible. The methanation reaction, Equation (11.4), and the water gas shift reaction, Equation (11.5), are both exothermic and reversible. The overall pre-reforming process is normally endothermic for natural gas and requires energy input. Steam-to-carbon ratios (moles of H_2O per mole of C atom) are not necessarily stoichiometric; they can vary from 0.3 to 2.0 on a molar basis and are adjusted based on the composition of the natural gas used as feed. Higher hydrocarbons produce a lower H_2:CO ratio than methane ($n + m/2$ mol H_2 per n mol of CO); thus, a higher fraction of methane in the natural gas results in a smaller pre-reformer and an overall higher H_2:CO ratio. As with all reforming processes that use steam, the process economics are improved by reducing the steam-to-carbon ratio. Lowering the steam-to-carbon ratio comes at the expense of lower conversion of the higher hydrocarbons, increased risk of coking and a lower H_2:CO ratio in the product.

A properly designed pre-reformer can shield the main reformer from variations in the feed composition.

11.5.2   Steam Reforming

Steam methane reforming (SMR) uses steam to convert methane to syngas over a catalyst (typically nickel) at temperatures $>800\,°C$. Pressures of 2 MPa and higher are common, but this is done to improve downstream operation. For example, higher pressure can improve the separation of hydrogen by pressure swing adsorption or increase the per pass syngas conversion during methanol synthesis. The SMR reaction is the opposite of methanation given by Equation (11.4) and is shown below, and can also be obtained by setting $n = 1$ in Equation (11.3):

$$CH_4 + H_2O \rightarrow CO + 3H_2, \quad \Delta H_{25\,°C} = +206\,\text{kJ mol}^{-1} \qquad (11.6)$$

The reaction is endothermic, requires large quantities of steam and is carried out at relatively high temperatures. Thus, the main challenge in SMR reactors is efficient heat generation and transfer to the reactor. Steam reforming of methane produces a high H_2:CO ratio in the syngas (typically H_2:CO>3). Although the stoichiometric requirement of steam is 1 mol per mole of methane, steam-to-carbon ratios of 2.5–3 are common in industrial practice.

A higher than stoichiometric steam-to-methane ratio is employed to reduce the possibility of carbon (coke) formation. Coke may be formed by the Boudouard reaction, Equation (11.7), and also by thermal cracking of the methane, Equation (11.8):

$$2CO \rightarrow CO_2 + C \tag{11.7}$$

$$CH_4 \rightarrow C + 2H_2 \tag{11.8}$$

In addition to the main steam reforming reaction, the water gas shift reaction, Equation (11.5), can also take place to some extent. However, a separate shift reactor (Section 11.5.5) is usually employed to manipulate the H_2:CO ratio for applications that target hydrogen generation.

For syngas applications such as methanol synthesis and Fischer–Tropsch synthesis, the H_2:CO ratio may have to be lowered. This can be achieved to a small extent by reducing the steam-to-carbon ratio in the feed. Other approaches include reverse water gas shift with CO_2 and by removing H_2 from the syngas after reforming or as it is formed during reforming, for example, using a membrane reactor.[25]

As with the pre-reformer, the steam-to-carbon ratio strongly affects the operating cost of the plant (based on energy required for steam generation), but lower steam-to-carbon ratios increase the risk of coking. Compared with other reforming technologies, steam reforming has a higher natural gas consumption, because additional natural gas is used as fuel to generate steam and to heat the reforming reactor itself. However, because heat is supplied externally (Figure 11.4), air can be used as oxidant without introducing inert gases (N_2 and Ar) into the syngas produced. It is also possible to use a different fuel than natural gas for supplying the heat.

11.5.3 Partial Oxidation

The conversion of methane to syngas using oxygen (or air) can be carried out using partial oxidation (POX). This is a non-catalytic process that is operated at 3–10 MPa and at very high temperatures ($>1200\,°C$). The POX reaction is shown in Equation (11.1) and it is repeated here for convenience: $CH_4 + \frac{1}{2}O_2 \rightarrow CO + 2H_2$.

This appears to be a very attractive process, since it produces H_2 and CO in a 2:1 ratio, which is ideal for many GTL applications.

In addition, the reaction is fast compared with the SMR reaction. However, POX also has some limitations: the absence of a catalyst means that very high temperatures are necessary and carbon formation is easier, and when oxygen is employed as oxidant, it means that an air separation unit (ASU) is required. Air separation is expensive and energy intensive.

Using air instead of oxygen is an alternative; however, this reduces reaction rates and thermal efficiency. It also presents much greater challenges for the downstream purification of the syngas, since there is a large amount of nitrogen present. The additional energy consumption in compressing air to partial oxidation (gasification) conditions is also substantial. Introducing nitrogen into the syngas lowers flexibility with respect to recycle streams, such as recycling unconverted synthesis gas. The possibility of forming nitrogen compounds at the high temperatures in the reactor cannot be discounted.

POX can also be carried out catalytically.[17] Noble metal and nickel-based catalysts have been employed for this purpose and result in lower operating temperatures, which in turn lowers the possibility of carbon formation.

11.5.4   Autothermal Reforming

An alternative to SMR and POX technologies is to combine them and perform them in a single reactor, as in the autothermal reforming (ATR) process.[26] The process is termed 'autothermal' since the exothermic partial oxidation (and combustion) reactions provide the energy that is required by the endothermic steam reforming reaction. This is distinct from processes that use a primary reformer (based on POX, for example) that is followed by a secondary reformer (based on SMR), and offers better energy utilisation.

The autothermal reactor is fed with oxygen (or air) and steam and operates at lower temperatures (1000–1100 °C) than the POX process. Commercially, ATRs are usually operated using oxygen and steam, which implies that they require an associated air separation unit. The arguments for choosing oxygen over air are similar to those for POX.

Since the partial oxidation process produces an H_2:CO ratio that is slightly lower than 2:1 and the SMR process produces a ratio higher than 2:1, a judicious mixture of the two should produce a process with tuneable H_2:CO ratio (by adjusting the exit temperature of the reformer). The ATR produces syngas with an H_2:CO ratio close to 2:1 at low steam-to-carbon ratios and relatively high exit temperatures; this tends to be more selective towards the POX reaction given by Equation 11.1. Changing the amounts of oxidant and steam in the feed can also be used to change the H_2:CO ratio. Other options for tuning the H_2:CO ratio include partial recycling of CO_2 or CO_2-rich tail gas from downstream of the shift reactor and separation units or by removal of hydrogen from the syngas using a membrane. The use of the CO_2 'dry' reforming reaction leads to a lower H_2:CO ratio:

$$CO_2 + CH_4 \rightarrow 2CO + 2H_2 \tag{11.9}$$

Combined reforming,[27] which uses a steam reformer followed by an autothermal reformer, can also be used to obtain the desired H_2:CO ratio. As mentioned earlier, this is not as effective in heat integration as ATR. The temperature of the steam reformer can then be used to manipulate the H_2:CO ratio.

11.5.5 Water Gas Shift Conditioning

For the case of hydrogen production using reforming (*e.g.* for fuel cell applications), it is typical to have a water gas shift (WGS) conditioning reactor downstream of the reformer to increase the H_2:CO ratio.[28] Also, if the reformer produces a different H_2:CO ratio than desired, Equation (11.5) can be used to manipulate the H_2:CO ratio.

Due to equilibrium, the forward reaction is thermodynamically favoured at lower temperatures; however, this slows the reaction kinetics. Typically, catalysts based on iron (90% Fe_2O_3 + 10% Cr_2O_3), copper (33% CuO + 34% ZnO + 33% Al_2O_3) and cobalt (sulfided 3.5% CoO + 12.5% MoO_3 on Al_2O_3) have been used for the WGS process.[29] Since the product stream from the reformer is at a high temperature, it is used to preheat the reformer feed. Since the stream is still at a relatively high temperature (450–550 °C), it is sent through a high-temperature shift reactor before being sent to a second, lower temperature (300–350 °C) shift reactor. The high-temperature shift reactor has lower conversion of CO, but the catalyst is designed to withstand the higher temperatures and protect the downstream low-temperature shift reactor, which enhances the CO conversion and hydrogen production.

Obviously, the inclusion of a shift reactor greatly influences the H_2:CO ratio and the CO_2 generated from the overall reforming process. The SMR and POX methods can produce CO_2:H_2 ratios of 0.25 and 0.33, respectively.[18] The net amount of CO_2 that is produced during syngas generation and conditioning is primarily a result of the energy requirements and the reaction stoichiometry, which depends on the H_2:CO ratio in relation to the H:C ratio of the feed.[30] The reforming and conditioning technology selection plays a minor role, mainly through the efficiency of the overall process.

11.6 SYNTHESIS GAS CONVERSION TECHNOLOGIES

The impression may have been created that once clean synthesis gas is produced, the selection of syngas conversion technology is independent of syngas production, since the syngas can be conditioned to meet the required H_2:CO ratio of the conversion process. This is not entirely true and the way in which the syngas was produced has an impact on the cost and efficiency of the syngas conversion technology. The three most important properties of the conditioned syngas are pressure, trace catalyst poisons and the inert content of the syngas.

The pressure at which the gas is available from the reformer, sets the upper limit of pressure at which the syngas conversion technology can operate, unless an additional syngas compressor is installed. In the case of methanol synthesis, a syngas compressor is usually required.[31] In the case of Fischer–Tropsch synthesis, synthesis is typically conducted at a pressure that does not require additional compression, even though a higher pressure has benefit.

Clean and conditioned syngas is never completely free of trace contaminants, even though such contaminants are typically present at sub-parts per million levels. Trace catalyst poisons will over time result in deactivation of the syngas conversion catalyst. If such deactivation results only in activity loss, it only affects the economics, but when it leads to changed selectivity, it can have an impact on the efficiency of the overall process.[32]

Inert materials in the feed have a more subtle effect and it has a similar impact to unselective syngas conversion.

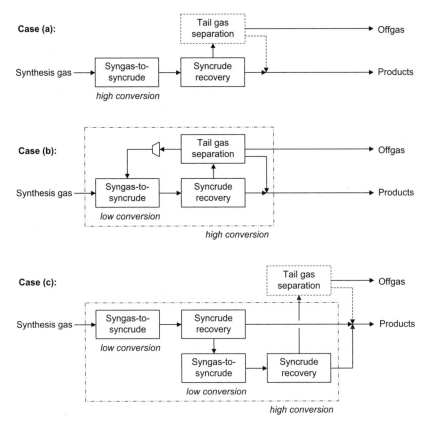

Figure 11.5   Basic gas loop designs to achieve high overall syngas conversion.

The per pass syngas conversion varies depending on the technology. It is usually limited to increase the selectivity to desirable products and to reduce the catalyst deactivation rate. The product after syngas conversion therefore still contains unconverted syngas. Since the syngas is valuable economically and environmentally (energy and carbon footprint), it is important to lose as little as possible of the unconverted syngas. Three basic strategies are employed to gain maximum benefit from the syngas (Figure 11.5):

(a) High per pass conversion in a single reactor.
(b) Low per pass conversion in a single reactor with recycle, resulting in a high overall conversion.

  (c) High overall conversion by more than one reactor in series
      with intermediate product recovery, where each individual
      reactor has a low per pass conversion.

  Hybrid configurations employing variations on these three basic
strategies can also be found in industrial practice.
  When the syngas conversion is high before recycling, the gas
loop design can be simplified by having less rigorous or no tail gas
separation. Although the unconverted syngas and unrecovered
light gases in the tail gas can be used as fuel gas, it would be wrong
to suggest that the overall conversion of syngas and tail gas
separation are not important. It makes no sense to substitute
methane (raw material feed) with syngas and light product gases
(value added products) for heating purposes.
  In any gas loop that involves syngas recycling, there is always a
fraction of the unconverted syngas that must be purged (offgas that
cannot be recycled). The size of the purge stream depends on the
separation efficiency during product recovery, where the unconverted
syngas is separated from the products and the inert materials that
were originally present in the syngas feed. The need for a purge
follows from the mass balance over the gas loop; for steady-state
operation, inert materials and products must leave the gas loop at the
same rate as they enter or are produced. Inert material in the syngas
feed is consequently detrimental to the gas loop in a similar way to
unrecovered and 'inert' gaseous products from syngas conversion.
  In order to improve sustainability, it is always preferable to have
high syngas conversion and good tail gas separation to limit the
amount of carbon in the offgas. When a tail gas recycling process is
employed, it is also important to limit the amount of inert material.
The main sources of carbon loss in the syngas conversion technology and its associated gas loop are as follows:

  • Methane production by the syngas conversion technology. It
    is pointless to have a process that converts methane via syngas
    into methane and consumes energy doing so. The methane
    selectivity of the syngas conversion technology must be as low
    as possible.
  • The product selectivity to other compounds than methane is
    also important. Not all products are equally useful and the
    usefulness of a product can be very application dependent.

For example, the production of ethane in a petrochemical facility is very valuable, because it can be converted with high selectivity into ethene, but in a fuels refinery it is not valuable and is often used as fuel gas. The synthesis gas conversion technology must be selected with the final application in mind and there is no single best technology for all applications.

- The syncrude recovery and tail gas separation steps determine how much of the carbon feed can be retained and refined to useful products. Unrecovered material cannot be refined to products.

- Any inert material that cannot be separated in the gas loop will increase the amount of carbon lost through the offgas purge. It does not matter whether the inert material entered with the syngas (*e.g.* N_2 and Ar) or whether it is inert material produced by syngas conversion (*e.g.* CH_4 and C_2H_6).

- The quality of energy that is produced during syngas conversion indirectly affects the efficiency of the process. Syngas conversion is very exothermic and this energy is normally employed to generate steam. The usefulness of the steam as energy source, either for heating or power generation, depends on the steam pressure (Table 11.7). When the syngas conversion technology operates at a higher temperature, the steam pressure that can be generated from the heat of reaction is higher.

11.6.1 Methanol Synthesis

Methanol is predominantly produced from methane through syngas conversion, although direct conversion by partial oxidation is also possible (Section 11.8.4).[33] The conversion can be

Table 11.7 Saturated steam pressure that can be obtained at different temperatures.

Temperature/°C	Saturated steam pressure/MPa
200	1.5
220	2.2
240	3.2
260	4.5
280	6.3
300	8.5

described as a partial hydrogenation of CO and it does not involve chain growth:

$$CO + 2H_2 \rightleftarrows CH_3OH, \quad \Delta H_{25°C} = -91 \, \text{kJ mol}^{-1} \quad (11.10)$$

If CO_2 is present in the syngas, it can also produce methanol and the reverse water gas shift may also occur if CO_2 is present:

$$CO_2 + 3H_2 \rightleftarrows CH_3OH + H_2O, \quad \Delta H_{25°C} = -41 \, \text{kJ mol}^{-1} \quad (11.11)$$

Historically, methanol production was associated with ammonia synthesis for the specific purpose of removing CO as methanol before ammonia synthesis from the nitrogen-containing synthesis gas.[34] The origin of purpose-designed methanol synthesis is credited to Badische Anilin- und Soda-Fabrik (BASF), where a synthetic process for methanol synthesis from synthesis gas was commercialised in the early 1920s. This process employed ZnO on a difficult to reduce oxide, Cr_2O_3, and synthesis was conducted at 320–400 °C and 25–35 MPa.[31] The ZnO/Cr_2O_3 catalyst formed the basis of various 'high-pressure' or 'classical' methanol synthesis technologies. Natta provided a detailed discussion of this type of catalysis.[35]

The development of 'low-pressure' or 'modern' methanol synthesis stemmed from the use of Cu–ZnO-based catalysts by Imperial Chemical Industries (ICI) in the 1960s. These catalysts were known before, but were very sensitive to catalyst poisons such as sulfur and halides. New technologies based on Cu-containing catalysts were enabled by better syngas cleaning, rather than an advance in catalysis.[31] Synthesis is conducted at 200–300 °C and 3.5–10 MPa. Currently, all industrial syngas-to-methanol conversion is performed using Cu-based catalysts.[36] These catalysts are also active for the water gas shift reaction shown in Equation (11.5). Catalyst deactivation is a major concern in this process and control of the temperature and efficient heat removal are critical in obtaining the maximum possible catalyst lifetime. Much of the technology associated with methanol synthesis is therefore related to reaction engineering and reactor design. Owing to the exothermic nature of methanol synthesis, heat management is very important. Industrial operation is dominated by gas-phase methanol synthesis in fixed-bed reactors. The main reactor technologies that are employed to control the reaction temperature are multiple catalyst beds with inter-bed quenching, and multitubular and radial flow reactors.[31] When a multitubular reactor is employed, 4 MPa steam can be co-produced.[37] Slurry bubble column liquid-phase methanol synthesis (LPMEOH process) over a powdered $Cu–ZnO/Al_2O_3$ catalyst was successfully demonstrated by Eastman Chemical. This technology was developed specifically to allow on-line catalyst replacement when using syngas with higher levels of catalyst poisons.[38]

Methanol synthesis is very selective and side reactions to produce dimethyl ether (DME), formaldehyde and heavier products are kinetically limited.[39] The conversion per pass is around 10%,[36] with the unconverted synthesis gas being recycled in a closed gas loop (Figure 11.5) to achieve a high overall conversion.

In some applications, methanol synthesis is combined with etherification, although etherification can also be a separate refining step. As separate step, DME is typically produced by the dehydration of methanol over zeolite or amorphous silica–alumina catalysts at around 325 °C and 2 MPa:

$$2CH_3OH \rightleftarrows CH_3OCH_3 + H_2O, \quad \Delta H_{25°C} = -22\,kJ\,mol^{-1} \quad (11.12)$$

It was pointed out that methanol synthesis and DME synthesis have very similar production costs, because the conversion of methanol to DME improves the equilibrium conversion of methanol, allowing higher per pass conversion.[40] Unless methanol is the desired end product, DME will in any case be a reaction intermediate during fuels refining (Section 11.7.1). In fact, producing hydrocarbons from DME instead of methanol is advantageous.[41]

11.6.2 Fischer–Tropsch Synthesis

Fischer–Tropsch synthesis is a conversion process that involves CO polymerisation in combination with hydrogenation. An in-depth discussion of Fischer–Tropsch synthesis can be found in books that appeared over an extended period of time, but are all still relevant today.[42–45] The reaction is often reported as a hydrocarbon synthesis:

$$nCO + 2nH_2 \rightarrow -(CH_2)_n- + nH_2O, \quad \Delta H_{25°C} = -140 - 160\,kJ\,mol^{-1}$$
$$(11.13)$$

The original syngas reaction reported by Franz Fischer and Hans Tropsch in 1923 was mainly an oxygenate synthesis. Oxygenates are always co-produced with hydrocarbons during the reaction and the main oxygenate classes found in Fischer–Tropsch syncrude are alcohols, carbonyls and carboxylic acids. Compounds with an oxygen-containing functionality are concentrated in the lighter syncrude fractions, and also in the aqueous product. Water is the main product from Fischer–Tropsch synthesis and light oxygenates, such as ethanol and acetic acid, preferentially dissolve in the water, rather than in the oil product.

The carbon number distribution as result of CO polymerisation can be described in the same way as in other polymerisation reactions and it is a function of the chain growth probability or α-value. The α-value relates the molar concentration of species with i carbon atoms (C_i) to the molar concentration of species with j carbon atoms (C_j) by the relationship

$$C_i/C_j = \alpha^{(i-j)} \tag{11.14}$$

In practice, there are deviations from this simple relationship for the C_1 and C_2 hydrocarbons, with methane being more and ethane/ethene being less than calculated from Equation (11.14). The α-value and thereby the actual carbon number distribution of syncrude are determined by many factors, such as the Fischer–Tropsch catalyst, reactor type and operating conditions, but mostly by the operating temperature. It is for this reason that Fischer–Tropsch processes are classified as either low-temperature Fischer–Tropsch (LTFT) synthesis or high-temperature Fischer–Tropsch (HTFT) synthesis. LTFT syncrude is rich in aliphatic hydrocarbons and about half of the syncrude product is a paraffinic wax. HTFT syncrude contains more oxygenates and unsaturated hydrocarbons, including aromatics, with little heavy material. The carbon number distributions obtained from these two types of syncrude are shown in Figure 11.6.

Anderson reviewed the metals capable of Fischer–Tropsch synthesis and these include, among others, Fe, Co, Ni, Ru and Rh.[44] Industrially only Fe and Co are employed for Fischer–Tropsch catalysts. The main technologies for Fischer–Tropsch synthesis are as follows:

(a) *Fe-based LTFT*. Three main technology types are found in industrial practice, namely the Arbeitsgemeinschaft Ruhrchemie-Lurgi (Arge) multitubular fixed-bed technology used by Sasol, the iron-based Sasol slurry bed process

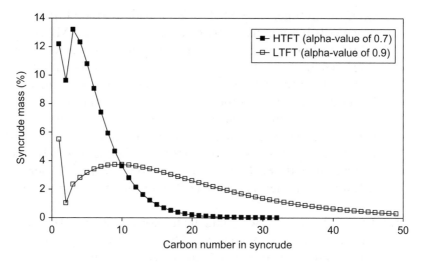

Figure 11.6 Carbon number distribution of a typical LTFT and a typical HTFT process.

(Fe-SSBP) and the high-temperature slurry Fischer–Tropsch process (HTSFTP) employed by Synfuels China. The Arge and SSBP technologies are operated at around 240 °C, whereas the HTSFTP technology is operated at 270 °C to improve the quality of steam that can be produced (Table 11.7). The Arge technology has been in operation since 1955 and it is a proven robust technology. Its main drawback is the onerous nature of catalyst replacement, which takes place every 70–100 days.[46] With slurry bed technology, on-line catalyst addition and removal are possible. The low cost and attrition resistance of precipitated Fe-LTFT catalysts make them ideal for slurry bed operation.

(b) *Co-based LTFT.* Two main technology types are found in industrial practice, namely the Shell middle distillate synthesis (SMDS) multitubular fixed-bed technology and the cobalt-based Sasol slurry bed process (Co-SSBP). Statoil also has a demonstration-scale cobalt-based slurry bed process in industrial operation at the PetroSA GTL facility. Around 220 °C is a typical operating temperature for these processes. Unlike Fe, Co is not water gas shift active, which can be beneficial or detrimental depending on the gas loop design. The main advantage of Co-LTFT over Fe-LTFT is catalyst lifetime. The SMDS fixed-bed Co-LTFT catalyst reportedly has a lifetime of at least 5 years,[47] and SMDS Co-LTFT technology has the same robustness as the Arge Fe-LTFT technology. The motivation for slurry bed technology in combination with Co-LTFT synthesis is based mainly on better heat transfer. Supported Co-LTFT catalysts are less attrition resistant and the poor initial performance of Sasol's Co-LTFT technology in the Oryx GTL facility was related to attrition problems.[48] The main drawbacks of Co-LTFT compared with Fe-LTFT are the more than an order of magnitude higher catalyst cost and the higher methane selectivity.

(c) *Fe-based HTFT.* Two main technology types are found in industrial practice, namely Sasol Synthol circulating fluidised bed technology and Sasol advanced Synthol (SAS) fixed fluidised bed technology. Both technologies operate at around 320 °C. In terms of steam quality and volumetric reactor productivity, Fe-HTFT is better than the LTFT technologies.[49] On the downside, HTFT has a higher methane selectivity and it produces a large fraction of C_2–C_3 products, that are advantageous only with appropriate downstream processing.

The syncrude composition obtained from each technology differs. The syncrude compositions in Table 11.8 give only an indication of what can typically be expected from each of the main technology classes. As mentioned before, the actual syncrude composition that is obtained in practice depends on many factors, such as the reactor design, operating conditions, catalyst formulation and state of catalyst deactivation.

11.7   REFINING TO COMMERCIAL TRANSPORTATION FUELS

It was pointed out in Section 11.3 that methane, methanol and DME can all be directly employed as 'special' transportation fuels. As such, no further refining is required for these special applications. Fischer–Tropsch syncrude has no equivalent opportunity for direct application and Fischer–Tropsch syncrude must always be refined in order to produce a useful product.

Table 11.8 Typical compositions of the syncrude from the main industrially applied Fischer–Tropsch technology classes.

Compounds	*Fischer–Tropsch syncrude/mass%*[a,b]		
	Fe-LTFT	*Co-LTFT*	*Fe-HTFT*
Gaseous products			
Methane	2.0	2.5	6.5
Ethene	0.5	<0.1	2.9
Ethane	0.5	0.5	2.3
Propene	1.6	0.9	5.9
Propane	0.4	0.4	0.8
Butenes	1.2	0.6	5.0
Butanes	0.4	0.4	0.7
Naphtha (C_5–C_{10})			
Olefins	3.6	3.5	13.3
Paraffins	1.5	5.3	2.2
Aromatics	0	0	0.9
Oxygenates	0.6	0.1	0.8
Distillate (C_{11}–C_{22})			
Olefins	2.7	0.5	2.5
Paraffins	6.4	9.2	0.5
Aromatics	0	0	0.4
Oxygenates	0.1	0	0.3
Residue/wax fraction			
Olefins	0.3	0	0.8
Paraffins	23.0	19.8	0.2
Aromatics	0	0	0.4
Oxygenates	0	0	0.1
Aqueous products			
Alcohols	1.9	0.6	2.3
Carbonyls	0	0	2.0
Carboxylic acids	0.2	0.1	0.7
Water	53.1	55.6	48.5

[a]These are the net products and exclude unconverted $H_2 + CO$ and $H_2 + CO_2$ from water gas shift conversion.
[b]Values that are zero do not imply the complete absence of such compounds.

The subsequent discussion focuses on the refining of methanol and Fischer–Tropsch syncrude from GTL facilities to produce 'standard' transportation fuels that are compatible with existing crude oil-based transportation fuel infrastructure. When considering 'standard' transportation fuels that are based mainly on liquid hydrocarbons, one must also differentiate between the production of on-specification fuels and the production of blending materials.

(a) *On-specification fuels.* These are fuels that conform to all fuel specifications for the specific jurisdiction where they are marketed. On-specification fuels can be sold directly to the consumer. Such fuels have a composition that falls within the range found in fuels produced by crude oil refining and are therefore for all practical intents and purposes indistinguishable from crude oil-derived products. Even though the refining effort required to produce on-specification transportation fuels from syncrude is less than that from crude oil,[50] the refinery is still complex and contains at least five different conversion units.

(b) *Blending materials.* These are fuels that are compatible with crude oil-derived transportation fuels. Blending materials must be mixed with other fuel components in order to produce on-specification transportation fuels and it cannot be sold directly to the consumer as a transportation fuel. In order to differentiate between on-specification fuels and blending materials, it is best to refer to the boiling range fraction of blending material, for example, distillate instead of diesel fuel and naphtha instead of motor gasoline. This distinction is not always clear in the GTL literature, which often refers to the distillate product from GTL as diesel fuel even though it is just a blending material. Furthermore, material only becomes useful for blending if it provides an advantage to the fuel blend or, at worst, if it is neutral and only increases the blending volume. A blending material by definition requires no further refining and must therefore have no undesirable properties that cannot be blended away. For example, a distillate rich in linear paraffinic material is waxy and has poor cold flow properties (solid waxy particles that form at cold temperatures). Even though the impact of the poor cold flow properties can be mitigated at low concentration, blending cannot change the freezing point of such molecules; the freezing point can only be changed by further refining.

From a sustainability perspective, it is important to refine methanol and Fischer–Tropsch syncrude only to the point where it becomes more efficient to produce on-specification fuels by blending instead of further refining. This is especially pertinent for diesel fuel production,[51] where the production of blending material is by far the preferred option. It will be shown that the same is not necessarily true for motor gasoline and jet fuel, where refining of methanol and Fischer–Tropsch syncrude to final on-specification fuels is efficient and that blending is an alternative that does not necessarily yield a sustainability gain. In GTL applications, the same applies to natural gas liquids that may be associated with the natural gas feed.

11.7.1 Methanol Refining to Transportation Fuels

The first step in the refining of methanol to produce 'standard' transportation fuels is methanol to hydrocarbon conversion.[14,15] The methanol is converted into a range of olefinic products over an H-ZSM-5 catalyst and the product bears some resemblance to an isomerised naphtha fraction of HTFT syncrude without the oxygenates:

$$n\text{CH}_3\text{OH} \rightarrow \text{C}_n\text{H}_{2n} + n\text{H}_2\text{O}, \quad \Delta H_{25°\text{C}} = -85\text{–}90 \text{ kJ mol}^{-1} \quad (11.15)$$

The actual conversion chemistry and catalysis are more complex.[52] The first step is dehydration of methanol to DME, Equation (11.12), which is followed by conversion of the DME to olefins. The olefins are the key intermediates that lead to subsequent aromatisation to produce paraffins and aromatics. During aromatisation, the hydrogen is rejected by transfer to olefins, as illustrated by the stoichiometry of the following pseudo-reaction:

$$6\text{C}_3\text{H}_6 \rightarrow \text{C}_9\text{H}_{12} + 3\text{C}_3\text{H}_8 \quad (11.16)$$

When either methanol or DME is used as feed material, the operating temperature must be sufficiently high to overcome the inhibitory effect of water.[53] The carbon number distribution is dictated by a limited equilibrium (the equilibrium does not extend to isomers within each carbon number). This restricts the formation of heavy products and kerosene is the heaviest product being produced. Unless methanol is first converted into olefins using a more olefin-selective catalyst, such as SAPO-34, and then refined, it is not suitable for distillate production. The most efficient refining pathway for transportation fuel production from methanol is the direct synthesis of motor gasoline.

Table 11.9 Typical product compositions from fixed-bed and fluidised-bed methanol-to-gasoline conversion.

	Product composition/mass%	
Compounds	Fixed-bed	Fluidised-bed
Water	56.59	55.87
H_2, CO and CO_2	0.20	0.22
Methanol and DME	0.04	1.20
Hydrocarbons	43.17	42.71
Hydrocarbon composition		
Methane	1.2	0.9
Ethene	0.0	5.8
Ethane	0.5	0.2
Propene	0.2	6.6
Propane	5.8	3.0
Butenes	1.1	5.4
n-Butane	3.3	1.3
Isobutane	8.3	13.0
C_5 and heavier aliphatics	49.8	36.5
Aromatics	29.9	27.3

The methanol-to-gasoline (MTG) process was developed to produce on-specification motor gasoline from methanol. Two technology variants exist, namely fixed-bed and fluidised-bed methods. Typical product compositions from each are given in Table 11.9.[54,55] These are average compositions and in both instances the operating conditions and catalyst age affect the actual composition that will be obtained in practice. It is noteworthy that the C_5 fraction is rich in isopentane (RON = 92, MON = 90), which is good for motor gasoline. Furthermore, the aromatic content is high, but not too high, and the aromatic fraction contains little benzene.

The fixed-bed design operates in swing mode; oxidative catalyst regeneration is required after each reaction cycle, which typically lasts 20 days. It employs two conversion steps in series. In the first reactor, methanol is etherified to DME and in the second, the DME is converted into hydrocarbons. Typical operating conditions are 315 °C inlet and 405 °C outlet for the first reactor and 360 °C inlet and 415 °C outlet for the second reactor. The pressure is 300 kPa and a 9:1 recycle ratio is employed.[16] The gasoline product from the fixed-bed process is shown in Table 11.10.[54] Depending on the location of the facility, the product may meet

Table 11.10 Straight-run motor gasoline quality obtained from fixed-bed methanol-to-gasoline conversion.

Property	Straight-run motor gasoline[a]
Research octane number, ASTM D2699	93.9–97.0
Motor octane number, ASTM D2700	83.2–86.0
Density/kg m^{-3}	739–767
Reid vapour pressure/kPa	79–96
Distillation profile, ASTM D86/°C	
IBP	27–29
T20	62–68
T50	105–123
T80	156–166
FBP	207–219
Composition, ASTM D1319/vol.%	
Paraffins	47.8–54.1
Olefins	5.5–17.8
Aromatics	29.4–46.7

[a]Straight-run motor gasoline implies that it is the raw product and beyond distillation it was not modified; the product included no blending components or additives.

motor gasoline specifications. However, a more robust refining strategy requires blending with alkylate (trimethylpentane-rich high octane number blending component), so that it can be marketed as a premium motor gasoline.[16] The alkylate helps to reduce the aromatic and olefin content, decrease the vapour pressure and increase the octane number slightly.

The fluidised-bed design employs only a single reactor with continuous catalyst regeneration. Conceptually it is much like a fluid catalytic cracker. The feed is introduced at 175 °C at the bottom of the reactor and the reaction heat increases the temperature to 400 °C. Although the reaction chemistry is the same as in the fixed-bed unit, the operating conditions and partial pressure of compounds are different. This causes the fluidised bed process to produce a product with higher yield of light olefins compared with the fixed-bed process.[55]

The higher C_3–C_4 olefin yield specifically is advantageous for refinery design, because it permits the inclusion of an aliphatic alkylation unit in the design. Aliphatic alkylation consumes olefins in a 1.1–1.4 ratio to isobutane.[56] The alkylate required for the robust production of on-specification motor gasoline can therefore be incorporated in the refinery. This has the added benefit of transforming part of the gaseous products into a liquid product, leading to a more efficient refinery design for motor gasoline production. Such a conceptual refinery design for methanol-based GTL employing the fluidised bed MTG process is shown in Figure 11.7. Depending on the octane requirements of the final product, it may also be necessary to include a C_6 hydroisomerisation unit, because the C_6 aliphatic fraction contains around 40% n-hexane (RON = 25, MON = 26).

The distillate fraction from such a process is essentially light kerosene and it is too aromatic to be employed for jet fuel or diesel fuel without extensive blending. The product also contains 2% durene (1,2,4,5-tetramethylbenzene) that imparts poor cold flow properties to the fuel. When the total product is used as motor gasoline, extensive vehicle performance testing indicated that this level of durene is acceptable in motor gasoline (crude oil motor gasoline contains 0.2–0.5% of durene).[16] However, the comparatively high level of durene makes the kerosene fraction a very poor jet fuel blending component.

Methanol is best refined to on-specification motor gasoline. It is also apparent that the refinery design required to produce on-specification motor gasoline from methanol (Figure 11.7) is less complex than a modern fourth-generation crude oil refinery, which typically requires hydrotreating, naphtha reforming, fluid catalytic cracking, naphtha hydroisomerisation and aliphatic alkylation units in order to produce on-specification motor gasoline.[11]

Associated natural gas liquids can be co-refined with the methanol in the same way, since H-ZSM-5 is also capable of converting paraffins at high temperature. This is a convenient upgrading pathway, but from a sustainability point of view it is not

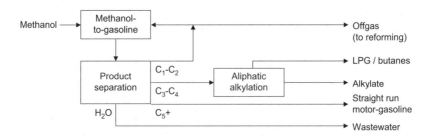

Figure 11.7   Flow diagram of the fuels refinery of a methanol-based GTL facility employing fluidised-bed MTG technology.

the most efficient pathway. The paraffinic naphtha and distillate fractions are easily converted, but the conversion is accompanied by a decrease in liquid yield. Unnecessarily losing liquid yield is contrary to the principle of sustainability in GTL conversion.

11.7.2 Fischer–Tropsch Syncrude Refining to Transportation Fuels

The success and simplicity of the methanol-to-gasoline process suggested that Fischer–Tropsch syncrude could also be refined in a similar way. Three approaches have been taken along these lines, mostly employing H-ZSM-5 as catalyst. The first approach involves the modification of the Fischer–Tropsch catalyst itself to include a zeolitic component (*e.g.* Fe/H-ZSM-5).[57] The second approach is to feed the zeolite as co-catalyst with the Fischer–Tropsch catalyst under synthesis conditions.[58] The third approach is analogous to the MTG process, namely to feed the total syncrude product to a conversion unit containing the zeolite.[59] In all cases, an aromatic naphtha is produced. At low per pass conversion, a very aromatic product can be obtained, but generally the conversion has a C_5 and heavier liquid yield of only around 50%. Compared with methanol as feed, the yield is lower and, considering that syncrude recovery and the gas loop design for Fischer–Tropsch synthesis is more complex than for methanol synthesis, this is not a preferred refining pathway. Should one want to produce motor gasoline exclusively, methanol is the preferred intermediate.

With sustainability in mind, the properties of Fischer–Tropsch syncrude should be exploited to produce products that are more difficult to produce in other ways. Here we focus only on transportation fuels; petrochemicals will be discussed separately (Section 11.8.3). Central to the efficient refining of syncrude and the principles of sustainability that were discussed earlier is the selection of appropriate refining techniques and catalysts.[11,60,61] Even though syncrude can be refined much like crude oil, it has few refining technology and catalyst combinations in common. In fact, refining syncrude like crude oil is inefficient and leads to complex refinery designs.[62]

On-specification motor gasoline and jet fuel can both be produced in >70% selectivity from HTFT and LTFT syncrudes, but the same is true of diesel fuel only if there is no minimum density requirement.[11,63] Extreme designs to maximise the yield of a specific fuel type are generally not the most efficient refining options, although such designs are instructive. For example, it highlights the versatility of Fischer–Tropsch syncrude, and also the potential flexibility to change the production focus in a refinery to meet changes in market demands.

Blending materials can also be produced from Fischer–Tropsch syncrude, but these make sense only when producing kerosene or distillate fractions for jet fuel or diesel fuel blending. The quality of straight-run naphtha or partially refined naphtha provides no blending advantage for modern-day motor gasoline. HTFT naphtha has a higher straight-run octane number than crude oil-derived naphtha on account of its high olefin content, but still requires refining. Historically, the HTFT naphtha from industrial GTL conversion in the American Hydrocol facility was refined using only bauxite treatment and could be marketed after addition of tetraethyllead.[11,62]

In GTL facilities where part of the feed to the refinery comes from associated natural gas liquids, it is very important to have refining pathways to deal with paraffins. The associated natural gas liquids resemble LTFT syncrude properties more closely than HTFT syncrude properties. However, there is a refining synergy between LTFT and HTFT syncrude and consequently an advantage in co-refining natural gas liquids with HTFT syncrude. It was also pointed out that paraffin refining is key to the efficient production of motor gasoline from Fischer–Tropsch syncrude.[11,63]

In order to transform Fischer–Tropsch syncrude into on-specification transportation fuels or blending materials, the refinery design must accomplish the following tasks:

(a) Adjust the carbon number distribution so that the refined products fall within the boiling range of transportation fuels (typically 20–360 °C).

(b) Introduce a high degree of branching in naphtha-range paraffins, because the octane number is strongly dependent on the degree of branching. Some branching must also be introduced in kerosene- and distillate-range paraffins to ensure adequate cold flow properties.

(c) Convert oxygenates to oxygenate classes that are acceptable in transportation fuels (alcohols and ethers) or convert oxygenates to hydrocarbons.

(d) Reduce the olefin content of the naphtha fraction to meet specifications. In the case of naphtha, olefin hydrogenation leads to a significant decrease in octane number and the refinery design must compensate for this. In the case of kerosene and distillate, hydrogenation improves the fuel properties. Whenever practical it is best to exploit the reactive nature of the olefins before hydrogenation.

(e) Produce additional aromatics, which are needed to produce on-specification motor gasoline and jet fuel, and also to avoid elastomer compatibility problems in the fuel systems of vehicles and aeroplanes. Elastomeric seals swell due to absorption of aromatics and polar compounds in the fuel and, when subsequently exposed to fuels with low aromatic content, the seals contract and leak. HTFT syncrude contains sufficient aromatics to avoid such problems, but not LTFT syncrude.

Figure 11.8 shows a generic Fischer–Tropsch refinery design that incorporates the co-refining of natural gas liquids. The cut points are only indicative. The quantity and quality of the different transportation fuel types depend on the Fischer–Tropsch syncrude. However, this type of design is capable of producing on-specification fuels for most jurisdictions. One can argue that liquid petroleum gas (LPG) is not widely used as a transportation fuel and that further refining of this cut is required. However, upgrading of LPG is a separate topic; there is little difference between synthetic, natural gas-derived and crude oil-derived LPG.

Some new industrial GTL facilities have been designed with a minimalist refinery. In these designs, crude oil technology-based hydrocracking is employed as the only unit to convert the syncrude into LPG, naphtha and distillate fractions. In such cases the distillate can be employed as a blending material, but the naphtha requires further refining to be useful and it cannot be considered a blending material. Such minimalist design makes poor use of the potential refining advantages offered by Fischer–Tropsch syncrude and it does not support the sustainability considerations discussed before.

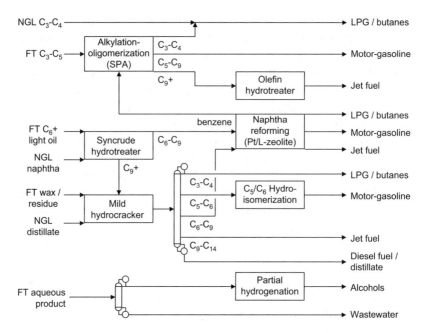

Figure 11.8   Flow diagram of a generic GTL Fischer–Tropsch (FT) refinery
that co-refines associated natural gas liquids (NGL).

Fischer–Tropsch refinery designs that are sensitive to sustainability principles are more complex. However, even in such refineries inadequate syncrude recovery in the gas loop and the lack of
refining pathways for the C_2 hydrocarbons and for the carboxylic
acids in the aqueous product erode sustainability. Cryogenic
separation and petrochemical use of the C_2 hydrocarbons and
carboxylic acids are obvious ways to improve sustainability. In
fact, Fischer–Tropsch syncrude has considerable petrochemical
potential that will be discussed in the next section.

11.8   PETROCHEMICALS AND LUBRICANTS FROM METHANE

Petrochemicals and lubricants are higher value products than
transportation fuels and can be produced from methane in two
different ways: indirect liquefaction and direct chemical conversion.

Indirect liquefaction, with synthesis gas as an intermediate
(Section 11.5), is already practised in industry. Synthesis gas itself is
a petrochemical intermediate (Section 11.8.1). The two main

industrial synthetic routes for synthesis gas conversion, namely methanol synthesis and Fischer–Tropsch synthesis, can produce products that are petrochemicals, in addition to intermediate products that may be converted into petrochemicals and lubricants (Sections 11.8.2 and 11.8.3). Of these, indirect conversion to Fischer–Tropsch syncrude is the only practical route to lubricants from methane.

Methane is an unreactive molecule, but there are processes that have been devised to convert methane directly into products. Oxygen is the most studied conversion agent for the direct oxidative conversion of methane to petrochemicals (Section 11.8.4). However, there are industrial processes that employ reagents other than oxygen for direct petrochemical production from methane too (Sections 11.8.5–11.8.7).

11.8.1   Petrochemicals *via* Synthesis Gas

Synthesis gas is employed as reagent in the production of many petrochemical products. Probably the largest use of syngas is for the production of hydrogen.[64] Hydrogen is ubiquitous in the refining and petrochemical industries and reforming of methane is the most efficient way to produce H_2 on a large scale at present.

Carbon monoxide is also employed in petrochemical production, on its own and in combination with hydrogen. The use of syngas to produce methanol and Fischer–Tropsch syncrude represents important routes to petrochemicals and will be discussed separately (Sections 11.8.2 and 11.8.3).

In addition to the aforementioned applications, some important petrochemical uses of syngas are as follows:

(a) Ammonia (NH_3) is one of the highest volume chemicals, mainly due to its use in the agricultural industry as a plant fertiliser and in the mining industry for the production of explosives. The synthesis gas is converted by the water gas shift reaction to produce H_2. The H_2 is then combined with N_2 from an air separation unit to produce ammonia.

(b) Hydroformylation, represented by Equation (11.17), forms the basis of many industrial processes and employs Co- and Rh-based homogeneous catalysts.[65] The aldehydes produced from the hydroformylation of alkenes are usually hydrogenated to the corresponding alcohols. Depending on the carbon chain length, these alcohols are employed as solvents, plasticisers and in the manufacture of detergents:

$$RCH = CH_2 + CO + H_2 \rightarrow RCH_2CH_2CHO \qquad (11.17)$$

(c) Phosgene ($COCl_2$), despite its hazardous nature, is still employed as a chemical intermediate in the manufacture of various products, including polyurethane, isocyanates ($RN{=}C{=}O$) and polycarbonates. Phosgene is produced by the free radical addition of Cl_2 to CO.[66]

(d) Higher alcohol ($C_nH_{2n+1}OH$) synthesis is more than just a variation on methanol synthesis, because it combines the principles of methanol synthesis with the chain growth properties of Fischer–Tropsch synthesis. Most catalysts for higher alcohol synthesis are based on alkaline ZnO with, among others, Cu, Cr_2O_3 and Co as the main elements. Methanol is invariably a major product, irrespective of the catalyst and conditions.[67,68] A typical product distribution from higher alcohol synthesis over a Cu–Co–ZnO/Al_2O_3 catalyst is provided in Table 11.11.[68]

11.8.2 Petrochemicals and Lubricants *via* Methanol Synthesis

Methanol itself is one of the highest volume petrochemicals, with a market size of over 40 Mt per year.[69] Methanol synthesis is therefore a petrochemical production process. It is also possible to convert methanol into other petrochemicals. Some of the main applications of methanol to petrochemical production are as follows:

Table 11.11   Liquid product from demonstration-scale higher alcohol synthesis from syngas over a Cu–Co–ZnO/Al_2O_3 catalyst.

Compound	Liquid product, water-free basis/mass%[a]
Methanol	63.5
Ethanol	24.9
Propanol	6.1
Butanol	2.3
C_5 and heavier alcohols	2.5
C_5 and heavier hydrocarbons	0.2
Esters	0.3
Other oxygenates	0.2

[a]Water content of the liquid product is 0.2%.

(a) Methanol-to-propylene (MTP) technology.[15,70,71] It can be based on H-ZSM-5- or SAPO-34-catalysed conversion of methanol into alkenes. SAPO-34 is more selective for olefin production. The conversion chemistry is similar to the original H-ZSM-5-based methanol-to-olefins (MTO) process,[14] but it is optimised for propene production.

(b) Methanol is used in a variety of etherification processes.[72] The etherification of alkenes with methanol, which is given by Equation (11.18), saw rapid growth in the 1990s due to the use of methyl *tert*-butyl ether (MTBE) in motor gasoline. Although this market has since declined, ethers are still employed both as fuel additives and as solvents. Etherification is usually conducted at 60–90 °C in the liquid phase over an acidic resin catalyst:

$$C_4H_8 + CH_3OH \rightarrow (C_4H_9)OCH_3 \qquad (11.18)$$

(c) Formaldehyde (methanal) can be produced from methanol by partial oxidation, given by Equation (11.19), and reduction, givn by Equation (11.20). Industrially, the main production route is by the Haldor Topsøe process for oxidative methanol conversion over an iron–molybdenum oxide catalyst at 400–425 °C:[66,73]

$$CH_3OH + \tfrac{1}{2}O_2 \rightarrow CH_2O + H_2O \qquad (11.19)$$

$$CH_3OH \rightarrow CH_2O + H_2 \qquad (11.20)$$

(d) Acetic acid is produced from methanol by carbonylation, as shown in Equation (11.21).[73,74] There are Ni, Co and Rh catalysed processes. Carbonylation of methanol can synergistically exploit the availability of CO in the syngas needed for methanol production:

$$CH_3OH + CO \rightarrow CH_3COOH \qquad (11.21)$$

11.8.3 Petrochemicals and Lubricants *via* Fischer–Tropsch Synthesis

Indirect liquefaction based on Fischer–Tropsch synthesis produces some molecules and syncrude distillation fractions that have petrochemical value. Petrochemicals can be found in all the main compound classes from Fischer–Tropsch synthesis, namely, alkanes (*e.g.* light paraffinic solvents and paraffin waxes), alkenes (*e.g.* ethene, propene and *n*-1-alkenes) and oxygenates (*e.g.* ethanol, acetone and acetic acid).

The type and concentration of each petrochemical depends on the Fischer–Tropsch technology (Section 11.6.2) and various industrial processes are in operation for petrochemical refining from Fischer–Tropsch syncrude. A more in-depth discussion of petrochemical and lubricant production from Fischer–Tropsch syncrude can be found in the literature.[11,75–77] The main petrochemical products derived from syncrude are the following:

(a) Light paraffins can be separated from hydrogenated Fischer–Tropsch oil. The degree of linearity (*n*-alkane content) depends on the Fischer–Tropsch technology employed for production, and industrially light paraffins are only produced from LTFT oil.

(b) Waxes can be recovered from the heavy fraction of LTFT syncrudes. The wax products are classified by congealing point and various wax grades are produced industrially from LTFT syncrude.

(c) Lubricant base oils are prepared industrially from LTFT waxes in an analogous fashion to the production of lubricant base oils from crude oil derived waxes. High viscosity index poly-α-olefin (PAO) lubricant base oils can also be prepared by oligomerisation of linear 1-alkenes.

(d) Light alkenes (ethene and propene) can be recovered by distillation from HTFT syncrude. Production can be supplemented by high-temperature thermal or catalytic cracking and alcohol dehydration. All of these approaches are or have been practised industrially.

(e) Linear *n*-1-alkene recovery and purification (1-pentene, 1-hexene and 1-octene) from HTFT syncrude are practised industrially. Additionally recovery and purification of C_7 and C_{12}/C_{13} *n*-1-alkenes are employed industrially in the production of petrochemicals from HTFT syncrude. Linear *n*-1-alkene production can also be supplemented by ethene oligomerisation.

(f) Light alcohols (methanol, ethanol, 1-propanol, 2-propanol and butanols) can be recovered from the Fischer–Tropsch aqueous product. Production can be supplement by partial hydrogenation of the aldehydes and ketones found in the same stream. It is also possible to produce ethanol and 2-propanol by alkene hydration. Separation and partial hydrogenation routes are practised industrially in Fischer–Tropsch refineries.

(g) Light carbonyls (acetaldehyde, acetone and butanone) can be recovered from the Fischer–Tropsch aqueous product. This is practised industrially.

(h) Light carboxylic acids (acetic and propanoic acid) can be recovered from the acid wastewater derived from the Fischer–Tropsch aqueous product. This is not practised industrially.

(i) Alcohols can be synthesised by hydroformylation of *n*-alkenes (Section 11.8.1). These are derived products that are produced industrially from syncrude. Hydroformylation has a specific advantage in a Fischer–Tropsch context due to the availability of syngas.

(j) Aromatics can be recovered from HTFT syncrude or produced by catalytic reforming or aromatisation in an analogous way to that used for aromatics production from crude oil. Industrially, aromatics production is limited to fuel use at present.

Other derived petrochemicals and lubricants can also be produced and in some instances there may be an advantage in producing them from syncrude.

11.8.4 Petrochemicals from Direct Oxidative Coupling of Methane

The free radical reaction network involved in the partial oxidation of methane is complex and model descriptions involving hundreds of elementary reactions can be found.[78] When looking at the reaction network, one cannot help but notice that partial oxidation of methane produces many useful intermediate products. Direct activation methane to produce higher value products is therefore an alluring prospect. In the literature there are reports on direct oxidative coupling of methane to produce

- ethene and ethane[79–82]
- liquid hydrocarbons[83]
- aromatics[84]
- methanol[85]
- formaldehyde.[86]

Thermodynamically, the direct oxidative conversion of methane is feasible. For example, the conversions to methanol and ethane are exothermic and spontaneous:[87]

$$CH_4 + \frac{1}{2}O_2 \rightarrow CH_3OH, \quad \Delta H_{400\,°C} = -125\,kJ\,mol^{-1},$$
$$\Delta G = -92\,kJ\,mol^{-1} \quad (11.22)$$

$$2CH_4 + \frac{1}{2}O_2 \rightarrow C_2H_6 + H_2O, \quad \Delta H_{800\,°C} = -88\,kJ\,mol^{-1},$$
$$\Delta G = -59\,kJ\,mol^{-1} \quad (11.23)$$

The practical problem is that there is a trade-off between selectivity and per pass conversion. As the oxidative conversion of methane increases, so does further oxidation to CO_x. There is consequently a limit to the overall yield of intermediate products. The upper limit for the yield of a continuous process for the oxidative coupling of methane to ethene and ethane is around 28%.[88] Is this good enough?

Kuo performed a comparative analysis of direct oxidative coupling of methane and GTL conversion using syngas-to-methanol technology.[87] The outcome of this study was twofold. First, oxidative coupling to produce ethene requires a C_2 and heavier hydrocarbon selectivity of at least 88% at a per pass conversion of 35% (*i.e.* a yield of 31%) to be competitive. The required ethene yield (31%) is above the theoretical limit (28%) previously cited[88] and, without evidence to the contrary, one has to conclude that oxidative coupling of methane to produce ethene does not seem to be a worthwhile pursuit. Second, it was found that direct partial oxidation to methanol becomes competitive with syngas-derived methanol synthesis when 90% methanol selectivity at 7.5% per pass conversion can be achieved or when 80% methanol selectivity at 15% per pass conversion can be achieved. This also seems to be beyond reach.[85]

It has to be concluded that the direct oxidative coupling of methane is unlikely to yield a process for the production of petrochemicals that is more efficient than indirect conversion via syngas.

11.8.5 Petrochemicals from Direct Methane Sulfurisation

Carbon disulfide (CS_2) is used as a solvent and chemical intermediate for the production of materials such as Rayon and

Cellophane. It can be produced by the direct conversion of methane with sulfur over an activated alumina or clay catalyst:[66,73,89]

$$CH_4 + 4S \rightarrow CS_2 + 2H_2S \tag{11.24}$$

The conversion is conducted in the vapour phase at around 675 °C and 0.2 MPa with a slight excess (5–10%) of methane. The CS_2 yield based on methane is 85–90% and slightly higher based on sulfur.[66,73]

11.8.6 Petrochemicals from Direct Methane Halogenation and Related Reactions

Halogenation of alkanes is literally a textbook free radical reaction. The halogens, F_2, Cl_2, Br_2 and I_2, have low homolytic bond dissociation energies (150–250 kJ mol^{-1} at 25 °C) and can readily be decomposed by heat or ultraviolet light. Methane can be repeatedly halogenated to produce a mixture of halomethanes and the corresponding hydrogen halide:

$$CH_4 + nX_2 \rightarrow CH_{4-n}X_n + nHX \tag{11.25}$$

Industrially, the most important halomethanes are the chloromethanes. Thermal chlorination is more widely used than light-induced chlorination. The process typically takes place at 350–370 °C and near atmospheric pressure.[73] The distribution of species can to some extent be controlled by the Cl_2:CH_4 ratio and per pass conversion. The reaction is straightforward and most of the process complexity arises from product separation and recycling of unconverted material.[89] All of the chloromethanes are useful petrochemicals and their main uses are as solvents, refrigerants and to produce silicones.[66,73] Halomethanes can also be employed as intermediates for further synthesis; for example, halogenation is used to activate methane for higher hydrocarbon synthesis.[90,91] This is not relevant to petrochemical synthesis, but it is an alternative pathway to fuels.

Chlorination also forms the basis for some of the other direct derivatisations of methane, such as sulfochlorination and oxychlorination:

$$CH_4 + SO_2 + Cl_2 \rightarrow CH_3SO_2Cl + HCl \tag{11.26}$$

$$CH_4 + O_2 + HCl \rightarrow CH_3OCl + H_2O \tag{11.27}$$

Table 11.12   Conversion of 40% CH_4, 40% HCl and 20% O_2 over a CuCl–KCl–LaCl$_3$ on SiO_2 catalyst at 330 °C, 0.1 MPa and residence times as indicated.

	Conversion and selectivity/mol% C		
Description	*23 s*	*38 s*	*77 s*
Methane conversion	18.4	22.3	42.7
Carbon selectivity[a]			
CH_3Cl	85.1	74.8	59.6
CH_2Cl_2	10.7	22.7	27.9
$CHCl_3$	0.6	1.5	4.6
HCOOH	3.1	0.3	0.8
CO_2	0.6	0.8	7.1

[a]CCl4 selectivity was 0.01% and CO was not detected.

Although sulfochlorination of methane is not of much industrial interest, it is an example of a derivative free radical reaction initiated by Cl_2.[92] Oxychlorination has been suggested as an alternative pathway for methane activation and the production of chloromethanes, by exploiting the production of water to provide the thermodynamic driving force. A catalytic cycle was proposed whereby the chloromethane is employed for higher hydrocarbon synthesis and by doing so the HCl is regenerated. Some performance data for the halogenation, which is of petrochemical interest, are given in Table 11.12.[93]

11.8.7 Petrochemicals from Direct Methane Conversion with NH_3 and HNO_3

Hydrogen cyanide (HCN) can be produced directly from methane and ammonia. The two main industrial production pathways for methane conversion are the oxidative Andrussow process in Equation (11.28) and the direct Degussa process in Equation (11.29):[66,73]

$$CH_4 + NH_3 + 1.5O_2 \rightarrow HCN + 3H_2O \qquad (11.28)$$

$$CH_4 + NH_3 \rightarrow HCN + 3H_2 \qquad (11.29)$$

The difference in these two approaches parallels the difference between oxidative and steam methane reforming (Section 11.5). In the oxidative Andrussow process, the heat of reaction generated by the co-production of water ($\Delta H = -473\,kJ\,mol^{-1}$) provides the energy for the otherwise endothermic conversion. In the Degussa process, the net reaction is endothermic ($\Delta H = +251 \, \text{kJ mol}^{-1}$) and heat must be supplied externally.

The catalysis of these two processes is completely different. In the Andrussow process, the reaction intermediate is NO, which is produced over a noble metal gauze catalyst, *e.g.* Pt, PtIr or PtRh.[66,73,94] The reaction is typically conducted at around 1100 °C and near atmospheric pressure. The yield of HCN based on ammonia and methane is 60 and 70%, respectively.[66] In the Degussa process, a Pt–Ru–Al alloy on a ceramic support is employed as catalyst and the reaction is conducted at around 1200 °C and near atmospheric pressure.[66,73] The yield of HCN is 83% based on ammonia.[66]

The production of nitromethane (CH_3NO_2) by direct nitration of methane with nitric acid is not practised on a large scale. It is more difficult to produce nitromethane from methane than to produce nitromethane as a side product from the nitration of propane or *n*-butane.[89,92] The nitration reaction is a free radical reaction and the thermodynamic driving force is provided by the co-production of water:

$$CH_4 + HNO_3 \rightarrow CH_3NO_2 + H_2O \qquad (11.30)$$

11.9 NATURAL GAS USE FOR A SUSTAINABLE FUTURE

Society's use of natural resources is unfortunately often driven by convenience and cost, rather than need and future sustainability. It is unlikely that we will be able to replenish natural gas at the rate that it is being consumed and over time this resource will be depleted to a level where need exceeds production. The present rate of natural gas use is therefore not sustainable. What, then, constitutes methane use for a sustainable future?

When considering natural gas as feed material, we should ask ourselves, 'What is the next best alternative and what is the impact of implementing the alternative?' For example, we can substitute domestic heating by natural gas with heating by electricity. Electricity is a very convenient energy carrier with existing infrastructure in place and no gaseous emissions during use. However, the ratio of energy input to electric power available at the point of use is only around 4:1. Is this degradation of energy efficiency for a heating application acceptable (*i.e.* consuming four times more energy for the same heat output)? Furthermore, does the existing electricity generation and distribution capacity permit such a substitution? It will quickly become apparent that it is difficult to find acceptable alternatives for some uses of natural gas, whereas others have acceptable alternatives.

The properties that make natural gas, and specifically methane, valuable are its high H:C ratio, low heteroatom content and fluid nature. It is these properties that are necessary in distributed applications, but merely convenient in large-scale stationary applications. It is also these properties that are valuable in the synthesis of products with high H:C ratio and low contaminant level. Responsible use of natural gas is at the core of sustainability, in order to meet the needs of the present, without compromising the ability of future generations to meet their own needs.

The following specific applications of natural gas are considered to be natural gas use for a sustainable future:

- Direct combustion to provide heat and/or power in small-scale distributed applications, such as heating in cold climates.
- Direct combustion in mobile applications, for example, compressed natural gas engine-powered transportation.
- Liquefaction of natural gas in remote locations to allow the transportation and the ultimate beneficial use of such natural gas. For this purpose, one can employ direct liquefaction to produce liquefied natural gas (LNG), or indirect liquefaction to produce on-specification transportation fuels, blending materials or petrochemicals.
- Production of fuels and petrochemicals that are more efficiently produced from natural gas than from other carbon sources.

REFERENCES

1. S. Stelter, *The New Synfuels Energy Pioneers. A History of Dakota Gasification Company and the Great Plain Synfuels Plant*, Dakota Gasification Company, Bismarck, ND, 2001.
2. D. L. King and A. de Klerk, *ACS Symp. Ser.*, 2011, **1084**, 1.
3. G. C. Andrews, *Canadian Professional Engineering and Geoscience: Practice and Ethics*, 4th edn, Nelson Education, Toronto, 2009, p 359.

4. J. G. Speight, *Natural Gas: a Basic Handbook*, Gulf Publishing, Houston, TX, 2007.
5. *Chemical Economics Handbook*, SRI International, Menlo Park, CA, 2010.
6. M. Schweigkoflerand and R. Niessner, *Environ. Sci. Technol.*, 1999, **33**, 3680.
7. C. A. Forest and P. A. Muzzell, Fischer–Tropsch fuels: why are they of interest to the United States military? *SAE Tech. Pap. Ser.*, 2005, 2005-01-1807.
8. G. J. Born and E. J. Durbin, The natural gas fueled engine, in *Methane: Fuel for the Future*, ed. P. McGeer and E. Durbin, Plenum Press, New York, 1982, pp. 101–112.
9. W. Morris, *Oil Gas J.*, 2007, **105**, 58.
10. G. A. Karim, Methane and diesel engines, in *Methane: Fuel for the Future*, ed. P. McGeer and E. Durbin, Plenum Press, New York, 1982, pp. 113–129.
11. A. de Klerk, *Fischer–Tropsch Refining*, Wiley-VCH, Weinheim, 2011.
12. W. L. Lom, *Liquefied Natural Gas*, Wiley, New York, 1974.
13. W. R. True, *Oil Gas J.*, 2011, **109**, 100.
14. C. D. Chang, *Catal. Rev. Sci. Eng.*, 1983, **25**, 1.
15. F. J Keil, *Microporous Mesoporous Mater.*, 1999, **29**, 49.
16. A. Y. Kam, M. Schreiner and S. Yurchak, The Mobil methanol-to-gasoline (MTG) process, in *Handbook of Synfuels Technology*, ed. R. A Meyers, McGraw-Hill, New York, 1984, pp. 2.75–2.111.
17. S. S. Bharadwaj and L. D. Schmidt, *Fuel Process. Technol.*, 1995, **42**, 109.
18. J. N. Armor, *Appl. Catal. A*, 1999, **176**, 159.
19. D. J. Wilhelm, D. R. Simbeck, A. D. Karp and R. L. Dickenson, *Fuel Process. Technol.*, 2001, **71**, 139.
20. J. R. Rostrup-Nielsen, *Catal. Today*, 2002, **71**, 243.
21. K. Aasberg-Petersen, T. S. Christensen, I. Dybkjær, J. Sehested, M. Østberg, R. M. Coertzen, M. J. Keyser and A. P. Steynberg, *Stud. Surf. Sci. Catal.*, 2004, **152**, 258.
22. W. Strauss, *Industrial Gas Cleaning*, 2nd edn, Pergamon Press, Oxford, 1975.
23. T. S. Christensen, *Appl. Catal. A*, 1996, **138**, 285.
24. K. Aasberg-Petersen, T. S. Christensen, C. S. Nielsen and I. Dybkjær, *Fuel Process. Technol.*, 2003, **83**, 253.

25. L. Barelli, G. Bidini, F. Gallorani and S. Servili, *Energy*, 2008, **33**, 554.
26. I. Dybkjær, *Fuel Process. Technol.*, 1995, **42**, 85.
27. T. Sundset, J. Sogge and T. Strom, *Catal. Today*, 1994, **21**, 269.
28. D. Mendes, A. Mendes, L. M. Madeira, A. Iulianelli, J. M. Sousa and A. Basile, *Asia-Pacific J. Chem. Eng*, 2010, **5**, 111.
29. A. Platon and Y. Wang, Water-gas shift technologies, in *Hydrogen and Syngas Production and Purification Technologies*, ed. K. Liu, C. Song and V. Subramani, Wiley, Hoboken, NJ, 2010, pp. 311–328.
30. A. de Klerk, *ACS Symp. Ser.*, 2011, **1084**, 215.
31. F. Marschner and F. W. Moeller, Methanol synthesis, in *Applied Industrial Catalysis*, Vol. 2, ed. B. E. Leach, Academic Press, New York, 1983, pp. 215–243.
32. A. de Klerk, *Prepr. Pap. Am. Chem. Soc., Div. Pet. Chem.*, 2010, **55**(1), 86.
33. G. A. Olah, A. Goeppert and G. K. Surya Prakash, *Beyond Oil and Gas: the Methanol Economy*, Wiley-VCH, Weinheim, 2006.
34. A. B. Stiles, *ACS Symp. Ser.*, 1983, **222**, 349.
35. G. Natta, Synthesis of methanol, in *Catalysis, Vol. III. Hydrogenation and Dehydrogenation*, ed. P. H. Emmett, Reinhold, New York, 1955, pp. 349–411.
36. P. J. A. Tijm, F. J. Waller and D. M. Brown, *Appl. Catal. A*, 2001, **221**, 275.
37. E. Supp and R. F. Quinkler, The Lurgi low-pressure methanol process, in *Handbook of Synfuels Technology*, ed. R. A Meyers, McGraw-Hill, New York, 1984, pp. 2.113–2.131.
38. R. Quinn, T. A. Dahl and B. A. Toseland, *Appl. Catal. A*, 2004, **272**, 61.
39. P. L. Rogerson, The ICI low-pressure methanol process, in *Handbook of Synfuels Technology*, ed. R. A Meyers, McGraw-Hill, New York, 1984, pp. 2.45–2.73.
40. T. H. Fleisch and R. A. Sills, *Stud. Surf. Sci. Catal.*, 2004, **147**, 31.
41. S. Lee, M. Gogate and C. J. Kulik, *Fuel Sci. Technol. Int.*, 1995, **13**, 1039.
42. H. H. Storch, N. Golumbic and R. B. Anderson, *The Fischer–Tropsch and Related Syntheses*, Wiley, New York, 1951.

43. P. H. Emmett (ed.), *Catalysis, Vol. IV. Hydrocarbon Synthesis, Hydrogenation and Cyclization*, Reinhold, New York, 1956.
44. R. B. Anderson, *The Fischer-Tropsch Synthesis*, Academic Press, Orlando, FL, 1984.
45. A. P. Steynberg and M. E. Dry Eds., *Fischer-Tropsch Technology. Studies in Surface Science and Catalysis*, Vol. 152, Elsevier, Amsterdam, 2004.
46. P. F. Mako and W. A. Samuel, The Sasol approach to liquid fuels from coal via the Fischer–Tropsch reaction, in *Handbook of Synfuels Technology*, ed. R. A Meyers, McGraw-Hill, New York, 1984, pp. 2.5–2.43.
47. R. Overtoom, N. Fabricius and W. Leenhouts, Shell GTL, from bench scale to world scale, in *Proceedings of the 1st Annual Gas Processing Symposium*, ed. H. E. Alfadala, G. V. R. Reklaitis and M. M. El-Halwagi, Elsevier, Amsterdam, 2009, pp. 378–386.
48. GTL: Oryx breakthrough and oil-price surge lift industry spirits, *Pet. Econ.*, 2008, **75**, 36.
49. M. E. Dry, The Fischer-Tropsch synthesis, in *Catalysis Science and Technology*, Vol. 1, ed. J. R. Anderson and M. Boudart, Springer, Berlin, 1981, pp. 159–255.
50. A. de Klerk, *Green Chem.*, 2007, **9**, 560.
51. A. de Klerk, A., *Energy Fuels*, 2009, **23**, 4593.
52. J. F. Haw, W. Song, D. M. Marcus and J. B. Nicholas, *Acc. Chem. Res.*, 2003, **36**, 317.
53. C. D. Chang, J. C. W. Kuo, W. H. Lang, S. M. Jacob, J. J. Wise and A. J. Silvestri, *Ind. Eng. Chem. Process Des. Dev.*, 1978, **17**, 255.
54. S. Yurchak, S. E. Voltz and J. P. Warner, *Ind. Eng. Chem. Process Des. Dev.*, 1979, **18**, 527.
55. D. Liederman, S. M. Jacob, S. E. Voltz and J. J. Wise, *Ind. Eng. Chem. Process Des. Dev.*, 1978, **17**, 340.
56. J.-F. Joly, Aliphatic alkylation, in *Petroleum Refining, Vol. 3. Conversion Processes*, ed. P. Leprince, Editions Technip, Paris, 2001, pp. 257–289.
57. C. D. Chang, W. H. Lang and A. J. Silvestri, *J. Catal.*, 1979, **56**, 268.
58. N. Guan, Y. Liu and M. Zhang, *Catal. Today*, 1996, **30**, 207.

59. R. L. Varma, N. N. Bakhshi, J. F. Mathews and S. N. Ng, *Ind. Eng. Chem. Res.*, 1987, **26**, 183.
60. A. de Klerk, *Green Chem.*, 2008, **10**, 1249.
61. A. de Klerk and E. Furimsky, *Catalysis in the Refining of Fischer–Tropsch Syncrude*, Royal Society of Chemistry, Cambridge, 2010.
62. A. de Klerk, Refining Fischer–Tropsch syncrude: perspectives on lessons from the past, in *Advances in Fischer–Tropsch Synthesis, Catalysts and Catalysis*, ed. B. H. Davis and M. L. Occelli, Taylor and Francis, Boca Raton, FL, 2009, pp. 331–364.
63. A. de Klerk, *Energy Environ. Sci.*, 2011, **4**, 1177.
64. K. Liu, C. Song and V. Subramani (eds), *Hydrogen and Syngas Production and Purification*, Wiley, Hoboken, NJ, 2010.
65. M. Beller, B. Cornils, C. D. Frohning and C. W. Kohlpaintner, *J. Mol. Catal. A*, 1995, **104**, 17.
66. S. Lee, *Methane and Its Derivatives*, Marcel Dekker, New York, 1997.
67. G. Natta, U. Colombo and I. Pasquon, Direct catalytic synthesis of higher alcohols from carbon monoxide and hydrogen, in *Catalysis, Vol. V. Hydrogenation, Oxo-Synthesis, Hydrocracking, Hydrodesulfurization, Hydrogen Isotope Exchange and Related Catalytic Reactions*, ed. P. H. Emmett, Reinhold, New York, 1957, pp. 131–174.
68. P. Courty, A. Forestiere, N. Kawata, T. Ohno, C. Riambault and M. Yoshimoto, *ACS Symp. Ser.*, 1987, **328**, 42.
69. S. Davis, *Chemical Economics Handbook Marketing Research Report: Petrochemical Industry Overview*, Stanford Research Institute, Menlo Park, CA, 2008.
70. J. Q. Chen, A. Bozzano, B. Glover, T. Fuglerud and S. Kvisle, *Catal. Today*, 2005, **106**, 103.
71. H. Koempel and W. Liebner, W., *Stud. Surf. Sci. Catal.*, 2007, **167**, 261.
72. P. Travers, Olefin etherification, in *Petroleum Refining, Vol. 3. Conversion Processes*, ed. P. Leprince, Editions Technip, Paris, 2001, pp. 291–319.
73. S. Matar and L. F. Hatch, *Chemistry of Petrochemical Processes*, Gulf Publishing, Houston, TX, 1994.
74. N. Rizkalla, *ACS Symp. Ser.*, 1987, **328**, 61.
75. M. E. Dry, *ACS Symp. Ser.*, 1987, **328**, 18.

76. J. H. Gregor, *Catal. Lett.*, 1990, **7**, 317.
77. A. P. Steynberg, W. U. Nel and M. A. Desmet, *Stud. Surf. Sci. Catal.*, 2004, **147**, 37.
78. K. van der Wiele, J. W. M. H. Geerts and J. M. N. van Kasteren, Elementary reactions and kinetic modeling of the oxidative coupling of methane, in *Methane Conversion by Oxidative Processes*, ed. E. E. Wolf, Van Nostrand Reinhold, New York, 1992, pp. 259–319.
79. M. M. Bhasin, *Stud. Surf. Sci. Catal.*, 1988, **36**, 343.
80. A. Ekstrom, The oxidative coupling of methane: reaction pathways and their process implications, in *Methane Conversion by Oxidative Processes*, ed. E. E. Wolf, Van Nostrand Reinhold, New York, 1992, pp. 99–137.
81. B. Zhang, J. Wang, L. Chou, H. Song, J. Zhao, J. Yang and S. Li, *Stud. Surf. Sci. Catal.*, 2007, **167**, 237.
82. N. Y. Usachev, V. V. Kharlamov, E. P. Belanova, T. S. Starostina and I. M. Krukovskii, *Russ. J. Gen. Chem.*, 2009, **79**, 1252.
83. J. H. Edwards and R. J. Tyler, *Stud. Surf. Sci. Catal.*, 1988, **36**, 395.
84. J. S. Espindola, N. R. Marcilio and O. W. Perez-Lopez, *Stud. Surf. Sci. Catal.*, 2007, **167**, 31.
85. H. D. Gesser and N. R. Hunter, The direct conversion of methane to methanol (DMTM), in *Methane Conversion by Oxidative Processes*, ed. E. E. Wolf, Van Nostrand Reinhold, New York, 1992, pp. 403–425.
86. T. Baldwin and R. Burch, *Appl. Catal.*, 1991, **75**, 153.
87. J. C. W. Kuo, Engineering evaluation of direct methane conversion processes, in *Methane Conversion by Oxidative Processes*, ed. E. E. Wolf, Van Nostrand Reinhold, New York, 1992, pp. 483–526.
88. Y. S. Su, J. Y. Ying and W. H. Green Jr, *J. Catal.*, 2003, **218**, 321.
89. R. N. Shreve and J. A. Brink Jr, *Chemical Process Industries*, 4th edn, McGraw-Hill, New York, 1977.
90. K.-J. Jens, *Stud. Surf. Sci. Catal.*, 1988, **36**, 491.
91. V. Degirmenci, D. Uner and A. Yilmaz, *Catal. Today*, 2005, **106**, 252.
92. F. Asinger, *Paraffins Chemistry and Technology*, Pergamon Press, Oxford, 1968.

93. C. E. Taylor, R. P. Noceti and R. R. Schehl, *Stud. Surf. Sci. Catal.*, 1988, **36**, 483.
94. J. K. Dixon and J. E. Longfield, Oxidation of ammonia, ammonia and methane, carbon monoxide and sulfur dioxide, in *Catalysis, Vol. VII. Oxidation, Hydration, Dehydration and Cracking Catalysts*, ed. P. H. Emmett, Reinhold, New York, 1960, pp. 281–345.

CHAPTER 12

Carbon Capture: Materials and Process Engineering

NICK H. FLORIN, NIALL MAC DOWELL, PAUL S. FENNELL
AND GEOFFREY C. MAITLAND*

Department of Chemical Engineering, Imperial College London,
South Kensington, London SW7 2AZ, UK
*Email: g.maitland@imperial.ac.uk

12.1 INTRODUCTION

The global challenge of achieving the major reductions in CO_2 emissions necessary to mitigate climate change cannot be over-stated. It will require a wide range of strategic actions, including enhancing the efficiency of energy transformation, improving end-use energy efficiency, changing public attitudes and behaviour to reduce demand and increasing the use of renewable energy sources and nuclear power. Despite the increased deployment of the latter over the next few decades, to meet the world's growing energy needs it will be necessary to continue to use fossil fuels well into the second half of this century. We must therefore capture as much as possible of the CO_2 released in their production and use (for power generation, industrial processes, *etc.*) and store it in suitable underground locations – so-called carbon capture and storage (or sequestration) (CCS).

Materials for a Sustainable Future
Edited by Trevor M. Letcher and Janet L. Scott
© The Royal Society of Chemistry 2012
Published by the Royal Society of Chemistry, www.rsc.org

The energy sector accounts for the largest share of anthropogenic greenhouse gas emissions, mostly CO_2 as a by-product of fossil fuel combustion (including coal, gas and liquid fuels).[1] Coal has the highest emissions per kWh of electricity produced of the conventional fossil fuels and is also the fuel whose rate of utilisation is increasing most rapidly. The dramatic increase in CO_2 emissions over the past two centuries shown in Figure 12.1 is a consequence of rapid industrialisation, ignited by the industrial revolution in the 'West' and sustained by its spread to the rest of the world, particularly in recent decades to India and China, with the additional demands of the electronic and globalisation revolutions of the late twentieth century.

Deployment of CCS technology is therefore important to help mitigate climate change in the near term. It can also help to maintain energy security during this period by enabling local fossil fuels to continue to be used to meet the growing demand for fuel, power and heat, which, during the next few decades, cannot be met by renewable and nuclear energy. In the longer term, CCS can support a transition to a sustainable energy future by continuing to curtail emissions from coal, oil and gas as the contribution from zero/low carbon emission energy sources gradually increases. This

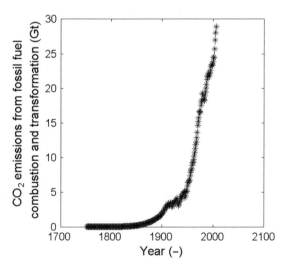

Figure 12.1 CO_2 emissions from fossil fuel combustion and fuel transformation.[2]

will require substantial investment; in order to achieve the increasingly accepted target of 50% reduction in global CO_2 emissions by 2050, about one-fifth of this (about $9\,Gt\,y^{-1}$) must come from the application of CCS to power generation, industry and fuel transformation, at a cost of about US\$316 $\times 10^{12}$ (US\$316 trillion).[3] In addition, it is possible that combustion of biomass in a CCS-enabled power station would allow 'negative' emissions of CO_2 to be effected.

With the exception of its use in enhanced hydrocarbon recovery, there is currently no significant market for the large volumes of CO_2 generated from fossil fuel usage (about $33\,Gt\,y^{-1}$). In the longer term, there is the possibility of using CO_2 as a renewable feedstock for the production of chemicals, plastics and fuels such as methanol,[4] but such processes are currently only at the stage of research feasibility and decades behind CCS in terms of proven technology. To have a significant effect on atmospheric concentrations, large volumes of CO_2 will need to be stored in safe geological structures, such as depleted oil and gas reservoirs, deep saline aquifers or unmined coal seams. The costs of separating, compressing, transporting and storing the CO_2 are major, so CCS deployment will only occur with changes in policy and regulatory frameworks which provide an incentive for investment or which introduce financial penalties or trading charges associated with emitting CO_2.

All the key technical elements for CCS are available today and in a few cases CCS is commercially viable, albeit only for niche applications such as enhanced oil and gas recovery, where additional revenue can be generated, or where unique taxes apply, *e.g.* Norway. The CCS process cycle is shown in Figure 12.2 for a range of CO_2 sources: from coal, gas and biomass used for power generation and industrial processes (*e.g.* cement plants and oil refineries).

12.2 CARBON CAPTURE OVERVIEW

While there are significant challenges associated with the transport and storage of CO_2, most of the cost and energy requirement (which will generate additional CO_2) of CCS is associated with CO_2 capture. It is here that the scope for materials development, to reduce costs and improve mitigation efficiency, is greatest. Many

Carbon Capture and Storage

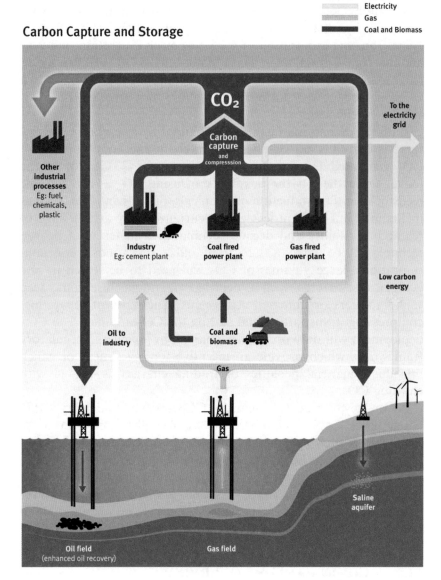

Figure 12.2   Diagram of CCS chain from capture to storage.[5]

potential capture technology options have been explored but only solvent absorption and potentially calcium looping are likely to be ready for commercial deployment in the near to medium term. Capture processes are of three main types: pre-combustion, post-combustion and oxyfuel combustion. The last is a relatively new but increasingly favoured option in which the fuel is combusted in nearly pure ($>95\%$) oxygen (with recycled CO_2 diluting this O_2 to moderate the flame temperature), resulting in a flue gas which is predominantly CO_2 and H_2O. The steam is readily condensed, leaving a relatively pure stream of CO_2 ready for immediate compression, transport and storage (see Figure 12.3). However, there is a significant energy penalty associated with standard cryogenic techniques[6] for the separation of air into its constituent parts, meaning that the efficiency penalty for an oxyfuel power station will be similar to that of a power station fitted with amine scrubbing,[7] unless a novel method, such as the use of high-temperature ceramic membranes, is utilised to produce the oxygen. However, it will be necessary to increase their current separation capacity of $5\,t\,d^{-1}$ by several orders of magnitude to achieve commercial requirements.

A number of alternative capture approaches are under development, such as more efficient solvents, other solid adsorbents, membranes, gas hydrates, cryogenic processes and biological capture. Each presents interesting materials challenges and may in time prove commercially viable. In the remainder of this chapter, we review the relative merits and the issues associated with the

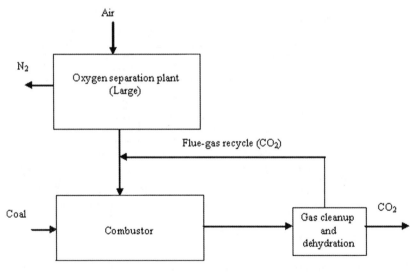

Figure 12.3 Oxy-combustion process.

design and selection of materials and processes for each of these capture approaches.

12.3   CARBON CAPTURE PROCESSES AND MATERIALS

12.3.1   Liquid-based Capture Processes

12.3.1.1   Post-combustion Chemical Absorption. Post-combustion capture is an 'end of pipe' technology involving the separation of CO_2 from a flue gas consisting mainly of nitrogen, water, CO_2 and other impurities such as oxides of sulfur (SO_x) or nitrogen (NO_x) and dust. This technology has the advantage that it is potentially suitable for retrofit without drastically affecting process operations, other than reducing the power output.

Chemical absorption capture technologies involve passing the flue gas to gas–liquid contacting and separating equipment where gas and liquid streams flow in a counter-current fashion in a vertical column, sufficient mixing and contacting being ensured by the inclusion of horizontal trays or packing material inside the column. A simplified flow diagram is given in Figure 12.4. These processes typically comprise two distinct unit operations – absorption and desorption (or solvent regeneration). In typical operations, the

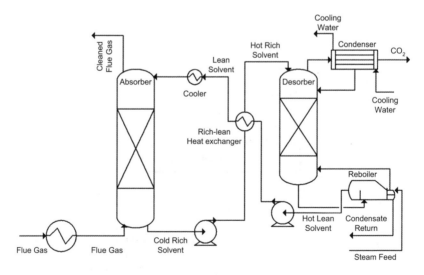

Figure 12.4   Simplified process flow diagram of an amine-based CO_2 capture process.

'lean' solvent stream is introduced to the top of the absorption column and flows vertically down the column over the packing material, absorbing its preferred components from the gas phase, which is introduced at the bottom of the column. This means that the mole fraction of CO_2 in the liquid phase is small. Although solvents which simply physically dissolve CO_2 (such as hydrocarbons and methanol) can be used, it is more usual to employ reactive or chemical absorption, since the gas uptake for such solvents can be considerably higher.

Once the CO_2-enriched solvent stream has reached the bottom of the column, it is sent to a solvent regeneration process, consisting of a further gas–liquid contacting column, with a reboiler at the bottom and a condenser at the top. The incoming liquid stream is heated in order both to break the chemical bonds formed in the reactive absorption process and to provide a vapour stream to act as a stripping fluid. The overhead condenser provides a reflux liquid stream to the column and ensures that the top-product stream is as pure as possible. The bulk of the energy penalty associated with chemisorption-based CO_2 capture processes arises from this solvent regeneration process.

For amines, the scrubbing column is kept at about 80 °C and 0.1 MPa and removes more than 90% of the CO_2. The solvent regeneration column is kept at about 120 °C and 0.2 MPa; here, heat transfer from the condensation of hot steam, diverted from the power cycle, releases CO_2 from the solvent, resulting in solvent regeneration (which is recycled) and a nearly pure (> 99%) stream of CO_2. The low temperature and pressure of this stage are largely determined by the relatively low thermal and oxidative degradation temperatures of most amines, meaning that the CO_2 stream is produced at a low pressure (0.2 MPa) relative to that required for transport and storage (~ 10 MPa), leading to considerable costs associated with the compression of the CO_2 stream. These processes are additionally expected to consume between 0.35 and 2.0 kg of solvent per tonne of CO_2 captured.[8] This obviously leads to significant costs associated with solvent make-up. Moreover, owing to the low CO_2 partial pressure usually in the inlet flue gas, relatively high solvent flow rates are required to achieve a high rate of CO_2 capture. The mass fraction of monoethanolamine (MEA) is restricted to around 30% to reduce corrosion in the columns; this corrosion is enhanced by oxidative degradation products.[4]

As with all current generation CO_2 capture technologies, chemisorption processes have the distinct disadvantage of their high cost, both capital and operational. It is estimated that the deployment of this technology will result in a reduction of the thermal efficiency of a modern power plant from ~45% to ~35%.[9] This efficiency penalty arises from the energy requirements of solvent regeneration (~4 GJ t^{-1} of CO_2 captured[10]), CO_2 compression and transport, and also that associated with transporting flue gases and solvents. This results in a reduction of ~20% in the electricity generated and means that more fuel must be burned and more CO_2 is released (although not emitted) in order to maintain the same power output.

Amine Solvents. Chemical absorption is a well-known technology, which has been widely deployed on a large scale across several industries.[11–13] Therefore, it seems that the major scope for reducing the costs associated with these processes lies in the selection and design of new, advanced sorbent materials, as it is the solvent which determines the thermodynamic and kinetic limits of the process. It is also the solvent chemistry that determines the type and seriousness of any deleterious environmental and public health impacts associated with fugitive emissions of organic solvents or of the products of solvent degradation or solvent-induced equipment corrosion.[14,15]

Amines have traditionally been the solvents of choice, with the primary alkanolamine, monoethanolamine (MEA), considered to be the benchmark solvent against which alternative solvents must be compared.[11] Alternative solvents include sterically hindered compounds such as 2-amino-2-methyl-1-propanol (AMP), secondary amines such as diethanolamine (DEA) and tertiary amines such as methyldiethanolamine (MDEA). Some amine-based processes have been commercialised on a large scale, namely the Kerr–McGee /ABB Lummus Crest Process,[11] Fluor Daniel's ECONAMINE process[16,17] and the Mitsubishi KS-1 process licensed by Kansai Electric Power and Mitsubishi Heavy Industries,[18] which uses a proprietary amine solvent, KS-1.

Much research in basic amine scrubbing is focused on the design of new solvent molecules and/or blends of existing solvents, together with the mitigation of problems such as corrosion. Blends are of interest because it is possible to combine the desirable characteristics of different solvents, for example, the high capture rate for primary and secondary solvents with the high ultimate CO_2 uptake for tertiary solvents.[4]

The type of amine is important because the CO_2 absorption chemistry varies with the degree of alkyl substitution (see, for example, references 19–24 and references therein for details). This can be understood by considering the absorption reaction chemistry. The principal reaction of interest between CO_2 and a primary or secondary amine (in water) is the formation of a carbamate, which occurs via the formation of a zwitterion and its subsequent base-catalysed deprotonation:[22,23]

$$CO_2 + 2R_1R_2NH \rightleftarrows \left[R_1R_2NCO_2^- + R_1R_2NH_2^+ \right] \qquad (12.1)$$

The other important reactions that contribute to the overall reaction rate are the reversion of carbamate to bicarbonate via a hydrolysis reaction [Equation 12.2] and the direct formation of bicarbonate [Equation 12.3]:

$$R_1R_2NCO_2^- + H_2O \rightleftarrows R_1R_2NH + HCO_3^- \qquad (12.2)$$

$$CO_2 + OH^- \rightleftarrows HCO_3^- \qquad (12.3)$$

In the case of non-sterically hindered compounds, the reversion of carbamate to bicarbonate by reaction (12.2) is only important at higher concentrations of CO_2. In the case of sterically hindered compounds, such as AMP, reaction (12.2) is particularly important as the presence of the 2-methyl group significantly reduces the stability of the carbamate bond, resulting in the preferred formation of bicarbonate, leading to the particularly high loading capacity (moles of CO_2 per mole amine) of this solvent.[25] Tertiary amines, such as MDEA, do not react directly with CO_2, but act as a base, catalysing the hydration of CO_2:[26]

$$CO_2 + H_2O + R_1R_2R_3N \rightleftarrows R_1R_2R_3NH^+ + HCO_3^- \qquad (12.4)$$

Amines are susceptible to degradation in the presence of O_2, SO_x and CO_2 found in flue gases, in addition to degrading thermally.[27–29] In addition, metallic packings used in the contacting columns can exacerbate this problem.[30,31] Oxidative degradation has been extensively investigated, particularly for MEA,[32,33]

DEA[26] and MDEA.[34] Here, the main degradation products are volatile compounds such as amines, aldehydes and carboxylic acids. While secondary amines are generally more stable than primary amines, the reaction paths involving flue gases are not well understood. Various studies have shown that there is a similarity in the degradation mechanisms and products in primary, secondary and tertiary amines.[28,29,35] Typically, the main products are amines, oxazolidinones and imidazolidinones.[36] Thus degradation processes under industrial conditions remain poorly understood,[37] so systematic improvement through new solvent design is difficult.

There are also environmental concerns around the use of amine solvents. A recent IEA report indicated that for every tonne of CO_2 captured, 0.0032 t of MEA will be emitted to the atmosphere.[38] Given that a typical 2.4 GW generator burning pulverised black coal produces ~ 30–50 t of CO_2 per minute,[39] the potential solvent losses to the atmosphere are highly significant. Most of the molecules considered for use in CO_2 capture applications are strongly polar and are therefore highly water soluble and quickly become part of the water cycle and biosphere. Recent results indicate that many of the degradation products can also be harmful to both human health and the environment.[39–41] Some amine degradation products include amides and aldehydes and also nitrosamines, which are potent carcinogens.[40] Thus amine solvent volatility is a very important characteristic in solvent selection and design. Environmental concerns are therefore a strong driver to explore alternative lower volatility, less degradable capture solvents.

Solvent Blends. Amine-based CO_2 capture is a very mature technology, with the first patent for the use of alkanolamines as absorbents for acidic gases being granted in 1930.[42] Consequently, it is a very well understood class of technologies and is considered to be economically feasible under certain conditions. Most of the uncertainties for CO_2 capture arise due to the application of this technology to flue gas. Blends of amines are considered to be an attractive option in developing improved sorbent materials because it is possible to combine the complementary attributes of different pure components; for example, it is possible to exploit the high absorption rates of primary and secondary amines and also the high capacity of tertiary amines.[43]

Sterically hindered compounds are considered to be particularly interesting owing to their tendency to form weak carbamate salts, leading to both higher carrying capacities for CO_2 (number of moles of CO_2 absorbed per mole of amine) and a lower enthalpy of regeneration (energy required per unit mass of CO_2 released).[44,45] However, lower absorption enthalpies lead to the generation of CO_2 at a lower pressure in the desorber[46] and consequently to higher subsequent pumping and compression costs. Such considerations underline the importance of simultaneously optimising the whole CO_2 capture process, cradle-to-grave, rather than focusing on one aspect, such as solvent properties or, worse, a subset of these. AMP has been regarded as the standard sterically hindered amine solvent and there has been considerable effort in developing predictive thermodynamic and process models describing the behaviour of AMP-based CO_2 capture processes.[47–50] Recently, other sterically hindered compounds, such as 2-amino-2-methyl-1,3-propanediol (AMPD), 2-amino-2-ethyl-1,3-propanediol (AEPD) and tri(dihydroxymethyl)aminomethane (THAM) have been examined. Blends with THAM show lower volatility and improved energetics compared with AMP blends,[51] and are promising alternative solvents for CO_2 capture. The proprietary industrial-scale sterically hindered solvent KS-1[18,52] is reported to be less susceptible to degradation and corrosion problems than MEA, but is currently considerably more expensive.[53]

The addition of other alkanolamines to an MEA solution frequently results in superior performance of the blend over that of MEA alone.[26,54–61] The substitution of a degradation-prone compound, such as MEA, by compounds less prone to degradation, such as NH_3, should allow higher amine concentrations to be used in processes compared with the current upper limit of 30%.[61] Some preliminary work on a blend of AMP and NH_3 has shown that it at least equals, if not surpasses, MEA solutions in terms of absorption capacity[62] and also in terms of reaction rate.[63]

There is some evidence to suggest that the addition of short-chain alkanols to amine–water blends increases the solubility of CO_2 in the liquid phase[64] while reducing the energy required for solvent regeneration. This is an interesting step towards the development of solvents based on physical interactions for the capture of CO_2. Physical solvents, such as methanol, have been used for the removal of CO_2 from natural gas streams,[65] but they are not generally effective on their own for CO_2 capture from the flue gases of coal-fired power plants, owing to the low CO_2 concentration and pressure of the gas stream.

Ammonia Solvent Processes. It has become apparent in recent years that one system which may give improved capture performance coupled with lower degradation and corrosion problems compared with amines is the parent molecule ammonia (NH_3). This has a favourable loading (moles of CO_2 per mole of NH_3) and energetics[66–69] compared with most amines. Ammonia can capture all the major acid gases (SO_x, NO_x, CO_2) from the flue gas of coal combustors; a single process to capture all acidic gases may reduce the total cost and complexity of any emission control system.[66,67] Moreover, ammonia is not expected to suffer from solvent degradation problems (caused by SO_2 and O_2 in addition to CO_2) or to cause equipment corrosion issues and could potentially reduce the energy requirements for CO_2 capture.[69] The underlying chemistry behind the resistance of NH_3 to degradation is unclear, but it may be due to the inherent simplicity of the NH_3 molecule, as alkanolamine sensitivity to O_2 is often attributed to the oxidation of the hydroxyl functional group to an acid.[36]

Alstom is developing the chilled ammonia process,[70] which reportedly uses only 15% of the amount of steam consumed by MEA for regeneration. However, electricity is required for refrigeration, so the overall efficiency is only marginally better than that of amine scrubbing. The process operates at 0–10 °C and so will be more efficient where cold cooling water is readily available. The process produces ammonium bicarbonate, which has the benefit that the regeneration can take place at elevated pressure (~ 3 MPa), reducing the energy penalty for CO_2 compression.[70] This process is undergoing pilot-plant scale trials at We Energies' Pleasant Prairie, WI, power plant and also AEP's Mountaineer Plant in New Haven, CT, USA.[71]

Commercial Developments. Several companies, such as Alstom, RWE and Endesa,[72,73] are actively engaged in developing improved, often proprietary, solvent options for post-combustion CO_2 capture. There are also collaborative research programmes such as the EU-funded CESAR project,[74] the principal aims of which are towards a breakthrough in the development of low-cost post-combustion CO_2 capture technology to reduce the costs of capture to €15 per tonne of CO_2. Sargas have developed a post-combustion technology based on wet potassium carbonate scrubbing, which is available in 100 MW modules for capturing SO_x, NO_x and CO_2. A CO_2 capture efficiency of more than 95% was reported based on results from pilot trials at the pressurised coal-fired Värtan combined heat and power plant in Stockholm, Sweden.[75] A 400 MW plant at Husnes, Norway, is planned,[76] with the captured CO_2 (2.6 Mt y^{-1}) to be used for enhanced oil recovery (EOR) in North Sea oil fields. This widespread focus on research, development and scale-up will probably result in post-combustion CO_2 capture technologies being the CO_2 emission mitigation technology of choice, at least during the initial stages of the large-scale deployment of CCS technology.

Ionic Liquids. Room temperature ionic liquids (ILs) are an exciting new class of materials which offer a promising alternative to the use of volatile organic compounds in gas scrubbing operations.[4] There has been an explosion of interest in ILs,[77] which are being increasingly studied and proposed for CO_2 capture applications. They offer the possibility of reducing the energy demand for CO_2 capture by as much as 16% relative to a 30% MEA-based solvent.[78] ILs are materials with low melting points and high boiling points (liquids below 100 °C), composed entirely of ions. They have been described as 'designer solvents' because of the ability to tailor the solvent's properties by appropriate combinations of cations and anions of different size and chemistry. The major advantage of ILs is that they are non-volatile under ambient conditions, minimising the risk of fugitive solvent losses; they have a wide liquid range, low flammability, good CO_2 solubility and thermal stability up to 300 °C,[79] offering flexibility in terms of process optimisation,[4] and can also integrate sulfur removal into one stage.[71] However, currently, ILs are prohibitively expensive and their manufacture is very complicated, hence significant cost reductions and process simplifications for large-scale production are required for economic viability. Future potential may arise from the possibility of combining ILs and amines.[80]

The use of ILs for CO_2 capture is most frequently proposed in a pressure-swing configuration,[80] where CO_2 is absorbed from flue gas before regeneration of the IL through a pressure decrease, with effectively zero solvent loss. The properties of ILs can be manipulated through ion selection: acid–base character, polarity, density, viscosity and thermal stability (to over 300 °C[79]) can all be tailored to the requirements of specific processes. Temperature-driven desorption is also possible. Physical absorption into the IL means that this desorption process will have a relatively low energy requirement.

ILs are inert to oxidation, even at high temperature,[79] allowing regeneration at higher temperatures and pressures with significantly reduced solvent degradation. High-pressure desorption could allow a significant saving in subsequent CO_2 compression cost. The substantial reduction or even elimination of solvent losses through thermal degradation, oxidative or chemical destruction and vapour loss means that the amount of solvent makeup needed for the process may be greatly reduced. Currently, ILs are vastly more expensive than traditional solvents; however, their low degradation and consequent required make-up rates potentially offset this. Hybrid technologies involving supported ionic liquid membranes (SLIMs) have been investigated. Here, by using a thin layer of IL supported by a membrane,[78] the IL advantages involving minimal solvent losses can be combined with a technique to overcome the potential limitations of the high liquid viscosity.

As discussed above, the key benefit of ILs is the 'tuneability' of their properties. For CO_2 capture, functionalised anions capable of reversible chemical reaction with CO_2 are the most frequent addition.[81] Such changes greatly increase the solubility of CO_2 in the IL but potentially reduce the ease of desorption. The selectivity towards CO_2 compared with other gases can also be significantly increased. In order to optimise ionic liquids for use in CO_2 capture processes, it will be necessary to acquire a detailed understanding of the thermophysical properties and phase behaviour of pure ionic liquids and their mixtures,[82,83] and how these relate to the ionic composition and interactions.

Some typical IL cations and anions are shown in Figure 12.5.[4] Classes of cation previously investigated for CO_2 capture ability include (a) 1,3-dialkylimidazolium[80,84] ([CmCnim] the most explored cation to date, owing to the low melting points and high thermal stability of the ILs formed, together with their relatively low viscosity

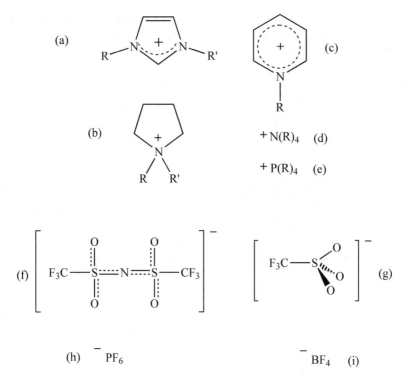

Figure 12.5 Ionic liquid cations and anions.[4]

and ready commercial availability[79]), (b) *N,N*-dialkylpyrrolidinium,[85] (c) *N*-alkylpyridinium,[86,87] (d) tetraalkylammonium[88,89] and (e) tetraalkylphosphonium.[85,88,90] Common anions used include (f) bis(trifluoromethyl)sulfonylimide (NTf2) (which gives the highest solubility of CO_2 of systems studied to date), (g) triflate (OTf), (h) hexafluorophosphate (PF_6^-) and (i) tetrafluoroborate (BF_4^-).

There are many different cation and anion functionalities available, leading to great design flexibility for ILs and their applicability in a range of applications. Although the choice of cation tends to have a smaller effect on the IL properties important for CO_2 capture than does the anion, there are some preferred options. Increasing CO_2 solubility results from the minimisation of unfavourable interactions between CO_2 and the positive charge centre, by using, for example, an imidazolium ring.[91] The addition of hydrophobic alkyl side groups to the imidazolium ring improves the absolute CO_2 solubility slightly, whereas the CO_2/N_2 selectivity can be improved by the addition of more polar functionalities to improve the interactions between ILs and CO_2. The increase can be by as much as 75%.[80]

The optimum IL class for CO_2 capture has been proposed to be [C1Cnim][NTf2]. This class has relatively good physical properties, CO_2 solubility and selectivity. A detailed consideration of the structures of the ILs studied for CO_2 capture and how specific ion combinations can improve the desired solvent characteristics can be found elsewhere.[4]

Hybrid solutions have been explored. IL–alkanolamine mixtures[92] combine CO_2 selectivity and stoichiometric uptake capacity (0.5 mol of CO_2 per mole of amine) with superior regeneration stability added by the IL. Another hybrid solution to raise the CO_2 loading capacity is to incorporate amine functionality on the IL cation,[81,92,93] although this also gives rise to large viscosity increases.

12.3.1.2 Pre-combustion Solvent Capture. Pre-combustion capture involves an initial gasification stage where the fuel is reacted with insufficient O_2 for complete combustion, producing a mixture known as synthetic gas or syngas, consisting mainly of carbon monoxide and hydrogen with some methane and CO_2. A series of reactions [water gas shift (WGS)] then converts these gases to a mixture of CO_2 and H_2. After separation from the CO_2, the H_2-rich fuel gas can be used to fire a gas turbine or run a fuel cell. Alternatively, liquid fuels can be produced from the syngas via Fischer–Tropsch catalytic processes to give flexible 'polygeneration'.[94] The conditions for CO_2 capture here are very different than in post-combustion capture because the gas is already at elevated pressure (2–7 MPa) and the CO_2 concentration is significantly higher (15–60% by volume).[53] Because of these conditions, a different range of solvents are used, known as physical solvents (such as methanol and *n*-alkanes). Generally, physical solvents combine less strongly with the CO_2, resulting in a lower energy penalty for desorption.

12.3.2 Solid Sorbents

12.3.2.1 Carbonate Looping Technology. The next major class of CO_2 capture technologies/materials is those based on adsorption and/or reaction with solid sorbents. Post-combustion capture using high-temperature solid sorbents is a promising technology and the leading candidate is post-combustion carbonate or calcium looping using calcium oxide derived from natural limestone. The process exploits the reversible gas–solid reaction between calcium oxide (CaO) and CO_2 to form calcium carbonate ($CaCO_3$), according to the equation

$$CaO(s) + CO_2(g) \rightleftarrows CaCO_3(s) \tag{12.5}$$

A simplified process flow diagram, as applied to post-combustion capture, is shown in Figure 12.6. In one vessel, the carbonator, the carbonation reaction between CO_2 and solid CaO separates CO_2 from a gas mixture, *e.g.* coal–combustion flue gas, resulting in the capture of CO_2 via the formation of solid $CaCO_3$. The $CaCO_3$ is then transferred to a second reactor, the calciner, where it is heated to reverse the reaction, releasing CO_2 and regenerating the CaO sorbent, which is recycled back into the carbonation reactor. A circulating fluidised-bed reactor (CFB) is considered the most suitable choice for the carbonation and calcination vessels owing to very good gas–solid contacting and temperature uniformity across the reactor bed.

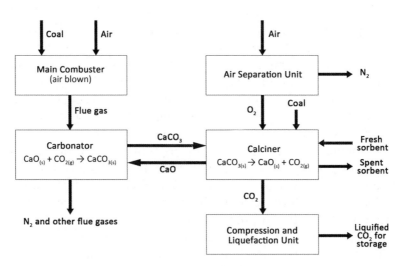

Figure 12.6 Simplified process flow diagram of calcium looping applied to post-combustion capture. After *e.g.* ref. 95.

Carbonation is an exothermic reaction and is typically carried out at an intermediate temperature of about 650 °C, balancing the decrease in equilibrium driving force for CO_2 capture according to Equation 12.5 and the enhanced reaction kinetics as the temperature increases. The carbonation reaction is characterised by an initial rapid chemical reaction-controlled phase followed by an abrupt transition to a slow diffusion-controlled phase. This transition is associated with the accumulation of a $CaCO_3$ product layer sufficient to impede significantly any further conversion, the rate-determining step becoming diffusion of CO_2 through the solid $CaCO_3$ product layer. As a result, the conversion of CaO to $CaCO_3$ is usually limited to about 70% in the first carbonation cycle of duration ~ 10 min. The extent of conversion decreases in subsequent cycles during repeated long-term carbonate looping, to $e.g.$ $<10\%$ after 30 cycles.[96]

By contrast, the endothermic calcination reaction proceeds rapidly to completion in a single step under a range of conditions above about 750 °C. In order to produce a pure stream of CO_2, then thermodynamic limitations dictate that under a high CO_2 partial pressure (>90 vol.%), calcinations must be conducted at 900–950 °C. To achieve this relatively high temperature, additional fuel must be combusted in the calcination vessel in pure O_2, which requires an air separation unit (though this is only about 30% of the size required for an oxyfuel-fired power station).[97] Critically, the energy penalty associated with the air separation is partially offset by the recuperation of heat from the hot CaO and CO_2 streams and heat produced from the exothermic carbonation reaction, all of which can used to generate additional steam. Hence the efficiency penalty associated with CO_2 capture from a power station using carbonate looping turns out to be extremely competitive (~ 3–4% for the capture stage and 3% for subsequent compression).

Carbonate looping technology offers four key advantages:[96] (i) low energy penalty; (ii) synergy with cement manufacturing; (iii) use of mature large-scale equipment, such as circulating fluidised beds (CFBs), which reduces scale-up risk; and (iv) cheap sorbent (natural limestone). Overall, this process could reduce the thermal efficiency de-rating associated with CO_2 capture to about 6–8%[97] compared with 8–10% for MEA scrubbing, representing a significant fuel and cost saving over the lifetime of a typical power station. Degradation of the sorbent over repeated cycles, particularly in the presence of ash and sulfur, is a potential problem.[96] However, owing to the low cost of the sorbent (crushed limestone), this is a minor concern. Significant research is being carried out to improve the long-term capacity of CaO-based sorbents.[98 – 100] Calcium looping technology underpins a range of advanced power schemes for the production of electricity and/or hydrogen, including combined shift–carbonation sorbent-enhanced reforming (SER).[101]

12.3.2.2 Large-scale Operations. At present, post-combustion carbonate looping is being tested in a 75 kW pilot plant at Canmet Energy (Ottawa, Canada),[102] and pilot-test facilities are also located at the Instituto Nacional del Carbón (INCAR, Spain),[103] Cranfield University (UK)[104] and the University of Stuttgart (Germany).[105,106] The Spanish utility Endesa and mining company Hunosa are constructing a 1.7 MW_{th} test facility for an EU-funded project known as CaOling[107] and a 1 MWth test facility, funded by the German government and also industry funders, is under development at the Technische Universität Darmstadt, Germany.[108] Cemex, which is the world's third largest cement manufacturer, have a pilot plant in Monterrey, Mexico;[109] cement manufacture is responsible for 7% of global industrial CO_2 production. There is a unique synergy between cement manufacture and carbonate looping because the exhausted sorbent (CaO) can be used instead of fresh limestone as a feedstock for cement manufacture,[110] thus reducing the direct CO_2 emissions of the entire cement manufacturing process by about 50%.

The significant cost advantage of post-combustion carbonate looping compared with MEA scrubbing arises because of the use of a cheap sorbent derived from natural limestone, *e.g.* priced at US$0.0015 per mole compared with US$0.544 per mole for MEA. This is particularly significant considering the degradation of the sorbent/solvent due to the presence of impurities (*e.g.* sulfur), necessitating a significant input of fresh material. The sale of exhausted sorbent to the cement industry would also improve the economics of carbonate looping technology.[97,110]

12.3.2.3 Future Materials Issues. As mentioned previously, the CaO sorbent derived from natural limestone loses its capacity to capture CO_2 after multiple capture-and-release cycles and large amounts of fresh limestone are required to maintain an acceptable CO_2 capture efficiency. The main factors influencing the decrease in capacity are sintering, attrition and chemical deactivation to the competing chemical reaction with sulfur dioxide (SO_2).

Sintering refers to changes in the pore shape and size distribution and grain growth, which reduce the total pore volume and reactive surface area. This process occurs during the heating of particles and the severity is increased by high temperatures, length of exposure to high temperature (as during calcination), and the presence of steam, CO_2 and impurities.[96] The process involves handling a large quantity of solids in an abrasive environment leading to attrition and the formation of fine particles susceptible to elutriation by the fluidising gas. Competing chemical reactions between CaO and fuel-bound impurities, most significantly sulfur, also affect the long-term CO_2 capture capacity of CaO sorbents. In the case of coal, where sulfur may be present at concentrations up to about 8 wt%,[111] $CaSO_4$ forms under the oxidising conditions in the carbonator and calciner, representing an 'irreversible' loss of sorbent that must be replenished. On the other hand, it has been suggested that the reaction could improve the economics of carbonate looping by eliminating the need for a separate flue gas desulfurisation unit.

Strategies for overcoming these issues and improving the reactivity of CaO for long-term carbonate looping include sorbent hydration,[112] doping with foreign ions,[113] thermal pretreatments,[114] the use of nanomaterials[115] and the use of inert porous supports with pelletisation.[116–119] It is difficult for some of these customised materials to compete with cheap and abundant limestones. Although sorbent hydration has been demonstrated as a successful strategy to regenerate exhausted sorbent particles, mechanical stresses associated with the formation of $Ca(OH)_2$, which has a higher molar volume than $CaCO_3$, make regenerated sorbent particles more susceptible to attrition.[112]

12.3.2.4 Chemical Looping Combustion. This approach eliminates direct contact between the fuel and air by using a metal oxygen carrier (MO), such as the oxides of iron, nickel or copper, to transfer the oxygen needed to combust the fuel. As in calcium looping, a typical process involves two fluidised beds. In this case, the first bed is used to oxidise the metal M in air or oxygen:

$$2M + O_2 \rightarrow 2MO \tag{12.6}$$

in an exothermic process that produces a hot flue gas that can be used to raise steam for power generation. The other bed reduces the oxide to its initial state by reaction with the hydrocarbon fuel:

$$C_nH_{2n} + 3nMO \rightarrow nCO_2 + nH_2O + 3M \tag{12.7}$$

producing a gas containing a high concentration of CO_2 and some steam, which is readily condensed to give a pure CO_2 stream for storage. Limestone can be added to remove sulfur impurities.

Unfortunately, the oxygen carriers tend to degrade during long-term cycling, a limitation that must be overcome to realise the potential for high overall efficiencies of $> 50\%$.[120] Chemical looping combustion has been investigated using gaseous and solid fuels, and also advanced H_2 production processes.[120–123] Chemical looping combustion with natural gas for electricity production is limited by the efficiency of the steam cycle and therefore, in general, has a lower overall efficiency compared with a CCGT plant with CO_2 capture unless the system is pressurised. The major materials challenges are in optimisation of the oxide(s) used (suitable contenders are oxides of Cu, Fe, Ni, Co), balancing raw material availability and cost, the energetics of the redox process and the long-term degradation rate of the oxygen carrier. A complementary process, chemical looping gasification, which uses metal oxides to provide the oxygen for (coal) gasification to syngas and then $CO_2 + H_2$, is also being explored. This claims to be able to achieve $> 90\%$ CO_2 capture with marginal increases in the cost of the generated electricity compared with traditional coal plants.[124–126]

12.3.2.5 Amine-impregnated Solid Sorbents. These sorbents are sometimes seen as the 'next generation' of amine-based sorbents for CO_2 capture. Eliminating water from the system greatly reduces the amount of energy required for regeneration. Today there are commercially available solid amine sorbents which are used to remove CO_2 in closed environments such as submarines and space shuttles.[127] Amines have been incorporated on a wide range of substrates;[128–130] however, this work has not progressed to testing in reactors capable of simulating realistic conditions and the capture capacity of CO_2 currently demonstrated with supported amines is low in comparison with alternative solid sorbents, *e.g.* calcium oxide (CaO). The US Department of Energy has recently reported a supported amine sorbent[129] which has a maximum theoretical capacity for CO_2 of 0.132 g of CO_2 per gram of sorbent, but this number decreases to 0.101 g of CO_2 per gram of sorbent under typical process conditions. A detailed discussion of this area was presented by Blamey *et al.*[112]

12.3.2.6 Structured Porous Solids – Molecular Cages. In addition to capturing CO_2 by adsorption on the active surface of porous solid materials, it can also be separated, and sometimes stored, using solids which contain cage-like structures that can act as host materials for guest gas molecules. One class of materials of this type are gas hydrates. These have ice-like structures in which hydrogen-bonded water molecules form a range of cavities of different size and shape, *e.g.* dodecahedra, which can host small gas molecules to form clathrate or hydrate materials (see Figure 12.7). These are stable typically in the range −10 to 20 °C at pressures up to 100 bar, depending on the nature and concentration of the guest molecules.

These compounds can be produced synthetically but also exist widely in Nature, where the appropriate temperature/pressure conditions occur beneath the Earth's surface – under the permafrost or in relatively shallow sub-sea sediments. The gas trapped in these natural hydrates is natural gas, predominantly methane, and constitutes a huge potential source of methane, containing upwards of 20×10^{12} m^3, well in excess of the gas stored in conventional oil and gas reservoirs and enough to meet demand for thousands of years at current consumption rates. The problem is that, because the gas hydrate is a critical structural component of the sediments, 'melting' it to produce CH_4 can sometimes lead to dramatic reductions in the mechanical strength of the sediment, leading to sub-sea landslides and possibly tsunami activity, alongside uncontrolled release of methane, which is 20 times worse than CO_2 in its climate-warming potential once released into the upper atmosphere. Research and development are being carried out to develop production routes to overcome these difficulties, but it is a challenging though high-reward endeavour.

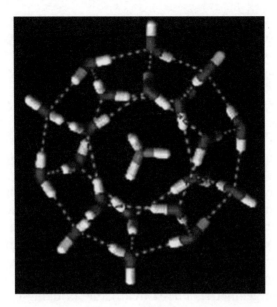

Figure 12.7 A pentagonal dodecahedral cavity containing a methane molecule as part of the methane hydrate sI structure. Reproduced from C. Koh, *Chem. Soc. Rev.*, 2002, **31**, 157–167.

The link to carbon capture rests on the fact that CO_2 forms gas hydrates which are significantly more thermodynamically stable than those of CH_4. This gives rise to the possibility of pumping gas streams containing CO_2 into methane hydrates, displacing the methane to produce large quantities of the cleanest fossil fuel, leaving the CO_2 captured and stored within the hydrate materials. Although such processes are thermodynamically favoured, kinetically they are very slow. Nevertheless, significant research is being carried out to identify conditions of pressure, temperature, co-solvent and co-guest molecules that may make such gas exchange processes practically viable.

Because of the high selectivity for guest molecules exhibited by gas hydrates, based on the available cage sizes and specific dispersion interactions between the guest gas molecules and the hydration cage, synthetic systems have attracted considerable attention for gas capture, purification of mixed gas streams, gas transport and long-term gas storage. Processes for capture of CO_2 from flue gas streams by creation of hydrate slurries have been described,[131] although none have reached commercialisation.

For example, Kang and Lee[132] described a small-scale process involving three hydrator–dissociator stages for CO_2 capture which exploits the effectiveness and selectivity of hydrates for separating both low and high concentrations of CO_2 from the usual flue gas components (N_2, H_2, H_2S). The relatively low energy and temperature requirements for the formation/dissociation of CO_2 hydrates compared with those for conventional solvent and oxide looping processes, combined with the possibility of carrying out the capture process at elevated pressure, make this approach attractive for further exploration.

Another class of materials which contains well-defined cavities whose size and molecular interactions can be varied and controlled are metal organic frameworks (MOFs). This new class of nanoporous materials has a range of potential applications in separation processes and catalysis, including gas capture and storage. They are hybrid materials synthesised from metal ions having well-defined coordination geometries linked by organic bridging ligands which self-assemble to form crystalline materials with very well-defined porous structures, high internal surface areas and desired chemical functionalities.[133] These characteristics make them possible candidates for CO_2 capture. Like gas hydrates, their potential storage capacity is high and the heat of adsorption of CO_2 is relatively low, making the recovery stage significantly less energy intensive than solvent or reactive adsorption processes. The number of combinations of metal ions and ligands is extremely high and hundreds of MOFs have been developed and studied in recent years. Molecular modelling is proving a very useful tool for screening existing and potential MOF structures for different applications, including CO_2 capture.

Studies to date suggest that the capture capacity of MOFs is very pressure dependent, with different structures required depending on the precise operating regime. The nature of the metal ion is critical; changing the metal dioxybenzenedicarboxylate MOFs (M-DOBDC) from Zn to Mg, Co or Ni results in large changes in CO_2 uptake.[134] Interestingly, adding an amine functionality to the organic linkers had only a small effect on CO_2 uptake, indicating that the dominant interaction is probably with the metal sites and their binding to the ligands. M-DOBDC MOFs have open metal sites that can interact with adsorbate molecules and Mg-DOBDC performs particularly well, with CO_2 uptakes a factor of five higher than for Zn on a molar basis and almost four times on a molar basis (0.721 molecules of CO_2 per metal atom under ambient conditions). MgO exothermically chemisorbs CO_2 to form $MgCO_3$ and it has been suggested[134] that the performance of Mg-DOBDC may be attributed to the relatively higher ionic character of the Mg–O bond promoting enhanced CO_2 physisorption. CO_2 is not chemisorbed by the Mg–O bonds in Mg-DOBDC, but the ionic character of this bond promotes a higher CO_2 uptake. Indeed, the uptake of CO_2 correlates well with decreasing M–O bond length in these MOF structures. Much research remains to be done on understanding and controlling the relationships between MOF structure and CO_2 uptake–release characteristics, and also the tolerance of the materials to thermal cycling, flue gas contamination and attrition during processing. However, they appear to have significant long-term potential as carbon capture materials.

A range of other high surface area porous materials have been investigated as potential CO_2 capture agents. These include zeolites and carbon nanotubes, sometimes functionalised with amine moieties, porous silicas impregnated with liquid amines such as polyethylenimine or tetraethylenepentamine and relatively simple, inexpensive salts and minerals, such as sodium carbonate. Adsorption–release kinetics are an important factor and surfactants have been shown to accelerate CO_2 transport in porous sorbents. The challenge is to identify cost-effective materials which combine high CO_2 uptake with low energy release and are capable of robust recycling under commercial conditions, which requires high tolerance to thermal/pressure cycling and to the contaminants always present in flue and well gases.

12.3.3 Other Separation Techniques

12.3.3.1 Membrane Separation Technology. Membrane separation and capture involve the selective permeation of gases through porous materials, driven by a pressure difference that is achieved by either compressing the gas upstream or creating a vacuum downstream. Three main types of material are used, polymeric, metallic and ceramic membranes, and the choice is strongly dependent on the temperature and gases involved in the particular application, *e.g.* separating CO_2 from N_2 in flue gas, O_2 separation from N_2 in air and H_2 separation from coal-derived synthesis gas. In the case of CO_2 separation from flue gas, which involves very large volumes of gas, a recent study highlighted the critical importance of increasing the membrane permeability to reduce the efficiency penalty associated with achieving the pressure gradient across the membrane.[135] For the application of O_2 separation from N_2 for oxy-combustion or gasification plants, the development of the ion transport membranes by Air Products represents a significant advance.[71] Major challenges which must be overcome to make this technology robust include the cost of membrane materials, membrane lifetimes and reliability issues due to exposure to particulates SO_x, NO_x and trace metals, demonstration of large-scale compression or vacuum equipment and efficient integration with power systems.

12.3.3.2 Biological Capture. Biological capture using micro-organism systems such as algae to remove CO_2 from industrial flue gas is also an active area of research. The approach involves using the waste heat and exhausted CO_2 from combustion, together with water and sunlight, to grow algae. This can take place in open ponds and tanks or in closed bioreactor systems, which give better process control and higher productivity, but a significant increase in capital cost. The large-scale cultivation of algae represents a valuable source of biomass, which may be may be converted to liquid fuels such as bio-oil and biodiesel. Solix (a start-up company funded by the US Department of Energy) contend that bio-oil would be competitive with an oil price of more than US$75 per barrel.[136] However, it is important to note that the re-use of CO_2 as a liquid biofuel only displaces the use of fossil transport fuels (the CO_2 from the power station is, of course, eventually emitted), hence limiting the CO_2 mitigation potential. Decarbonisation of the transport sector by electrification of vehicles, using electricity from power plants fitted with CCS, would result in considerably lower emissions. Biological capture processes suffer from inherent scale-up issues because of the limited rate at which algae can grow and challenges associated with bioreactor design. The use of highly productive genetically engineered algal strains is expected to enhance greatly the capture capacities and rates for these systems.

Another biological capture option being pursued involves enzymes. One such process uses carbonic anhydrase to capture and release CO_2 in a similar way to the mammalian respiratory system. The enzyme is contained in a hollow-fibre membrane and at the laboratory scale can capture CO_2 with 90% efficiency.[71] Significantly, the CO_2 can then be regenerated at ambient pressure and temperature with a very low heat of absorption and consequent energy penalty. Its efficiency derives from the ability of the enzyme to enhance dramatically the rate of CO_2 hydration to carbonic acid and recent results suggest that the costs of scaled-up technology would be significantly lower than those of the existing MEA solvent process.[137] There are several materials challenges still to be addressed, such as controlling surface fouling, pore wetting and maintaining enzyme activity beyond a few months.

12.3.3.3 Cryogenic Separation. This is an alternative approach for gas–gas separation, exploiting the different boiling temperatures and partial pressures of the gases in a mixture to separate them into distinct phases by cooling or pressurisation. Cryogenic air separation is the main method used currently for the separation of O_2 from N_2 by cooling the air to about $-196\,°C$, at which point N_2 becomes a liquid. For CO_2 separation, CO_2 can be frozen at $-75\,°C$ and atmospheric pressure or condensed to a supercritical fluid when pressurised beyond its critical point at $31\,°C$ and 74 bar. The major problem for cryogenic separation is the high energy consumption and cost associated with compression and cooling. For the separation of CO_2 from a flue gas stream, where it is often less than 15% of the total gas stream, much energy is consumed in compressing and cooling the remaining 85% of the gas stream. Another challenge is the removal of water which is necessary before cooling to avoid the formation of ice.[138] Owing to the considerable energy penalty of these processes, there is significant scope for fundamental research to reduce this or develop less energy-intensive alternatives, such as gas membranes.

12.3.3.4 Building CO_2 into Materials. Carbon dioxide can be used in principle as a renewable C_1 feedstock for producing fine and commodity chemicals.[139,140] However, relatively few large-scale processes utilise CO_2 as a feed, although this use has a number of advantages:[4]

- CO_2 is a cheap and abundant feedstock and in certain applications can replace toxic chemicals, such as phosgene, while avoiding the depletion of fossil fuel resources.
- It can be transformed into totally new materials for which conventional petrochemical synthesis routes are not preferable or available.
- These new materials include novel polymers, enabling captured CO_2 to be stored permanently within solid materials.
- Adding value to CO_2 by transforming it into other products, such as polymers, chemicals and fuels, could improve the economics of carbon capture and storage.[71]

With carbon in its highest oxidation state ($+4$), CO_2 corresponds to the lowest energy state of all carbon-containing binary neutral species. Carbon dioxide and water are the main products of most combustion processes and metabolic pathways and most useful energy-releasing processes. However, the central carbon of CO_2 has a strong affinity towards nucleophiles and electron-donating reagents because of the electron deficiency of the carbonyl carbon. The bonding of a third atom such as O to this central carbon is exothermic and as anhydrous carbonic acid it reacts rapidly with basic compounds. Hence carbonates, formed by reaction with hydroxyl ions, are even lower in energy than CO_2 and the natural weathering of silicates is also an exothermic process, albeit extremely slow.

On the other hand, the reduction of CO_2 to lower oxidation states requires significant energy. The use of electrochemical or photoelectrochemical reduction to convert CO_2 to formic acid, formaldehyde, methanol or methane has been intensively studied since the early 1970s.[141,142] Alternatively, CO_2 can be reduced by highly reactive organic or organometallic compounds to produce a wide range of chemicals.[143-145] Currently, about 150 Mt of CO_2 is used for chemical synthesis, the main products being urea,[146] inorganic carbonates (such as sodium carbonate through the Solvay process),[147] methanol and cyclic and polycarbonates.[144] However, the length of time that the CO_2 is locked away is often an issue (for example, urea-based fertilisers immediately release CO_2

when applied to the soil),[148] and the lifecycle CO_2 emissions associated with the formation of the highly reactive organic or organometallic compounds must be carefully accounted for.

The potential for expanding commercial processes for using CO_2 as a feedstock[139,142–144] is in producing three main classes of products: fine chemicals, including urea, carboxylic acids and carbonates; fuels or commodity chemicals such as methanol and formic acid; and plastics such as polycarbonates and polyurethanes. The most important current processes are the synthesis of organic carbonates and conversion to urea using the Bosch–Meiser process. Urea synthesis (see Figure 12.8) is an overall exothermic process where dry-ice and liquid ammonia are reacted exothermically to form ammonium carbamate, with subsequent endothermic decomposition to form water and urea.[149] Of course, the production of methanol as a fuel from CO_2 is highly questionable on the grounds of thermodynamics. It is often supposed that there will be large quantities of 'free' renewable energy to conduct such reactions, although of course there will always be an opportunity cost – for example, the electricity used (and the capital employed for the building of the methanol plant) could also have been employed to build some form of pumped storage. There are numerous other disadvantages, with recent calculations putting the efficiency of transformation of 'spare' electricity into motive force in a vehicle at around 12%, as opposed to $> 50\%$ if the electricity were stored in a pumped storage facility and then used to charge an electric vehicle.[150]

CO_2-derived Plastics. A long-term goal is to capture and store CO_2 as a constituent of solid materials, where it remains sequestered indefinitely, at least until the materials come to be recycled.

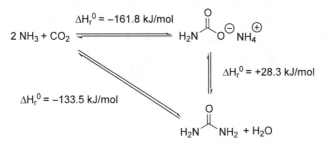

Figure 12.8 Thermodynamics of the industrial synthesis of urea.[4]

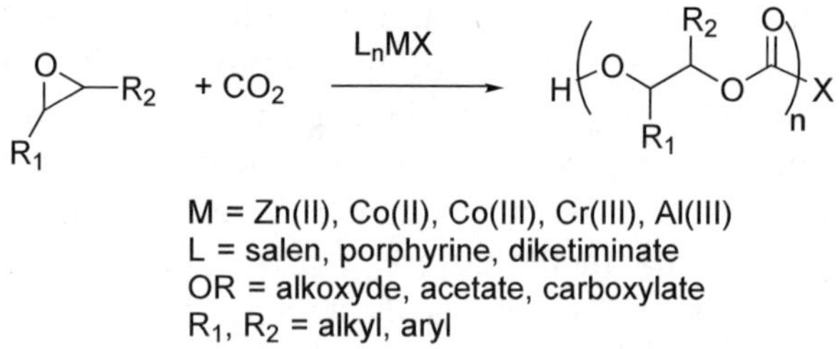

$M = Zn(II), Co(II), Co(III), Cr(III), Al(III)$
$L = salen, porphyrine, diketiminate$
$OR = alkoxyde, acetate, carboxylate$
$R_1, R_2 = alkyl, aryl$

Figure 12.9   A general scheme for the copolymerisation of CO_2 and epoxides.[4]

The copolymerisation of CO_2 with epoxides has attracted much attention[151,152] (see Figure 12.9). Compared with the established commercial syntheses, which involve polycondensation reactions between the corrosive and highly toxic nerve gas phosgene and diols, using methylene chloride as a solvent, or alternatively transesterification processes of cyclic carbonates with diols [such as the widely used aromatic bisphenol-A (BPA)], this process is both more sustainable and safer. Another option that has been explored is the synthesis of polyurethanes from supercritical CO_2 and aziridines.[153]

Polycarbonates, currently produced at the 4 Mt y^{-1} scale,[154] are thermosetting polymers with physical properties that allow them to be processed in bulk, and they find application in a wide range of uses, *e.g.* safety helmets and children's toys. It is possible to produce a variety of aliphatic polycarbonates by sequential copolymerisation of CO_2 and epoxides;[155,156] these may fulfil the increasing demand for sustainably produced packaging and construction materials. The thermomechanical properties of the aliphatic polycarbonates and impact aromatic (BPA-based) polycarbonates can also potentially be improved.

The major challenge for making these polymerisation processes commercially viable is the development of more effective catalysts. A wide range of systems have been explored and found to have moderate activity:[157] chromium, cobalt, zinc, aluminium and lanthanide alkoxide, halide and acetate complexes have proved fairly effective,[157–164] with some systems active at low CO_2 pressure.[157,165,166] Catalysts involving two metallic centres have emerged as candidates for reaching the level of activity (turnover numbers up to 700, active at 1 bar) needed for commercial processes.[166–169] Most studies have involved the epoxides cyclohexene oxide and propylene oxide; copolymerisation of carbon dioxide and a wider range of epoxides, such as substituted epoxides or vinyl ethers, needs to be explored to widen the range of materials and potential applications, in order to meet the demands of the emerging markets outlined above (*e.g.* materials for sustainable construction or packaging). In the medium term, the conversion of CO_2 into fuels and materials could provide an economic stimulus for CO_2 capture (although with regard to fuels, see the discussion above concerning thermodynamics and energy efficiency) and in the longer term it could make a contribution itself as a capture/storage route.

12.4 INTEGRATION OF CAPTURE PROCESSES AND MATERIALS WITH CO_2 SOURCE PROCESSES

The large-scale deployment of CCS technology requires the collaboration between a diverse range of stakeholders, including: power plants, energy-intensive industries such as cement, iron and steel plants, CO_2 capture and compression plants, energy networks and associated infrastructure, CO_2 transport networks and associated infrastructure and CO_2 storage sites. The integration of such a complex process chain is a major challenge, with new technical and economic uncertainties that must be negotiated to achieve a successful CCS project. Equally, there are significant opportunities for optimising overall cost efficiencies with consideration of the integration of individual processes within the chain. This section discusses important examples and opportunities for process system integration and illustrates how optimising the design of separation and capture materials is an integral part of the overall process integration.

12.4.1 Integration of CO_2 Capture Plant with Power Production

A key consideration for any CO_2 capture technology is the parasitic energy penalty that is imposed on the power or industrial plant where it is applied. An integrated development of the solvent, the capture plant and the power cycle is essential for optimising the overall energetic performance.[4] For post-combustion capture systems, where the CO_2 capture plant can be 'bolted on' to the end of a combustion exhaust pipe, the integration may be focused mainly on minimising heat requirements for solvent regeneration; however, for pre-combustion systems where the fuel conversion is directly integrated with the CO_2 separation, the opportunity for process integration (and simplification) is significant. For example, the use of pre-combustion calcium looping can eliminate the need for separate shift reactors by incorporating the CO_2 separation with the WGS reaction.[101] A similar process simplification is possible using H_2-selective membranes.[170,171] Furthermore, these processes can potentially produce CO_2 at elevated pressure (15–20 bar), limiting the energy penalty associated with CO_2 compression for transportation.

12.4.2   Integration of CO_2 Capture Plant with Industrial Processes

In the energy-intensive industrial sectors, *e.g.* chemical and petrochemical, iron and steel and cement, CCS is an essential technology for achieving large-scale emission abatement. In general, good integration of industrial plants and power plants is important for minimising CO_2 capture costs because capture and compression plant will increase on-site industrial power requirements. Depending on the quantity and purity of CO_2 and in some cases where CO_2 separation is integral to the industrial process, there are relatively low-cost opportunities for CCS. When applied to the production of hydrogen, ethylene, ammonia and natural gas processing, the main additional operating cost component is associated with compression. CCS integrated with natural gas processing represents an early example of a fully integrated CCS project, *e.g.* Statoil's Sleipner project in Norway has been operational since 1996.[172]

Different CO_2 capture technologies are suited to different industrial processes. For example, post-combustion and oxy-combustion capture technology are best suited for CO_2 capture from cement manufacture for capturing CO_2 produced during the pre-calcination of limestone ($\sim 50\%$), and also CO_2 produced from burning fuel in the kiln. Pre-combustion capture is the most suitable for iron and steel in terms of CO_2 capture efficiency considering the composition of blast furnace gas, with 20–28%

CO.[173,174] The conversion of CO to CO_2 with steam via the WGS reaction increases the calorific value of the blast furnace gas that may be used to generate steam or electricity, potentially offsetting emissions associated with auxiliary operations on-site.[174] Post-combustion capture from blast furnace gas would achieve $<50\%$ capture owing to the carbon content remaining in the form of CO. The synergy between calcium looping and cement manufacture was discussed earlier.

12.4.3 Network Design and Operation

The integration of power or industrial plant and CO_2 capture plant has implications for the design and dynamic operations of the energy or CO_2 network. Specifically, how does the integration of a power plant with a CO_2 capture plant affect the plant's operability within the network? Chalmers *et al.*[175] considered desirable operating modes, defining the 'dynamic flexibility' of power plants with CO_2 capture, including: quick start-up/shut-down, quick change in output, efficient operation at part load, capture-plant by-pass ability and fuel flexibility. These characteristic are different for different power plants and different capture technologies. For example, Scottish Power's post-combustion carbon capture plant design at Longannet Power Station (UKCCS Demonstration Competition, now withdrawn) allowed for a CO_2 capture rate from 75 to 250 t h^{-1} and the capability to turn injection rates down to 60% of peak flow. Design for flexible operations also necessitated the ability to vent CO_2 from the capture plant during start-up as well as some capacity to 'line-pack' the pipeline.[176] The quality of the CO_2 stream exiting the capture plant also has downstream consequences for a CO_2 transport network. CO_2 quality, which is dependent on the power/industrial plant and the capture technology, is important because the presence of impurities (H_2O, N_2, Ar, O_2, SO_2, H_2S, NO_x, CO, CH_4) affects CO_2 properties, including thermophysical properties (*e.g.* vapour–liquid equilibrium) and transport properties (*e.g.* viscosity, thermal conductivity, diffusivity). These properties influence the design and operation of pumps/compressors and pipelines and are directly relevant to defining feasible, safe and cost-efficient operating conditions. Despite the importance of these properties, there is a paucity of experimental data, particularly for liquid-phase CO_2 mixtures.[177] Although there may be no significant technical barriers in terms of the CO_2 capture technology to produce a highly pure stream of CO_2, the associated increase in energy for purification, and hence cost, maybe unaffordable.

When the total power output correlates with CO_2 capture efficiency, there are new implications for optimising cost efficiency and the cost-optimal capture efficiency likely varies with plant scale, industrial sector and the degree of integration between the power and industrial plant.[178,179] For instance, utilising 'waste heat' (in the form of hot water or air) for solvent regeneration or preheating can significantly improve overall system efficiency by maximising fuel utilisation. The combined production of heat and power (CHP) is not a new concept; however, the exploitation of this mature technology in the context of CCS projects can mitigate costs and emissions associated with the increased on-site power demands.

12.5 CONCLUSION

The application of carbon capture and storage, using closest-to-market technology, would currently de-rate the thermal efficiency of a power plant by about 8–10%, equivalent to a reduction in power output of 20%. To achieve the same power output, more fuel must be consumed, more CO_2 is produced (although not emitted), more waste is generated and the potential for adverse environmental impacts is increased. This efficiency penalty correlates directly with the long-term energy requirements and cost of capture. New capture materials and processes which dramatically reduce these critical parameters will be essential to the long-term commercial viability of CCS. Figure 12.10 illustrates a likely trajectory towards cost reduction arising from improved overall efficiency and increased process integration. Thus, capture systems deployed in 2030 and beyond may look very different to those deployed in the first raft of full-scale demonstration projects. Therefore, to avoid lock-in to sub-optimal approaches, it is essential to encourage diverse academic research to explore new options and focused industrial R&D to transfer the most promising capture materials into more advanced technologies.

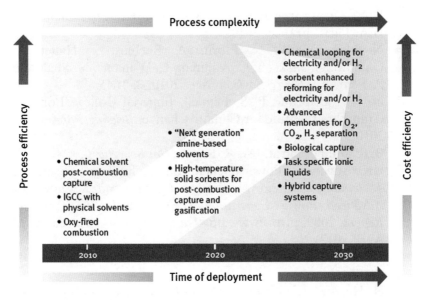

Figure 12.10 Likely adoption trajectories, future capture plants and cost efficiencies. After ref. 71.

Acknowledgements

The authors thank many colleagues at Imperial College London for valuable discussions and collaborations that have informed the contents of this chapter, especially Claire Adjiman, Antoine Buchard, Amparo Galindo, Jason Hallett, George Jackson, Martin Trusler, Danlu Tong, Nilay Shah and Charlotte Williams.

REFERENCES

1. United Nations Framework Convention on Climate Change (UNFCCC), *National Greenhouse Gas Inventory Data for the Period 1990–2007. Note by the Secretariat*, FCCC/SBI/ 2009/12, http://unfccc.int/2860.php (last accessed 4 April 2012).
2. Carbon Dioxide Information Analysis Centre (CDIAC), *Production of CO₂ from Fossil Fuel Burning by Fuel Type*, 1860–1982 (NDP–006), http://cdiac.ornl.gov/ftp/ndp006/ ndp006.pdf (last accessed 29 April 2012).

3. IEA, *Energy Technology Perspectives. Executive Summary*, IEA, Paris, France, 2010.
4. N. Mac Dowell, N. Florin, A. Buchard, J. Hallet, A. Galindo, G. Jackson, C. Adjiman, C. Williams, N. Shah and P. Fennell, *Energy Environ. Sci.*, 2010, **3**, 1645.
5. N. H. Florin and P. S. Fennell, Imperial College London, Grantham Institute for Climate Change, *Briefing Paper No. 3* 2010.
6. B. R. Dunbobbin and W. R. Brown, *Gas Sep. Purif.*, 1987, **1**, 23.
7. UK Advanced Power Generation Technology Forum, *A Technology Strategy for Carbon Capture and Storage, UK Advanced Power Generation Technology Forum Technical Report*, 2009, http://www.apgtf-uk.com/ (last accessed on 28 April 2012).
8. D. W. Bailey and P. H. M. Feron, *Oil Gas Sci. Technol.*, 2005, **60**, 461.
9. H. Herzog, *The Capture, Utilisation and Disposal of Carbon Dioxide from Fossil Fuel-Fired Power Plants*, Department of Energy Technical Report DOE/ER-30194, US Department of Energy, Washington, DC, 1993.
10. C. Alie, L. Backham, E. Crosietand and P. L. Douglas, *Energy Convers. Manage.*, 2005, **46**, 475.
11. A. L. Kohl and R. B. Nielsen, *Gas Purification*, 5th edn Gulf Publishing, Houston, TX, 1997.
12. D. S. Arnold, D. A. Barrett and R. H. Isom, *Oil Gas J.*, 1982, **80**, 130.
13. S. Reddy, J. Scherffius, S. Freguia and C. Roberts, presented at the 2nd Annual Confernece on Carbon Sequestration, Alexandria, VA, 2004.
14. S. M. Murphy, A. Sorooshian, J. H. Kroll, N. L. Ng, P. Chhabra, C. Tong, J. D. Surratt, E. Knipping, R. C. Flagan and J. H. Seinfeld, *Atmos. Chem. Phys.*, 2007, **7**, 2313.
15. Q. G. J. Malloy, L. Qi, B. Warren, D. R. Cocker, M. E. Erupe and P. J. Silva, *Atmos. Chem. Phys.*, 2009, **9**, 2051.
16. M. T. Sander and C. L. Mariz, *Energy Convers. Manage.*, 1992, **33**, 341–348.
17. D. G. Chapel, C. Mariz and L. Ernest J., Recovery of CO_2 from flue gases: Commercial trends, *Annual Meeting of the Canadian Society of Chemical Engineering*, Saskatoon, Canada, 1999.

18. T. Mimura, S. Shimojo, T. Suda, M. Ijima and S. Mitsuoka, *Energy Convers. Manage.*, 1995, **36**, 397.
19. G. Astarita and D. W. Savage, *Chem. Eng. Sci.*, 1980, **35**, 1755.
20. H. Hikita, S. Asai, H. Ishikawa and M. Honda, *Chem. Eng. J.*, 1977, **13**, 7.
21. P. V. Danckwerts, *Chem. Eng. Sci.*, 1979, **34**, 443.
22. S. S. Laddha and P. V. Danckwerts, *Chem. Eng. Sci.*, 1981, **36**, 479.
23. D. E. Penny and T. J. Ritter, *J. Chem. Soc., Faraday Trans. 1*, 1983, **79**, 2103.
24. P. M. M. Blauwhoff, G. F. Versteeg and W. P. M. van Swaaij, *Chem. Eng. Sci.*, 1984, **39**, 207.
25. G. Sartori and D. W. Savage, Sterically hindered amines for CO_2 removal from gases, *Ind. Eng. Chem. Fundam.*, 1983, **22**, 239–249.
26. M. Edali, A. Aboudheir and R. Idem, *Int. J. Greenhouse Gas Control*, 2009, **3**, 550.
27. M. L. Kennard and A. Meisen, *Hydrocarbon Process., Int. Ed.*, 1982, **59**, 103.
28. A. Chakma and A. Meisen, *J. Chromatogr. A*, 1988, **457**, 287.
29. A. Chakama and A. Meisen, *Can. J. Chem. Eng.*, 1997, **75**, 861.
30. M. Attalla, presented at the 5th Trondheim Conference on CO_2 Capture, Transport and Storage, 2009.
31. N. Kladkaew, R. Idem., P. Tontiwachwuthikul and C. Saiwan, *Ind. Eng. Chem. Res.*, 2009, **48**, 10169.
32. A. Bello and R. O. Idem, *Ind. Eng. Chem. Res.*, 2006, **45**, 2569.
33. T. Supap, R. Idem, P. Tontiwachwuthikul and C. Saiwan, *Ind. Eng. Chem. Res.*, 2006, **45**, 2437.
34. J. Critchfield and J. Jenkins, *Pet. Technol. Q.*, 1999, 87.
35. P. D. Clark, N. I. Dowling and P. M. Davis, Introduction to sulfur chemistry and sulfur handling, presented at the Sulfur 2004 Conference, Barcelona, 2004.
36. H. Lepaumier, D. Picq and P. L. Carrette, *Ind. Eng. Chem. Res.*, 2009, **48**, 9061.
37. A. Meisen and X. Shuai, *Energy Convers. Manage.*, 1997, **38**, S37.
38. IEA Greenhouse Gas R&D Programme, *Improvement in Power Generation with Post-Combustion Capture of CO_2*, Technical Report PH4/33, 2004, p. 46.

39. P. Jackson and M. Attalla, *Energy Proc.*, 2011, **4**, 2277.

40. J. J. Renard, S. E. Calidonna and M. V. Henley, *J. Hazard. Mater.*, 2004, **108**, 29.

41. B. Thitakamol, A. Veawab and A. Aroonwilas, *Int. J. Greenhouse Gas Control*, 2007, **1**, 318.

42. R. R. Bottoms (Girdler Corp.), Separating acid gases, *US Patent* 1 783 901, 1930.

43. H. Bosch, G. F. Gersteeg and W. P. M. Van Swaaij, *Chem. Eng. Sci.*, 1989, **44**, 2745.

44. G. Sartori and D. W. Savage, *Ind. Eng. Chem. Fundam.*, 1983, **22**, 239.

45. G. Sartori, W. S. Ho, D. W. Savage, G. R. Chludzinski and S. Wlechert, *Sep. Purif. Rev.*, 1987, **16**, 17.

46. J. Oexmann and A. Kather, *Int. J. Greenhouse Gas Control*, 2010, **4**, 36.

47. P. Tontiwachwuthikul, A. Meisen and C. J. Lim, *Chem. Eng. Sci.*, 1992, **47**, 381.

48. J. Gabrielsen, M. L. Michelsen, E. H. Stenby and G. M. Kontogeorgis, *Ind. Eng. Chem. Res.*, 2005, **44**, 3348.

49. J. Gabrielsen, M. L. Michelsen, E. H. Stenby and G. M. Kontogeorgis, *AIChE J.*, 2006, **52**, 3443.

50. J. Gabrielsen, H. F. Svendsen, M. L. Michelsen, E. H. Stenby and G. M. Kontogeorgis, *Chem. Eng. Sci.*, 2007, **62**, 2397.

51. J. K. You, H. Park, S. H. Yang, W. H. Hong, W. Shin, J. K. Kang, K. B. Yi and J. N. Kim, *J. Phys. Chem. B*, 2008, **112**, 4323.

52. T. Mimura, S. Satsumi, M. Iijima and S. Mitsuika, presented at the 4th International Conference on Greenhouse Gas Control Technologies, 1998.

53. J. Davison and K. Thambimuthu, *Proc. Inst. Mech. Eng., Part A*, 2009, **223**, 201.

54. P. Mandal, M. Guha, A. K. Biswas and S. S. Bandyopadhyay, *Chem. Eng. Sci.*, 2001, **56**, 6217.

55. B. P. Mandal and S. S. Bandyopadhyay, *Chem. Eng. Sci.*, 2006, **61**, 5440.

56. A. Setameteekul, A. Aroonwilas and A. Veawab, *Sep. Purif. Technol.*, 2008, **64**, 16.

57. M. Nainar and A. Veawab, *Energy Proc.*, 2009, **1**, 231.

58. A. Dey and A. Aroonwilas, *Energy Proc.*, 2009, **1**, 211.

59. A. Aroonwilas and A. Veawab, *Energy Proc.*, 2009, **1**, 4315.

60. R. Sakwattanapong, A. Aroonwilas and A. Veawab, *Energy Proc.*, 2009, **1**, 217.
61. W. J. Choi, J. B. Seo, S. Y. Jang, J. H. Jung and W. J. Oh, *J. Environ. Sci.*, 2009, **21**, 907.
62. H. Lee, W. J. Choi, S. J. Moon, S. H. Ha, I. G. Kim and K. J. Oh, *Korean J. Chem. Eng.*, 2008, **25**, 279.
63. W. J. Choi, B. M. Min, B. H. Shon, J. B. Seo and K. J. Oh, *J. Ind. Eng. Chem.*, 2009, **15**, 635.
64. A. Archane, L. Gicques, E. Provost and W. Furst, *Chem. Eng. Res. Des.*, 2008, **86**, 592.
65. Linde, Rectisol Wash, http://www.linde-engineering.com/en/process_plants/hydrogen_and_synthesis_gas_plants/gas_processing_plants/rectisol_wash/index.html (last accessed 29 April 2012).
66. H. Bai, P. Biswas and T. Keener, *Ind. Eng. Chem. Res.*, 1994, **33**, 1231.
67. H. Bai and A. C. Yeh, *Ind. Eng. Chem. Res.*, 1997, **36**, 2490.
68. C. Yeh and B. Hsunling, *Sci. Total Environ.*, 1999, **228**, 121.
69. K. P. Resnik, J. T. Yeh and H. W. Pennline, *Int. J. Environ. Tech. Manage.*, 2004, **4**, 89.
70. R. Rhudy and S. Black, Chilled ammonia process update, in *10th International Network for CO_2 Capture*, Lyon, France, 24–25 May 2007.
71. J. D. Figueroa, T. Fout, S. Plasynski, H. McIlvried and R. D. Srivastava, *Int. J. Greenhouse Gas Control*, 2008, **2**, 9.
72. C. Larsson, presented at the 5th Trondheim Conference on CO_2 Capture, Transport and Storage, 2009.
73. Hitachi Power Europe, *CO_2 Capture*, www.hitachi-power.com/en/co2-capture.html (last accessed 4 April 2012).
74. Cesar, www.co2cesar.eu/site/en/about.php (last accessed 4 April 2012).
75. M. Bryngelsson and M. Westermark, *Energy Proc.*, 2009, **1**, 1403.
76. Sargas, http://www.sargas.no/Operations/husnes.html (last accessed 4 April 2012).
77. J. P Hallett and T. Welton, *Chem. Rev.*, 2011, **5**, 3508.
78. M. Hasib-ur-Rahman, M. Siaj and F. Larachi, *Chem. Eng. Process.*, 2010, **49**, 313.
79. P. Wasserscheid and T. Welton, *Ionic Liquids in Synthesis*, Wiley-VCH, Weinheim, 2008.

80. J. E. Bara, T. K. Carlisle, C. J. Gabriel, D. Camper, A. Finotello, D. L. Gin and R. D. Noble, *Ind Eng. Chem. Res.*, 2009, **48**, 2739.
81. J. H. J. Davis, *Chem. Lett.*, 2004, **33**, 1072.
82. I. Bandres, S. Meler, B. Giner, C. P and C. Lafuente, *J. Solution Chem.*, 2009, **38**, 1622.
83. I. Bandres, F. M. Royo, I. Gascon, M. Castro and C. Lafuente, *J. Phys. Chem. B*, 2010, **114**, 3601.
84. J. L. Anthony, S. N. V. K. Aki, E. J. Maginn and J. F. Brennecke, *Int. J. Environ. Technol. Manage.*, 2004, **4**, 105.
85. J. L. Anthony, J. L. Anderson, E. J. Maginn and J. F. Brennecke, *J. Phys. Chem. B*, 2005, **109**, 6366.
86. J. L. Anderson, J. K. Dixon and J. F. Brennecke, *Acc. Chem. Res.*, 2007, **40**, 1208.
87. Y. Hou and R. E. Baltus, *Ind. Eng. Chem. Res.*, 2007, **46**, 8166.
88. P. K. Kilaru, R. A. Condemarin and P. Scovazzo, *Ind. Eng. Chem. Res.*, 2008, **47**, 900.
89. A. Finotello, J. E. Bara, S. Narayan, D. Camper and R. D. Noble, *J. Phys. Chem. B*, 2008, **112**, 2335.
90. P. K. Kilaru and P. Scovazzo, *Ind. Eng. Chem. Res.*, 2008, **47**, 910.
91. P. A. Hunt, B. Kirchner and T. Welton, *Chem. Eur. J.*, 2006, **12**, 6762.
92. D. Camper, J. E. Bara, D. L. Gin and R. D. Noble, *Ind. Eng. Chem. Res.*, 2008, **47**, 8496.
93. H. Shirota, J. F. Wishart and E. W. Castner Jr, *J. Phys. Chem. B*, 2007, **111**, 4819.
94. R. H. Williams, E. D. Larson, G. Liu and T. G. Kreutz, *Energy Proc.*, 2009, **1**, 4379.
95. T. Shimizu, T. Hirama, H. Hosoda, K. Kitano, M. Inagaki and K. Tejima, *Trans. Inst. Chem. Eng.*, 1999, **77A**, 62.
96. J. Blamey, E.J. Anthony, J. Wang and P. S. Fennell, *Prog. Energy Comb. Sci.*, 2010, **36**, 260.
97. J. C. Abanades, E. J. Anthony, J. Wang and J. E. Oakey, *Environ. Sci. Technol.*, 2005, **39**, 2861.
98. V. Manovic and E. J. Anthony, *Environ. Sci. Technol.*, 2008, **42**, 4170.
99. V. Manovic and E. J. Anthony, *Environ. Sci. Technol.*, 2007, **41**, 1420.

100. N. H. Florin, J. Blamey and P. Fennell, *Energy Fuels*, 2010, **24**, 4598.
101. D. Harrison, *Ind. Eng. Chem. Res.*, 2008, **47**, 6486.
102. D. Y. Lu, R. W. Hughes and E. J. Anthony, *Fuel Process. Technol.*, 2008, **89**, 1386.
103. J. C. Abanades, M. Alonso, N. Rodriguez, B. Gonzalez, F. Fuentes, G. Grasa and R. Murillo, presented at *In Situ Carbon Removal 4*, Imperial College, London, 2008.
104. O. Masek, A. Bosoaga and J. Oakey, presented at *In Situ Carbon Removal 4*, Imperial College, 2008.
105. C. Hawthorne, A. Charitos, C. A. Perez-Pulido, Z. Bing and G. Scheffknecht, presented at the 9th International Conference on Circulating Fluidized Beds, Hamburg, 2008.
106. A. Charitos, C. Hawthorne, A. Bidwe, L. He and G. Scheffknecht, presented at the 9th International Conference on Circulating Fluidized Beds, Hamburg, 2008.
107. CaOling, *Development of Postcombustion CO$_2$ Capture with CaO in a Large Testing Facility*, www.caoling.eu (last accessed 4 April 2012).
108. J. Strohle, A. Galloy and B. Epple, *Energy Proc.*, 2009, **1**, 1313.
109. A. Roder, CEMEX – Climate Change Strategy and CCS, UK Technical Report, Imperial College, London, 2008.
110. C. Dean, D. Dugwell and P. S. Fennell, *Energy Environ. Sci.*, 2011, **4**, 2050.
111. M. Smith, Properties and Behaviour of SO$_2$ Adsorbents for CFBC, Technical Report, IEA Clean Coal Centre, London, 2007.
112. J. Blamey, D. Y. Lu, D. Dugwell, E. J. Anthony and P. S. Fennell, *Ind. Eng. Chem. Res.*, 2011, **50**, 10329.
113. H. Lu and P. G. Smirniotis, *Ind. Eng. Chem. Res.*, 2009, **48**, 5454.
114. V. Manovic and E. J. Anthony, *Environ. Sci. Technol.*, 2008, **42**, 4170.
115. N. H. Florin and A. T. Harris, *Chem. Eng. Sci.*, 2009, **64**, 187.
116. Z. S. Li, N. S. Cai, Y. Y. Huang and H.-J. Han, *Energy Fuels*, 2005, **19**, 1447.
117. V. Manovic and E. J. Anthony, *Environ. Sci. Technol.*, 2009, **43**, 7117.
118. J. S. Dennis and R. Pacciani, *Chem. Eng. Sci.*, 2009, **64**, 2147.

119. N. Florin, J. Blamey and P. S. Fennell, *Energy Fuels*, 2010, **24**, 4598.

120. A. Lyngfelt, B. Leckner and T. Mattisson, *Chem. Eng. Sci.*, 2001, **56**, 3101.

121. J. S. Dennis, S. A. Scott and A. N. Hayhurst, *J. Energy Inst.*, 2006, **79**, 187.

122. S. A. Scott, J. S. Dennis, A. N. Hayhurst and T. Brown, *AIChE J.*, 2006, **52**, 3325.

123. M. Ryden, A. Lyngfelt and T. Mattisson, *Fuel*, 2006, **85**, 1631.

124. H.-J. Ryu and G.-T. Jin., *Korean J. Chem. Eng.*, 2007, **24**, 527.

125. G.-T. Jin, H.-J. Ryu, S.-H. Jo, S.-Y. Lee, S. R. Son and S. D. Kim, *Korean J. Chem. Eng.*, 2007, **24**, 542.

126. C. D. Bohn, C. R. Müller, J. P. Cleeton, A. N. Hayhurst, J. F. Davidson, S. A. Scot and J. S. Dennis, *Ind. Eng. Chem. Res.*, 2008, **47**, 7623.

127. A. Lin, F. Smith, J. Sweterlitsch, J. Graf, T. Nalette, W. Papale, M. Campbell and S.-D. Lu, *Testing of an Amine-Based Pressure-Swing System for Carbon Dioxide and Humidity Control*, NASA Technical Report (ID 20070020003), 2007, http://ntrs.nasa.gov/archive/nasa/casi.ntrs.nasa.gov/20070020003_2007014376.pdf (last accessed 29 April 2012).

128. T. C. Drage, J. M. Blackman, C. Pevida and C. E Snape, *Energy Fuels*, 2009, **23**, 2790.

129. M. L. Gray, K. J. Champagne, D. Fauth, J. P. Baltrus and H. Pennline, *Int. J. Greenhouse Gas Control*, 2008, **2**, 3.

130. S. Chuang, presented at the Annual NETL CO_2 Capture Technology for Existing Plants R&D Meeting, 24–26 March 2009.

131. P. Englezos and J. D. Lee, *Korean J. Chem. Eng.*, 2005, **22**, 671.

132. S. P. Kang and H. Lee, *Environ. Sci. Technol.*, 2000, **34**, 4397.

133. J. R. Long and O. M. Yaghi, *Chem. Soc. Rev.*, 2009, **38**, 1201.

134. A. O. Yazaydin, R. Q. Snurr, T.-H. Park, K. Koh, J. Liu, M. D. LeVan, A. I. Benin, P. Jakubczak, M. Lanuza, D. B. Galloway, J. J. Low and R. R. Willis, *J. Am. Chem. Soc.*, 2009, **131**, 18198.

135. T. C. Merkel, H. Lin, X. Wei and R. Baker, *J. Membr. Sci.*, 2010, **359**, 126.

136. J. Sheehan, T. Dunahay, J. Benemann and P. Roessler, *A Look Back at the U.S. Department of Energy's Aquatic Species Program: Biodiesel from Algae*, Technical Report NREL TP-580-24-190, National Renewable Energy Laboratory, Golden, CO, www.nrel.gov/docs/legosti/fy98/24190.pdf (last accessed 4 April 2012).

137. W. C. Yang and J. Ciferno, *Assessment of Carbozyme Enzyme-Based Membrane Technology for CO_2 Capture from Flue Gas*, DOE/NETL 401/072606, National Renewable Energy Laboratory, Golden, CO, 2006.

138. Stanford University Global Climate and Energy Project, *An Assessment of Carbon Capture Technology and Research Opportunitie*s, Global Climate and Energy Project, Stanford, CA, 2005.

139. H. Arakawa, M. Aresta, J. N. Armor, M. A. Barteau, E. J. Beckman, A. T. Bell, J. E. Bercaw, C. Creutz, E. Dinjus, D. A. Dixon, K. Domen, D. L. DuBois, J. Eckert, E. Fujita, D. H. Gibson, W. A. Goddard, D. W. Goodman, J. Keller, G. J. Kubas, H. H. Kung, J. E. Lyons, L. E. Manzer, T. J. Marks, K. Morokuma, K. M. Nicholas, R. Periana, L. Que, J. Rostrup-Nielson, W. M. H. Sachtler, L. D. Schmidt, A. Sen, G. A. Somorjai and P. C. Stair, *Chem. Rev.*, 2001, **101**, 953.

140. M. Aresta, *ACS Symp. Ser.*, 2003, **852**, 2.

141. E. Benson, C. P. Kubiak, A. J. Sathrum and J. M. Smieja, *Chem. Soc. Rev.*, 2009, **38**, 89.

142. J. P. Collin and J. P. Sauvage, *Coord. Chem. Rev.*, 1989, **93**, 245.

143. H. Gibson, *Chem. Rev.*, 1996, **96**, 2063.

144. T. Sakakura, J. C. Choi and H. Yasuda, *Chem. Rev.*, 2007, **107**, 2365.

145. P. G. Jessop, F. Joo and C. C. Tai, *Coord. Chem. Rev.*, 2004, **248**, 2425.

146. *Ammonia and Urea Production, New Zealand Institute of Chemistry*, http://nzic.org.nz/ChemProcesses/production/1A.pdf (last accessed 29 April 2012).

147. D. S. Kostick, 2009 *Minerals Yearbook*, United States Geological Survey, US Government Printing Office, Washington, DC, 2009.

148. C. S. Snyder, T. W. Bruulsema, T. L. Jensen and P. E. Fixen, *Agric. Ecosyst. Environ.*, 2009, **133**, 247.

149. M. Bernard and M.-M. Borel, *Bull. Soc. Chim. Fr.*, 1968, **6**, 2362.

150. P. S. Fennell, N. Florin, M. Al-Jeboori and I, MacKenzie. Carbon capture technology/future research needs, presented at What's the Future for Carbon Capture and Storage? Evaluating the Latest Regulatory, Financial And Project Developments, I. Mech. E. Symposium, London, October 2011.

151. W. Coates and D. R. Moore, *Angew. Chem. Int. Ed.*, 2004, **43**, 6618.

152. J. Darensbourg, *Chem. Rev.*, 2007, **107**, 2388.

153. O. Ihata, Y. Kayaki and T. Ikariya, *Chem. Commun.*, 2005, 2268.

154. ICIS.com Website, *Polycarbonate Uses and Market Data* http://www.icis.com/Articles/2007/11/06/9076146/polycarbonate-uses-and-market-data.html (last accessed 29 April 2012).

155. M. Byrne, S. D. Allen, E. B. Lobkovsky and G. W. Coates, *J. Am. Chem. Soc.*, 2004, **126**, 11404.

156. C. Koning, J. Wildeson, R. Parton, B. Plum, P. Steeman and D. J. Darensbourg, *Polymer*, 2001, **42**, 3995.

157. R. Moore, M. Cheng, E. B. Lobkovsky and G. W. Coates, *J. Am. Chem. Soc.*, 2003, **125**, 11911.

158. J. Darensbourg and J. Yarbrough, *J. Am. Chem. Soc.*, 2002, **124**, 6335.

159. B. Li, G. P. Wu, W. M. Ren, Y. M. Wang, D. Y. Rao and X. B. Lu, *J. Polym. Sci., Part A: Polym. Chem.*, 2008, **46**, 6102.

160. L. Shi, X. B. Lu, R. Zhang, X. J. Peng, C. Q. Zhang, J. F. Li and X. M. Peng, *Macromolecules*, 2006, **39**, 5679.

161. K. L. Peretti, H. Ajiro, C. T. Cohen, E. B. Lobkovsky and G. W. Coates, *J. Am. Chem. Soc.*, 2005, **127**, 11566.

162. T. Aida, M. Ishikawa and S. Inoue, *Macromolecules*, 1986, **19**, 8.

163. D. M. Cui, O. Nishiura, M. Tardif and Z. M. Hou, *Organometallics*, 2008, **27**, 2428.

164. Z. Zhang, D. Cui and X. Liu, *J. Polym. Sci., Part A: Polym. Chem.*, 2008, **46**, 6810.

165. K. Nakano, M. Nakamura and K. Nozaki, *Macromolecules*, 2009, **42**, 6972.

166. M. R. Kember, P. D. Knight, P. T. R. Reung and C. K. Williams, *Angew. Chem. Int. Ed.*, 2009, **48**, 931.
167. M. R. Kember, A. J. P. White and C. K. Williams, *Inorg. Chem.*, 2009, **48**, 9535.
168. P. G. Jessop, F. Joo and C. C. Tai, *Coord. Chem. Rev.*, 2004, **248**, 2425.
169. M. A. Pacheco and C. L. Marshall, *Energy Fuels*, 1997, **11**, 2.
170. M. D Dolan, R. Donelson and N. C. Dave, *Int. J. Hydrogen Energy*, 2010, **35**, 10994.
171. C. A. Scholes and K. H. Smith, S. E. Kentish and G. W. Stevens, *Int. J. Greenhouse Gas Control*, 2010, **4**, 739.
172. Statoil, *Sleipner West*, http://www.statoil.com/en/ TechnologyInnovation/NewEnergy/Co2Management/Pages/ SleipnerVest.aspx (last accessed 4 April 2012).
173. G. Dolf, *Energy Convers. Manage*, 2003, **44**, 1027.
174. T. Kuramochi, *Prog. Energy Comb. Sci*, 2012, **38**, 87.
175. H. Chalmers, M. Leach and J. Gibbins, *Energy Proc.*, 2011, **4**, 2596.
176. *UK Carbon Capture and Storage Demonstration Competition.* FEED Close Out Report. Report SP-SP 6.0 - RT015, ScottishPower CCS Consortium. http://www.decc.gov.uk/ assets/decc/11/ccs/sp/sp-sp-6.0-rt015-feed-close-out-report-final. pdf (last accessed 29 April 2012).
177. H. Li, Ø. Wilhelmsen, Y. Lv, W. Wang and Y. Yan, *Int. J. Greenhouse Gas Control*, 2011, **5**, 1119.
178. A. B. Rao and E. S. Rubin, *Ind. Eng. Chem. Res.*, 2006, **45**, 2421.
179. N. M. Mac Dowell and N. Shah., in *Computer Aided Chemical Engineering*, ed. M. C. G. E. N. Pistikopoulos and A. C. Kokossis, Elsevier, Amsterdam, 2011, pp. 1205–1209.

CHAPTER 13

Carbon Dioxide Utilisation in the Production of Chemicals, Fuels and Materials

MICHELE ARESTA*[1] AND PAOLO STUFANO[2]

[1]CIRCC, Consorzio Interuniversitario Reattività Chimica e Catalisi, Via Celso Ulpiani 27, 70126 Bari, Italy; [2]Department of Chemistry, Campus Universitario, University of Bari, 70126 Bari, Italy
*Email: m.aresta@chimica.uniba.it

13.1 INTRODUCTION

Carbon dioxide is produced in several anthropomorphic activities at a rate of $\sim 35\,Gt\,y^{-1}$. The main sources (Figure 13.1) are the combustion of fossil carbon (production of electrical energy, transport, heating, industry), the utilisation of biomass (combustion to obtain energy, fermentation) and the decomposition of natural carbonates (mainly in the steel and cement industries).

As the natural system (Figure 13.2) is not able to buffer this release by dissolving CO_2 into oceans (or water basins in general) or by fixing it into biomass or inert carbonates, CO_2 is accumulating in the atmosphere with serious concerns about its influence on climate change.

Materials for a Sustainable Future
Edited by Trevor M. Letcher and Janet L. Scott
Published by the Royal Society of Chemistry, www.rsc.org

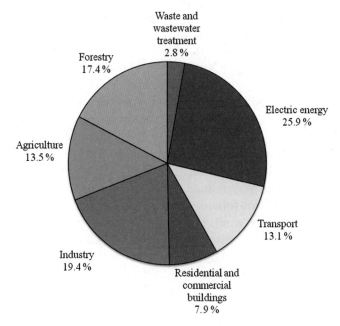

Figure 13.1 Anthropogenic sources of carbon dioxide.

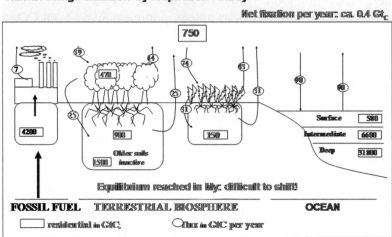

Figure 13.2 The carbon cycle and the excess anthropogenic CO_2.

This has made it imperative to find solutions that will avoid the atmospheric concentration of CO_2 increasing well beyond the present 391 ppm (the pre-industrial era value was 275 ppm). The growth in the energy demand by humanity makes the solution not simple because, according to most scenarios, at least 80% of the total energy will still be produced from fossil carbon in the coming 30 years or so. This adds urgency to implementing technologies that may reduce both the amount of CO_2 released to the atmosphere and the utilisation of fossil carbon. Furthermore, in addition to more efficient technologies (in the production and use of energy), other routes must be discovered that may reduce either the production of CO_2 or its emissions to the atmosphere. Among the former, perennial energy sources (such as sun, wind, hydro, geothermal) are being exploited. The reduction of the release of CO_2 to the atmosphere is based on its capture from its point sources (power, industrial, fermentation, cement plants) by using liquid or solid sorbents or membranes.[1]

Such captured CO_2 can be either disposed of in geological cavities and aquifers or recycled. The former option corresponds to the CCS technology (CO_2 Capture and Storage)[2] and the latter to the CCU technology (CO_2 Capture and Utilisation).[3] CCS is believed to be able in general to manage larger amounts of CO_2 than CCU. The latter, on the other hand, is able to recycle carbon, reducing the extraction of fossil carbon. CCS is energy demanding and economically unfavourable, whereas CCU may or may not require energy (depending on the nature of the species derived from CO_2) and is economically viable, as all compounds derived from CO_2 or any use of CO_2 will have an added value. A concern about the utilisation of CO_2 lies in the amount of energy eventually necessary that cannot be derived from fossil carbon. This has so far prevented the utilisation of CO_2. However, in a changing paradigm of deployment of primary energy sources, if the development of perennial sources becomes more widespread, the conversion of CO_2 into chemicals and fuels may become economically convenient and energetically feasible. A key role in this direction will be played by the wind and by the sun. The former can be coupled with electricity generation in the conversion of CO_2, whereas the latter can be used in a direct (photochemical, thermal) or indirect (photoelectrochemical) conversion of CO_2.

The products obtainable from CO_2 are of various nature: fine chemicals, intermediates, fuels (Figure 13.3).[3,4]

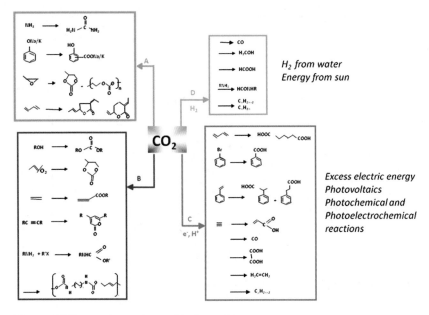

Figure 13.3   Products obtainable from CO_2.

In the latter class of compounds (Figure 13.3, right), the CO_2 carbon atom is in a lower oxidation state (n_{ox}) than in CO_2 itself. Figure 13.4 shows the variation of the standard free energy of formation (ΔG°_f) of C_1 species against the carbon oxidation state. The reduction of CO_2 thus requires an input of energy and the amount depends on Δn_{ox}.

As mentioned above, such energy must come from energy sources other than fossil carbon. Other products may involve the entire CO_2 molecule incorporated into an organic substrate (Figure 13.3, left): the energy requirement of this class of compounds is much lower and the relevant reactions can be carried out in a 'business as usual' mode. This chapter discusses the conversion (direct or indirect) of CO_2 into chemicals, fuels and materials.

13.2   CARBONATES

Carbonates, organic or inorganic, molecular or polymeric, or extended ionic lattices, are characterised by the –OC(O)O– moiety, either ionic or covalently bound to organic groups (Figure 13.5).

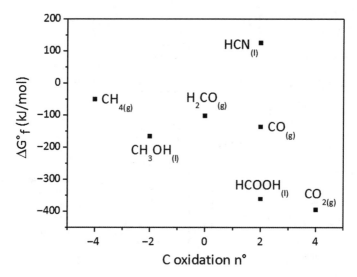

Figure 13.4   Gibbs free energy of formation *versus* the oxidation state of carbon in C_1 molecules.

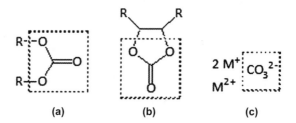

Figure 13.5   Carbonates: organic (a, b) and inorganic (c).

The building-up of the 'CO_3' moiety from CO_2 and a co-reagent such as an oxide [Equation (13.1)] or an alcohol [Equation (13.2)] is exothermic, as shown by the ΔH_f of the compounds (Table 13.1).

$$CO_2 + CaO \rightarrow CaCO_3 \qquad (13.1)$$

$$CO_2 + 2ROH \rightleftharpoons (RO)_2CO + H_2O \qquad (13.2)$$

Conversely, due to the entropy decrease in the reaction because gaseous CO_2 is fixed into a solid or a liquid, the Gibbs free energy change is more positive. This makes the synthesis of organic carbonates from alcohols and CO_2 thermodynamically unfavourable

Table 13.1 Enthalpy and Gibbs free energy of carboxylation reactions.

Reaction	$\Delta H°_f/kcal\ mol^{-1}$	$\Delta G°_f/kcal\ mol^{-1}$
$CaO + CO_2 \rightarrow CaCO_3$	-42.58	-31.2
$2MeOH + CO_2 \rightarrow (MeO)_2CO$	-4.00	$+6.1$
$2EtOH + CO_2 \rightarrow (EtO)_2CO$	-3.80	na
$2PhOH + CO_2 \rightarrow (PhO)_2CO$	$+12.09$	na

and the equilibrium concentration of the product is fairly low (often less than 1% at the temperature of operation).

13.2.1 Organic Molecular Compounds

Industrially, organic carbonates are easily prepared at room temperature by reacting phosgene ($COCl_2$) with alcohols [Equation (13.3)] in the presence of a base such as NaOH. Such reaction does not require any catalyst. As phosgene has been banned in several countries because of its high toxicity and the fact that large amounts of chlorinated waste are produced in reaction (13.3), either as by-products or as solvents, substitutes for phosgene are being investigated. CO_2 [Equation (13.2)], urea [Equation (13.4)] and CO/O_2 [Equation (13.5)] are good candidates. The last system is not relevant to the content of this chapter and is not discussed.

$$2ROH + COCl_2 + 2B \rightarrow (RO)_2CO + 2B \cdot HCl \qquad (13.3)$$

$$2ROH + (H_2N)_2CO \rightarrow (RO)_2CO + 2NH_3 \qquad (13.4)$$

$$2ROH + CO + \tfrac{1}{2}O_2 \rightarrow (RO)_2CO + H_2O \qquad (13.5)$$

Carbonates of Group 1 and 2 elements react with RX (X = Cl, Br, I). In addition, organic carbonates can be formed by reacting alkylating agents such as alkyl halides (RX) with inorganic carbonates [Equation (13.6)] at a temperature of ~400 K under phase-transfer conditions to produce symmetrical and unsymmetrical dialkyl carbonates.

$$2RX + MCO_3 \rightarrow (RO)_2CO + MX_2 \qquad (13.6a)$$

$$2RX + M_2CO_3 \rightarrow (RO)_2CO + 2MX \qquad (13.6b)$$

Alkyl halides can alkylate alcohols in pressurised CO_2 in the presence of strong bases[5] in ionic liquids.[6] The existing trend to reduce or exclude the use of halogenated organics does not make reaction (13.6) an effective alternative to phosgene. The catalytic conversion of alcohols and CO_2 is by far the most attractive route to organic acyclic carbonates.

Equation (13.2) represents the direct use of CO_2, whereas Equation (13.4) is an example of an indirect use of CO_2: ammonia, formed in the reaction, can be recovered and reacted with CO_2 to produce urea (which can be considered as an activated form of CO_2) or used in other applications. Another way to make linear carbonates is by transesterification, which converts a carbonate into another type of carbonate. In particular, Equation (13.7) shows the conversion of a cyclic into an acyclic carbonate.

$$2ROH + \quad \underset{O}{\overset{O}{\bigcirc}}{=}O \rightarrow (RO)_2CO + HOCH_2CH_2OH \qquad (13.7)$$

Cyclic carbonates can in turn be produced by reacting epoxides with CO_2 [Equation (13.8)] or by reaction of diols with urea [Equation (13.9)].

$$\underset{RCH-CH_2}{\overset{O}{\triangle}} + CO_2 \rightarrow \quad \overset{R}{\underset{O}{\bigcirc}}{=}O \qquad (13.8)$$

$$HOCH_2CH_2OH + H_2NC(O)NH_2 \rightarrow \quad \underset{O}{\overset{O}{\bigcirc}}{=}O + 2NH_3 \qquad (13.9)$$

The new synthetic routes based on CO_2, urea, epoxides and transesterification are more environmentally friendly than reaction (13.3) as they do not make use of toxic substances such as $COCl_2$ and do not produce harmful waste. In some cases [*e.g.* Equation (13.8)] there are no co-products. This justifies the industrial interest in developing new processes that may avoid phosgene.

13.2.1.1 Synthesis of Acyclic Carbonates via Carboxylation of Alcohols. The formation of a carbonate from an alcohol and CO_2 should in principle occur according to Equations (13.10) – (13.13), where A is an acid and B is a base. Therefore, bifunctional catalysts are good candidates for the synthesis of linear carbonates, a reaction that has the thermodynamic limitation discussed above.

$$ROH + B \rightarrow RO^- + BH^+ \tag{13.10}$$

$$RO^- + CO_2 \rightarrow ROC(O)O^- \tag{13.11}$$

$$ROH + A \rightarrow R^+ + AOH^- \tag{13.12}$$

$$RC(O)O^- + R^+ \rightarrow ROC(O)OR \tag{13.13}$$

At the reaction temperature (often above 400 K), the equilibrium position of Eq. 13.2 is shifted to the left and the equilibrium concentration of the carbonate may be as low as $<1\%$. Such conditions are not suitable for developing an industrial process and the shift of the equilibrium to the right is a key issue for the exploitation of reaction (13.2). Attempts have been made based on different strategies, such as water elimination, increasing the CO_2 concentration and the using various alcohol:CO_2 molar ratios.

In water removal, either inorganic or organic water traps have been used. The former are molecular sieves that require a low temperature (250 K or lower[7]) for the efficient trapping of water and avoiding carbonate decomposition that may occur at the reaction temperature due to the acidic –OH groups present on silica that may protonate the carbonate and decompose it back to alcohol and CO_2. The use of such water traps is not energetically favourable owing to the oscillating temperature (above 400 K for the carbonate formation to occur and below 250 K for trapping water) and to the energy necessary for dewatering the molecular sieves for their recycling. Moreover, pulverisation and the loss of inorganic material are possible owing to their physical stress.

Organic traps such as aldols,[8] ketals,[9] orthoformates[10] or cyanides[11] have also been used. The issue here is that the resulting compounds are soluble in the reaction medium and this will increase the cost of processing the reaction mixture for product separation and regeneration of the organic trap. Among organic traps, carbodiimides (RN=C=NR) are interesting molecules as their hydrated form [RHN–C(O)–NHR, a urea] is not soluble in the reaction medium and can be easily separated. It has been shown[12] that CyN=C=NCy (DCC) is not a simple water trap, but it acts as an 'organic catalyst' as it is able under very mild reaction conditions (330 K and 0.2 MPa CO_2; metal systems require temperatures higher than 410 K and high pressures) to promote the formation of dimethyl carbonate (DMC), diethyl carbonate (DEC) and diallyl carbonate (DAC) from the relevant alcohols and CO_2 in quantitative yields. DFT calculations have demonstrated that in the presence of DCC, the formation of DMC has much more favourable thermodynamics as $\Delta G°$ is equal to − 146 kJ mol^{-1} (see Scheme 13.5) Such a dramatic change with respect to the metal-catalysed carboxylation of methanol is due to the simultaneous formation of dicyclohexylurea (DCU), the hydrated form of DCC. The dehydration of DCU^{13} carried out with quantitative yields closes the cycle (Scheme 13.1) and makes the process based on DCC alone of potential usefulness in the synthesis of niche carbonates or, in the case the alcohol, brings moieties sensitive to temperature.

As mentioned above, the thermodynamics of Equation 13.2 are made unfavourable by the use of gaseous CO_2 as reagent. Therefore, working under pressurised CO_2 or even better in supercritical CO_2 (sc-CO_2) would improve the thermodynamics of the reaction. In fact, increasing the pressure from 0.1 to 30 MPa positively affects the conversion, but it is only under very severe pressure (> 300 MPa) that the reaction is shifted to the right. Therefore, in order to push the equilibrium to the right, the reaction should be carried out in either (i) liquid alcohol pressurised with CO_2 or (ii)

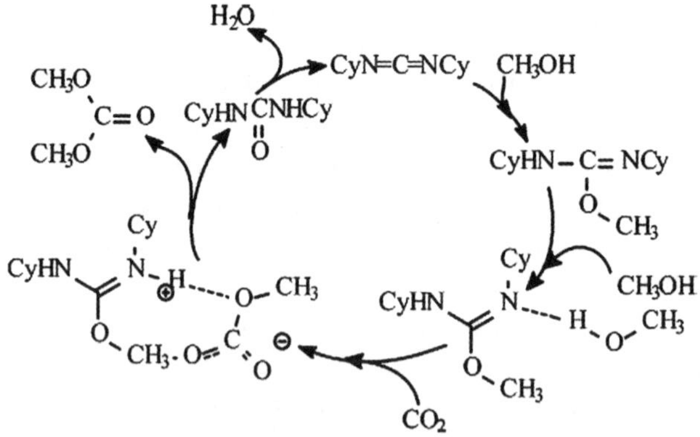

Scheme 13.1   Synthesis of organic linear carbonates catalysed by DCC.

sc-CO_2. Case (i) is governed by the solubility of CO_2 in the co-reagent alcohol; case (ii) improves the thermodynamics from the point of view of the entropy of CO_2, but requires that alcohol and CO_2 are in a single phase in order to react better. In fact, the use of sc-CO_2 produces a better conversion of alcohols.

Recent reviews on the synthesis of linear carbonates based on homogeneous catalysts or organic catalysts are available.[14] Here we discuss most recent results relevant to attempts to improve the conversion yield of alcohols. It is worth emphasising that much attention has been paid in recent years to the synthesis of dimethyl carbonate (DMC) from methanol and CO_2, but recently the synthesis of diethyl carbonate (DEC)[15,16] has been a target. Such interest is due to the fact that DMC and DEC have similar properties and uses. DEC prepared from bioethanol gives a 'bio-derived' label to DEC and features as 'green' and 'sustainable' in the processes in which it is used and the products derived from it.

For the synthesis of DMC and DEC, homogeneous, heterogeneous and heterogenised catalysts have been used or else organic catalysts.[12] The reaction mechanism has been elucidated for a few of them, namely the homogeneous catalysts $[Nb(OR)_5]_2$[17] (Scheme 13.2), $^nBu_2Sn(OMe)_2$[18a] (Scheme 13.3), the heterogeneous catalysts ZrO_2[19] and CeO_2/Al_2O_3[20] (Scheme 13.4) and the organic catalyst DCC[16] (Scheme 13.5).

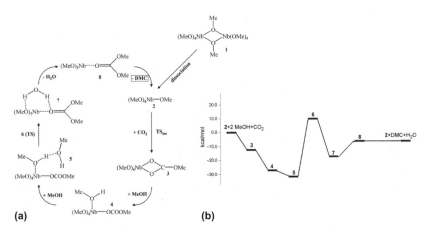

Scheme 13.2 Reaction pathway (a) and energetic profile (b) for the $[Nb(OCH_3)_5]_2$-catalysed conversion of methanol to DMC.

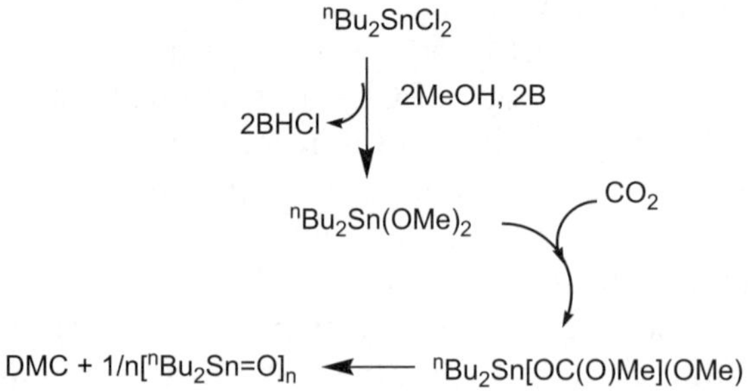

Scheme 13.3 Mechanism of the carboxylation of methanol promoted by homogeneous di-*n*-butyltin(IV) methoxide.

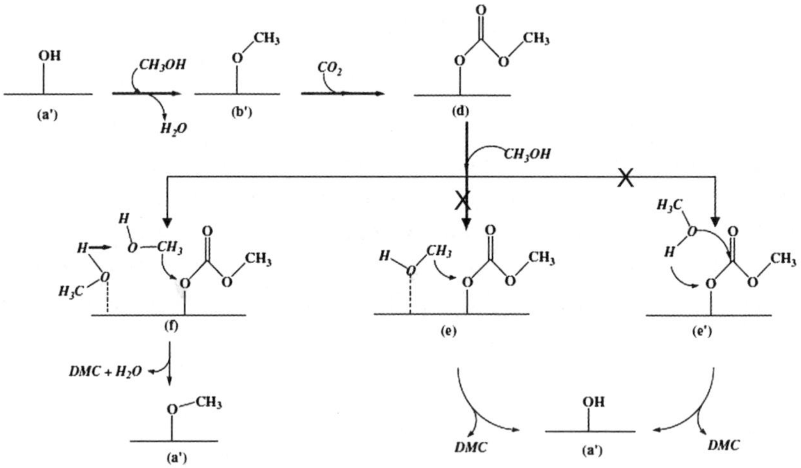

Scheme 13.4 Putative steps for the formation of DMC from methanol and CO_2 *via* gas-phase methanol attack on surface-bound $OC(O)OCH_3$. Gas-phase species are in italics.

It is clear from a comparative analysis of Schemes 13.2–13.5 that a different mechanism is operating for the different catalysts. In particular, in the case of the Sn catalyst (Scheme 13.3), 2 mol of methanol are both activated through a base mechanism to give the $Sn(OMe)_2$ moiety that undergoes CO_2 insertion to produce $Sn(OMe)OC(O)OMe$. Then, an intramolecular methyl transfer from Sn–OMe to $SnOC(O)OMe$ produces DMC and a by-product oligomeric Sn compound $[^nBu_2SnO]_n$ that needs to be converted

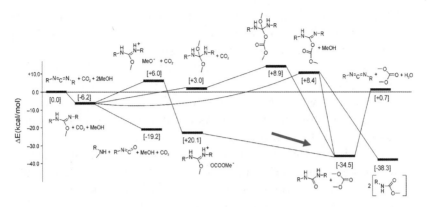

Scheme 13.5 Energy profile for the DCC-promoted carboxylation of methanol.

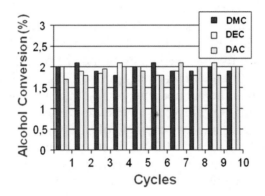

Figure 13.6 Recycling of Nb catalysts in the synthesis of carbonates.

into a monomer so that the catalysis may continue, as shown in studies by Ballivet-Tkatchenko *et al.*[18b]

In the case of the Nb catalyst (Scheme 13.2), 2 mol of methanol are activated through a base-plus-acid mechanism and DMC and water are formed out of the coordination sphere of the metal, regenerating the catalyst that can maintain the same activity after several cycles (Figure 13.6). Anyway, at the end of each cycle it must be recovered as lengthy contact with water must be avoided as it may convert into $(RO)_3NbO$, inactive in catalysis. The Sn-like mechanism is prevented with Nb by a higher transition state energy (Scheme 13.2). The organic catalyst DCC follows the same reaction path as the Sn complex and produces urea as the final form of DCC.

Heterogeneous catalysts (based on W and Ce oxides) show the same mechanism of activation of methanol and insertion of CO_2 common to Sn and Nb. For Al_2O_3 (3%)–CeO_2, the closure of the cycle has been shown to proceed according to what is found for Nb (Scheme 13.4).

The main drawbacks and limitations for such reactions are due to the deactivation of the catalysts for many reasons. The mixed oxides[21,22] seem to be more resistant than single oxide systems and operating times of weeks have been reported.[23] The elimination of water may improve the lifetime of the catalyst and the conversion yield, which always remains low at just a few percent. The use of membranes instead of water traps may help in setting a continuous process.[23] However, the low conversion yield problem still persists, making this synthetic procedure uneconomic.

Sn and Nb complexes have been tethered on a polystyrene matrix[24] and show a much longer life [and, thus, turnover number (TON)] than the corresponding soluble forms. A similar behaviour has been shown by the Sn complex supported on either SBA-15[25a] or SBA-16.[25b] SnO_2 and ZrO_2 supported on silica[25c] also show a better TON than the bulk parent species, but Si–SnO_2 has a worse selectivity than SnO_2 alone, as ethers are formed.

In an attempt to improve the yield of carbonate, ZrO_2–MgO has been used in the synthesis of DMC in the presence of butene oxide. The latter is reported to act as a water trap producing butenediol. It is worth recalling that, as will be discussed later, oxiranes easily react with CO_2 to produce cyclic carbonates and the latter are known to react with alcohols to produce acyclic carbonates and diols. In the reaction above, an apparent activation energy for the synthesis of DMC of 62 kJ mol^{-1} has been calculated.

The same oxirane as used in the synthesis of DEC[26] in the presence of CeO_2 has been shown to improve the conversion of ethanol to 15.8 %, but the selectivity towards DEC was only 10%. Mixed oxides[27] with variable Ce:Zr ratio have been used in the synthesis of DMC. The best performance was shown by 1:1 Ce–Zr oxides calcined at 1273 K: in the presence of 1,1,1-trimethyoxymethane a conversion of 7.8% of methanol was observed at 373 K.

Alkoxy-ionic liquids (AILs) have been used as solvents with ZrO_2–MgO catalysts to effect a conversion of methanol of 12% with 90% selectivity at 400 K and 7.8 MPa; the AILs were recycled.[28] Graphene[29] nanosheets have also been used to support

Cu–Ni catalysts in the synthesis of DMC. Heteropolymetallates[30] also catalyse the formation of DMC from methanol and CO_2, the selectivity of the reaction being strongly temperature dependent: with increasing temperature, methyl formate and dimethoxymethane are easily formed.

13.2.1.2 Synthesis of Carbonates via *Transesterification.* The market for acyclic carbonates could expand much beyond its present size if new phosgene-free processes were to be developed. The DMC market and that of other alkyl carbonates (DEC mainly) may easily expand to the tens of Mt y^{-1} scale. DMC and DEC can be used as gasoline components or as solvents, reagents and monomers for polymers and DAC is already used for the synthesis of specialty polymeric materials (production of lenses). Therefore, finding new phosgene-free routes is essential for such market expansion. The direct carboxylation of alcohols with CO_2, as discussed above, at present is limited by low conversion (a few percent). If more drastic conditions are implemented in an attempt to increase the conversion yield, a reduction in selectivity is observed, which unfortunately increases the processing costs. Finding alternative routes that may be characterised by a conversion yield of the alcohol of a few tens of percent would be most welcome. The transesterification of cyclic carbonates and the reaction of alcohols with urea represent two approaches that may contribute to implementing new sustainable synthetic routes.

The transesterification of ethene carbonate [Equation (13.7)] with methanol produces DMC and ethene glycol ($HOCH_2$-CH_2OH). The latter must be recovered and reacted with urea to recover the ethene carbonate. This recycling is essential in order to avoid wasting the ethene glycol.

A reactive distillation approach[31] has been used in the transesterification of DMC with ethanol to produce a mixture of methyl ethyl carbonate and diethyl carbonate. The chemical equilibrium and reaction kinetics of such a reactive system have been investigated experimentally and theoretically and a kinetic model has been derived.[31b]

DEC has been produced by using the same reactive distillation technique in the transesterification of ethene carbonate with ethanol in presence of sodium ethoxide,[32a] reaching a 91% conversion yield. Zn–Y oxides have been shown to catalyse the transesterification of ethene carbonate with methanol.[32b] The Y content and the acid–base properties of the oxides play a key role in the conversion yield.

13.2.1.3 Synthesis of Carbonates via Alcoholysis of Urea. The alcoholysis of urea is now a process being investigated with much attention all around the world. It was shown, almost a decade ago,[33] that the process proceeds in two steps, (i) formation of a urethane H_2NCOOR [Equation (13.14)] and (ii) formation of DMC [Equation (13.15)], that require different operating conditions. The second step is much more energy demanding than the first. Either a single [ZnO] catalyst[34] or two different catalysts[33] have been used.

$$(H_2N)_2CO + ROH \rightarrow ROC(O)NH_2 + NH_3 \qquad (13.14)$$

$$H_2NCOOR + ROH \rightarrow (RO)_2CO + NH_3 \qquad (13.15)$$

The process presents some problems of separation of products with low-boiling alcohols; reactive distillation is a technique that may help to solve the problems. The reaction of urea and methanol using ZnO has been studied in detail,[34] and the effects of temperature, space–time and urea loading in the feed have been investigated. These studies confirm that the second stage is the rate-determining step. FTIR studies have shown that urea is decomposed into isocyanic acid (HCNO),[35] and with ZnO forms zinc isocyanate, which has been proposed as being the real catalyst. Furthermore, mixed oxides, Zn/FeO_x, have shown interesting properties of stability and recyclability.[36] Hydrotalcites[37] have also been used and the relevant Fe–Zn materials have been shown to be the most active catalysts. The route based on urea has the advantage of having the highest alcohol conversion rate ($\sim 25\%$.) coupled with easy of availability of urea that is produced on a scale of 120 Mt y^{-1} with a predicted annual increase of 7%.

13.2.1.4 Synthesis of Cyclic Carbonates. The carboxylation of epoxides [Equation (13.6)] to produce pentaatomic cyclic carbonates has been known since 1943.[38–40] It is now produced in several plants. The reaction may produce polymers[41] in addition to cyclic carbonates, depending on the catalyst used and the reaction conditions.[42–52]

Several different catalysts have been used, such as organic catalysts,[53] metal halides,[54] transition metal complexes,[55] classical Lewis acids[44] and metal phthalocyanines.[56] Heterogeneous catalysts.[57,58] including metal oxides,[59-63] are active catalysts characterised by a longer life with respect to homogeneous systems. Amides, such as dimethylformamide (DMF) and dialkylacetamides (DAAs), used as solvents, promote the carboxylation of epoxides,[62] albeit to a limited extent. Optically active carbonates have been produced using metal oxides that catalyse the carboxylation of optically active epoxides with total retention of configuration.[63] Racemic mixtures of the epoxides have been carboxylated with an enantiomeric excess (*ee*) of 22% using Nb(IV) complexes with optically active (N, O, P as donor atoms) ligands. Deanchoring of the ligand from the metal centre causes the low *ee*.[63] Ionic liquids are good media for the fast carboxylation of oxiranes and microwaves even promote the formation of carbonates in ILs.[64] Several papers and reviews on the carboxylation of oxiranes have recently been published.[65-70]

The advantages of using sc-CO_2 alone as a solvent[71] or in combination with ILs[72,73] have also been described.

The oxidative carboxylation of olefins appears to be an interesting approach to the synthesis of cyclic carbonates from cheap and easily available reagents such as olefins, CO_2 and O_2 [Equation (13.16)].

$$RCH=CH_2 + \tfrac{1}{2}O_2 + CO_2 \rightarrow \quad \text{(structure)} \qquad (13.16)$$

The direct oxidative carboxylation of olefins couples two processes: (i) the epoxidation of the olefins and (ii) the carbonation of the epoxides, an approach that avoids the separate production of the epoxide.

Only a few examples have been reported of the direct carbonation of olefins,[74-80] which has the drawback of the addition of dioxygen across the C–C double bond of the olefin with formation of aldehydes or the relevant acids. A recent review surveyed the area.[81] Using homogeneous Rh complexes, the active species in the epoxidation of the olefin has been shown to be the peroxocarbonate. Using $^{16(18)}O–^{18(16)}O$ peroxo groups, it has been possible to show that the metal-bound O-atom of the peroxo group is transferred to an oxophile.[82]

Several other oxidants have been used for the oxidation of olefins, such as organic hydroperoxides and hydrogen peroxide,[65] but the goal must be the direct use of dioxygen under controlled conditions.

In order to reduce the cleavage of the olefin double bond by dioxygen, a two-step reaction has been investigated in which a metal oxide transfers one lattice oxygen to an olefin and is then reoxidised using dioxygen.[83] When the reaction is carried out in the presence of CO_2, the oxygen transfers to the olefin to produce the cyclic carbonate.

The reaction of polyols with urea is a recent strategy to prepare cyclic carbonates. Efficient catalysts have been used for the synthesis of glycerol carbonate,[84a] which has been used as a platform molecule for the synthesis of several chemicals, including epichlorohydrin.[84b]

13.2.2 CO_2 as Comonomer in the Synthesis of Polycarbonates

As already mentioned, the carboxylation of epoxides may produce polycarbonates. Al–porphyrin complexes[85] were the first to be used for this purpose, followed by Zn–compounds.[86] The pioneering discoveries by Inoue,[87] who found a recent implementation in a plant, and the studies by Rokicki and Kuran[88] have made the copolymerisation of CO_2 and epoxides fairly popular and, as a result, it has been studied in detail by several researchers around the world with the aim of mastering the alternative insertion of the comonomers for making regular copolymers. Darensbourg *et al.*[89] and Kember *et al.*[90] have published recent reviews on epoxide–CO_2 copolymerisation, and this is not repeated here. More recent results are discussed below. Different polymers have different applications and Table 13.2 presents some applications of specific polycarbonates.

The amount of polycarbonates used around the world has grown with a quasi-exponential trend in the last few years, as shown in Figure 13.7. In 2011, over 4 Mt y^{-1} of polycarbonates were used in fields such as the automotive, construction, medical care, packaging and electronics industries and also to make optical discs. Their use in construction, optical discs and electronics covers over 90% of the total use.

Figure 13.8 shows some applications of polycarbonates in architecture where they fulfil completely different roles, from

Table 13.2 Polycarbonates and their uses.

Type of polycarbonate	Application	Market (Mt y^{-1})
Bisphenol-A	Optical media, materials for construction, safety goggles and protective visors and helmets, suitcase shells, bottles and packaging, medical and healthcare, electronics automotive	3.5
Ethene	Ceramics binder, adhesive coatings, electronics, thin films	0.25
Propene	Packaging, coatings, electronics, ceramics binder	0.4
Styrene	Materials for construction, coatings	0.15

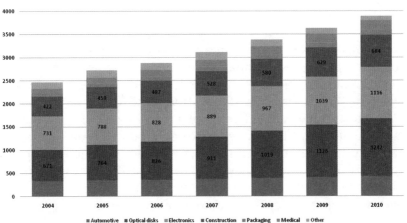

Figure 13.7 Yearly growth of the world consumption of polycarbonates (kt y^{-1}).

insulating to light spreading, decoration and protection from external sun, in addition to their architectural function.

The use of a variety of epoxides characterised by different molecular structures may be of interest as polymers with different structural properties may be obtained. In this area, there is still much work to be done to find new routes for the production of epoxides and to find new catalysts for the production of regular

Figure 13.8   Utilisation of polycarbonates in architecture. Top, from the left: façade of the Malakoff metro station, Paris, France; Expo 2010, Shanghai, China; Sports Centre, Bakio, Spain. Bottom, from the left: Pompeu Fabra University, Barcelona, Spain; Barbie store, Shangai, China; Sung Kyun Kwan University, Seoul, Korea.

copolymers. A key issue is the production of epoxides. A clean route to epoxides is essential for expanding the use of such synthetic routes. As indicated above, the use of hydrogen peroxide, which represents the cleanest industrial route for the conversion of olefins to epoxides, should be supplanted with that of dioxygen. In fact, the quantity of hydrogen peroxide produced today is around 500 kt y^{-1}, very much less than what would be required for the synthesis of epoxides if they were to be used to synthesise polycarbonates. Using dioxygen, there would be no limit to the production of epoxides or to the amount of polycarbonates produced ($>$ Mt y^{-1}).

Let us consider now some recent advances in research in this specific field. Bifunctional Co(III)–salen complexes have been used as promoters of the alternating copolymerisation of epoxides and CO_2,[91–94] with a turnover frequency (TOF) of 15 000 h^{-1}, and working with a fairly high substrate:catalyst ratio (50 000), with an interesting TOF of 6105 h^{-1}.[94] Double metal cyanide systems also show interesting activity[95] (TOF 3856 h^{-1}). Polyindene carbonates have been prepared by using indene oxide and CO_2 in the presence of Co(III)–salen complexes.[96] The complex has a molecular mass of up to 7100 Da and a glass transition temperature of 407 K. Such a polymer is stable up to >520 K. 'SalenCrCl'[97] complexes have been demonstrated to be able to produce either cyclic carbonates or polycarbonates. New metal-free catalysts have also been developed[98] for the synthesis of defined functionalised polymers. Functionalised bisphenol-A polycarbonates (BPA-PC[99]) have been prepared using siloxanes to produce silicone–BPA-PC. Sulfonated polycarbonates have been obtained using BPA and the sodium salt of sulfobenzoic acid phenyl ester.[100] The synthesis of polyol–aliphatic polycarbonates in a two-stage process[101] produces polyurethane-free elastomers for biomedical applications.

An alternative to the epoxide–CO_2 copolymerisation is transesterification that produces carbonates that can be polymerised. Metal–organic-frameworks[102] based on Zn–1,4-benzenedicarboxylate show interesting activity as catalysts in the production of polycarbonate diol from diphenyl carbonate and 1,6-hexanediol.

A new approach is now emerging, namely the use of enzymes as catalysts. Dialkyl carbonates and diols or polyols and a diester can produce a polycarbonate in the presence of lipases.[103] Diethyl carbonate, 1,8-octanediol and tris(hydroxymethyl)ethane were polymerised in the presence of a lipase B extracted from *Candida antarctica* to produce a polycarbonate polyol.[104] A recent review[105] summarised the behaviours of enzymes in ring-opening polymerisation and discussed the mechanism that is essentially based on the activation of monomers.

13.3 CARBAMATES

13.3.1 Synthesis of Molecular Carbamates

The interaction of carbon dioxide with amines is the basis of several processes, not only in the field of synthetic chemistry[106] but also in advanced technological applications such as CO_2 capture from flue gases using monoethanolamine (MEA) or di/polyamines.[107,108]

The simplest species is carbamic acid, H_2NCOOH, a labile compound that easily decomposes back to NH_3 and CO_2. Only recently have examples of such a class of compound been isolated (see below) by reacting CO_2 with secondary amines [Equation (13.17)], with suitable structural features.[109] In general, in such reactions the relevant ammonium carbamates are formed [Equation (13.18)].

$$RR'NH + CO_2 \rightarrow \frac{1}{2}[RR'NCO_2H]_2 \qquad (13.17)$$

$$2RR'NH + CO_2 \rightarrow (RR'NH_2)^+ {}^-OOCNRR' \qquad (13.18)$$

The carbamate moiety 'RR'NCO$_2$' can be bonded to *an electrophilic centre* through either an *ionic or a covalent linkage*. In Nature, carbamates are involved either in the CO_2 activation by RuBisCO in the Calvin cycle for photosynthesis or for physiological removal of N-containing compounds.

13.3.1.1 Direct Synthesis of Molecular Carbamates from CO$_2$. The formation of the carbamate 'RHNCOO' moiety can be promoted by metals or metal salts [Equation (13.19)] *via* the formal insertion of CO_2 in the M–N bond of amides resulting from the reaction of the metal centre with the amine employed:[110]

$$RR'NH + L + MBPh_4 + CO_2 \rightarrow M^+ {}^-O_2CNRR' + [HL]BPh_4 \quad (13.19)$$

where, $R' = H$, alkyl, $L = RR'NH$; $R = $ aryl, $R' = H$, $L = NR'_3$ ($R' = $ alkyl); $M = $ Li, Na, K.

Carbamic acid, NH_2CO_2H, has been detected and found to be stable as a solid in the zwitterionic form, $^+NH_3COO^-$, at low temperature.[111–113] Carbamic acids formed from primary and secondary amines are labile and easily convert back to CO_2 and the original amine. Nevertheless, solid carbamic acids have been isolated and characterised by X-ray diffraction using dibenzylamine or a Co–aminophosphane complex.[109,114] The solids show H-bonded dimeric structures [Equation (13.17)], which are most probably responsible for their stability. Also, some aminosilanes were found to produce dimeric carbamic acids $[(RO)_3SiCH_2CH_2NHCH_2CH_2NHCOOH]_2$ at 273 K.[115] Theoretical studies, supported by experimental data, gave an activation enthalpy of 40 kJ mol^{-1} for the formation of ammonium salts from CO_2 and amines.[116]

Carbamate salts such as $RR'NCOOM$ (M = Group 1 cation or ammonium ion) generated from CO_2 and amines have attracted the interest of many researchers as they could be used to transfer the entire 'RR'NCO$_2$' to suitable organic substrates or even be converted into isocyanates and ureas. The latter approach represents an attractive and innovative alternative to the problematic phosgene-based technologies. The reactivity and the application of organic carbamates from amines and CO$_2$ has been extensively investigated over the past years and were reviewed recently.[106]

Carbamate sources such as alkylammonium and p-block element carbamates, directly generated from CO$_2$ as discussed above, are usefully reacted with suitable organic electrophiles to produce stable organic carbamates, which find application in pharmacology,[117] as agrochemicals or in synthetic chemistry as protecting groups,[118,119] or precursors not only for ureas and isocyanates but also for polymers.[120] The reaction of carbamate sources 'RR'NCO$_2$' with an electrophile can produce not only the target carbamate ester, but also undesired side products if the electrophile attacks the N-atom, causing CO$_2$ elimination at the same time.[106,121]

Here we focus on the different classes of electrophiles and the relevant concepts employed for the synthesis of stable carbamate esters. For additional specific examples, one can refer to an extensive review published recently by Chaturvedi.[122]

One of the first reports on carbamate synthesis was by Yoshida and Inoue, who reacted CO$_2$, an amine and a vinyl ether to produce the relevant carbamates.[123] Subsequently, the same group investigated the reaction with alkyl halides in a one-pot reaction.[124] Alkyl halides have often been used as alkylating agents for carbamates. Despite the great number of metal carbamates isolated, only a few examples of transfer of a carbamic group from a metal centre to an alkyl halide or sulfide have been documented.[110,125,126] The general reaction between CO$_2$, amines and alkyl halides to produce the relevant carbamates is illustrated in Scheme 13.6.

The modulation of the electrophilicity of the cation in ammonium or metal carbamates is a crucial factor for the selection of the centre involved in the reaction.[127] Complexing agents such as crown ethers and cryptands have been successfully used to drive the electrophilic attack of the alkyl halides selectively on to the O-atom of the carbamate.[128]

Another improvement of this strategy has been achieved using sterically hindered strong organic bases under mild conditions. Amidines, pentaalkylguanidines such as CyTMG

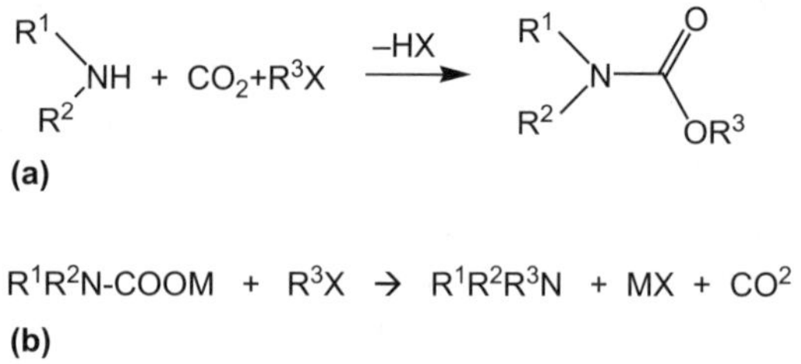

(a)

$$R^1R^2N\text{-}COOM + R^3X \rightarrow R^1R^2R^3N + MX + CO^2$$

(b)

Scheme 13.6 Synthesis of carbamates from CO_2, amines and alkyl halides.

(cyclohexyltetramethylguanidine),[129] phosphazenes[129–131] and DBU (1,8-diazabicyclo[5.4.0]undec-7-ene) have been successfully used for this purpose. Interesting results have also been obtained using different systems such as tetraethylammonium superoxide,[132] basic resins,[133] Triton B (benzyltrimethylammonium hydroxide),[134] and inorganic bases such as K_2CO_3 in the presence of catalytic amounts of $(Bu_4N)I$.[135]Kong *et al.*[136] recently reported the use of poly(ethylene glycol) (PEG) as solvent and phase-transfer catalysts (PTCs) in the synthesis of carbamates from CO_2, amines and alkyl halides in the presence of K_2CO_3. They also reported a positive effect of PEG in increasing the selectivity, avoiding the alkylation of both the amine and the carbamate.[136]

Alcohols have been advantageously used in place of alkyl halides, which is equivalent to considering $X = OH$ in Scheme 13.6. This route seems to be more eco-friendly as it implies no halogen atoms in the process and gives only water as co-product of carbamates. Unfortunately, this reaction suffers from thermodynamic and kinetic limitations, thus requiring, fairly drastic conditions. Several strategies have been investigated in order to overcome these problems, such as the use of 1,2-amino alcohols or their formal dehydrated form, aziridines, to produce cyclic oxazolidin-2-ones in sc-CO_2.[106] Some recent work has been reported on the synthesis of methyl benzylcarbamate in high yields from benzylamine, methanol and CO_2 catalysed by CeO_2, suggesting the formation of dimethyl carbonate as intermediate.[137] Another recent paper, by Li *et al.*,[138] described the synthesis of methyl-, ethyl- and *n*-butyl carbamates from CO_2, ammonia, and the relevant alcohols catalysed by V_2O_5. In this case, despite the optimal selectivity (98%), the yields were lower than 25%.[138]

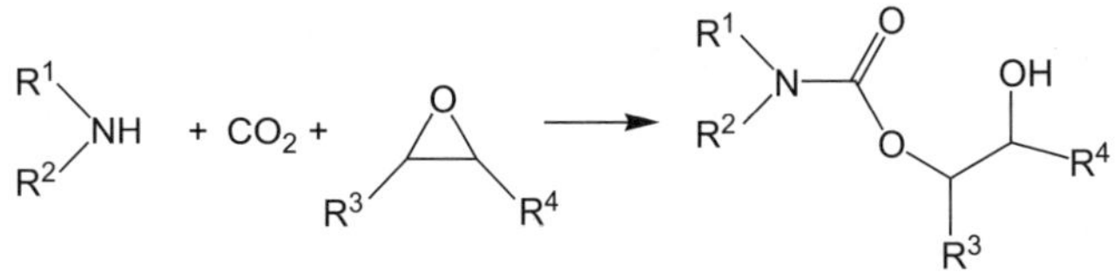

Scheme 13.7 Synthesis of carbamates from CO_2, amines and epoxides

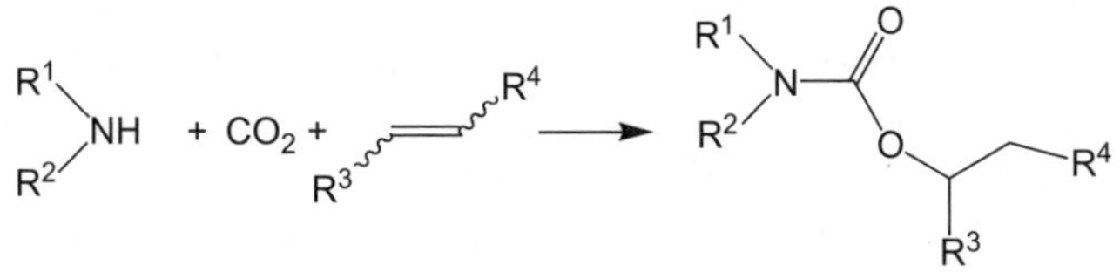

Scheme 13.8 Synthesis of carbamates from CO_2, amines and alkenes.

Even epoxides are suitable substrates for the formation of carbamates by reaction with CO_2 and amines (Scheme 13.7).

In the first example reported by Yoshida and Inoue, $Ti(NMe_2)_4$ was reacted with CO_2 and 1,2-epoxycyclohexane.[139] Other metal amides have been employed in different studies, such as $TiCp(NMe_2)_3$, $W(NMe_2)_6$[140] and $EtZn(NPh_2),Et_2Al(NPh_2)$,[141] and also a variety of different epoxides, such as propylene and styrene oxide,[142] and chloromethyloxirane,[143] producing cyclic carbamates as the final product. α-Haloacylphenones have also been used as oxirane precursors,[144] and oxetanes were successfully employed to produce monocarbamates of 1,3-propanediols.[145]

The reaction of transfer of the carbamate moiety to C–C double and triple bonds has been documented. Actually, only a few reports concern the reaction with olefins (Scheme 13.8). The first was the case of the vinyl ether reported above. Non-activated double bonds require the coordination to a metal centre employed as catalyst. Commonly, Pd-coordinated olefins have been used, as in the case of norbornene, dicyclopentadiene and 1,5-cyclopentadiene, reacted with $PdCl_2$.[146] Pd(0)–phosphine complexes catalyse the formation of allylic carbamates and, more recently, $Pd(PPh_3)_4$ has been used to prepare vinyloxazolidinones. In both cases DBU was used to achieve good yields.[147,148]

1-Alkynes have been used for the preparation of vinyl carbamates (Scheme 13.9) with $Ru_3(CO)_{12}$ as catalyst.[149] Several

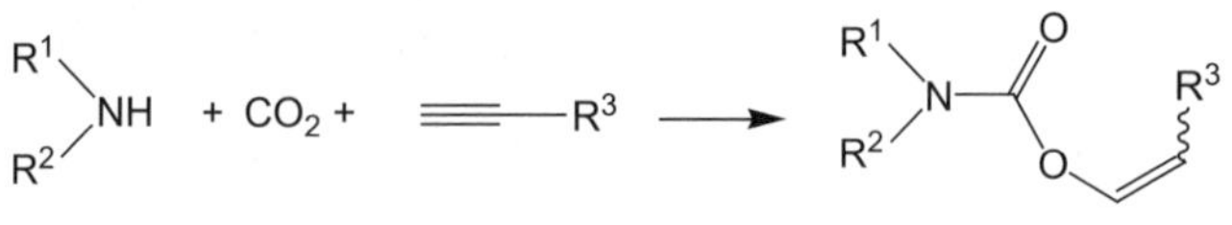

Scheme 13.9 Synthesis of carbamates from CO_2, amines and alkynes.

different mononuclear ruthenium(II) complexes have been used, showing better performance and even higher activity in sc-CO_2 with respect to organic solvents.[150] The mechanism proposed for the Ru-catalysed processes involves an Ru–vinylidene intermediate which promotes the transfer of the carbamate moiety to the vinylidene moiety.[151] Non-ruthenium catalysts for the activation of terminal alkynes have been poorly investigated and only ReBr(CO)$_5$ was reported to catalyse the synthesis of N,N-diethylcarbamates in very good yields.[152]

Propargyl alcohols have been reacted successfully with primary amines and CO_2 to produce cyclic carbamates, also in sc-CO_2, in the presence Cu(I) halides.[153] A recent example was reported by Della Ca' et al. for the guanidine-catalysed synthesis of carbamates and carbonate in sc-CO_2.[154] Propargylamines and CO_2 produce 5-methylene-2-oxazolidinones in high yields in the presence of strong organic bases under mild conditions.[155] Later this reaction was carried out in sc-CO_2 with solid bases (alumina, hydrotalcites) or supported organic bases.[156]

A different approach to the synthesis of carbamates is represented by the electrochemical reduction of CO_2 in conventional solvents[157] or ionic liquids, in the presence of alkyl halides and aromatic or aliphatic amines.[157–159] Alternatively, the base used to promote the carbamate formation from CO_2 can be regenerated electrochemically, reducing O_2 to superoxide.[158]

13.3.1.2 Indirect Synthesis of Carbamates. All the methods described above represent a direct route for the synthesis of carbamates from CO_2 and amines. Nevertheless it is also possible to use an indirect route, reacting carbonates, derived from CO_2, with amines, and Scheme 13.10 shows the reaction of primary amines with dimethyl carbonate (DMC), which can be produced by methanol carboxylation. This strategy represents an alternative green process to the use of phosgene as acylating agent for amines and has attracted the interest of several

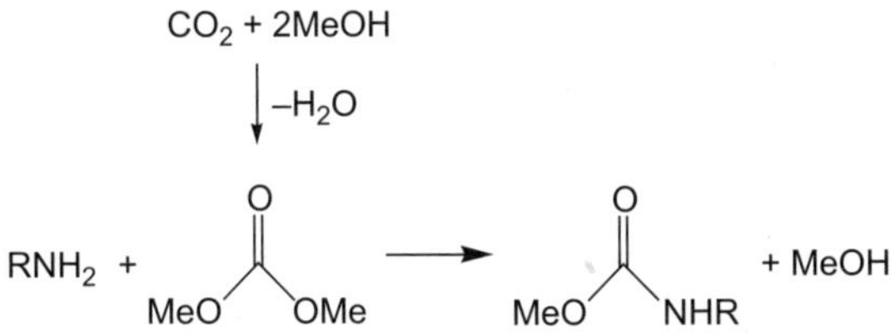

$$CO_2 + 2MeOH$$

$$\downarrow -H_2O$$

RNH$_2$ + [MeO–C(=O)–OMe] → [MeO–C(=O)–NHR] + MeOH

Scheme 13.10 'Indirect' route for the synthesis of carbamates from CO_2 and amines.

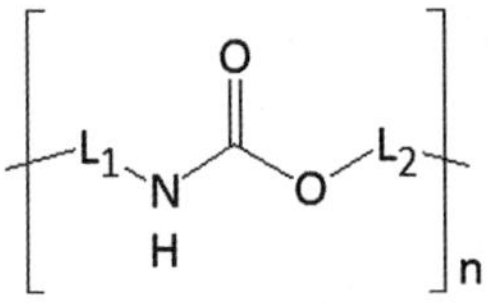

$$\left[-L_1-\underset{H}{N}-\overset{O}{\underset{\|}{C}}-O-L_2- \right]_n$$

Figure 13.9 Polyurethane.

research groups. The use of carbonates for the synthesis of carbamates has recently been reviewed.[122,160]

In the synthesis of urethanes (or carbamates characterised by an –NH$_2$ moiety, H$_2$N–COOR), Guo *et al.*[161] used urea and organic carbonates with an La$_2$O$_3$/SiO$_2$ catalyst in the absence of additional solvents and obtained excellent yields of up to 95%.

13.3.2 Production of Polyurethanes

The carbamate or urethane moiety is the fundamental constituent part of polyurethanes (Figure 13.9). The urethane moieties are connected by different linkers (aliphatic, aromatic, *etc.*) depending on the reactants used.

The first polyurethane, namely Perlon U, was synthesised in 1937 by Bayer by the reaction of 1,6-diisocyanatohexane and 1,4-dihydroxybutane,[162] and until recently the industrial production of polyurethanes was based on diisocyanates and polyols. Depending on the number of different functional groups and structures of the substrates used in the polycondensation, the polymers can be either linear or branched, in a two- or three-dimensional network. Moreover, the stoichiometric ratio between the reagents employed is crucial for the structure of the final product. Factors that drive the polyaddition process are temperature, the catalyst, additional

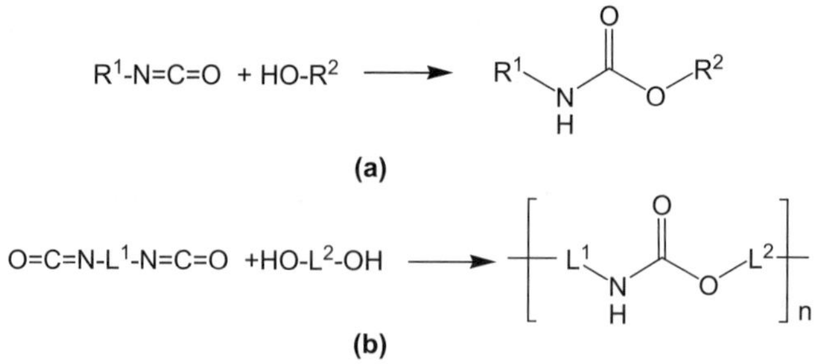

R^1-N=C=O + HO-R^2 $\longrightarrow$ [structure: R^1-NH-C(=O)-O-R^2]

(a)

O=C=N-L^1-N=C=O +HO-L^2-OH $\longrightarrow$ [polymer structure: L^1-NH-C(=O)-O-L^2]$_n$

(b)

Scheme 13.11   (a) Condensation of an isocyanate with an alcohol; (b) synthesis of polyurethanes by condensation of diisocyanates and polyalcohols.

agents to control the polymerisation process and reactor feeding and volumes. A more detailed insight into the practical aspects of polyurethanes is available in a review by Kròl.[163]

As expected, the fundamental reaction in the synthesis of polyurethanes is the condensation between an isocyanate and an alcohol (Scheme 13.11a)

Reaction (a) is reversible and isocyanates can be produced from carbamates of primary amines (see below). By reacting a diisocyanate with a polyol (diols or even a polyether, aliphatic or aromatic, derived for example from the known bisphenol A,[164,165] it is possible to produce polymeric products. It must be emphasised that side reactions can occur, involving in particular the isocyanate functionality that can easily react with both electrophilic and nucleophilic species.[166,167] Isocyanates can also dimerise, trimerise and polymerise, even if the latter process requires alkaline conditions and low temperatures, not used in the polyurethane synthesis. Trimerisation could be also desirable in order to have hyperbranched structures with improved thermal stability.[163]

The most commonly used diisocyanates are the aliphatic 1,6-diisocyanatohexane, used by Bayer, and the aromatic toluene diisocyanate (TDI), a mixture of 2,4- and 2,6-isomers, 4,4'-methylenebis(phenyl isocyanate) (MDI) and 1,5-naphthylene diisocyanate (NDI). In the last decade, cyclic isocyanates such as isophorone diisocyanate (IPDI) and 4,4'-methylenebis(cyclohexyl isocyanate) (HMDI) have also been utilised successfully for the production of polymers less susceptible to photodegradation (Figure 13.10).[163]

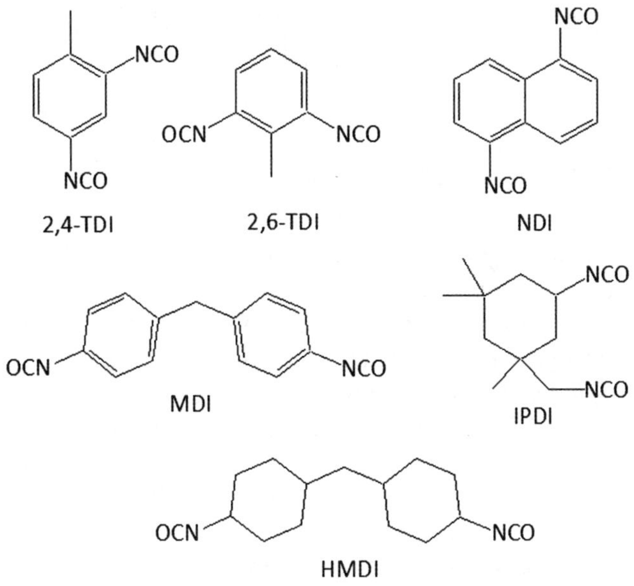

Figure 13.10 Aromatic and cyclic diisocyanates used for polyurethanes synthesis.

The synthesis of isocyanates is a crucial aspect of polyurethane production, as discussed above. They are synthesised using phosgene but, as already mentioned, could be obtained using CO_2, *via* a carbamate intermediate, as illustrated earlier. In fact, isocyanates can be obtained by thermal decomposition of carbamate esters, a reaction that is easier when R^2 is CH_3 [Equation (13.20)]:

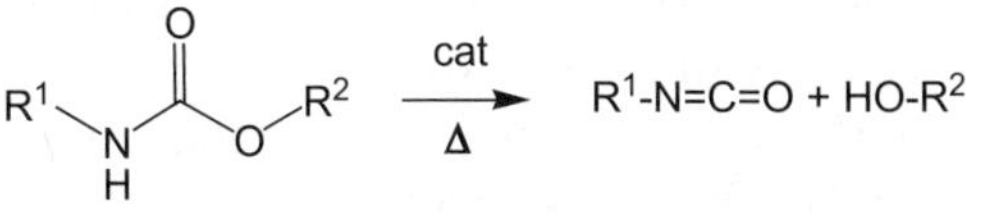

$$R^1-\underset{H}{\underset{|}{N}}-\overset{O}{\overset{\|}{C}}-O-R^2 \xrightarrow[\Delta]{cat} R^1-N=C=O + HO-R^2 \qquad (13.20)$$

The process is usually carried out at high temperature in the presence of a powdered metal system as catalyst, such as boron, bismuth, germanium oxide, manganese, molybdenum, tungsten, zinc, zirconium, *etc.*[168–170] In addition, trapping agents for alcohols to avoid recombination with the isocyanate are employed, with the possibility of using milder conditions.[171,172] The thermal decomposition of carbamates is an alternative synthetic route to the phosgenation of amines, obviously if carbamates are produced *via* a phosgene-free approach. Other methods involve the reductive

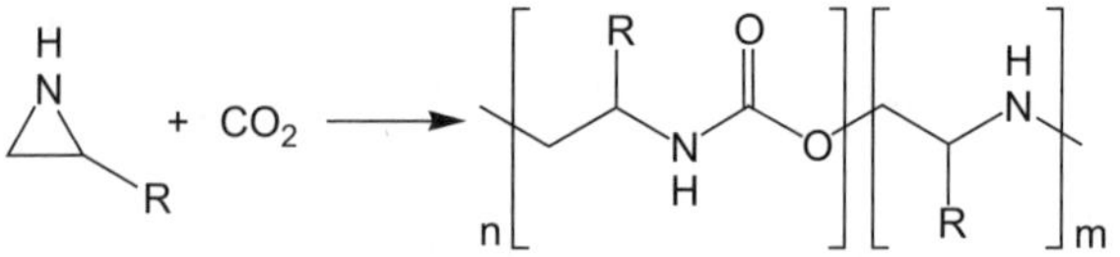

Scheme 13.12   Synthesis of polymers derived from CO_2 and aziridines.

carbonylation of nitroaromatic compounds and the oxidative carbonylation of amines.[173,174]

A direct use of CO_2 for polyurethane synthesis is represented by its copolymerisation with aziridines or azetidines.[175] Both cyclic urethanes and polymers were obtained by reaction of CO_2 and aziridines,[176–178] in the first attempts, and the polymeric products were characterised by a mixed structure derived from both the copolymerisation and the aziridine homopolymerisation (Scheme 13.12).

The urethane content in the copolymer shown in Scheme 13.12 was improved by increasing the CO_2 pressure or working in supercritical conditions and adding *N,N*-dimethylacetamide, as reported by Ihata *et al.*[179] Polyurethanes have a wide range of applications, in particular for the production of elastomeric materials used for protective coatings, for example for automobile seats, for construction, as plastic equipment and as textile fibres.[163] The modulation of the mechanical properties of these materials is a fundamental question and it can be addressed by manipulating the chemical structures of the individual constituents, *i.e.* isocyanates and polyols. Polyethers, for example, give rise to more flexible polymers than polyesters.[180] On the other hand, symmetrical diisocyanates, such as MDI and HMDI, make materials with high mechanical strength.[181] Other post-synthetic modifications can be applied to regulate mechanical properties.

Another huge area of application of polyurethanes is the biomedical field.[182] Polyurethanes are used as biocompatible materials for cardiac valves, regenerative membranes and tissues.[163,183,184] Among the newest and outstanding technologies is the production of polyurethane ionomers that can be easily dispersed in an aqueous phase and used as 'green' varnishes for different substrates,[185] or combined with other polymers to produce water-resistant adhesives and polymeric coatings.[163,186] Other applications of polyurethanes include the production of very hard polymer–ceramic composites, films used for capacitors and materials with interesting opto-electronic properties.[163,187] Finally, the utilisation of biogenic building blocks for synthesising polyurethanes opens the door to more easily biodegradable materials; this area is now creating great interest.[163,188,189]

13.4 THE ENERGY ISSUE IN THE PRODUCTION OF FUELS FROM CO_2

The conversion of CO_2 into fuels requires either hydrogen or energy in the form of heat or electrons. Actually, 90% of hydrogen is produced from fossil carbon, according to the water gas shift reaction [Equation (13.21)] and wet reforming of methane [Equation (13.22)].

$$C + H_2O_{(vap)} \rightarrow CO + H_2, \quad \Delta H^\circ_{298K} = 131 \, kJ \, mol^{-1} \qquad (13.21)$$

$$CH_4 + H_2O_{(vap)} \rightarrow CO + 3H_2, \quad \Delta H^\circ_{298K} = 206 \, kJ \, mol^{-1} \qquad (13.22)$$

Both reactions are strongly endoergic and occur at high temperature (>800 °C). Obviously, such routes cannot be taken into consideration for producing H_2 for CO_2 conversion! A new approach is needed that does not make use of fossil carbon, namely the use of water (water splitting). Various techniques can be considered to this end, such as (i) the use of excess electrical energy for H_2O electrolysis or for the direct reduction of CO_2 in water, (ii) the thermal scission of water using solar energy, (iii) the application of solar energy for photochemical or photoelectrochemical water splitting or for the direct conversion of CO_2 in water, and (iv) the exploitation of thermodynamic cycles.[190]

As an example of the latter, in Table 13.3 the sulfur–iodine cycle is reported: the net reaction of such a cycle is given in Equation (13.23):

$$H_2SO_4 \rightarrow H_2 + SO_2 + O_2 \qquad (13.23)$$

However, hydrogen and O_2 are really produced from water.

The simplest route for converting CO_2 into energy products would be its deoxygenation to produce CO [Equation (13.24)], and

Table 13.3 Sulfur–iodine cycle.

T/K	Reaction
1123	$2H_2SO_4(g) \rightarrow 2SO_2(g) + 2H_2O(g) + O_2(g)$
723	$2HI \rightarrow I_2(g) + H_2(g)$
393	$I_2 + SO_2(aq) + 2H_2O \rightarrow 2HI(aq) + H_2SO_4(aq)$

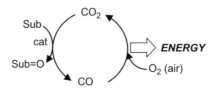

Scheme 13.13 Use of CO_2 as oxidant in a cycle that produces energy. Sub is an O-acceptor substrate with a market size of Mt y^{-1}.

either a purely thermal or a radiative route (UV, IR)[191] can be used.

$$CO_2 \rightarrow CO + \tfrac{1}{2}O_2, \quad \Delta G_{1000\,K} = 190.5\,kJ\,mol^{-1} \qquad (13.24)$$

If energy is produced from fossil carbon, it makes no sense to use such technology. Today we are living in a change of paradigm in the production of energy and new technologies are ready that are not based on the use of fossil carbon and can make the process in Equation (13.24) feasible, *e.g.* the use of concentrators of solar power (CSPs).[192] Temperatures of up to 1500 K can be reached that may be used in CO_2 dissociation. The CO formed can be used for producing energy in direct combustion with oxygen.

It is worth emphasising that when CO_2 is coordinated to a metal system the deoxygenation may occur at room temperature[193] or even just above 14 K in a gas matrix[194] with oxygen transfer to a substrate. Such 'catalytic deoxygenation' would become of interest if a cyclic 'reduction of CO_2–oxidation of CO' were implemented on a large scale with the production of useful materials (*i.e.* Sub = O, formed upon reduction of CO_2 to CO) and energy (during the reoxidation of CO to CO_2) (Scheme 13.13).

Obviously, such application is limited by the marketable volume of the species 'Sub = O' produced from the O-acceptor; their market should be comparable to that of the CO produced. The

CO_2 reduction can be further continued and a number of interesting chemicals can be obtained according to the catalysts and conditions used [Equations (13.25)–(13.29)].

$$CO_2 + H_2 \rightarrow HCOOH \tag{13.25}$$

$$CO_2 + 2H_2 \rightarrow H_2COH_2O \tag{13.26}$$

$$CO_2 + 3H_2 \rightarrow CH_3OH + H_2O \tag{13.27}$$

$$CO_2 + 4H_2 \rightarrow CH_4 + 2H_2O \tag{13.28}$$

$$(n+2)CO_2 + [3(n+2)+1]H_2 \rightarrow CH_3(CH_2)_nCH_3 + 2(n+2)H_2O \tag{13.29}$$

However, if hydrogen becomes available, CO_2 can be reduced to a variety of compounds each having a large market and interesting and unique properties for application in different sectors. Hydrogen can be produced from water using the CSP approach: the water splitting is also an endoergonic reaction ($\Delta G^\circ = 237.2$ kJ mol^{-1}) [Equation (13.30)].

$$H_2O \rightarrow H_2 + \tfrac{1}{2}O_2 \tag{13.30}$$

The CO_2–H_2-derived products given in Equations (13.25)–(13.29) find application on a large scale. For example, formic acid (HCOOH) can be used both in the chemical industry (~ 1 Mt y^{-1}) or as an H_2 vector/storage (several Mt y^{-1}), considering that Equation (13.15) can easily be reversed. Formaldehyde (HCHO) can be used as a comonomer for large-scale polymers (formaldehyde thermo-resins used as insulators, several Mt y^{-1}). Methanol can find use in many different applications (over 100 Mt y^{-1} in a mature market), such as: fuel cells and combustion engines in cars. Methanol is also a very versatile bulk chemical that may be used for producing acetic acid (CH_3COOH), ethene and derived polymers such as LDPE or HDPE, ethers and hydrocarbons. The synthesis of methanol from CO_2 is very well known and mastered at the plant level; the reaction mechanism is well understood and the efficiency of the catalysts, although good, could be further improved. The conversion of CO_2 occurs very selectively (100%) under relatively mild conditions (420 K) with energy recovery and reuse. The same is true for the conversion of methanol into other chemicals or fuels.[195]

Thus, if hydrogen is available it will find a use in recycling fairly large volumes of CO_2 producing chemicals and fuels [Equations (13.25)–(13.29)] or H_2 vectors and reducing the need to extract fossil carbon. The reaction of CO_2 with hydrogen is, thus, a '*multipurpose*' application that can give several positive answers.

13.4.1 CO_2 Conversion into Fuels as a Store of Electrical Energy

Electrochemical and electrocatalytic methods are considered with particular attention these days as they can really contribute to recycling large volumes of CO_2. It is obvious that one cannot produce electricity from fossil fuels expressly for such an application. Selected options that may have a potential application will be considered in the follow-up of this section.

Excess electrical energy (*e.g.* energy produced during the night or holidays when lower power is demanded by the grid) could be conveniently converted into chemical energy (*i.e.* fuels) to be used in cars (substituting fuels produced from fossil carbon) or for regenerating electrical energy during the peak hours. A similar practice is already implemented in several power stations which pump water uphill during the night and use the falling water during the day for electricity production during peak hours.

However, the conversion of electrical energy into fuels by reduction of CO_2 also gives an answer to an old issue: the storage of electrical energy. The solution today is to store it in batteries which have a fairly low energy density by volume or by mass. Figure 13.11 gives an idea of how the storage of energy in batteries compares with chemicals that can be derived from CO_2.

By converting CO_2 into chemicals such as methanol or gasoline, we would increase the energy density by a factor ranging from 10 to even 100, depending on the kind of battery considered.

Another possible approach is to use perennial energy, such as wind power, to run an electrolyser. Figure 13.12 shows how one can couple a wind tower with CO_2 reduction.

Today, very often in many areas wind towers are not in operation because the grid is saturated and cannot accept further electrical energy. Why not keep them in operation permanently and use the excess electrical energy they generate for the reduction of CO_2

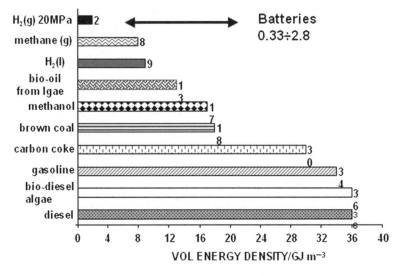

Figure 13.11 Volume energy density of different classes of chemicals.

Figure 13.12 How to couple wind energy with CO_2 reduction.

produced nearby or, even better, captured from the atmosphere? It is worth emphasising that if non-fossil energy is used for CO_2 entropy reduction (concentration/separation from N_2 and O_2), then atmospheric CO_2 (and water) could conveniently be used for conversion into fuels. In a situation in which an effective electrochemical conversion of CO_2 and H_2O is identified (see below), the production of fuels from CO_2 becomes of great interest from both the storage of energy and the recycling of carbon points of view.

Considering that electrolysers may have an efficiency of 70–80% and the selectivity of conversion of H_2 and CO_2 into methanol is close to 100%, despite the loss of energy for the compression of H_2 (~ 3.6 kWh kg^{-1})[196] to the pressure required for the synthesis of methanol (3.0–5.0 MPa), the efficiency of the whole process will anyway remain significantly high. To improve the efficiency of the conversion of electrical energy into chemical energy, the electrolysis of water could be carried out under pressure so as to produce H_2 ready for use; such a process still requires development. Anyway, the reduction of CO_2 with H_2 is a remarkable process to take into consideration for short-term implementation on a large scale.

Merging H_2 production and CO_2 reduction may give the answer to two problems that we have today, namely the storage of H_2 and the reduction of CO_2 emission. It is also worth noting that this approach produces fuels that can be used in a 'business as usual' (BAU) mode. Moreover, the recycling of carbon will reduce the extraction of equivalent (if not higher!) amounts of fossil carbon.

13.4.2   The Direct Reduction of CO_2 to Fuels in Water

An alternative to the production of hydrogen from water is the direct reduction of CO_2 in water to produce C_1 or C_n molecules. The electroreduction of CO_2 in water is strongly dependent on the electrode used. Table 13.4 lists a number of electrodes and the

Table 13.4   Electrodes and products of reduction of CO_2 in water.

Electrode	Products
Cu	C_2H_4 (32–80%)
	C_nH_{2n+2}
Zn, Au, Ag, p-InP, p-GaAs, Pt–Pd–Rh	CO
	CO + HCOOH
RuO_x on conductive diamond, B-doped-C	MeOH, EtOH, $C_nH_{2n+1}OH$

products obtained. The yield is also given in parentheses in one specific case. In any case the electrodes are affected by the electrolysis and severe losses are observed.

In order to improve the efficiency of the process and the life of the electrodes, the *electrocatalytic* approach can be used. In this case, a catalyst is used that avoids the reactions that occur at the electrodes, thus extending their life. Under homogeneous catalysis, soluble catalysts are used that catalyse the reaction without contact with the electrodes. In this way, the electrodes are only involved in the transfer of electrons to and from the catalysts and are not the site where the reactions occur.[197] A particular case is the deposition on the electrode of electrocatalysts which protect the electrode surface from chemical attack.

Aspects of paramount importance in the electrocatalytic reduction of CO_2 to fuels are the kinetics of reactions and the electron transfer processes in which the catalyst is involved; this requires that the catalyst must have energy levels that match the reduction potential of CO_2 to the desired species. Homogeneous catalysts are particularly suitable for adaptation to the different potentials required as the properties of the metal system can be finely tuned through the ligands.

13.5   OTHER MONOMERS PRODUCED FROM CO_2

Materials can be obtained from CO_2 using a synthetic strategy different from those presented above. Below, two examples are discussed, in which CO_2 is first converted into a molecular species that is then used as monomer or comonomer for polymers. Two molecules are discussed: one is urea, which has been made from CO_2 for over 150 years and used so far essentially as an agrochemical; the other is ethane, which is produced on a very large scale (36 Mt y^{-1}) from fossil carbon and used for making polymers and chemicals. The fixation of CO_2 in aquatic biomass (see Chapter 8) used as a source of biogenic materials is also mentioned.

13.5.1   Urea

Urea, $H_2NC(O)NH_2$, has been produced from CO_2 since 1869. It is used today largely as a soil additive, for delivering nitrogen to soils poor in N, for the production of vegetable proteins. The global production of urea is around 220 Mt y^{-1}, implying that about 110 Mt y^{-1} of CO_2 are used in its synthesis [Equation (13.31)].

$$2NH_3 + CO_2 \rightarrow H_2NC(O)NH_2 + H_2O \qquad (13.31)$$

Owing to the expansion of the use of urea in developing countries, a growth of production of the order of 7% per year is predicted over the next 20 years. In addition to such utilisation that follows a BAU trend, urea has, for the past 40 years, been known as a ligand for metal systems,[198a] and as such is currently attracting attention as a chemical reagent.[84]

In addition, urea has been used as monomer for oligomers and polymers and urea-derived materials have also been used for many years as flame retardants.[198b] Urea polymers with ethanedial, formaldehyde and propanal are used as coatings for cardboard in food storage.[198c] Polymers are used as pervaporative membranes,[198d] and polymers with formaldehyde (Polynoxylin) are used as non-toxic flame-resistant agents in hot-press moulding. Another use of urea polymers is as slow urea-releasing agents in soil.

13.5.2 Olefins

Ethene, $CH_2 = CH_2$, has for a very long time been produced by oil cracking. It is a major product of the petrochemical industry with a production rate of 36 Mt y^{-1}. Ethene is a key platform molecule for the chemical and polymers industry. Interestingly, ethene can be produced from CO_2 by electrochemical reduction in water using copper electrodes. The actual selectivity is around 80%. Should this limit be increased, the production of ethane from CO_2 could become economically and environmentally important. Several materials can be derived from ethene, such as low-density polyethylene (LDPET) and high-density polyethylene (HDPET) and other polymers in which ethene is a comonomer.

13.5.3 Enhanced Fixation into Aquatic Biomass

The fixation of CO_2 into aquatic biomass (microalgae) under non-natural conditions (concentration of CO_2 in the gaseous phase up to 150 times higher than natural conditions) makes the production of such biomass attractive for CO_2-enhanced fixation. The biomass can in turn be used for the production of fuels, chemicals and materials. This aspect is discussed in Chapter 8.

13.6 PRODUCTION OF INORGANIC CARBONATES

The carbonation of natural silicate minerals [Equations (30 and 31)] or industrial slag (with an estimated potential of sequestration of 180 Mt y^{-1} of CO_2) or mine tailings rich in Group 2 metal oxides, such as magnesium oxide (MgO) and calcium oxide (CaO), is an interesting option for long term storage of CO_2. The risk of leakage of CO_2 from carbonates is practically close to zero. Minerals such as serpentine and olivine contain high concentrations of MgO, while pyroxenes and amphiboles are a potential source for both CaO and MgO. Carbonation of serpentine [Equation (13.32)] and wollastonite [Equation (13.33)] represent interesting routes to long-term sequestration of CO_2. Such an option is appealing for those countries that do not have suitable sites for underground storage and are far away from deep seas and aquifers.[199–201]

$$Mg_3Si_2O_5(OH)_4(s) + 3CO_2(g) \rightarrow 3MgCO_3(s) + 2SiO_2(s) + 2H_2O$$

$$(13.32)$$

$$CaSiO_3(s) + CO_2(g) \rightarrow CaCO_3(s) + SiO_2(s) \qquad (13.33)$$

An important feature of carbonation of natural rocky materials is the solid mining and handling. A 1 t amount of CO_2 will require 2.5–3 t of magnesium silicate mineral: for CO_2 derived from the combustion of fossil carbon, this implies ~ 8 t of mineral per tonne of coal. Altogether, this will result in a mining activity similar to typical commercial mining of coal or metal-containing ore. As mentioned above, the carbonation reaction of such basic oxides is exothermic. The use of the natural rocky materials as such requires a very long time for the reaction to occur; therefore, in order to shorten this time, it is wise to extract the basic component that has to be reacted. The production of Mg or Ca hydroxides/oxides from silicates is an endoergonic process, while the reaction of the hydroxides with CO_2 releases energy that should be recovered if the whole process must feature a quasi-zero energy input.

An alternative to such processes is the carbonation of slag from the iron and steel sector. Slag by-products need to be processed in a wet, aqueous process that involves extraction of calcium followed by carbonation with CO_2 and recovery of the solvent. In this case, the total energy balance is strongly influenced by the solvent (water) that needs to be processed before discharge. Working in a cycle is the best option that allows the reuse of water. In all cases, it is necessary to use flue gas with the minimum cleaning treatment (elimination of excess NO_x and SO_x).

Slag by-products from iron- and steelmaking may produce valuable precipitated calcium carbonate (PCC). Such carbonates have special applications and have an economic value. In addition to the above industrial sectors, the pulp and paper industry may also be a provider of waste or by-products rich in basic oxides that can be converted to added-value materials.

The key issues in such an approach are (i) the development of low-energy and low-environmental impact extraction or activation of the reactive component MgO or CaO from silicate mineral and (ii) speeding up the carbonation kinetics of basic oxide produced in whatever way.

Technical solutions to perform the carbonation process on a large scale have been intensively searched for since the mid-1990s. Whereas most early and subsequent research was concentrated on methods using aqueous solutions, most recent approaches address gas–solid methods. The latter require higher temperatures than the former [say 770 K, where the rate of the gas–solid carbonation using MgO or $Mg(OH)_2$ appears to become significant], which all together are more energy demanding owing to the use of water and chemicals for making available the Group 2 cations Mg^{2+} or Ca^{2+} for carbonation.

In the open literature, hardly any data can be found on the kinetics of the carbonation of magnesium silicates or oxides and most of the reports from the late 1990s are qualitative. Furthermore, even under conditions of 34 MPa and 780 K and after several hours, a significant fraction of the sub-millimetre size particles were reported not to have carbonated. New technologies such as the high-pressure, high-temperature fluidised-bed operation at 880 K and 10 MPa seem to improve the kinetics. Some old and more recent data are summarised in Table 13.5.

The wet route based on the use of solvents/reagents that may dissolve cations which are then precipitated raises the problem of handling huge volumes of liquids/reagents that require *ad hoc* treatment and must be recycled.

Table 13.5 Magnesium–silicate mineral carbonation.

Mineral	Temperature/K	Pressure/MPa	Particle size/m	Time/h	Conversion/%
Serpentinite	773	34	50–100	2	25–30
Serpentinite	423–473	5.7–30	75–125	3.5	<2
$Mg(OH)_2$	838	5.2	20	0.5	90
$Mg(OH)_2$	648	5.2	50–100	12	16.5
MgO from $Mg(OH)_2$	473	0.12	~20	3	<5

An issue is the use of the carbonates produced. Their use depends on their purity, which strongly depends on the source of Group 2 cations. Slag or mining retailing can be accompanied by a variety of other elements difficult to separate from Ca or Mg that will give a low quality of the carbonates produced, preventing their use in, for example, the paper industry, which requires pure carbonates.[199] Special PCC had a price of €100 per tonne in 2096[199] and could be obtained if the Group 2 cations are extracted from the liquid solution after dissolution of the slag or natural minerals, but this will increase the cost and the impact of the process. Slag-derived impure carbonates could be used as building materials (bricks) or recycled to the steel or cement industry, saving natural carbonates. In general, such products also find an application as flame retardants. Finely divided $MgCO_3$ could be also used as a soil additive. Finally, Ca and Mg carbonates can be returned to natural sites, but this may be opposed by local groups who see it as a blight on the landscape.

13.7 CONCLUSION

Carbon dioxide can be used in a number of added-value applications for the production of chemicals, fuels and materials that store the entire CO_2 molecule or its reduced forms. Materials have a special interest during their lifetime; while all molecular compounds made out of CO_2 will eventually release CO_2, many materials can be made that do not release CO_2.

The most useful materials are polycarbonates, polyurethanes and inorganic carbonates. The polycarbonates have an ever growing market, which at present is rated at >4 Mt y^{-1}. The market for polyurethanes is of comparable size and their application range is growing. Inorganic carbonates have the greatest potential for new applications, but are the most difficult to make owing to the complex process of making available Group 2 cations for CO_2 fixation. Should a process be found that makes it easy to extract Group 2 metal oxides from natural sources or slag and the carbonation process, the 'inorganication' of CO_2 would acquire an important role and significant volumes of CO_2 could be recycled as inorganic carbonates.

REFERENCES

1. M. Aresta (ed.), *Carbon Dioxide Recovery and Utilisation*, Kluwer, Dordrecht, 2003.
2. B. Metz, O. Davidson, H. de Coninck, M. Loos and L. Meyer (eds), *IPCC Special Report on CO$_2$ Capture and Storage*, Cambridge University Press, Cambridge, 2005.
3. M. Aresta (ed.), *Carbon Dioxide as Chemical Feedstock*, Wiley-VCH, Weinheim, 2010.
4. M. Aresta, in *Developments and Innovation in Carbon Dioxide (CO$_2$) Capture and Storage Technology*, ed. M. M. Maroto-Valer, Woodhead Publishing, Cambridge, 2009, **2**, 377–410.
5. W. McGhee and D. Riley, *J. Org. Chem*, 1995, **60**, 6205.
6. Y. R. Jorapur and D. Y. Chi, *J. Org. Chem.*, 2005, **70**, 10774.
7. J. C. Choi, L. N. He, H. Yasuda and T. Sakakura, *Green Chem.*, 2002, **4**, 230.
8. N.S. Isaacs, B. O'Sullivan and C. Verhaelen, *Tetrahedron*, 1999, **55**, 11949.
9. (a) T. Sakakura, Y. Saito, J. C. Choi, T. Masuda, T. Sako and T. J. Oriyama, *Org. Chem.*, 1999, **64**, 4506; (b) T. Sakakura, Y. Saito, J. C. Choi and T. Sako, *Polyhedron*, 2000, **19**, 573.
10. T. Sakakura, Y. Saito, M. Okano, J. C. Choi and T. J. Sako, *Org. Chem.*, 1998, **63**, 7095.
11. M. Honda, S. Kuno, N. Begum, K. Fujimoto, K. Suzuki, Y. Nakagawa and K. Tomishige, *Appl. Catal. A: Gen.*, 2010, **384**, 165.
12. M. Aresta, A. Dibenedetto, E. Fracchiolla, P. Giannoccaro, C. Pastore, I. Papai and G. Schubert, *J. Org. Chem.*, 2005, **70**(16), 6177.
13. (a) M. Aresta, A. Dibenedetto and P. Stufano, *Italian Patent*, MI2009A001221, 2009; (b) M. Aresta, A. Dibenedetto,

P. Stufano, M. Aresta, S. Maggi, I. Pàpai, T. A. Rokob and B. Gabriele, *Dalton Trans.*, 2010, **39**, 6985.

14. (a) D. Ballivet-Tkatchenko and A. Dibenedetto, in *Carbon Dioxide as Chemical Feedstock*, ed. M. Aresta, Wiley-VCH, Weinheim, 2010, p. 169; (b) E. Leini, P. Maki-Arvela, V. Eta, D. Yu. Murzin, T. Salmi and J. P. Mikkola, *Appl. Catal. A: Gen.*, 2010, **383**, 1; (c) N. Keiler, G. Rebmann and V. Keller, *J. Mol. Catal. A: Chem.*, 2010, **317**(1–2), 1.

15. N. Yamazaki, S. Nakahama and F. Higashi, *Ind. Eng. Chem. Prod. Res. Dev.*, 1979, **18**, 249.

16. J. Kizlink and I. Pastucha, *Collect. Czech. Chem. Commun.*, 1994, **59**, 2116.

17. M. Aresta, A. Dibenedetto and C. Pastore, *Inorg. Chem.*, 2003, **42**, 3256.

18. (a) D. Ballivet-Tkatchenko, J. H. Z. dos Santos, K. Philippot and S. Vasireddy, *C. R. Chim.*, 2011, **14**(7), 780; (b) D. Ballivet-Tkatchenko, O. Dauteau and S. Steizman, *Organometallics*, 2000, **19**(45), 63.

19. K. T. Jung and A. T. Bell, *Top. Catal.*, 2002, **20**(1–4), 97.

20. M. Aresta, A. Dibenedetto, C. Pastore, A. Angelini, B. Aresta and I. Pàpai, *J. Catal.*, 2010, **269**(1), 44.

21. M. Aresta, A. Dibenedetto, C. Pastore, C. Cuocci, B. Aresta, S. Cometa and E. De Giglio, *Catal. Today*, 2008, **137**, 125.

22. (a) K. Tomishige and K. Kunimori, *Appl. Catal. A: Gen.*, 2002, **237**, 103; (b) W. Wang, S. Wang, X. Ma and J. Gong, *Catal. Today*, 2009, **148**, 323.

23. A. Dibenedetto, M. Aresta, A. Angelini, J. Ethirj, C. Pastore and B. M. Aresta, *Chem. Eur. J.*, 2012, in press.

24. (a) M. Aresta, A. Dibenedetto, F. Nocito and C. Pastore, *Inorg. Chim. Acta*, 2008, **361**, 3215; (b) M. Aresta, A. Dibenedetto, F. Nocito, A. Angelini, B. Gabriele and S. De Negri, *Appl. Catal. A: Gen.*, 2010, **387**, 113.

25. (a) B. Fan, J. Zhang, R. Li and W. Fan, *Catal. Lett.* 2008, **121**, 297; (b) B. Fan, H. Li, W. Fan, J. Zhang and R. Li, *Appl. Catal. A: Gen.*, 2010, **372**, 94. (c) M. P.Kalhor, H. Chermette and D. Ballivet-Tkatchenko, *Polyhedron*, 2012, **32**(1), 73.

26. E. Leino, P. Mäki-Arvela, K. Eränen, M. Tenho, D. Y. Murzin, T. Salmi and J. P. Mikkola, *Chem. Eng. J*, 2011, **176–177**, 124.

27. Z. F. Zhang, Z. W. Liu, J. Lu and Z. T. Lu, *Ind. Eng. Chem. Res.*, 2011, **50**, 1981.
28. V. Eta, P. Mäki-Arvela, E. Salminen, T. Salmi, D. Yu., Murzin and J. P. Mikkola, *Catal. Lett.*, 2011, **141**(9), 1254.
29. J. Bian, X. W. Wei, L. Wang and Z. P. Guan, *Chin. Chem. Lett.*, 2011, **22**, 57.
30. A. Aouissi, A. W. Apbiett, Z. A. Al-Othman and A. Al-Amro, *Transition Met. Chem.*, 2010, **35**, 927.
31. (a) T. Keller, J. Holtbruegge and A. Górak, *Chem. Eng. J.*, 2012, **180**(15), 309; (b) T. Keller, J. Holtbruegge, A. Niesbacht and A. Górak, *Ind. Eng. Chem. Res.*, 2011, **50**(19), 11073.
32. (a) P. Qiu, L.Wang, X. Jiang and B. Yang, *Energy Fuels*, 2012, **26**, 1254; (b) L. Wang, Y. Wang, S. Liu, L. Lu, X. Ma and Y. Deng, *Catal. Commun.*, 2011, **16**(1), 45.
33. M. Aresta, A. Dibenedetto, C. Devita, O. A. Bourova and O. N. Chupakhin, *Stud. Surf. Sci. Catal.*, 2004, **153**, 213.
34. J. Zhang, F. Wang, W. Wei, F. Xiao and Y. Sun, *Korean J. Chem. Eng.*, 2010, **27**(6), 1744.
35. H. Wang, M. Wang, W. Zhao, W. Wei and Y. Sun, *React. Kinet. Mech. Catal.*, 2010, **99**, 381.
36. D. Wanga, X. Zhang, Y. Gao, F. Xiao, W. Wei and Y. Sun, *Catal. Commun.*, 2010, **11**, 430.
37. D. Wang, X. Zhang, W. Zhao, W. Peng, N. Zhao, F. Xiao, W. Wei and Y. Sun, *J. Phys. Chem. Solids*, 2010, **71**, 427.
38. M. A. Pacheco and C. L. Marshall, *Energy Fuels*, 1997, **11**, 2.
39. S. J. Ainsworth, *Chem. Eng. News*, 1992, **70**, 9.
40. K. Weissermel and H. J. Arpe,*Industrial Organic Chemistry*, 3rd edn, VCH, Weinheim, 1997, p. 162.
41. (a) G. W. Coates and D. R. Moore, *Angew. Chem. Int. Ed.*, 2004, **43**, 6618; (b) D. J. Darensbourg, *Chem. Rev.*, 2007, **107**, 2388; (c) D. J. Darensbourg, R. M. Mackiewicz, A. L. Phelps and D. R. Billodeaux, *Acc. Chem. Res.*, 2004, **37**, 836.
42. N. Limura, M. Takagi, H. J. Iwane and H. Ookago, *Kokai Tokkyo Koho, Japanese Patent*. 07267944, 1995.
43. K. Inoe and H. Oobkuko, *Kokai Tokkyo Koho, Japanese Patent*, 07206846, 1995.
44. K. Inoe and H. Oobkubo, *Kokai Tokkyo Koho, Japanese Patent*, 07206847, 1995.

45. K. Inoe and H. Oobkubo, *Kokai Tokkyo Koho, Japanese Patent*, 07206848, 1995.
46. M. Inaba, K. Hasegawa and H. Nagaoka, *Kokai Tokkyo Koho, Japanese Patent*, 09067365, 1997.
47. S. Ichikawa and H. Iwane, *Kokai Tokkyo Koho, Japanese Patent*, 09235252, 1997.
48. M. Tojo and S. Fukuoka, *Kokai Tokkyo Koho, Japanese Patent*, 03120270,1991.
49. L. I. Bobyleva, S. I. Kryukov, B. N. Bobylev, A. G.Liakumovich, A. A. Surovstev, O. P. Karpov, R. A. Akhmedyanova and S. A. Koneva, *Russian Patent*, 1781218, 1992.
50. F. J. Mais, H. J. Buysch, C. Mendoza-Frohn and A. Klausener, *European Patent*, 543249, 1993.
51. W. Kuran and T. Listos, *Macromol. Chem. Phys.*, 1994, **195**, 977.
52. T. Sakai, N. Kihara and T. Endo, *Macromolecules*, 1995, **28**, 4701.
53. T. Sakai, Y. Tsutsumi and T. Ema, *Green Chem.*, 2008, **10**, 337.
54. M. Sone, T. Sako and C. Kamisawa, *Kokai Tokkyo Koho, Japanese Patent*, 11335372, 1999.
55. D. J. Darensbourg and M. W. Holtcamp, *Coord. Chem. Rev.*, 1996, **153**, 155.
56. E. T. Marquis and J. R. Sanderson, *US Patent*, 5 283 365, 1994.
57. Y. Li, X. Q. Zhao and Y. J. Wang, *Appl. Catal. A: Gen.*, 2005, **279**, 205.
58. X. Zhang, W. Wei and Y. Sun, presented at the International Conference on Carbon Dioxide Utilisation – ICCDU VIII, Oslo, 2005.
59. T. Yano, H. Matsui, T. Koike, H. Ishiguro, H. Fujihara, M. Yoshihara and T. Maeshima, *Chem. Commun.*, 1997, 1129.
60. K. Yamaguchi, K. Ebitani, T. Yoshida, H. Yoshida and K. Kaneda, *J. Am. Chem. Soc.*, 1999, **121**, 4526.
61. M. Aresta and A. Dibenedetto, presented at the 221st ACS National Meeting, San Diego, 2001, Abstract 220.
62. M. Aresta, A. Dibenedetto, L. Gianfrate and C. Pastore, *J. Mol. Catal. A: Gen.*, 2003, **245**, 204.
63. M. Aresta, A. Dibenedetto, L. Gianfrate and C. Pastore, *Appl. Catal. A: Gen*, 2003, **255**, 5.

64. J. R. Machac Jr, E. T. Marquis and S. A. Woodrum, *US Patent*, 6 395 103, 2002.

65. M. G. Clerici, G. Belussi and U. Romano, *J. Catal.*, 1991, **129**, 159.

66. (a) D. Ballivet-Tkatchenko and A. Dibenedetto, in *Carbon Dioxide as Chemical Feedstock*, ed. M. Aresta, Wiley-VCH, Weinheim, 2010, p. 169; (b) A. Decortes, A. M. Castilla and A.W. Klei, *Angew. Chem. Int. Ed.*, 2010, **49**, 9822; (c) S. N. Riduan and Y. Zhang, *Dalton Trans.*, 2010, **39**, 3347.

67. (a) J. Song, B. Zhang, T. Jiang, G. Yang and B. Han, *Front. Chem. China*, 2011, **6**, 21; (b) M. Ulusoy, A. Kilic, M. Durgun, Z. Tasci and B. Cetinkaya, *J. Organomet. Chem.*, 2011, **696**, 1372; (c) S. Liang, H. Liu, T. Jiang, J. Song, G. Yang and B. Han, *Chem. Commun.*, 2011, **47**, 2131; (d) A. Buchard, M. R. Kember, K. G. Sandeman and C. K. Williams, *Chem. Commun.*, 2011, **47**, 212; (e) A. Shibata, I. Mitani, A. Imakuni and A. Baba, *Tetrahedron Lett.*, 2011, **52**, 721; (f) J. E. Dengler, M. W. Lehenmeier, S. Klaus, C. E. Anderson, E. Herdtweck and B. Riege, *Eur. J. Inorg. Chem.*, 2011, 336; (g) A. Kilic, M. Ulusoy, M. Durgun, Z. Tasci, I. Yilmaz and B. Cetinkaya, *Appl. Organomet. Chem*, 2010, **24**, 446; (h) M. Ulusoy, O. Sahin, A. Kilic and O. Buyukgungor, *Catal. Lett.*, 2010, **141**, 717.

68. J. L. Wang, C. X. Miao, X. Y. Dou, J. Gao and L. N. He, *Curr. Org. Chem.*, 2011, **15**(5), 621.

69. W. Dai, S. Luo, S. Yin and C. Au, *Front. Chem. Eng. China*, 2010, **4**(2), 163.

70. W. L. Dai, S. L. Luo, S. F. Yin and C. T. Au, *Appl. Catal. A: Gen.*, 2009, **366**(1), 2.

71. H. Kawanami and Y. Ikushima, *Chem. Commun.*, 2000, 2089.

72. D. Jairton, F. D. S. Roberto and A. Z. S. Paulo, *Chem. Rev.*, 2002, **102**, 3667.

73. H. Kawanami, A. Sasaki, K. Matsui and Y. Ikushima, *Chem. Commun.*, 2003, 896.

74. S. E. Jacobson, *European Patent Application*, 117147, 1984.

75. M. Aresta, E. Quaranta and A. Ciccarese, *J. Mol. Catal.*, 1987, **355**, 41.

76. M. Aresta and A. Dibenedetto, *J. Mol. Catal.*, 2002, **399**, 182.

77. M. Aresta, A. Ciccarese and E. Quaranta, *C1 Mol. Chem.*, 1985, **1**, 267.

78. M. Aresta, E. Quaranta and I. Tommasi, *New J. Chem.*, 1994, **18**, 133.
79. M. Aresta, A. Dibenedetto and I. Tommasi, *Eur. J. Inorg. Chem.*, 2001, 1801.
80. M. Aresta, A. Dibenedetto and I. Tommasi, *Appl. Organomet. Chem.*, 2000, **14**, 799.
81. J. San, L. Liang, J. Sun, Y. Jiang, K. Lin, X. Xu and R. Wang, *Catal. Surv. Asia*, 2011, **15**, 49.
82. M. Aresta, I. Tommasi, A. Dibenedetto, M. Fouassier and J. Mascetti, *Inorg. Chem.*, 2002, **330**(1), 63.
83. A. Dibenedetto, M. Aresta, C. Fragale, M. Distaso, C. Pastore, A. M. Venezia, C. Liu and M. Zhang, *Catal. Today*, 2008, **137**, 44.
84. (a) M. Aresta, A. Dibenedetto, F. Nocito and C. Ferragina, *J. Catal.*, 2009, **268**, 106; (b) A. Dibenedetto, A. Angelini, M. Aresta, J. Ethiraj, C. Fragale and F. Nocito, *Tetrahedron*, 2011, **67**, 1308.
85. (a) H. Sugimoto and Y Inoue, *J. Polym. Sci. A: Polym. Chem.*, 2004, **42**, 5561; (b) H. Sugimoto, H. Ohtsuka and S. Inoue, *Stud. Surf. Sci. Catal.*, 2004, **153**, 243.
86. M. Super, E. Berluche, C. Costello and E. Beckman, *Macromolecules*, 1997, **30**, 368.
87. S. Inoue, in *Carbon Dioxide as a Source of Carbon: Chemical and Biochemical Uses. NATO ASI Series, Ser. C*, Vol. 206, ed. M. Aresta and G. Forti, Reidel, Dordrecht, 1987, p. 331.
88. A. Rokicki and W. Kuran, *J. Macromol. Sci., Rev. Macromol. Chem.*, 1981, **135**, C21.
89. D. J. Darensbourg, J. R. Andreatta and A. I. Moncada, in*Carbon Dioxide as Chemical Feedstock*, ed. M. Aresta, Wiley-VCH, Weinheim, 2010, p. 213.
90. M.R. Kember, A. Buchard and C.K. Williams, *Chem. Commun.*, 2011, **47**, 141.
91. H. Li and Y. Niu, *Polym. J.*, 2011, **43**, 121.
92. A. Gosh, P. Ramidi, S. Pulla, S. Z. Sullivan, S. L. Collom, Y. Gartia, P. Munshi, A. S. Biris, B. C. Noll and B. C. Berry, *Catal. Lett.*, 2010, **137**, 1.
93. W. M. Ren, X. Zhang, Y. Liu, J. F. Li, H. Wang and X. B. Lee, *Macomolecules*, 2010, **43**, 1396.
94. B. E. Kim, J. K. Varghese, Y. G. Han and B. Y. Lee, *Bull. Korean Chem. Soc.*, 2010, **31**, 829.

95. K. Lee, J. J. Y. Ha, C. Cao, D. W. Park, C. S. Ha and I. Kim, *Catal. Today*, 2009, **148**, 389.

96. D. J. Darensbourg and S. J. Wilson, *J. Am. Chem. Soc.*, 2011, **133**(46), 18610.

97. D. J. Darensbourg and A. I. Moncada, *Macromolecules*, 2010, **43**(14), 5996.

98. F. Suriano, O. Coulembier, J. L. Hedrick and P. Dubois, *Polym. Chem.*, 2011, **2**, 528.

99. M. M. Islam, D. W. Seo, H. H. Jang, Y. D. Lim, K. M. Shin and W. G. Kim, *Macromol. Res.*, 2011, **19**(12), 1278.

100. M. Colonna, C. Berti, E. Binassi, M. Fiorini, S. Sullalti, F. Acquasanta, S. Karanam and D. J. Brunelle, *React. Funct. Polym.*, 2011, **71**(10), 1001.

101. E. Fov, J. B. Farrell and C. L. Higginbotham, *J. Appl. Polym. Sci.*, 2009, **111**(1), 217.

102. L. Wang, B. Xiao, G. Y. Wang and J. Q. Wu, *Sci. China Chem.*, 2011, **54**(9), 1468.

103. R. A. Gross and Z.-Z. Jiang, *US Patent*, 7 951 899, 2011.

104. C. Liu, Z. Jiang, J. Decatur, W. Xie and R. A. Gross, *Macromolecules*, 2011, **44**(6), 1471.

105. Y. Yang, Y. Yu, Y. Zhang, C. Liu, W. Shi and Q. Li, *Process Biochem.*, 2011, **46**(10), 1900.

106. E. Quaranta and M. Aresta, in *Carbon Dioxide as Chemical Feedstock*, ed. M. Aresta, Wiley-VCH, Weinheim, 2010, p. 121.

107. M. Aresta, *Carbon Dioxide Recovery and Utilisation*, Kluwer, Dordrecht, 2003.

108. P. D. Vaidya and E. Y. Kenig, *Chem. Eng. Technol.*, 2007, **30**, 1467.

109. M. Aresta, D. Ballivet-Tkatchenko, D. Belli Dell'Amico, M. C. Bonnet, D. Boschi, F. Calderazzo, R. Faure, L. Labella and F. Marchetti, *Chem. Commun.*, 2000, 1099.

110. (a) M. Aresta and E. Quaranta, in *Proceedings of Internetional Conference on Carbon Dioxide Utilisation*, Bari, 1993, p. 63; (b) D. Belli Dell'Amico, F. Calderazzo, F. Labella, F. Marchetti and G. Pampaloni, *Chem. Rev.*, 2003, **103**, 3857.

111. J. K. Terlouw and H. Schwarz, *Angew. Chem. Int. Ed. Engl.*, 1987, **26**, 805.

112. N. V. Kaminskaia and N. M. Kostic, *Inorg. Chem.*, 1997, **36**, 5917.

113. M. Remko, K. R. Liedl and B. M. Rode, *J. Chem. Soc., Faraday Trans.*, 1993, **89**, 2375.
114. M. H. Jamroz, J. C. Dobrowolski, J. E. Rode and M. Borowiak, *J. Mol. Struct.*, 2002, **618**, 101.
115. A. Dibenedetto, M. Aresta, C. Fragale and M. Narracci, *Green Chem.*, 2002, **4**, 439.
116. M. H. Jamroz, J. C. Dobrowolski and M. A. Borowiak, *J. Mol. Struct.*, 1997, **404**, 105.
117. S. Ray, S. R. Pathak and D. Chaturvedi, *Drugs Future*, 2005, **30**, 161.
118. R. C. Gupta (ed.), *Toxicology of Organophosphate and Carbamate Compounds*, Elsevier Academic Press, Burlington, MA, 2006.
119. T. W. Green and P. G. M. Wuts, *Protective Groups in Organic Synthesis*, Wiley, New York, 2007.
120. (a) C. Wu, H. Cheng, R. Liu, Q. Wang, Y. Hao, Y. Yu and F. Zhao, *Green Chem.*, 2010, **12**, 1811; (b) J. Li, X. G. Guo, L. G. Wang, Y. Z. Ma, Q. H. Zhang, F. Shi and Y. Q. Deng, *Sci. China Chem.*, 2010, **53**, 1534; (c) S. L. Peterson, S. M. Stucka and C. J. Dinsmore, *Org. Lett.*, 2010, **12**, 1340.
121. A. Dibenedetto, M. Aresta and P. Stufano, in *Comprehensive Inorganic Chemistry II*, ed. J. Reedijk, Elsevier, Amsterdam, 2012, in press.
122. D. Chaturvedi, *Tetrahedron*, 2012, **68**, 15.
123. Y. Yoshida and S. Inoue, *Chem. Lett.*, 1977, 1375.
124. (a) Y. Yoshida, S. Ishii and T. Yamashita, *Chem. Lett.*, 1984, 1571; (b) Y. Yoshida, S. Ishii, M. Watanabe and T. Yamashita, *Bull. Chem. Soc. Jpn.*, 1989, **62**, 1534.
125. T. Tsuda, H. Washida, K. Watanabe, M. Miva and T. Saegusa, *J. Chem. Soc., Chem. Commun.*, 1978, 815.
126. Y. Yoshida, S. Ishii, M. Watanabe and T. Yamashita, *Bull. Chem. Soc. Jpn.*, 1989, **62**, 1534.
127. M. Aresta and E. Quaranta, *Tetrahedron*, 1992, **21**, 1515.
128. M. Aresta and E. Quaranta, *Italian Patent*, 1237207, 1993.
129. W. D. McGhee, D. Riley, C. Kevin, Y. Pan and B. Parnas, *J. Org. Chem.*, 1995, **60**, 2820.
130. M. Shi and Y. M. Shen, *Helv. Chim. Acta*, 2001, **84**, 3357.
131. W. D. McGhee and J. J. Talley, *US Patent*, 5 302 717, 1994.
132. K. N. Singh, *Synth. Commun.*, 2007, **37**, 2651.

133. D. Chaturvedi, N. Mishra and V. Mishra, *Chin. Chem. Lett.*, 2006, **17**, 1309.
134. D. Chaturvedi and S. Ray, *Monatsh. Chem.*, 2006, **137**, 459.
135. D. Chaturvedi, A. Kumar and S. Ray, *Synth. Commun.*, 2002, **32**, 2651.
136. D. L. Kong, L. N. He and J. Q. Wang, *Synth. Commun.*, 2011, **41**, 3259.
137. M. Honda, S. Sonehara, H. Yasuda, Y. Nakagawa and K. Tomishige, *Green Chem.*, 2011, **13**, 3406.
138. J. Li, X. Qi, L. Wang, Y. He and Y. Deng, *Catal. Commun.*, 2011, **12**, 1224.
139. Y. Yoshida and S. Inoue, *Bull. Chem. Soc. Jpn.*, 1978, **51**, 559.
140. Y. Yoshida and S. Inoue, *Polym. J.*, 1980, **12**, 763.
141. Y. Yoshida, S. Ishii, A. Kawato, T. Yamashita, M. Iano and S. Inoue, *Bull. Chem. Soc. Jpn.*, 1988, **61**, 2913.
142. (a) Y. Yoshida and S. Inoue, *Chem. Lett.*, 1978, 139; (b) Y. Yoshida and S. Inoue, *J. Chem. Soc., Perkin Trans. 1*, 1979, 3146.
143. T. Asano, N. Saito, S. Ito, K. Hatakeda and T. Toda, *Chem. Lett.*, 1978, 311.
144. T. Toda, *Chem. Lett.*, 1977, 957.
145. S. Ishii, M. Zhou, Y. Yoshida and H. Noguchi, *Synth. Commun.*, 1999, **29**, 3207.
146. W. D. McGhee and D. P. Riley, *Organometallics*, 1992, **11**, 900.
147. W. D. McGhee, D. P. Riley, M. E. Christ and K. M. Christ, *Organometallics*, 1993, **12**, 1429.
148. M. Yoshida, Y. Ohsawa, K. Sugimoto, H. Tokuyama and I. Masataka, *Tetrahedron Lett.*, 2007, **48**, 8678.
149. D. Sasaki and P. H. Dixneuf, *J. Chem. Soc., Chem. Commun.*, 1986, 790.
150. M. Rohr, C. Geyer, R. Wandeler, M. S. Schneider, E. F. Murohy and A. Baiker, *Green Chem.*, 2001, **3**, 123.
151. R. Mahè, Y. Sasaki, C. Bruneau and P. H. Dixneuf, *J. Org. Chem.*, 1989, **54**, 1518.
152. J. L. Jiang and R. Hua, *Tetrahedron Lett.*, 2006, **47**, 953.
153. H. Jiang, J. Zhao and A. Wang, *Synthesis*, 2008, 763.
154. N. Della Ca', B. Gabriele, G. Ruffolo, L. Veltri, T. Zanetta and M. Costa, *Adv. Synth. Catal.*, 2011, **353**, 133.

155. M. Costa, G. P. Chiusoli, D. Taffurelli and G. Dalmonego, *J. Chem. Soc., Perkin Trans. 1*, 1998, 1541.
156. R. Maggi, C. Bertolotti, E. Orlandini, C. Oro, G. Sartori and M. Selva, *Tetrahedron Lett.*, 2007, **48**, 2131.
157. M. Feroci, M. Orsini, L. Rossi, G. Sotgiu and A. Inesi, *J. Org. Chem.*, 2007, **72**, 200.
158. M. Feroci, M. A. Casadei, M. Orsini, L. Palòombi and A. Inesi, *J. Org. Chem.*, 2003, **68**, 1548.
159. S. Ikeda, T. Takagi and K. Ito, *Bull. Chem. Soc. Jpn.*, 1987, **60**, 2517.
160. M. Carafa and E. Quaranta, *Mini-Rev. Org. Chem.*, 2009, **6**, 168.
161. X. Guo, J. Shang, J. Li, L. Wang, Y. Ma, F. Shi and Y. Deng, *Synth. Commun.*, 2011, **41**, 1102.
162. O. Bayer, W. Siefken, H. Rinke, L. Orthner and H. Schild, *German Patent*, DRP 728981, 1937.
163. P. Kròl, *Prog. Mater. Sci.*, 2007, **52**, 915.
164. H. Szewczyk, E. Dziwinski and P. Kròl, *J. Chromatogr.*, 1988, **446**, 109.
165. J. Pielichowski and P. Kròl, *Polimery*, 1988, **33**, 182.
166. J. H. Saunders and K. C. Frisch, in *Polyurethanes, Chemistry and Technology, Part I. Chemistry*, ed. F. L. Malabar, Imterscience, New York, 1983.
167. A. Kaji, Y. Arimatsu and M. Murano, *J. Appl. Polym. Sci. Part A: Polym. Chem.*, 1992, **30**, 287.
168. S. Okuda, *Japanese Patent*, JP57158746, 1983; *Chem. Abstr.*, 1983, **98**, 144386b.
169. T. Yagii, *US Patent*, 5 789 614, 1998.
170. L. Gerhard, *German Patent*, 19541384, 1997.
171. V. L. K. Valli and H. Alper, *J. Org. Chem.*, 1995, **60**, 257.
172. D. C. D. Butle and H. Alper, *Chem. Commun.*, 1998, 2575.
173. F. Paul, *Coord. Chem. Rev.*, 2000, **203**, 269.
174. (a) P. Giannoccaro, A. Dibenedetto, M. Gargano, E. Quaranta and M. Aresta, *Organometallics*, 2008, **27**, 967; (b) P. Giannoccaro, D. Cornacchia, S. D'Oronzo, E. Mesto, E. Quaranta and M. Aresta, *Organometallics*, 2006, **25**, 2872.
175. D. J. Darensbourg, J. R. Andreatta and A. I. Moncada, in *Carbon Dioxide as Chemical Feedstock*, ed. M. Aresta, Wiley-VCH, Weinheim, 2010, p. 213.
176. K. Soga, S. Hosoda, H. Nakamura and S. Ikeda, *J. Chem. Soc., Chem. Commun.*, 1976, **16**, 617.

177. K. Soga, W. Y. Chiang and S. Ikeda, *J. Polym. Sci. Polym. Chem. Ed.*, 1974, **12**, 121.
178. S. Inoue, *Chemtech*, 1976, 588.
179. O. Ihata, Y. Kayaki and T. Ikariya, *Angew. Chem. Int. Ed.*, 2004, **43**, 717.
180. D. J. Liaw, *J. Appl. Polym. Sci.*, 1997, **66**, 1251.
181. P. Kròl and J. Wojturska, *Polimery*, 2002, **47**, 6.
182. M. Szycher, in *An Introduction to Biomaterials*, ed. J. O. Hollinger, CRC Press, Boca Raton, FL, 2011, p. 281.
183. L. Poussard, F. Burel, J. P. Couvercelle, Y. Merhi, M. Tabrizian and C. Bunel, *Biomaterials*, 2004, **25**, 3473.
184. J. H. Wang and C. H. Yao, *J. Biomed. Mater. Res.*, 2000, **51**, 761.
185. J. P. Santerre and J. L. Brash, *Ind. Eng. Chem. Res.*, 1997, **36**, 1352.
186. P. Kròl, B. Kròl, S. Pikus and K. Skrzypiec, *Colloid Surf. A: Physicochem. Eng. Aspects*, 2005, **259**, 35.
187. M. T. Sebastian and H. Jantunen, *Int. J. Appl. Ceram. Technol.*, 2010, **7**, 415.
188. M. Zhang, Y. H. Zhou, X. H. Yang and L. H. Hu, *Adv. Mat. Res.*, 2011, **250**, 974.
189. M. J.-L. Tschan, E. Brulé, P. Haquette and C. M. Thomas, *Polym. Chem.*, 2012, **3**, 836.
190. L. C. Brown, G. E. Besenbruch, K. R. Scultz, A. C. Marshall, S. K. Showalter, P. S. Pickard and J. F. Funk, *General Atomics Report*, GA-A23944, 2002, p. 1.
191. M. Aresta and G. Forti (eds), *Carbon Dioxide as a Source of Carbon: Chemical and Biochemical Uses. NATO ASI Series, Series C*, Vol. 206, Reidel, Dordrecht, 1987.
192. C. J. Winter, R. L. Sizmann and L. L. Vant-Hull (eds), *Solar Power Plants, Fundamentals, Tecnology, Systems, Economics*, Springer, New York, 1991.
193. M. Aresta, C. F. Nobile, V. G. Albano, E. Forni and M. Manassero, *J. Chem. Soc., Chem. Commun.*, 1975, 636.
194. J. Mascetti, in *Carbon Dioxide as Chemical Feedstock*, ed. M. Aresta, Wiley-VCH, Weinheim, 2010, 55–88.
195. G. A. Olah, A. Goeppert and G. K. S. Prakash (eds), *Beyond Oil and Gas: the Methanol Economy*, Wiley-VCH, Weinheim, 2010.
196. R. Drnevich, presented at the Strategic Initiatives ror Hydrogen Delivery Workshop, Tonawanda, NY, 7 May 2003.

197. E. Barton Cole and A. B. Bocarsly, in *Carbon Dioxide as Chemical Feedstock*, ed. M. Aresta, Wiley-VCH, Weinheim, 2010, 291–316.
198. (a) D. Theophanides and P. D. Hardey, *Cord. Chem. Rev.*, 1987, **76**, 237; (b) A. Basch, B. Nachamowitz, S. Hasenfratz and M. Lewin, *J. Polym. Sci., Polym. Chem. Ed.*, 1979, **17**(1), 39; (c) National Occupational Health and Safety Commission, *Exposure Standards for Atmospheric Contaminants in the Occupational Environment*, 2nd edn, Australian Government Publishing Service, Canberra, 1991; (d) C.-t. Zhao and M. Norberta de Pinho, *Polymer*, 1999, **40**(22), 6089.
199. R. Zevenhoven, S. Eloneva and S. Teir, *Catal. Today*, 2006, **115**, 73.
200. S. Eleneva, *Lic. Tech. Thesis*, Helsinki University of Technology, Espoo, 2008.
201. S. Teir, *PhD (Eng) thesis*, Helsinki University of Technology, Espoo, 2008.

CHAPTER 14

Carbon Dioxide in the Manufacture of Plastics

MATTHEW D. JONES

Department of Chemistry, University of Bath, Claverton Down,
Bath BA2 7AY, UK
Email: m.jones2@bath.ac.uk

14.1 INTRODUCTION

This chapter focuses on the preparation of sustainable 'green' polymers specifically using CO_2 as a resource, which in the modern era is becoming an ever-important area of contemporary polymerization science.[1] It is generally acknowledged that CO_2 is the ultimate C_1 feedstock. Carbon dioxide is the most oxidized form of carbon – and CO_2 is the product of the majority of energy-releasing (combustion) processes. This high oxidation state results in CO_2 being extremely thermodynamically stable and therefore any CO_2 utilization may require significant energy input to initiate a reaction. Not only is CO_2 a sustainable chemical, it is universally acknowledged that CO_2 is the main cause of anthropogenic global warming[2] – with the UK alone emitting over 200 Mt y^{-1} of CO_2 in 2008.[3] However, a significant challenge is not only to use this 'waste' material but to capture it in a viable manner – major

Materials for a Sustainable Future
Edited by Trevor M. Letcher and Janet L. Scott
© The Royal Society of Chemistry 2012
Published by the Royal Society of Chemistry, www.rsc.org

advances in this area are dealt with in Chapters 12 and 13 – and to convert the captured CO_2 to a useful material or 'lock' the CO_2 away in polymeric materials. CO_2 is abundant, nontoxic, biorenewable and inexpensive, hence there is currently a huge desire in both academia and industry to find viable uses for this material.

The utility of renewable feedstocks is of paramount importance in the chemist's armory to achieve Paul Anastas's so-called '*12 Principles of Green Chemistry*', which are summarized as follows:[4]

1. Prevent waste.
2. Be atom efficient.
3. Avoid production of toxic/harmful by-products.
4. Products should be designed efficiently.
5. Avoid the use of solvents where possible.
6. Reduce energy requirements and the environmental impact of any process.
7. Use renewable feedstocks.
8. Avoid unnecessary protection chemistry whenever possible.
9. Catalytic reagents are superior to stoichiometric reagents.
10. The final products are not harmful to the environment.
11. Develop real-time analytical methods to allow the detection of hazardous materials.
12. All materials should minimize the risk of explosion or harm to health/environment.

Without doubt, one of the global challenges for the twenty-first century is the development of sustainable and low-carbon technologies.[5] With the rise in oil prices in recent years (which are predicted to rise even further in coming years), technology that was too costly is now becoming more and more cost-effective and has helped to ignite and rejuvenate research in this area.[6] Concurrently, organizations and governments (for example, SuSChem organization in Europe) have published action plans supporting the use of renewable materials in chemical synthesis.[5a,7] There are various reasons that prevent this technology from being utilized on an industrial scale, one of the main ones being the high cost involved in processing the renewable feedstocks into chemicals – as this technology is competing with well-established and proven technology developed over the last century for the petrochemical industry. The majority of the polymeric materials, which we rely upon as a nation, are based on crude oil-derived monomers and therefore now is the time to address this scenario before it is too late. As such, there has been a flurry of publications and patents in this area in recent years, and this chapter aims to highlight why this has been the case and to emphasise some important catalytic developments in the area.

14.2   POLYMERS

One of the most common polymerisations involving CO_2 is its copolymerisation with epoxides (such as cyclohexene oxide and propylene oxide) to produce polycarbonates) (Figure 14.1).[1b] Polycarbonates prepared *via* this methodology have been proposed as alternatives for commodity applications such as packaging, engineering polymers and elastomers.[1b] Moreover, traditionally these polymers are prepared using phosgene ($COCl_2$), which is highly harmful to health and to the environment, hence there is a tremendous drive to reduce its use and find environmentally benign methods for the production of polycarbonates. This is one of the core principles of the '*green chemistry movement*'. This is typically achieved by the polymerisation of bisphenol-A with $COCl_2$ (Figure 14.2).[8]

For the reasons previously described, this chapter focuses on recent catalytic developments for the conversion of epoxides into

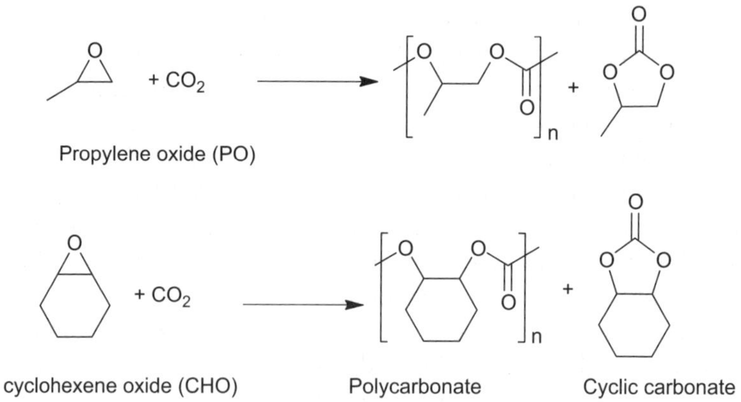

Figure 14.1   Copolymerisation of CO_2 and epoxides (cyclohexene oxide or propylene oxide) to produce a polycarbonate or cyclic carbonate.

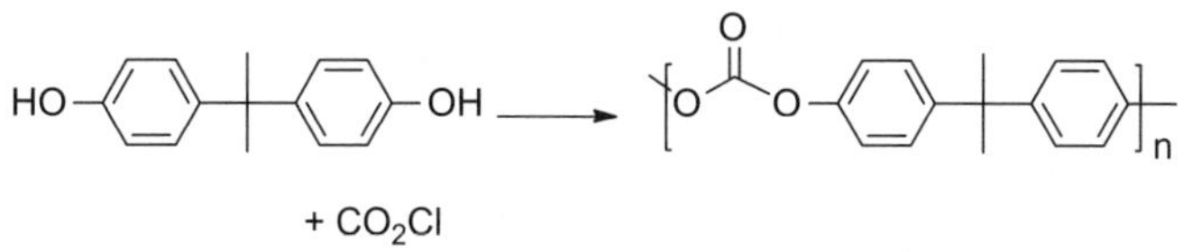

Figure 14.2 Preparation of polycarbonates using bisphenol-A and phosgene.

polycarbonates, a more environmentally benign approach to this industrially important polymeric material.[1b,c] The two must common epoxides encountered in the literature are cyclohexene oxide (CHO) and propylene oxide (PO); however, there is currently a desire to utilise non-petrochemically derived epoxides, and this is discussed further. Furthermore, both of these monomers have at least one chiral centre; therefore, by a judicious choice of the catalyst, it is possible to alter the stereochemistry (and hence properties) of the final polymer. If CHO is utilised as the monomer, the polymer has a high T_g (glass transition temperature) and good tensile strength and it also is degradable.[3] The two main by-products are cyclic carbonates and polyethers; typically, ^{1}H NMR spectroscopy is utilised to determine the selectivity of each polymerisation experiment. It is possible to minimise these by-products by a judicious choice of catalyst and reaction conditions. Cyclic carbonates are an important industrial chemical in their own right, since they were commercialised in the mid-1950s.[9] The current uses of such materials are as polar aprotic solvents, electrolytes for lithium ion batteries and intermediates in a variety of chemical processes.[9b] Furthermore, cyclic carbonates can actually undergo ring-opening polymerisation (a mechanism analogous to that for the polymerisation of lactide) to achieve polycarbonates.[10] For example, Jones and co-workers used Lewis acidic metal complexes to catalyse the polymerisation of 1,3-dioxan-2-one to afford polycarbonates.[11] For completeness, a further method to prepare polycarbonates is *via* the coupling of CO_2 with oxetanes (four-membered cyclic ethers); in certain cases, polymerisation proceeds *via* the formation of a six-membered cyclic carbonate that is subsequently ring opened to generate the polycarbonate.[12] Although not the main thrust of this chapter, it must be noted that there have been many elegant examples, in particular by North and co-workers in the UK, for the conversion of the epoxide into the

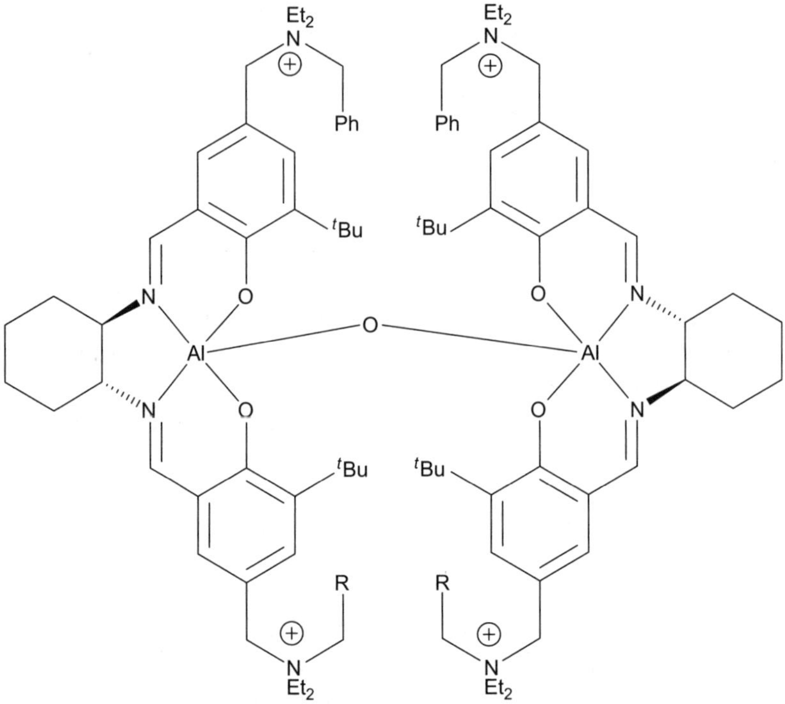

Figure 14.3   Bimetallic Al(III) complex prepared by North and co-workers for the production of cyclic carbonates.[9a] R = Ph or Merrifield resins.

cyclic carbonate. In this work, North pioneered the use of heterogeneous bimetallic aluminium salen complexes (Figure 14.3) for use under flow conditions.[9a,13] The utility of salen ligands is commonplace for such processes. It is hypothesised that such heterogeneous systems can be added to power stations to convert the exhaust gases to cyclic carbonates, hence locking away the CO_2 from the atmosphere. However, the main aim of this chapter is to examine the reaction of CO_2 and epoxides to form polycarbonates (Figure 14.1).

The majority of polymerisations are performed at high pressures (usually around 20 bar CO_2) and at elevated temperatures, typically in the range 60–100 °C.[1b,c,3] However, there is growing body of literature advocating the application of ambient pressure and temperature conditions (this is dependent upon having a highly active catalyst).[3] It is also a common observation that the activity of the

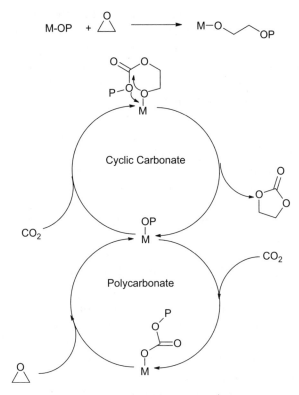

Figure 14.4 Proposed mechanism for the production of polycarbonates from CO_2 and epoxides. The upper cycle also indicates how cyclic carbonates can be formed.[1c]

system can be enhanced by the addition of a co-catalyst, for example, bis(triphenylphosphoranylidene)ammonium chloride (PPNCl). The generally accepted mechanism for the polymerisation of CO_2 and epoxides is shown in Figure 14.4.

The first step (initiation) involves the ring opening of the epoxide to form a metal alkoxide. At this point there are three possible processes that can occur:

1. CO_2 inserts into the metal alkoxide, forming a metal carbonate, and back-biting occurs with the elimination of a cyclic carbonate (upper pathway).
2. CO_2 inserts into the metal alkoxide. In this scenario, chain propagation occurs by the insertion of an epoxide into the

metal carbonate linkage and so the polymerisation can continue, forming a perfectly alternating copolymer, until the process is terminated (lower pathway).

3. Alternatively, an epoxide can insert into the metal–alkoxide bond, generating another metal–alkoxide linkage; this process will form a polyether (not shown in Figure 14.4).

The relative ratios of products can be easily determined by analysis of the ^{1}H NMR spectrum of the polymeric product. When using the racemic version of PO, it is possible to produce either atactic (random), isotactic (same stereochemistry) or syndiotactic (alternate stereochemistry) polycarbonate material. Furthermore, with PO it is also possible to observe regioselectivity with substituents on alternate carbon atoms adding in a head-to-tail (H-T) or with the substituents on consecutive carbon centres in head-to-head (H-H) or tail-to-tail (T-T) configurations (Figure 14.5).[3,14] These changes can affect the crystallinity of the polymer and hence the thermal and mechanical properties of the final polymer. The amount of each linkage can be determined by ^{13}C{^{1}H} NMR spectroscopy and often in the case of polymerisation of CHO the polymer can be hydrolysed and the enantioselectivity of the diol can be determined.[3]

The first production of polycarbonates was by Inoue *et al.* in 1969.[15] They showed that it was possible to copolymerise CO_2 with

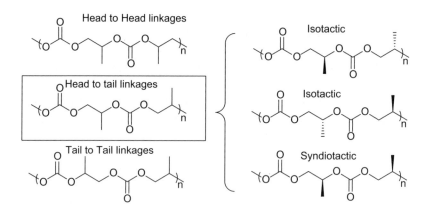

Figure 14.5   Regio- and stereoregular polymers formed with PO. Adapted from ref. 2.

PO utilising Zn(II) as the initiator. The catalyst was prepared *via* the reaction of a 1:1 mixture of $ZnEt_2$ and H_2O, producing a heterogeneous catalyst. Further advances in this chemistry came with the development of stable Zn(II) dicarboxylic acid systems by Soga *et al.* in 1981.[16] They found that a zinc–glutarate complex was the most active species. The concentration of active zinc on the surface was low ($< 5\%$); therefore, to achieve significant conversions it was necessary to use relatively high loadings of catalyst. Further development in the use of heterogeneous catalysts came with the development of double metal cyanides for the polymerisation. For example, in 2006 Coates and co-workers prepared a series of Ni(II), Pd(II) and Pt(II) catalysts with Co(II) two-dimensional double metal cyanide (DMC) initiator – $Co(H_2O)_2$ $[M(CN)_4] \cdot 4H_2O$ – for the polymerisation.[17] A summary of their polymerisation results is given in Table 14.1. They found (from ^1H NMR spectroscopic analysis) that these double metal cyanides were able to copolymerise CO_2 and PO randomly, with both ether and carbonate linkages observed in the spectrum.

Darensbourg and co-workers also prepared DMC initiators for the polymerisation. In this case they utilised an Fe(II)–Zn(II) system,[18] which showed modest activity in the polymerisation of cyclohexene oxide and CO_2 to afford polycarbonate and cyclic

Table 14.1 Results for the polymerisation of *rac*-PO and CO_2; in all cases [PO]:[Co] = 2530.

Catalyst	Temperature/ °C	Pressure/ bar	Yield/%^{a}	f_{CO_2}^{b}	TOF^c/h^{-1}	M_n^{d}/g mol^{-1}	PDI^d
CoNi(CN)$_4$	130	54.4	5.57	0.20	1860	74 300	3.1
	110	54.4	5.39	0.22	1770	84 100	2.9
	90	54.4	4.77	0.27	1510	86 000	2.8
	70	54.4	3.79	0.31	1170	152 000	3.7
	50	54.4	1.29	0.36	390	163 000	5.8
	70	81.7	1.83	0.38	540	152 000	4.3
	70	68.1	2.57	0.35	7800	233 000	4.8
	70	40.8	3.92	0.27	1250	116 000	3.5
	70	27.2	3.74	0.23	1220	111 000	2.6
CoPd(CN)$_4$	90	54.4	1.47	0.43	18	25 600	3.6
CoPt(CN)$_4$	90	54.4	1.11	0.44	13	27 900	3.7

^{a}Isolated yield.
^{b}Carbonate fraction determined by ^1H NMR spectroscopy.
^{c}Turnover frequency.
^{d}Determined by gel permeation chromatography (GPC).

carbonates. After these early forays into the area, there has been a tremendous drive in the last 20 years to prepare homogeneous catalysts for this process. The following sections detail some specific examples of initiators for the process based on Co(III), Cr(III), Zn(II) and Al(III) systems that are effective at ambient pressures and finally recent advances in the area of novel epoxides for this process.[1b,c,3]

14.3   INITIATOR SYSTEMS

14.3.1   Cobalt Initiators

The use of Jacobsen's Co(III) complex for the production of polycarbonates was investigated by Lu and co-workers (Figure 14.6).[19] Initially, with $X = O_2CCCl_3$ and $^nBu_4NBr$ as the co-catalyst, only a 6% selectivity to polypropylene carbonate (PPC) was observed, the major product being the cyclic carbonate. Interestingly, they observed that if X or the anion of the quaternary

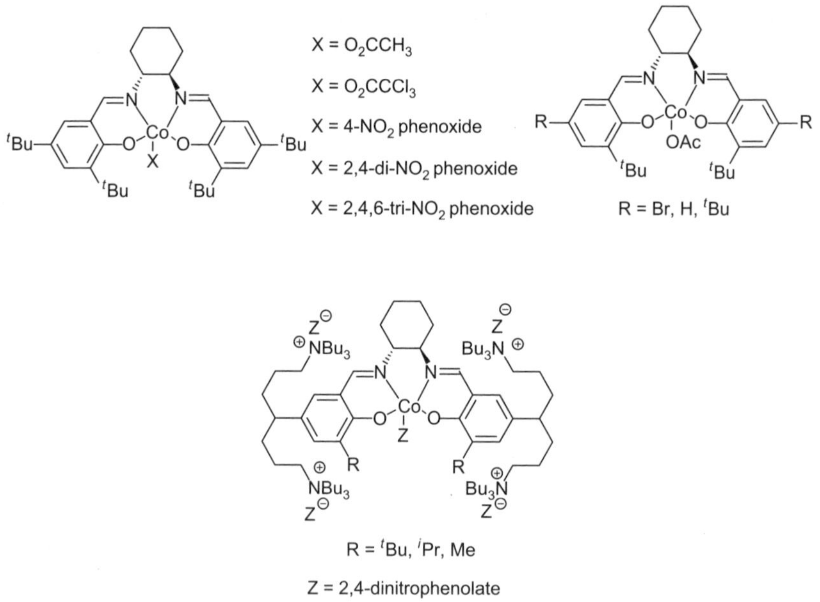

Figure 14.6   Co(III) system prepared by Lu and co-workers (top left), Coates and co-workers (top right) and Lee and co-workers (bottom).[19c,20]

Table 14.2 CO_2/PO polymerisation using the catalysts shown in Figure 14.6; the catalyst:co-catalyst:epoxide ratio was 1:1:2000, temperature $= 25\,°C$, pressure $= 2$ MPa.

X^a	Y^a	TOF^b/h^{-1}	$Selectivity^c$	Carbonate linkages^d	$M_n^e/g\,mol^{-1}$	PDI^e
O_2CCH_3	Br	228	3	>99	–	–
	Br	264	6	>99	–	–
2,4-$NO_2C_6H_3O$	Br	289	78	>99	23 500	1.29
	Cl	257	99	>99	30 400	1.36
	I	272	69	>99	22 100	1.34
	OAc	167	99	>99	18 200	1.28
2,4,6-$NO_2C_6H_2O$	Cl	248	99	>99	27 500	1.43

^{a}See Figure 14.6 for X, Y is the counter ion for $^nBu_4NY$ co-catalyst.
^{b}Turnover frequency.
^{c}% PPC as determined by ^1H NMR spectroscopy.
^{d}Determined by ^1H NMR spectroscopy.
^{e}Determined by GPC.

Table 14.3 Selected data for the polymerisation of CO_2 and *rac*-PO for Coates and co-workers' Co(III) catalysts shown in Figure 14.6, at a temperature of $25\,°C$.

Catalyst, $R =$	Time/h	[PO]: [cat]	Pressure/ MPa^a	TOF^b/h^{-1}	$Selectivity^c$	Carbonate linkages^c	$M_n^d/g\,mol^{-1}$	PDI^d
Br	3	500	5.51	81	>99	95	15 300	1.22
	3	500	4.13	19	>95	94	3100	2.60
	3	200	5.51	51	>99	95	8200	1.25
	8	2000	5.51	38	>99	95	21 700	1.41
H	3	200	5.51	51	>99	96	6600	1.21
	3	500	5.51	66	>99	96	9000	1.31
^tBu	3	200	5.51	42	>99	99	5700	1.28
	3	500	5.51	59	>99	99	8100	1.57

^{a} Pressure of CO_2 (1 psi $= 6.89 \times 10^3$ Pa).
^{b}Turnover-frequency of PO to PPC.
^{c}Determined by ^1H NMR spectroscopy.
^{d}Determined by GPC in THF.

amine salt (Y) was varied, then it was possible to alter the product selectivity significantly (Table 14.2).

Coates and co-workers have also utilised such Co(III) complexes (Figure 14.6) for the controlled production of polycarbonates, one of the first examples of such (Table 14.3).[20a] In this case they varied the nature of the group *para* to the phenoxide moiety. With all R groups almost perfectly alternating polymers (epoxide $= rac$-PO) were prepared and no cyclic material observed. For R = ^{t}Bu the head-to-tail selectivity was the highest at 80%; it was 70 and 60% when R = Me and H, respectively. Moreover, when the enantiomerically pure form of the catalyst for R = ^{t}Bu was employed with (S)-PO as the monomer, a polymer with a head-to-tail selectivity of 93% was isolated. This Co(III) complex also preferentially polymerised (S)-PO over (R)-PO with $k_S/k_R = 2.8$. The authors compared their catalyst with the Zn(BDI)OAc system, which afforded a higher turnover frequency (TOF) of 184 h^{-1} but a significantly lower selectivity towards PPC at 87% compared with the Co(III) systems.

Lee and co-workers prepared a series of Co(III) initiators for the controlled polymerisation of CO$_2$ and epoxides and investigated the effect of the *ortho* substituent on the overall catalytic performance.[20b] Furthermore, they observed that the initiators were recyclable. The initiators were screened for the polymerisation of propylene oxide and CO$_2$ with a [PO]:[cat] ratio of 25 000 at 80 °C and CO$_2$ pressures in the range 2.0–1.7 MPa. In this example, it was observed that the largest TOFs were achieved with a methyl group in the *ortho* position. For example, with a methyl group a TOF of 26 000 h^{-1} was observed, compared with 1300 h^{-1} with a *tert*-butyl group. In all cases, very high selectivity towards the polycarbonate was observed over formation of cyclic carbonate and all polymers were strictly alternating copolymers with no ether linkages. It was also noted that the catalyst (as the ammonium salt) could be easily removed from the isolated polymer by passing a solution of the polymer through a short silica gel pad, affording a colourless polymer. If the catalyst did not have the ammonium salts bound, then after flash chromatography the polymer was still highly coloured. It was also possible to recover the catalyst from the silica gel and reuse it with an almost identical TOF compared with the virgin catalyst.

In addition to these Jacobsen-style ligands based on salens, systems based on porphyrins have also been exploited under mild conditions by Sugimoto and Kuroda[21] (Figure 14.7).

If the polymerisation of CHO and CO$_2$ was performed at 80 °C, a pressure of 50.5 bar of CO$_2$ and a CHO:(TPP)CoCl ratio of 500 without the addition of co-catalyst, then only small amounts of polyether were observed. However, with the addition of 0.5 equiv. of DMAP (*N,N*-dimethylaminopyridine), 99% polycarbonate was

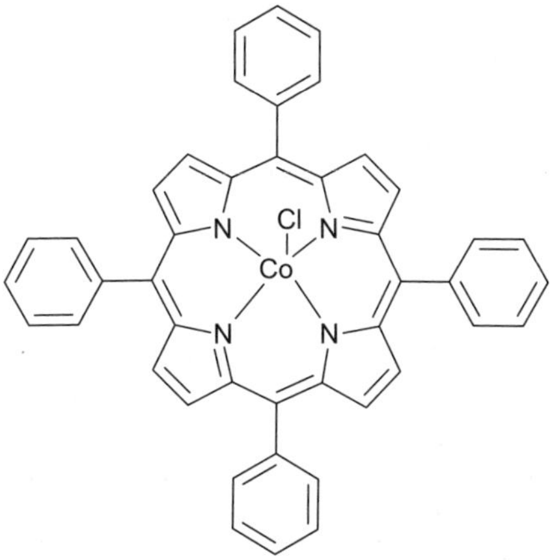

Figure 14.7 Structure of (TPP)CoCl porphyrin complexes employed by Sugimoto and Kuroda.[21]

observed without production of the cyclic species. It was observed that the M_n of the polymer increased in a linear fashion with respect to conversion with a narrow molecular weight distribution.

14.3.2 Chromium Initiators

Chromium–salen complexes have also been utilised with great success for copolymerisation. A significant contributor to this area was Darensbourg, whose group prepared an extensive series of Cr(III) complexes for this process[22] (Figure 14.8). A significant advantage of these systems is in the synthetic variability of the ligand itself, with four sites that can potentially be altered, thus, allowing an in-depth study of the effect of sterics and electronics on the polymerisation; see Table 14.4 for selected polymerisation data for cyclohexene oxide.

It can be seen that sterically hindering groups bound to the diimine backbone have a detrimental effect on the production of copolymers, with lower carbonate incorporation in the final material. This steric bulk presumably hinders the interaction of the epoxide with the catalyst and therefore its ring opening. It is also noted that the nature of the X anion is also important. When X = N_3, the catalyst is always more active than its chloride counterpart. Furthermore, it is also apparent that increasing the electron-donating effect of the substituents on the phenolate ring

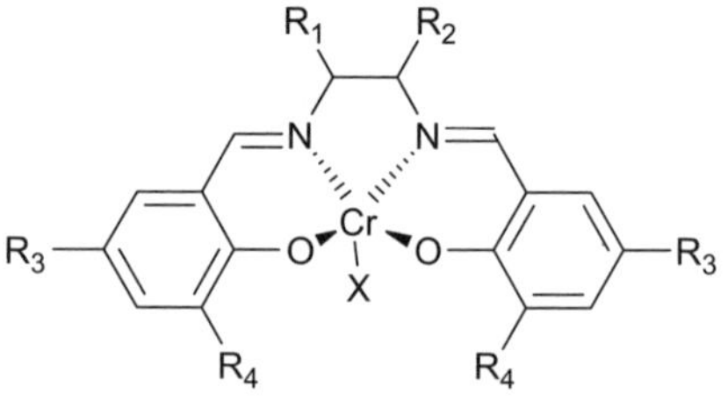

Figure 14.8 Cr(III) complexes for the copolymerisation of CHO and CO_2.[22]

Table 14.4 Copolymerisation of CHO and CO_2 using the (salen)CrX complexes shown in Figure 14.8[a].

R_1	R_2	R_3	R_4	X	TOF^b/h^{-1}	$Carbonate/\%^c$
'Bu	'Bu	'Bu	'Bu	Cl	0.8	48
	C_6H_{11}	'Bu	'Bu	Cl	1.5	61
	CH_3	'Bu	'Bu	Cl	7.3	90
H	H	'Bu	'Bu	Cl	35.7	99
	H	'Bu	'Bu	N_3	46.9	>99
	H	H	H	Cl	11.2	92
	H	OMe	H	Cl	0.6	36
	H	OMe	'Bu	Cl	56.7	>99
	H	OMe	'Bu	N_3	62.7	>99
$-C_4H_4-$		H	H	Cl	15.9	97
$-C_4H_4-$		'Bu	'Bu	Cl	36.2	99
$-C_4H_4-$		'Bu	'Bu	N_3	42.9	>99
$-C_4H_4-$		OMe	'Bu	Cl	45.7	>99
$-C_4H_4-$		OMe	'Bu	N_3	65.9	>99
$-C_4H_4-$		Cl	Cl	Cl	3.8	15

[a]Conditions: 55 bar CO_2, 20 mL of CHO, 50 mg of catalyst, 2.25 equiv. of *N*-methylimidazole, time 24 h.
[b]Moles of epoxide consumed per mole of Cr per hour.
[c]Estimated by ^{1}H NMR spectroscopy.

(incorporation of OMe moieties) has a beneficial effect on the TOF. It was noted that the addition of *N*-methylimidazole was crucial for successful polymerisation. On optimisation of the loading for the $R_1 = R_2 = H$, $R_3 = R_4 = {}'Bu$, that the optimum co-catalyst loading was found to be 5 mol%. Under these conditions, a turnover number (TON) of 1141 and 97% carbonate linkages were observed. As the amount of co-catalyst increased, the percentage of carbonate linkages decreased.

Further investigations into Cr(III) complexes for this polymerisation were carried out by Luinstra *et al.*[23] They used a combination of DFT (density functional theory) calculations and experiments to probe the mechanism of the formation of aliphatic polycarbonates in more detail.

14.3.3 Zinc Initiators

One of the earliest examples of the use of discrete Zn(II) catalysts for the polymerisation was undertaken in 1999 by Darensbourg *et al.*[24] A series of Zn(II) phenoxides were prepared by reaction of $Zn[N(SiMe_3)_2]_2$ with substituted phenols in THF (THF was observed to coordinate to the zinc centre). Various 2,4,6-trisubstituted phenoxides were employed, with methyl-substituted systems affording the highest yield of polycarbonate. Following this initial work on relatively simple complexes, a plethora of Zn(II) catalysts has been reported in the last 10 years.[25]

A popular trend in this area is to use bimetallic complexes; for example, Ding and co-workers prepared dizinc complexes as shown in Figure 14.9.[26]

Coates and co-workers in 2001, inspired by their success with such complexes for the polymerisation of *rac*-lactide, utilised Zn(II)–β-diiminate complexes (Figure 14.10) for the copolymerisation.[1a]

Further to the structures shown in Figure 14.10, they also prepared an acetate-bridged complex. The polymerisations were similar when the bridging groups were OAc, OMe, O^{i}Pr and $N(SiMe_3)_2$; this presumably suggests that the propagating species is the same in all cases. For example, after 2 h at 50 °C with a [CHO]:[Zn] ratio of 1000:1 with neat CHO at a pressure of 690 kPa (100 psi) of CO_2, when Ar = 2,6-diethylphenyl with bridging group = OAc a polymer with $M_n = 22.0 \text{ kg mol}^{-1}$ [polydispersity index (PDI) = 1.12] and TON = 470 was observed, whereas with an OMe bridging group $M_n = 23.7 \text{ kg mol}^{-1}$ (PDI = 1.14) and TON = 477 were found. In all cases, there was a high proportion (>86%) of carbonate linkages and atactic polymers were formed (as determined by $^{13}C\{^1H\}$ NMR spectroscopic analysis).

14.3.4 Ambient-pressure Initiators

Although there have been many elegant approaches to the preparation of polycarbonates, the examples discussed so far require a

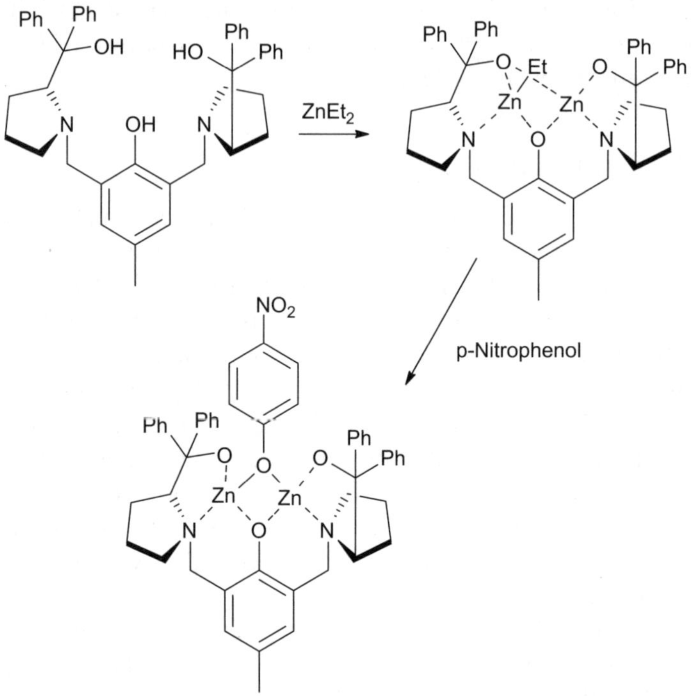

Figure 14.9 Bimetallic Zn(II) complexes for the polymerisation of CO_2 and CHO. Each Zn(II) centre has a coordinated molecule of THF which has been removed for clarity.[26]

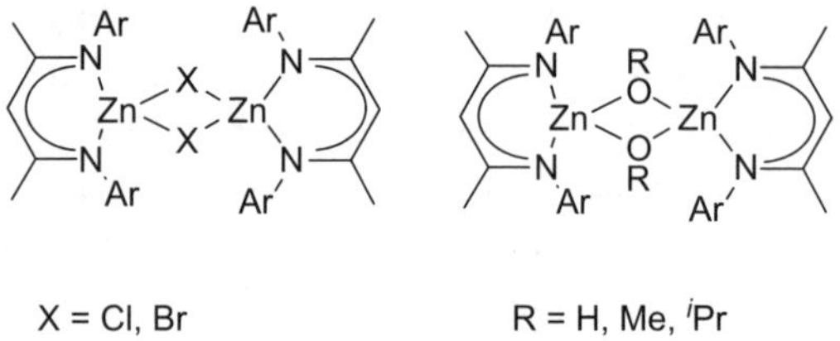

X = Cl, Br   R = H, Me, ^{i}Pr

Figure 14.10 Zn(II) β-diiminate complexes utilised for the copolymerisation of CHO and CO_2.[1a]

relatively high pressure of CO_2. There is an exigent need for the production of initiators that are capable of producing poly-carbonates at relatively low pressures. One such example was reported by Williams and co-workers, who prepared macrocyclic dimeric Zn(II), Co(II) and Fe(III) complexes (Figure 14.11).[27]

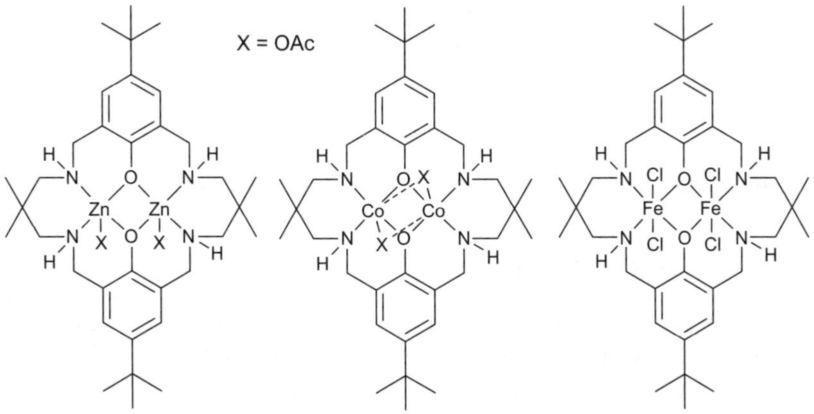

Figure 14.11 Dimeric Zn(II), Co(II) and Fe(III) complexes prepared by Williams and co-workers for the ambient-temperature copolymerisation of CO_2 and CHO.[27]

Table 14.5 Copolymerisation of CHO and CO_2 using the macrocyclic complexes shown in Figure 14.11[a].

Metal	Temperature/°C	Time/h	TON^b	TOF^c/h^{-1}	Polymer/%[d]	$M_n^e/g\ mol^{-1}$	PDI^e
Zn	80	2	340	172	>99	5100	1.26
Co	100	1	410	410	>99	4200	1.29
Co	60	7	310	45	>99	4900	1.04

[a]Reaction conditions: neat CHO, temperature 80 °C, pressure 1 bar CO_2.
[b]TON (turnover number) = number of moles of CHO consumed per mole of catalyst.
[c]TOF = TON per hour.
[d]Determined by ^{1}H NMR spectroscopy.
[e]Determined by GPC.

Furthermore, trimetallic cobalt and zinc complexes were prepared and also showed excellent activity.

Williams and co-workers also utilised Fe(II) and Co(III) complexes of the same ligand and found these to be excellent catalysts at ambient pressure, a comparison of the these complexes are shown in Table 14.5:

The use of catalysts to convert CO_2 and epoxides into cyclic carbonates at ambient pressure has also been reported. For example, Decortes and Kleij used a series of Zn(II)–salphen catalysts, with high yields being observed with a wide variety of epoxide starting materials.[28]

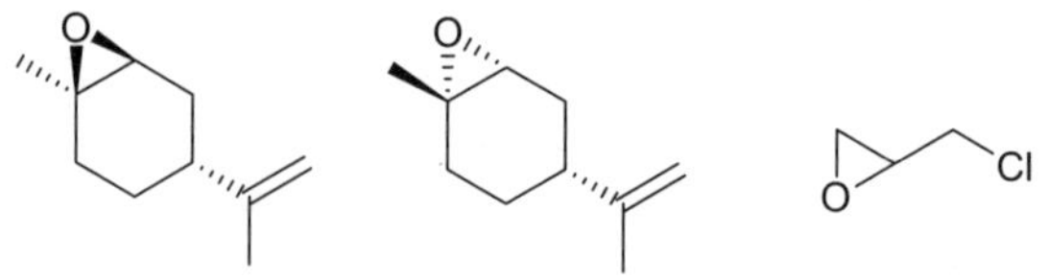

Figure 14.12   Two enantiomers of limonene oxide and epichlorohydrin.

14.4   OTHER EPOXIDES

The polymerisation is not limited to cyclohexene and propylene oxide. For example, epichlorohydrin and limonene oxide (an epoxide derived from citrus fruits) have both been employed (Figure 14.12).[29] The low cost of (R)-limonene oxide and its structural similarity to cyclohexene oxide make it an attractive choice as a sustainable comonomer.

Coates and co-workers were able to produce high molecular weight alternating polycarbonate which is highly regio- and stereoregular.[29a] The catalyst used was Zn(II) β-diiminate, with limonene oxide. Epichlorohydrin is an important industrial chemical and has many used in the chemical industry and organic synthesis. Perfectly alternating polycarbonates were produced with Co(III) complexes (Figure 14.6).[29b]

14.5   OTHER POLYMERIC MATERIALS

The area is not limited to polycarbonates. For example, Oi and co-workers showed that it is possible to prepare poly(alkyl alkynoates) from CO_2 using diynes and dihalides as the comonomer.[30] Organic carbamates ($R^1NHCO_2R^2$) are an important class of industrial chemicals used in a variety of processes, pharmaceuticals, pesticides and the production of intermediates for fine chemical processes. Moreover, carbamates can be used as the precursors for polyurethanes. It has been shown that the reaction of amines with CO_2 and alkyl halides is an environmentally friendly method for the production of such materials using solid-supported catalysts, such as MCM-41-type systems.[31]

14.6   CONCLUSION

There have been many remarkable advances in this area in recent years, specifically in terms of conversion and selectivity (regio- and stereo-). This method is truly showing promise over the bisphenol-A/COCl$_2$ route for the production of polycarbonates. The success of this polymerisation is obviously dependent upon the choice of catalyst – with the advent of bimetallic systems and intelligent ligand design, the future for this class of polymer can only be positive. The ability to perform this polymerisation under an ambient pressure of CO$_2$ should now make this process even more suited to industrial applications, as it will no doubt improve the energy balance of the process. A significant challenge is this area is the utilisation of sustainable epoxides that can be used to produce 100% sustainable polycarbonates with the appropriate physical properties.

REFERENCES

1. (a) M. Cheng, D. R. Moore, J. J. Reczek, B. M. Chamberlain, E. B. Lobkovsky and G. W. Coates, *J. Am. Chem. Soc.*, 2001, **123**, 8738; (b) D. J. Darensbourg, *Chem. Rev.*, 2007, **107**, 2388; (c) D. J. Darensbourg, R. M. Mackiewicz, A. L. Phelps and D. R. Billodeaux, *Acc. Chem. Res.*, 2004, **37**, 836; (d) D. J. Darensbourg and J. C. Yarbrough, *J. Am. Chem. Soc.*, 2002, **124**, 6335.
2. (a) P. M. Cox, R. A. Betts, C. D. Jones, S. A. Spall and I. J. Totterdell, *Nature*, 2000, **408**, 184; (b) T. C. Johns, J. M. Gregory, W. J. Ingram, C. E. Johnson, A. Jones, J. A. Lowe, J. F. B. Mitchell, D. L. Roberts, D. M. H. Sexton, D. S. Stevenson, S. F. B. Tett and M. J. Woodage, *Climate Dyn.*, 2003, **20**, 583.
3. M. R. Kember, A. Buchard and C. K. Williams, *Chem. Commun.*, 2011, **47**, 141.
4. (a) P. T. Anastas and M. M. Kirchhoff, *Acc. Chem. Res.*, 2002, **35**, 686; (b) P. T. Anastas, M. M. Kirchhoff and T. C. Williamson, *Appl. Catal. A: Gen.*, 2001, **221**, 3; (c) P. T. Anastas and J. B. Zimmerman, *Environ. Sci. Technol.*, 2003, **37**, 94A.
5. (a) P. Gallezot, *Green Chem.*, 2007, **9**, 295; (b) M. Schlaf, *Dalton Trans.*, 2006, 4645.
6. (a) M. Eissen, J. O. Metzger, E. Schmidt and U. Schneidewind, *Angew. Chem. Int. Ed.*, 2002, **41**, 414; (b) A. J. Ragauskas, C. K. Williams, B. H. Davison, G. Britovsek, J. Cairney, C. A. Eckert, W. J. Frederick, J. P. Hallett, D. J. Leak, C. L. Liotta,

J. R. Mielenz, R. Murphy, R. Templer and T. Tschaplinski, *Science*, 2006, **311**, 484.

7. (a) M. Mours, *ChemSusChem*, 2008, **1**, 59; (b) M. Poliakoff, J. M. Fitzpatrick, T. R. Farren and P. T. Anastas, *Science*, 2002, **297**, 807.

8. S. Fukuoka, M. Kawamura, K. Komiya, M. Tojo, H. Hachiya, K. Hasegawa, M. Aminaka, H. Okamoto, I. Fukawa and S. Konno, *Green Chem.*, 2003, **5**, 497.

9. (a) M. North, R. Pasquale and C. Young, *Green Chem.*, 2010, **12**, 1514; (b) A. A. G. Shaikh and S. Sivaram, *Chem. Rev.*, 1996, **96**, 951.

10. H. R. Kricheldorf, M. Berl and N. Scharnagl, *Macromolecules*, 1988, **21**, 286.

11. E. L. Whitelaw, M. D. Jones, M. F. Mahon and G. Kociok-Kohn, *Dalton Trans.*, 2009, 9020.

12. (a) D. J. Darensbourg and A. I. Moncada, *Inorg. Chem.*, 2008, **47**, 10000; (b) D. J. Darensbourg and A. I. Moncada, *Macromolecules*, 2009, **42**, 4063; (c) D. J. Darensbourg and A. I. Moncada, *Macromolecules*, 2010, **43**, 5996; (d) D. J. Darensbourg, A. I. Moncada, W. Choi and J. H. Reibenspies, *J. Am. Chem. Soc.*, 2008, **130**, 6523.

13. (a) W. Clegg, R. W. Harrington, M. North and R. Pasquale, *Chem. Eur. J.*, 2010, **16**, 6828; (b) J. Melendez, M. North and R. Pasquale, *Eur. J. Inorg. Chem.*, 2007, 3323; (c) M. North, P. Villuendas and C. Young, *Chem. Eur. J.*, 2009, **15**, 11454; (d) M. North and C. Young, *ChemSusChem*, 2011, **4**, 1685; (e) M. North and C. Young, *Catal. Sci. Technol.*, 2011, **1**, 93.

14. M. H. Chisholm and Z. P. Zhou, *J. Am. Chem. Soc.*, 2004, **126**, 11030.

15. S. Inoue, H. Koinuma and T. Tsuruta, *J. Polym. Sci. Part B: Polym. Lett.*, 1969, **7**, 287.

16. K. Soga, E. Imai and I. Hattori, *Polym. J.*, 1981, **13**, 4070.

17. N. J. Robertson, Z. Q. Qin, G. C. Dallinger, E. B. Lobkovsky, S. Lee and G. W. Coates, *Dalton Trans.*, 2006, 5390.

18. (a) D. J. Darensbourg, M. J. Adams and J. C. Yarbrough, *Inorg. Chem.*, 2001, **40**, 6543; (b) D. J. Darensbourg, M. J. Adams, J. C. Yarbrough and A. L. Phelps, *Inorg. Chem.*, 2003, **42**, 7809.

19. (a) W. M. Ren, X. Zhang, Y. Liu, J. F. Li, H. Wang and X. B. Lu, *Macromolecules*, 2010, **43**, 1396; (b) X. B. Lu and

Y. Wang, *Angew. Chem. Int. Ed.*, 2004, **43**, 3574; (c) W. M. Ren, Z. W. Liu, Y. Q. Wen, R. Zhang and X. B. Lu, *J. Am. Chem. Soc.*, 2009, **131**, 11509.

20. (a) Z. Q. Qin, C. M. Thomas, S. Lee and G. W. Coates, *Angew. Chem. Int. Ed.*, 2003, **42**, 5484; (b) S. Sujith, J. K. Min, J. E. Seong, S. J. Na and B. Y. Lee, *Angew. Chem. Int. Ed.*, 2008, **47**, 7306.
21. H. Sugimoto and K. Kuroda, *Macromolecules*, 2008, **41**, 3127.
22. D. J. Darensbourg, R. M. Mackiewicz, J. L. Rodgers, C. C. Fang, D. R. Billodeaux and J. H. Reibenspies, *Inorg. Chem.*, 2004, **43**, 6024.
23. G. A. Luinstra, G. R. Haas, F. Molnar, V. Bernhart, R. Eberhardt and B. Rieger, *Chem. Eur. J.*, 2005, **11**, 6298.
24. D. J. Darensbourg, M. W. Holtcamp, G. E. Struck, M. S. Zimmer, S. A. Niezgoda, P. Rainey, J. B. Robertson, J. D. Draper and J. H. Reibenspies, *J. Am. Chem. Soc.*, 1999, **121**, 107.
25. (a) D. J. Darensbourg, J. R. Wildeson, J. C. Yarbrough and J. H. Reibenspies, *J. Am. Chem. Soc.*, 2000, **122**, 12487; (b) D. J. Darensbourg, J. C. Yarbrough, C. Ortiz and C. C. Fang, *J. Am. Chem. Soc.*, 2003, **125**, 7586; (c) R. Eberhardt, M. Allmendinger, M. Zintl, C. Troll, G. A. Luinstra and B. Rieger, *Macromol. Chem. Phys.*, 2004, **205**, 42; (d) T. Sarbu and E. J. Beekman, *Macromolecules*, 1999, **32**, 6904; (e) I. Kim, M. J. Yi, S. H. Byun, D. W. Park, B. U. Kim and C. S. Ha, *Macromol. Symp.*, 2005, **224**, 181; (f) J. S. Kim, H. Kim, J. Yoon, K. Heo and M. Ree, *J. Polym. Sci. Part A: Polym. Chem.*, 2005, **43**, 4079; (g) W. J. van Meerendonk, R. Duchateau, C. E. Koning and G. J. M. Gruter, *Macromol. Rapid Commun.*, 2004, **25**, 382.
26. Y. L. Xiao, Z. Wang and K. L. Ding, *Chem. Eur. J.*, 2005, **11**, 3668.
27. (a) A. Buchard, M. R. Kember, K. G. Sandeman and C. K. Williams, *Chem. Commun.*, 2011, **47**, 212; (b) M. R. Kember, A. J. P. White and C. K. Williams, *Inorg. Chem.*, 2009, **48**, 9535; (c) M. R. Kember, A. J. P. White and C. K. Williams, *Macromolecules*, 2010, **43**, 2291.
28. A. Decortes and A. W. Kleij, *ChemCatChem*, 2011, **3**, 831.
29. (a) C. M. Byrne, S. D. Allen, E. B. Lobkovsky and G. W. Coates, *J. Am. Chem. Soc.*, 2004, **126**, 11404; (b) G. P. Wu,

S. H. Wei, W. M. Ren, X. B. Lu, T. Q. Xu and D. J. Darensbourg, *J. Am. Chem. Soc.*, 2011, **133**, 15191.

30. (a) Y. Fukue, S. Oi and Y. Inoue, *J. Chem. Soc., Chem. Commun.*, 1994, 2091; (b) S. Oi, K. Nemoto, S. Matsuno and Y. Inoue, *Macromol. Rapid Commun.*, 1994, **15**, 133; (c) M. S. Super and E. J. Beckman, *Trends Polym. Sci.*, 1997, **5**, 236.

31. R. Srivastava, D. Srinivas and P. Ratnasamy, *Tetrahedron Lett.*, 2006, **47**, 4213.

CHAPTER 15

Carbon Dioxide as a Sustainable Industrial Solvent to Replace Organic Solvents

STEVEN M. HOWDLE,* STEFAN POLLAK,
NATASHA A. BIRKIN AND MARIE WARREN

School of Chemistry, University of Nottingham, University Park,
Nottingham NG7 2RD, UK
*Email: steve.howdle@nottingham.ac.uk

15.1 INTRODUCTION

Solvents are essential to almost every sector of the chemistry-using industries – from the food industry where they are indispensable for extraction and fractionation processes or to remove undesired impurities or pesticides through to technical cleaning applications where solvents are used to remove fats, oils, cutting emulsions and other contaminations from a substrate. Solvents are also used to transport additives into a substrate, for example in dyeing or impregnation processes. In the traditional chemistry sector, solvents are needed to dissolve reactants and thereby bring them into a liquid state, aiding mass transfer and also removing reaction products or simply aiding in heat transfer.

Materials for a Sustainable Future
Edited by Trevor M. Letcher and Janet L. Scott
© The Royal Society of Chemistry 2012
Published by the Royal Society of Chemistry, www.rsc.org

A wide range of solvents are essential, from water to organics such as hexane, benzene, toluene, acetone and ethanol, which are the most common industrial solvents. With respect to sustainable, environmentally friendly and safe processing, these solvents function effectively, but also have some disadvantages. Organic solvents are flammable and their handling must be managed with extreme care at the large scale. Most organic solvents are volatile organic compounds (VOCs) with a considerable vapour pressure under atmospheric conditions and whose vapour may affect human health. In particular, residues of organic solvents in foodstuffs that have not been fully removed after processing can be hazardous to consumer health. Importantly, most organic solvents are derived from crude oil; thereby the supply and the price depend directly upon fossil fuel fluctuations. After use, solvents must be disposed of safely or recycled to prevent wastewater and environmental pollution. Hence there is certainly a need for alternative, more environmentally acceptable solvents which might provide 'green' alternatives to classical solvents.

Dense gases offer a promising alternative. There is significant use of propane in the petrochemical industry, but in this chapter we limit discussion to carbon dioxide (CO_2), the most common gas investigated as a benign alternative solvent. Although its ability to act as a solvent is poor to non-existent in the gaseous state, the solvent properties improve considerably when CO_2 is in the liquid or supercritical ($scCO_2$) state.[1] Although CO_2 is not a strong solvent, it has many advantages that make it interesting as a replacement solvent.[2] It is non-flammable, non-carcinogenic, non-mutagenic, thermodynamically stable and inexpensive and has a very low toxicity.[3] The threshold limit value (TLV) of CO_2 for an 8 h work day is 5000 ppm (0.5%). Since CO_2 turns into its gaseous state under ambient conditions, it can easily be separated from any liquid or solid substance by lowering the pressure below the vapour pressure. After CO_2 treatment, no solvent residues need to be removed from the product and no liquid solvent waste remains to be treated after processing.[4]

There is currently significant focus upon global warming and climate change, and carbon dioxide is strongly implicated. Indeed, there are now significant global research efforts targeted at better efficiency, alternative energy sources and even carbon capture and storage in order to mitigate the release of carbon dioxide from combustion processes. Significant research funding is being directed at this problem worldwide because hard targets have been set, *e.g.* by 2020 the emissions of carbon dioxide must be reduced by 20% compared with 1990. Indeed, the European Union, Canada and Japan aspire to a 50% cut to be achieved by 2050. Carbon capture and storage (CCS) technologies are being developed to store huge amounts of carbon dioxide in subterranean reservoirs such as depleted gas deposits, salt domes, aquifers or even on the ocean floor.[5] However, even through these measures, the problem of a massive excess of carbon dioxide will not be solved, but will be postponed at best. This means that alternative uses for this carbon dioxide should be found. As a feedstock it will be available in almost unlimited amounts and at a very low cost – but the catalytic breakthroughs to activate carbon dioxide have not yet been developed. However, there are other potential uses: 'Why not use supercritical carbon dioxide as a solvent for processing?' One barrier to commercialisation is that the use of liquid or supercritical carbon dioxide requires high-pressure technology. Thus capital costs and the necessity for energy expenditure for compression have been cited as key economic hurdles. Although in most known production processes the production costs per unit product decrease with increasing throughput, the expenses for high-pressure equipment, *e.g.* because of the increasing thicknesses of vessels, rise over-proportionally to the volume. Another reason is that despite significant research activities, there is still not enough knowledge available about the phase behaviour and the chemistry of dense carbon dioxide mixtures.[2] Moreover, for reactions, the use of $scCO_2$ as a solvent does not mean just replacing a conventional solvent, but requires careful investigation and understanding of the process and its mechanisms.[4]

15.1.1 The Supercritical State

Most substances can appear in a solid, liquid or gaseous state. By varying the pressure and/or the temperature, matter can be converted from one state into another. Under phase equilibrium conditions, two or even three states can exist in parallel. A well-known example of the equilibrium of two phases is boiling water and an ice cube floating in its liquid phase. In these two-phase systems under equilibrium conditions, pressure and temperature depend on each other. If one is changed, the other may not remain unchanged or one of the phases will disappear. This phase behaviour of pure substances can be illustrated in a phase diagram, as shown in Figure 15.1 for carbon dioxide.

The line where liquid and gas can exist simultaneously under equilibrium conditions is called the boiling line or dew line. This boiling line has two characteristic endpoints. On one side, it ends with the triple point; under triple conditions, all three states can appear at the same time. On the other side, it ends with the critical point, which is defined by the critical temperature T_c and the critical pressure P_c. Approaching this point on the boiling line, the densities of the two phases become identical and the phases become indistinguishable at the critical point. A substance with a temperature higher than T_c *and* a pressure higher than P_c is defined as being in a supercritical state. Under supercritical conditions, no separation of gas and liquid can be observed and the substance appears as one continuous fluid.

The supercritical state offers interesting opportunities for many process engineering applications. Although the density of a

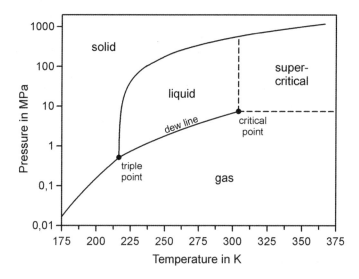

Figure 15.1 Pressure–temperature (*p*–*T*) diagram for carbon dioxide. The critical point marks the end of the dew line. In the supercritical region above this point, the liquid and gas phases are no longer distinguishable.

supercritical fluid (SCF) can begin to approach that of the liquid state, the fluid has an extremely low viscosity and almost no surface tension. Low viscosity, low surface tension, high diffusivity and the absence of a liquid–gas phase boundary make it attractive for extraction and impregnation applications where massive solid structures need to be penetrated and where good transport properties are required. For chemical reactions, the supercritical state is interesting because of the excellent transport properties. Supercritical fluids are miscible with any other gas and many non-polar liquids and can therefore be used to overcome transport limitations between reactants. As far as practical application is concerned, carbon dioxide is particularly interesting because of its easily accessible critical parameters: $T_c = 31.1\,°C$ and $P_c = 73.9$ bar. In comparison, water, which is also used for supercritical fluid applications, has $T_c = 374\,°C$ and $P_c = 220.64$ bar.

15.2 EXTRACTION

Extraction is a fractionation process that uses an extracting agent to separate a substance from a liquid (liquid–liquid extraction) or a solid substrate (solid–liquid extraction). A substance can only be extracted if it is soluble in or miscible with the extracting agent, at least to a certain extent. Industrial processes consist of at least two steps; an extraction step and a separation step. In the extraction step, substrate and extracting agent are brought into contact by mixing or percolation. The thermodynamic conditions have to be selected to allow for the desired component to dissolve in the extracting agent. Subsequently, the 'loaded' extracting agent is separated from the substrate.[6] In a second step, the extracting agent (solvent) and the extracted substance (solute) are separated by manipulating the thermodynamic conditions.

Because of the pressure dependence of its solvent properties, dense carbon dioxide is an interesting solvent for fractionation processes. Although CO_2 in its gaseous state is not a good solvent at all, the dissolving power increases considerably in the liquid and the supercritical states. Hence the solvent properties can easily be modified by changing the temperature and pressure. The temperature can be kept at a moderate level, which allows the treatment of thermally sensitive substances. Extractions can generally be divided into two groups. In the first group, the extract is of interest. This can be the case with the extraction of oils, flavours or pharmaceutical substances from herbs or plants. In the other group, not the extract but the purified remainder is interesting. Cleaning processes such as dry cleaning, degreasing, the removal of undesired flavours and the extraction of nicotine from tobacco are just a few examples. In the ideal case, both extract and remainder are of commercial interest, for example in the extraction of caffeine from coffee or tea.[7]

15.2.1 Extraction of Natural Substances

Extraction is still one of the most common industrial applications for carbon dioxide. Mainly high-value products are extracted, for example aromas and flavouring components from herbs and spices, lipids from oleiferous seeds and plants and active substances such as antioxidants for natural medicine.[6] Another important branch utilising supercritical fluid extraction (SFE) is the production of hop extract for the brewing industry. Natural extracts have several advantages compared with the raw crops. Through extraction, a uniform product is created, which may be much longer lived, easier to process and will have known properties to allow for consistent product quality. Moreover, extracts have a much smaller volume than the raw material and are therefore easier to store and to transport. SFE is operated by a wide range of providers, from major corporations to small- and medium-sized enterprises, and these are found all over the world. For natural materials, the relatively moderate conditions required for SFE are beneficial for the extraction of thermally sensitive components.[8] Table 15.1 gives an overview of just a few of the companies offering SFE products.

Most of the natural products obtained from SFE are further processed by the food, pharmacy or cosmetics industry and in all cases the purity of the products and the absence of any organic solvents after the extraction are the most important advantages of the CO_2 technology. The process is usually operated in a batch mode as solids, which are not easy to handle continuously, are treated.[6] This is further described below for decaffeination.

The increasing popularity of the supercritical fluid technique is creating a market for suppliers of process equipment. For example, the Natex, Austria,[9] Separex, France,[10] and CF Technologies, USA,[11] focus on the engineering and production of the machinery

Table 15.1 Companies offering supercritical fluid extraction products.

Company[a]	Location	Product examples
Evonik Industries http://extraction.evonik.com	Münchsmünster and Trostberg, Germany	Flavours: fruit, coffee, tea, ginger. Pesticide removal, sterilisation, defatting, oil and hop extraction
Flavex www.flavex.de	Rehlingen, Germany	Spices, herbs, aromas, anti-oxidants, essential oils for the food industry, cosmetics and natural medicine
RAPS GmbH www.raps-industrieservice.de	Kulmbach, Germany	Extraction of spices and aromas. Powdering of liquid flavours and aromas (CPF process, see Section 15.5.2).
B&K Technology Group www.bkherb.com	Xiamen, China	Wide range of herbal extracts (> 600), natural food colours (> 70), preservatives, stimulants
Ecomaat www.organicroseoil.eu	Mirkovo, Bulgaria	Rose extract, essential oils for the pharmaceutical, cosmetics and flavouring industries
CO_2 Extracts www.co2extract.com	New Delhi, India	Essential oils and flavours from flowers, herbs and spices for the pharmaceutical and food industries
Nutrizeal http://nutrizeal.com	Nelson, New Zealand	Bioactive substances for the pharmaceutical and cosmetics industries, hop extract
NATECO$_2$ www.nateco2.de	Wolnzach, Germany	Seed oils and hop extract. Largest producer of CO_2 hop extract with an annual throughput > 8000 t of hops
Baarth-Haas Group www.barthhaasgroup.com	Nuremberg, Germany	Hop extract, different essential oils and aromas from hops

[a]All Internet references last accessed 1 July 2011.

for supercritical fluid processes. Figure 15.2 shows the scheme of a commercial extraction plant. The substance to be extracted from is charged batchwise into the extractor vessel where it is flowed through by compressed carbon dioxide. The CO_2-soluble essence is extracted and carried to the next vessel, the separator. Here, the pressure and temperature conditions are changed in such a way that the extract precipitates. In many cases the pressure is lowered, but not below the vapour pressure in order to avoid an expensive

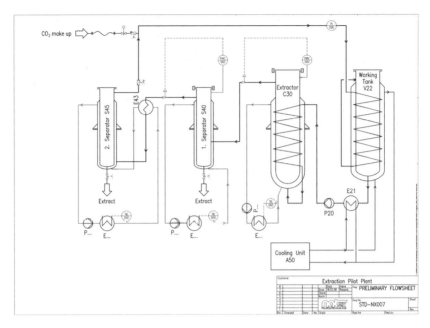

Figure 15.2   Scheme of a high-pressure extraction plant with one extractor, two
separator vessels and a reservoir tank. The scheme was kindly
provided by the manufacturer, Natex Prozesstechnologie
GesmbH, Ternitz, Austria.

recompression step. The extract is withdrawn from the separator
while the pure CO_2 is pumped to a storage tank. From here the gas
is recompressed and recirculated to the extractor vessel via a
downstream compressor to facilitate further extraction.

15.2.2   Cleaning and Purification

Cleaning and purification are extraction processes in which the
extract is not recovered but the treated substrate is the desired
product. Usually the extract is disposed of after being separated
from the solvent. Carbon dioxide is used as an extracting agent for
cleaning applications, although the technique is not as widespread
and developed as the extraction of valuable products.

Supercritical CO_2 can be used to remove pesticides or other
undesired impurities from foods such as rice and crops and from
other natural substances such as cork. The French Oeneo Bouch-
age Group operates an $scCO_2$ extraction plant for the purification

of cork in San Vicente de Alcantara, Spain. In three vessels with a capacity of 3800 L each, undesirable flavours, microorganisms and pollutants are extracted. The main target of the process is tri-chloroanisole (TCA), which causes an undesired cork taint. The cork is boiled and ground into flour before it is treated in the supercritical extraction process. The purified grains are then baked together using a binding agent to form ~500 million traditional bottle stoppers.[12]

SFE is also used for the extraction of nicotine from tobacco. There is not much freely available information about these industrial processes, but a description of the method can be found in several patents.[13] Dense gases might offer opportunities in terms of cleaning applications in the high-tech and electronics sectors. In Belgium, IMEC constructed significant operations to harness $scCO_2$ for the cleaning of high-purity silicon wafers for integrated circuit manufacture, but it appears that the levels of residual impurities were too high to pursue the technology further. CO_2 can also be used as a substitute for organic solvents in dry cleaning processes. In this sector, off-the-shelf equipment for the commercial cleaning of clothes and fabrics is available.[14]

15.2.3 Decaffeination of Coffee and Tea

The decaffeination of coffee and tea is a favourable case of an extraction process, as both extract and remainder are of commercial interest. The extracted caffeine is resold for pharmaceutical use or as an ingredient of beverages. The most common methods for decaffeinating coffee beans are the Swiss Water process and decaffeination with $scCO_2$. In all processes, the green beans are decaffeinated before roasting. In the Swiss Water process, the caffeine is extracted with hot water and then removed from the water with activated carbon. To avoid the extraction of flavours, the process water is presaturated with aroma compounds by initially extracting one batch of 'sacrificial' beans that are subsequently disposed of. The pretreated water is reused and no more aromas are extracted from further batches.[15]

In 1970, Kurt Zosel of the Max Planck Institute for Coal Research patented a method for decaffeinating green coffee beans with $scCO_2$.[16] Like most of the processes for the extraction from solid substrates, the decaffeination of coffee is run batchwise. The caffeine is extracted at pressures of ~ 120–180 bar and temperatures of 40–$80\,^{\circ}C$ with a small amount of water as a co-solvent. Prior to the decaffeination, the coffee beans are soaked in water to swell and to open the pores for the extraction.[17] In the original patent, the caffeine was separated from the CO_2 by absorption on activated carbon. To allow for the recovery and further use, the caffeine is now washed out in a separation column with a counter-current water flow.[18] During this separation step, the CO_2 is kept at an elevated pressure. A full separation of caffeine and CO_2 could also be achieved by decompressing the gas below its vapour pressure and thereby turning it into the vapour state, but this strategy would increase the cost of the recompression step required before the gas can be reused.

The industrial use of carbon dioxide for decaffeination purposes began in the late 1970s. Café Hag, Germany,[19] started the decaffeination of coffee in 1979 in a large-scale plant in Bremen. After it was sold to Kraft Foods, further plants were erected in Houston, TX, and Bremen.[20] Another large supplier of coffee using CO_2 for decaffeination purposes is the Maximus Coffee Group, also in Houston, TX.[21]

15.3 IMPREGNATION

Impregnating means bringing a functional substance into a solid substrate and depositing it throughout its inner structure. Impregnation is used, for example, to protect natural substances against degradation and decay, to dye fibres and fabrics, to strengthen the structure or to change material properties, *e.g.* by making the substrate hydrophobic. The impregnation agent can be either a liquid or a dissolved solid. Liquids can be applied directly by immersing the substrate in the impregnation agent or with the help of a solvent acting as an entrainer. Solvents are often used to lower the viscosity and the interfacial tension of the impregnation agent and, thereby, to help to penetrate the substrate. In the case of a solid impregnation agent, a solvent is compulsory. In a solvent-based impregnation, the impregnation agent is dissolved in a carrier medium and later released again in the substrate to be treated. The release can be triggered by changing the thermodynamic or the chemical conditions. Therefore, impregnation is a transport process contrary to extraction. $ScCO_2$ is used in a growing number of applications focused upon impregnation because of its outstanding ability to penetrate almost any structure.

15.3.1 Textile Dyeing

Most established commercial dyeing processes are water based. Water is needed in many steps of the procedure and leads to wastewater that is mainly contaminated with dyestuff that is poorly degradable and hazardous for the environment. For every 1 kg of dyed material, 30–400 L of water are polluted and need treatment. $ScCO_2$ with a small content of water as co-solvent can be used as the dyeing medium and can very easily be separated from the dyestuff after the process, therefore avoiding huge amounts of wastewater. Polyester, cotton and even silk can be dyed with SCF technology.[22] The main requirement is the solubility of the dyestuff and $scCO_2$ does dissolve several low-polarity dyestuffs and moreover has the ability to swell and penetrate the fabrics.[23] Non-polar fibres such as polyesters can be treated with non-polar and non-reactive dyes. Admixed with the CO_2, the dyestuff is transported to the surface and diffuses into the swollen material. On depressurisation, the CO_2 exits the fibre, leaving the dye molecules behind. In the case of polar fibres such as cotton and silk, a chemical bond between the dye and fibre is needed, so polar and reactive dyes have to be used. As these are not soluble in $scCO_2$, additional surfactants are required. For these reasons, the dyeing of non-polar fibres is the more developed technique[24] and there is a need for further research, especially on reactive dyeing.[22] Van der Kraan *et al.* reported good results in dyeing polyester, nylon, silk and wool using disperse dyes with reactive vinyl sulfone or dichlorotriazine groups.[25] Looking for carbon dioxide as a solvent in industrial-scale dyeing applications, one will merely find a few suppliers of the process equipment, which indicates that the technology is just getting started. The Dutch DyeCoo Textile Systems,[26] founded in 2008, is the first company to specialise in supercritical dyeing equipment. So far, the company offers a polyester dyeing machine for batches of up to 125 kg of fabrics. The dyeing process is operated at 120 °C and 250 bar and takes only half of the time required by the classical method. Moreover, the solubilisation agents required to dissolve the hydrophobic dyes in water-based processes are not required in CO_2. A larger scale plant and a

machine for cotton dyeing will be available in the future from the
same manufacturer.[27] Likewise, the Japanese process technology
supplier Hisaka[28] offers technical solutions for the supercritical
treatment of fabrics.

15.3.2 Wood Impregnation

Supercritical wood impregnation is an application fully harnessing
the low viscosity and the negligible surface tension of SCFs and
their ability to penetrate complex structures. Wood as a natural
substance is prone to insects, rot and decay when exposed to
environmental conditions. In order to use wood as a durable
building material, it needs treatment with fungicides or biocides to
suppress biodegradation. Classical impregnation processes are
based on the application of fungicides by vacuum and/or pressure.
The most common active substances are metal salts, for example
chromated copper arsenate (CCA), which has become a subject of
public discussion because of its possible effect on the environment.
For smaller structures, organic biocides such as tebuconazole and
propiconazole are used.[29] The high internal resistance of wood to
fluid flow makes it difficult to penetrate the structure fully with
liquid impregnation agents and leads to an uneven distribution of
fungicides in the wood, meaning that untreated areas may be
exposed after machining. In supercritical wood impregnation, CO_2
is applied as a carrier medium for organic biocides and even thick
structures can be impregnated completely.[30]

Since 2002, a Danish joint venture called Supertræ[31] has been
running the first supercritical wood impregnation plant on a
commercial scale in Hampen, Denmark. Three vessels with a
volume of 8000 L each allow for the treatment of wooden components with a maximum length of 6.6 m. The annual capacity of
the plant is in the range 40 000–60 000 m^3. For impregnation, the
conditions are adjusted to $\sim$150 bar and 40–60 °C. Carbon dioxide
is then circulated through a small mixing vessel containing the
fungicide. After 2–5 h, the system is depressurised. As the wood
provides a certain flow resistance even to the $scCO_2$, the venting
has to be done under controlled conditions, otherwise pressure
differences occur which could cause serious damage to the material.[32] During the depressurisation, the fungicide precipitates from
the CO_2 and is deposited in the wood structure.[33] The mechanical

characteristics and optical appearance of the wood are not affected by the procedure. After the impregnation, residual fungicide is separated from the CO_2 in a separator system. Both CO_2 and fungicide are collected after the process and can be recycled.[30]

15.4 REACTIONS IN CARBON DIOXIDE

A large number of chemical reactions in SCFs, particularly in $scCO_2$, have been described in numerous scientific publications and reviews. Frequently described reactions in $scCO_2$ are hydrogenation, hydroformylation, oxidation and polymer synthesis.[2,4] Many reactions are conducted heterogeneously as the reactants are in different states (*e.g.* liquid and gaseous) and/or the reaction is performed over a solid catalyst. The rate of reaction is mainly determined by how fast the reactants can be transported to each other or to the catalyst and how fast the products can be removed. Therefore, solvents are often used as reaction media to enhance the transport properties and to enable the reaction to proceed. As the supercritical gas is miscible with all other gaseous reactants and also with several liquids, the number of phase boundaries can be reduced in order to allow for a faster reaction. Liquid CO_2 or $scCO_2$ as a benign alternative to environmentally harmful solvents shows excellent heat and mass transfer properties and therefore is considered a highly attractive solvent for chemical reactions.[34] Although there is significant promise, so far industrial activity in this field is still comparatively low.

15.4.1 Hydrogenation

Hydrogenation is a chemical reaction in which hydrogen is added to a molecule, mostly at high temperatures and in the presence of a catalyst. The most common application is the saturation of $C=C$ double bonds in hydrocarbons. During exothermic hydrogenation, the double bond between two molecules is opened and a hydrogen atom is added to each carbon atom. A very well-studied reaction is the hydrogenation of fatty acids in the fat hardening process. Fatty acids with no double bond in the carbon chain are called saturated.

In the traditional process, a substrate (*e.g.* oil) is mixed with hydrogen and a solid catalyst. As the hydrogen is not fully miscible with the substrate, a three-phase system is formed. The process is usually conducted batchwise at elevated temperatures and at near-atmospheric pressures. The hydrogen has to be dissolved in the liquid and to be transported to the heterogeneous catalyst, where the reaction takes place. To obtain adequate yields in a reasonable time, the system has to be strongly agitated. Nevertheless, the low solubility of hydrogen in the oil phase and the high transport resistance are limiting factors that control the rate of reaction.[35]

SCF technology provides a real opportunity to overcome these limitations. In the presence of SCFs, the miscibility behaviour of substances that are hardly miscible with each other under atmospheric conditions might be significantly improved. If, under appropriate conditions, hydrogen, lipid and SCF form a single-phase mixture, the rate-limiting step of transporting mass from the gaseous hydrogen phase to the lipid phase is eliminated. Combined with the fact that diffusion coefficients in dense gases are much higher than those in liquids, an improved mass transfer and therefore higher reaction rates can be expected in SCFs.[36]

There are various publications showing that the process is generally feasible. In the past, efforts were made to establish SCF-aided hydrogenation processes by Thomas Swan in the UK and Härröd Research in Sweden. Hoffmann-La Roche in Switzerland operated a CO_2-assisted hydrogenation process, but has recently sold these activities to DSM Switzerland.

15.4.2 Polymer Synthesis

There are different methods for the polymerisation of monomers. In bulk polymerisation, no medium other than the pure monomer, an initiator and perhaps a catalyst is present during the reaction. The removal of heat from the exothermic reaction is the main issue with this method. Other methods, such as solution polymerisation (with or without precipitation), emulsion polymerisation or suspension polymerisation require an additional medium wherein either the monomer or both the monomer and polymer are stabilised. During emulsion and suspension polymerisation, reactants and reaction aids are dispersed and stabilised (but not dissolved) in a continuous, mostly aqueous phase. By contrast, during solution polymerisation, the reaction medium acts as a solvent in which initiator, catalyst, monomer and polymer are all dissolved. The dilution by a solvent particularly improves the removal of heat from the reaction but is also limited because of the generally poor solubility of long-chain molecules in the reaction medium. Only a few polymers are soluble enough in reaction media to be viable for solution polymerisation. Precipitation polymerisation is a special case of solution polymerisation where only the monomer stays in solution while the polymeric product precipitates.[37]

As most long-chain polymers show very poor solubility, CO_2 is not an ideal medium for solution polymerisation. Only amorphous fluoropolymers and polysiloxanes show significant solubility at high molecular weights. Hence CO_2 is a promising solvent for the polymerisation of fluoroolefins, which are usually synthesised either in water using non-degradable surfactants or in environmentally challenging chlorofluorocarbons (CFCs).[38] The most common fluoroolefin is tetrafluoroethylene (TFE), which is a reactant for the synthesis of the well-known Teflon [polytetrafluoroethylene (PTFE)]. In addition to the environmental concerns regarding CFCs or surfactants in the reaction medium, most fluoroolefin monomers are highly flammable or even explosive. This factor was an important aspect in the development of the well-publicised commercial process by DuPont in North Carolina for fluoropolymer resins with a process based on dense-phase CO_2 as solvent.[39,40]

In other cases, $scCO_2$ is used as a feedstock and is incorporated into a polymer. For example, Novomer in the USA has commercialised academic breakthroughs to produce poly(propylene carbonate) and poly(ethylene carbonate) polymers through 'green' chemistry based on the use of CO_2 (and CO).[41,42] A similar approach has been adopted by Bayer. Such processes permanently bind CO_2 into a polymer matrix and provide an alternative route to valuable commodity materials.[43,44]

15.5 CARBON DIOXIDE AS A PROCESS AID

All of the processes described above involve a chemical reaction or at least the transport of a substance into or out of a substrate. In some applications, however, carbon dioxide is used to change material properties without adding or removing anything or even supporting any chemical changes. During such processing, carbon dioxide simply interacts physically with a substance to change its appearance or properties. Two good examples of this kind of modification are the foaming of polymers and particle production. In both cases, CO_2 is utilised to change the material properties for a short time during the processing and to change the morphology of the material. For example, if liquid or $scCO_2$ is dissolved in a polymer, the low viscosity strongly influences the polymer properties and lowers the melting and/or the glass transition temperature, *i.e.* CO_2 is behaving as a plasticiser. Therefore, solids can be liquefied below their actual melting or glass transition temperature by exposing them to the compressed gas.

15.5.1 Polymer Foaming and Modification

Foamed polymers, usually polyurethane, are used extensively for insulation purposes in the building sector. Polymer foams are also applied in the automotive industry, in refrigeration engineering, for mattresses and furniture, for packaging, *etc.* Usually, polymer foams are produced with the help of an organic blowing agent that is admixed with the polymer during processing and then evaporated to form a porous structure. The blowing agent remains in the cavities or escapes to the atmosphere. In this context, the use of carbon dioxide as a replacement for organic solvents is considered to be an excellent example of green chemistry.[2]

The MuCell process by Trexel (USA) utilises either $scCO_2$ or nitrogen as a blowing agent to create microporous structures in polymers. The gas is admixed with the pressurised polymer. On depressurisation during the moulding or extrusion process, the gas expands and the polymer is foamed before solidification.[45] Bayer holds a patent for carbon dioxide-assisted polyurethane foaming[46] and is devoting new research efforts to developing a new, CO_2-based production process for polyurethane nanofoams.[47]

Additionally, $scCO_2$ foaming techniques can be utilised to produce polymer scaffolds for tissue engineering applications. Tissue engineering is the replacement or repair of diseased or damaged tissues such as bone.[48] Certain organs or tissues cannot adequately heal themselves and require treatment to restore their function. Transplantation is a common therapy. However, shortage of donor tissues means that not all patients receive the treatment they need.[49] A common method is to grow the tissue on porous three-dimensional polymer scaffolds. These scaffolds are used as cell supports to provide mechanical strength and structural guidance.[50]

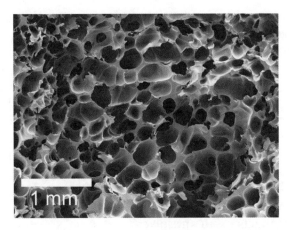

Figure 15.3 SEM image of a microporous PLA foam prepared using $scCO_2$.

They often contain drugs, nutrients and growth factors to promote cell differentiation and vascularisation of the developing tissue. Amorphous or semi-amorphous polymers are liquefied under supercritical conditions and on venting of the CO_2 from the system the gas expands leaving a porous network within the polymer on solidification. The polymer can be placed in moulds to take different shapes depending on the application. Additionally, the porosity of the polymer scaffold can be controlled by altering the processing conditions.[49] Although there has been no implementation of any commercial products, much research is being directed at developing foamed polymer scaffolds using $scCO_2$ techniques.[50,51] Figure 15.3 shows a scanning electron microscopy (SEM) image of poly(lactic acid) (PLA) made by SCF foaming.

15.5.2 Particle Production

A fairly well-investigated but industrially not very widespread SCF technology is particle or powder production by high-pressure spraying processes. Different methods have already proved their feasibility, such as PGSS (particles from gas-saturated solutions),[52] RESS (rapid expansion of supercritical solutions), CPF (concentrated powder form)[53] and PCA (precipitation with a compressed fluid antisolvent).[54] These processes allow for the production of powders from melts or even liquids. All of them can be run batchwise or continuously.

Even though the procedures appear to be very similar, carbon dioxide plays a completely different role in each method. It is only in the RESS process that CO_2 really acts as a solvent requiring the substrate to be micronised to dissolve in the $scCO_2$ phase. The PGSS and the CPF processes differ from RESS in that a much smaller amount of gas per kilogram of product is used. Here the CO_2 acts as a solute and dissolves in the substrate of interest. Finally, the PCA method is used to create particles from a substance that is initially dissolved in an organic solvent. This organic solution is injected into a vessel containing $scCO_2$; if the organic solvent has a significant interaction with CO_2, then its dissolving power is decreased significantly and the solute precipitates. In this case, the CO_2 acts as an antisolvent.

The PGSS process allows the production of particles from melts or from dissolved solids. In the former case, a solid is melted or liquefied upon exposure to liquid CO_2 or $scCO_2$. The absorption of the gas in the solid leads to liquefaction or plasticisation. This process is often applied to sensitive substances that would degrade if heated above their melting temperature. The gas-saturated mixture is depressurised through a small nozzle of typically 0.1–1 mm diameter, where it is atomised by the rapidly expanding gas into droplets of 5–1000 µm. In most cases, the pressure upstream of the nozzle is in the range 100–300 bar. The post-expansion pressure can be ambient pressure or slightly above. Because of the Joule–Thomson phenomenon, the droplets are instantly cooled below their glass transition or melting temperature. They solidify and a powder is formed. In a variant of this process known as PGSS drying, the feed substance is liquefied with the aid of an organic solvent. In this case, the solvent is separated after the depressurisation together with the $scCO_2$. Therefore, thermodynamic conditions have to be appropriately selected to keep the solvent in a gaseous condition.[55]

In the RESS process, the substance to be micronised is dissolved in $scCO_2$. In contrast to the PGSS process, the gas does not act as a solute but as a solvent. As even in the supercritical state the solubility of most substances in $scCO_2$ is rather low, high throughputs of gas are needed and the process is expensive and reserved for high-value products. The supercritical solution is depressurised through a nozzle. The dissolving power of CO_2 decreases dramatically at this point and the solute precipitates and forms small particles. The process is often combined with a prior extraction step, where the substance to be micronised is extracted from a substrate, *e.g.* vegetable raw materials.[55,56]

To produce a powder from a liquid, a solid powdery carrier has to be added. Such carriers can be starch, sugars, polymers or celluloses. In much the same way as the PGSS process, the CPF process mixes a liquid with compressed $scCO_2$ and then depressurises through a nozzle. Downstream of the depressurisation, the carrier material is injected into the jet of droplets by a number of auxiliary nozzles that are arranged around the main nozzle. The liquid is bound to the carrier by adhesion or soaked into the pores by capillary forces, and powders with 60% liquid content can be produced while the powder remains in an apparently dry, free-flowing condition.[57]

Only a few of the above processes have made their way into industry. Fraunhofer Umsicht in Oberhausen, Germany, is running a scale-up of the PGSS process in a commercial dimension (Figure 15.4). The plant is designed for a maximum throughput of 150 kg h^{-1} and used for research purposes and also for industrial contract manufacturing. The CPF technology is used commercially by Raps in Kulmbach, Germany,[58] to produce powders from liquid aromas, spices and flavours.

15.6 FUTURE TRENDS

There are significant opportunities for further utilisation of carbon dioxide as a green solvent or processing aid. What is clear is that even if the technical realisation of a process is possible, economic and environmental considerations have to justify the costs of expensive high-pressure equipment. However, increasingly draconian legislation is beginning to exert pressure to push down the levels of volatile organic residues permissible in everyday products, and this could be a driving force. More importantly, the use of $scCO_2$ can lead to products that are difficult or impossible to access via conventional routes and this could be a more significant driver. Below, three examples are given which may provide interesting opportunities for possible future commercial developments for $scCO_2$.

Figure 15.4   Plant for PGSS (particles from gas-saturated solutions), PGSS
drying and CPF (concentrated powder form) at Fraunhofer
Umsicht in Oberhausen, Germany, who kindly provided the
photograph.

15.6.1   Tanning of Leather

In the European Union alone, more than 300 million m^2 of leather
are produced per year.[59] In a time-consuming and complex tanning
process, raw hides are converted into durable, imperishable leather.
For each tonne of product, 20–40 t of wastewater are produced,
and ~25% of this wastewater is contaminated with metallic or
organic tanning agents.[60] Moreover, a considerable amount of the
global leather output is produced in emerging countries[61] where
proper wastewater treatment cannot be ensured. Hence new processes could help to reduce the impact on the environment.

Developing a carbon dioxide-based tanning process would
reduce significantly the need for water.[62] The CO_2-assisted tanning

process has already been trialled successfully on the pilot-plant industrial scale at Fraunhofer Umsicht. Considerable reductions in both the processing time and the wastewater effluent have been achieved with this technique. The skins are treated in a 1700 L tanning vessel using classical tanning agents but in the presence of compressed carbon dioxide at 30–60 bar. The carbon dioxide swells the raw hides and enhances the mass transfer of the chrome (or other tanning agents) into the skin. The tanning time can be reduced from 28 to ∼5 h.[60]

15.6.2 Medical Applications

15.6.2.1 Drug-encapsulated Particles. Critical Pharmaceuticals is a biotechnology company specialising in drug delivery technologies.[63] They have developed a novel encapsulation technology, CriticalMix, also based on the PGSS process (Figure 15.5).[64] This one-step process produces injectable microparticles for the sustained release of challenging drugs including proteins and peptides at ambient temperature and in the absence of solvents. The $scCO_2$ plasticises a biodegradable polymer and allows effective addition and mixing of potent therapeutics drugs into the polymer. After mixing, the pressure is released in a controlled fashion to yield drug-loaded microparticles suitable for subcutaneous injection.

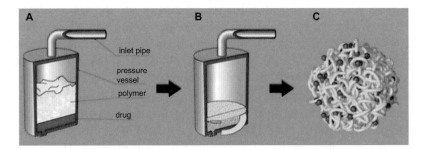

Figure 15.5   The CriticalMix process: (A) The polymer and drug are loaded into a pressure vessel and the vessel is pressurised with CO_2 to reach supercritical conditions. (B) The polymer becomes liquefied, which allows the drug to be mixed into the polymer, and is then sprayed through a nozzle. (C) The polymer solidifies, forming a particle trapping the drug inside.[63]

This technology has increasing relevance as advances in bio-technology have led to acceleration of the use of protein-based drugs. However, their fragile nature negates the traditional oral route of administration. Moreover, such drugs may require daily injections and these are inconvenient and painful for the patient and can lead to poor patient compliance. Therefore, producing a controlled release system where a drug is entrapped within a bio-degradable polymer reduces the need for repeated injection and leads to a more efficacious and cost-effective treatment. Spray drying, melt extrusion and emulsion methods are conventional techniques used for encapsulation of drugs within polymers. Unfortunately, these techniques are not suitable for protein-based drugs which are sensitive to heat or organic solvents and the pre-sence of surfactants and interfaces, all of which can lead to the denaturing of the proteins.[65] In such cases, $scCO_2$ shows great promise and has been demonstrated to retain the activity of very delicate protein-based drugs such as human growth hormone through processing into controlled release devices.[66,67]

15.6.2.2 Particle Micronisation. By contrast, others have concentrated on micronisation of the drug itself. CritiTech Pharma produce submicron drug particles using a process based on the PCA technology (Figure 15.6). Ultrasonic energy produced by a converging/diverging nozzle or an electromechanical oscillator is used to shatter droplets into even finer droplets. These droplets are sprayed into a flowing stream of $scCO_2$, which allows for rapid mass transfer of solvent into the stream of $scCO_2$. This rapid mass transfer forces precipitation or crystallisation to occur prior to the coalescence of droplets. It is this combination of techniques that allows CritiTech to produce submicron particles of drugs and other chemicals.[68]

Crystec Pharma have also developed an anti-solvent SCF par-ticle production process. The drug is dissolved in a suitable solvent and this solution and the SCF are passed simultaneously through a proprietary nozzle arrangement into a particle formation vessel. Rapid extraction of the solvent into the SCF occurs, creating a high level of supersaturation of the material of interest in the diminishing level of solvent causing rapid precipitation of solid particles.[69]

Figure 15.6 An SCF rig for making fine-particle drugs based on PCA technology.[68]

It is estimated that around 40% of the new drug entities produced by pharmaceutical companies have water solubility issues.[70] A potential solution is to reduce the particle size of the drug, increasing the surface area and thus enchaining the dissolution rate and bioavailability.[71] Conventional techniques, including fine grinding mills to micronise the drug, have shown some success in improving the dissolution rate.[72] However, the particle size remains relatively large and there can be variations in the particle size distribution. $ScCO_2$ processes, in particular supercritical antisolvent techniques, can offer an alternative route to reduce the particle size of these drugs.

15.6.3 Polymer Synthesis – New Opportunities

The synthesis of polymers in $scCO_2$ has been the subject of a great deal of research in recent years. The main challenge in employing $scCO_2$ as an alternative reaction medium for polymerisations is that carbon dioxide is good at solvating most non-polar and some polar molecules of low molecular mass, but it is a weak solvent for the majority of high molecular weight polymeric materials.[73] Despite this, scCO$_2$ has already been shown to be a promising reaction medium for the synthesis of polymers.

The poor solubility of most polymers actually makes scCO$_2$ an ideal medium for heterogeneous techniques such as dispersion polymerisation.[74] Here, a steric stabiliser, also referred to as a surfactant, facilitates the formation of a colloidal dispersion and prevents aggregation of particles from occurring.[75,76] So far, only fluorinated and siloxane-based polymers have shown sufficient solubility in scCO$_2$ to act as surfactants and this is perhaps one of the key reasons why scCO$_2$ processes have not been adopted commercially.

Considerable work has been focused on identifying and developing hydrocarbon stabilisers as alternatives for use in dispersion polymerisation in the last decade or so. Beckman and co-workers have developed poly(ether–carbonate) copolymers which are able to act as efficient, non-fluorous CO$_2$-philes and will readily dissolve at low CO$_2$ pressures,[77] and sugar acetate structures have also been proposed as potential CO$_2$-philes.[78] The synthesis of poly(vinyl alkylate) stabilisers via RAFT polymerisation and their successful application in the dispersion polymerisation of N-vinylpyrrolidone in scCO$_2$ was recently reported, resulting in a powder product with high conversion and spherical microparticles. This work highlighted the potential to make scCO$_2$ polymerisation processes much more accessible and inexpensive in the future through the use of hydrocarbon stabilisers.[79–81]

Emulsion polymerisation has also attracted interest in the field of scCO$_2$ utilisation and hydrocarbon stabiliser synthesis. A microemulsion is a thermodynamically stable dispersion of two immiscible fluids stabilised by surfactants at the interface.[82,83] A number of groups have focused on scCO$_2$ microemulsions as a reaction medium for chemical processes and polymer synthesis and the development of surfactants suitable for such applications.

CO$_2$-in-water (c/w) emulsion techniques have exploited the fact that CO$_2$ and water represent a non-volatile, non-toxic reaction medium and the droplet phase of the emulsion is easily removed by simply depressurising the system, removing the usual energy-intensive drying steps and allowing a route to the production of novel materials.[84] With the use of inexpensive surfactants such as OVAc-X, porous organic materials can be synthesised via this method and could potentially find application in areas such as the development of emulsion technologies for producing biodegradable scaffolds.[85] Water-in-CO_2 (w/c) microemulsions have also received particular attention, as they provide a water core that will act as a nanoreactor for the synthesis of inorganic nanoparticles.[86] A variety of materials have been produced using this method, including silver-based nanoparticles, which are attractive owing to their antibacterial properties, potentially useful for biomedical applications.[87,88]

Metal–polymer nanocomposites have been a subject of research, using scCO$_2$ to synthesise polymer microspheres with silver nanoparticles on the surface of the polymer bead, effectively creating a polymer powder with the potential for catalytic, antibacterial and biosensing applications.[89] CO_2 has also been utilised to produce polymers on the nanoparticle scale. Zetterlund and co-workers have also recently reported the development of a novel heterogeneous polymerisation technique which relies on CO_2-assisted particle formation and the homogenous expansion limit (HEL). The approach permits the formation of polymeric nanoparticles with diameters <20 nm.[90]

As research on polymerisations in scCO$_2$ continues to advance, new, inexpensive and more active CO_2-philic materials for use in CO_2 will be developed. This will open up significantly the opportunity to tailor industrial polymerisation processes to utilise scCO$_2$.

15.7 CONCLUSION

There are a wide range of industrial processes that already utilise the unique properties of scCO$_2$. In this chapter we have outlined those that are well known and also hinted at those that are perhaps on the cusp of commercial development. What is clear is that increasing legislation and the tightening of limits on volatiles will drive further development. In addition, there are new processes under development where the use of CO_2 as a solvent can lead to materials that would be difficult or impossible to achieve using conventional routes, and this may well be a much stronger driver towards the uptake of these new technologies. Moreover, the global focus on carbon capture and sequestration means that there will be significant volumes of readily available CO_2 for use in future processing.[91]

REFERENCES

1. W. Leitner, Designed to dissolve, *Nature*, 2000, **405**, 129.
2. E. J. Beckman, Supercritical and near-critical CO_2 in green chemical synthesis and processing, *J. Supercrit. Fluids*, 2004, **28**, 121.
3. Z. Knez, High pressure process technology – quo vadis?, *Chem. Eng. Res. Des.*, 2004, **82**, 1541.
4. W. Leitner and P. G. Jessop (eds), *Handbook of Green Chemistry, Vol. 4: Supercritical Solvents*, Wiley-VCH, Weinheim, 2010.
5. Anonymous, *Gasette – E.ON Ruhrgas Magazin*, 2008, **1**, http://www.eon-gas-storage.de/cps/rde/xbcr/SID-EC4E9270-60811A99/eon-gas-storage/090629_CCS-gasette_1-2008_Sonderdruck_%282%29.pdf.
6. G. Brunner, Supercritical fluids: technology and application to food processing, *Journal of Food Engineering*, 2005, **67**, 2.
7. T. Gamse, Industrial applications and current trends in supercritical fluid technology, presented at the VI Symposium: Contemporary Technologies and Economic Development, Leskovac, Serbia and Montenegro, 21–22 October 2005.
8. E. Ramsey, Q. Sun, Z. Zhang, C. Zhang and W. Gou, Mini-review: green sustainable processes using supercritical fluid carbon dioxide, *J. Environ. Sci.*, 2009, **21**, 720.
9. Natex Prozesstechnolgie, *Dense Gas Technology (CO$_2$)*, http://www.natex.at (last accessed 31 May 2011).
10. Separex, *Supercritical and High Pressure Technology*, http://www.separex.fr (last accessed 31 May 2011).
11. CF Technologies, *Process Innovators in Critical Fluid Technology*, http://www.cftechnologies.com (last accessed 14 June 2011).
12. DIAM, *Preserve*, http://www.diam-cork.com (last accessed 3 June 2011).
13. R. Prasad, Process and apparatus for the semicontinuous extraction of nicotine from tobacco, *European Patent*, EP323699A2, 1988.
14. Washpoint, http://www.washpoint.com (last accessed 3 June 2011).
15. Swiss Water Decaffeinated Coffee Co., *Science of Decaffeination*, http://www.swisswater.com/trade/the-swiss-water-experience/science-of-decaffeination (last accessed 29 June 2011).

16. K. Zosel, Decaffeinating green coffee with moist carbon dioxide in the supercritical state, *German Patent*, DE2005293A, 1970.
17. P. Munshi and S. Bhaduri, Supercritical CO_2: a twenty-first century solvent for the chemical industry, *Curr. Sci.*, 2009, **97**, 63.
18. G. Brunner, Counter-current separations, *J. Supercrit. Fluids*, 2009, **47**, 574.
19. Kraft Foods, http://www.kraftfoodscompany.com/eu/en/brands/brands_az/brands-h/index.aspx (Kraft, Café Hag, last accessed 28 April 2012).
20. M. SIMS, *Decaffeinating with Carbon Dioxide, Coffee Tea Trade J.*, 1990, **162**(9), 8–10.
21. Maximus Coffee Group, http://www.maximuscoffee.com/bulk-services/natural-decaffeination/sparkling-water-process (Maximus Coffee Group, last accessed 28 April 2012).
22. M. van der Kraan, Process and equipment development for textile dyeing in supercritical carbon dioxide, *Dissertation*, University of Delft, 2005.
23. M. Banchero, S. Sicardi, A. Ferri and L. Manna, Supercritical dyeing of textiles – from the laboratory apparatus to the pilot plant, *Text. Res. J.*, 2008, **78**, 217.
24. D. Knittel, W. Saus and E. Schollmeyer, Water-free dyeing of textile accessories using supercritical carbon dioxide, *Indian J. Fibre Text. Res.*, 1997, **22**, 184.
25. M. van der Kraan, M. V. Fernandez Cid, G. F. Woerlee, W. J. Veugelers and G. J. Witkamp, Dyeing of natural and synthetic textiles in supercritical carbon dioxide with disperse reactive dyes, *J. Supercrit. Fluids*, 2007, 40, 470.
26. DyeCoo Textile Systems, *Water-free and Environmentally Friendly Dyeing of Textiles*, http://www.dyecoo.com/profile.html (last accessed 23 June 2011).
27. J. Scrimshaw, CO_2 dyeing gets commercial roll-out, *Int. Dyer*, August 2010, 6.
28. Hisaka, *Supercritical Dyeing and Treatment*, http://www.hisaka.co.jp/english/vital/textile/index.cgi?c = zoom&pk = 33 (last accessed 23 June 2011).
29. E. Lack and H. Seidlitz, Supercritical CO_2 impregnation, presented at the 12th European Meeting on Supercritical Fluids, 9–12 May 2010, Graz, Austria.

30. S. B. Iversen, T. Larsen, O. Henriksen and K. Felsvang, The world's first commercial supercritical wood treatment plant, presented at the 6th International Symposium on Supercritical Fluids, 28–30 April 2003, Versailles, France.
31. Superwood, http://www.superwood.dk (last accessed 22 June 2011).
32. A. W. Kjellow and O. Henriksen, Supercritical wood impregnation, *J. Supercrit. Fluids*, 2009, **50**, 297.
33. A. W. Kjellow, O. Henriksen, J. C. Sørensen, M. Johannsen and C. Felby, Partitioning of organic biocides between wood and supercritical carbon dioxide, *J. Supercrit. Fluids*, 2010, **52**, 1.
34. W. Leitner, Supercritical carbon dioxide as a green reaction medium for catalysis, *Acc. Chem. Res.*, 2002, **35**, 746.
35. M. Härröd, M.-B. Macher, S. van den Hark and P. Møller, Hydrogenation at supercritical conditions, in *Proceedings of the 6th Meeting on Supercritical Fluids Chemistry and Materials*, Nottingham, 1999, p. 253.
36. A. Baiker, Supercritical fluids in heterogeneous catalysis, *Chem. Rev.*, 1999, **99**, 453.
37. W. Kaiser, *Kunststoffchemie für Ingenieure*, 2nd edn, Hanser, Munich, 2007.
38. S. L. Wells and J. M. DeSimone, Die CO_2-Technologie: ein wichtiges Instrument für die Lösung von Umweltproblemen, *Angew. Chem.*, 2001, **113**, 534.
39. DuPont, *DuPont Fluoropolymers Made with Supercritical CO_2 Polymerization Technology*, http://www2.dupont.com/Teflon_Industrial/en_US/products/supercritical_co2.html (last accessed 16 June 2011).
40. C. D. Wood, A. I. Cooper and J. M. DeSimone, Green synthesis of polymers using supercritical carbon dioxide, *Curr. Opin. Solid State Mater. Sci.*, 2004, **8**, 325.
41. G. Coates, Alternating copolymerization of propylene oxide and carbon dioxide with highly efficient and selective (salen) Co(III) catalysts: effect of ligand and cocatalyst variation, *J. Polym. Sci. Part A: Polym. Chem.*, 2006, **44**, 5182.
42. S. D. Allen, C. M. Byrne and G. W. Coates, Carbon dioxide as a renewable C1 feedstock: synthesis and characterisation of polycarbonates from the alternating copolymerisation of epoxides and CO_2, in *Feedstocks for the Future: Renewables for*

the *Production of Chemicals and Materials*, ed. J. Bozell and M. Patel, ACS Symposium Series, Vol. 921, American Chemical Society, Washington, DC, 2005, pp. 116–129.

43. M. Peters, B. Köhler, W. Kuckshinrichs, W. Leitner, P. Markewitz and T. E. Müller, Chemical technologies for exploiting and recycling carbon dioxide into the value chain, *ChemSusChem*, 2011, **4**, 1216.
44. Novomer, *High Performance, Environmentally Responsible Polymers*, http://www.novomer.com (last accessed 1 November 2011).
45. Trexel, *Trexel MuCell Processes*, http://www.trexel.com (last accessed 15 June 2011).
46. H.-M. Sulzbach, H. Steilen, R. Raffel, R. Eiben and W. Ebeling, Process for foam production using dissolved under pressure carbon dioxide, *US Patent*, 6 127 442, 2000.
47. Bayer MaterialScience, *Press Release*, Double the insulating performance, reduced energy consumption, Bayer, Leverkusen, 2010.
48. R. A. Quirk, R. M. France, K. M. Shakesheff and S. M. Howdle, Supercritical fluid technologies and tissue engineering scaffolds, *Curr. Opin. Solid State Mater. Sci.*, 2004, **8**(3–4), 313.
49. M. Sokolsky-Papkov, K. Agashi, A. Olaye, K. Shakesheff and A. J. Domb, Polymer carriers for drug delivery in tissue engineering, *Adv. Drug Deliv. Rev.*, 2007, **59**(4–5), 187.
50. J. J. A. Barry, M. Silva, V. K. Popov, K. M. Shakesheff and and S M. Howdle, Supercritical carbon dioxide: putting the fizz into biomaterials, *Philos. Trans. R. Soc. London A*, 2006, **364**(1838), 249.
51. L. J. White, S. M. Howdle and K. M. Shakesheff, Controlling the architecture and mechanical properties of supercritical fluid foamed poly D,L-lactic acid scaffolds, *Tissue Eng., Part A*, 2008, **14**(5), 102.
52. E. Weidner, Z. Knez and Z. Novak, Process for preparing particles or powders, *World Patent*, WO9521688, 1994.
53. B. Weinreich, R. Steiner and E. Weidner, Method for producing a powder product from a liquid substance of a mixture of substances, *World Patent*, WO9917868, 1999.
54. H. Kröber, Partikelherstellung mit überkritischen Fluiden – Ein Vergleich zwischen PCA- und RESS-Prozess, *Chem. Ing. Tech.*, 2005, **77**(3), 248.

55. E. Weidner, High pressure micronization for food applications, *J. Supercrit. Fluids*, 2009, **47**, 556.

56. M. Türk, Formation of small organic particles by RESS: experimental and theoretical investigations, *J. Supercrit. Fluids*, 1999, **15**, 79.

57. S. Grüner, F. Otto and B. Weinreich, CPF-technology – a new cryogenic spraying process for pulverization of liquid, in *Proceedings of the 6th International Symposium on Supercritical Fluids*, Versailles, 2003, p. 1935.

58. RAPS, www.raps-industrieservice.de (last accessed 7 July 2011).

59. Confederation of National Associations of Tanners and Dressers of the European Community (COTANCE), *The European Tanning Industry Sustainability Review*, COTANCE, Brussels, 2002.

60. M. Renner, E. Weidner and G. Brandin, High-pressure carbon dioxide tanning, *Chem. Eng. Res. Des.*, 2009, **87**, 987.

61. K. Joseph and N. Nithya, Material flows in the life cycle of leather, *J. Cleaner Prod*, 2009, **17**, 676.

62. E. Weidner and H. Geihsler, Verfahren zur Zurichtung von tierischen Häuten oder Fellen, *German Patent*, DE19507572A1, 1995.

63. Critical Pharmaceuticals, www.criticalpharmaceuticals.com (last accessed 29 September 2011).

64. M. J. Whitaker, J. Y. Hao, O. R. Davies, G. Serhatkulu, S. Stolnik-Trenkic, S. M. Howdle and K. M. Shakesheff, The production of protein-loaded microparticles by supercritical fluid enhanced mixing and spraying, *J. Controlled Release*, 2005, **101**(1–3), 85.

65. C. E. Upton, C. A. Kelly, K. M. Shakesheff and S. M. Howdle, One dose or two? The use of polymers in drug delivery, *Polym. Int.*, 2007, **56**(12), 1457.

66. F. Jordan, A. Naylor, C. Kelly, S. M. Howdle, A. Lewis and L. Illum, Sustained release hGH microsphere formulation produced by a novel supercritical fluid technology: *in vivo* studies, *J. Controlled Release*, 2010, **141**, 153.

67. C. A. Kelly, S. M. Howdle, A. Naylor, G. Coxhill, L. C. Tye, L. Illum and A. L. Lewis, Stability of human growth hormone in supercritical carbon dioxide, *J. Pharm. Sci.*, 2012, **101**, 56.

68. CritiTech, http://www.crititech.com (last accessed 29 September 2011).
69. CrystecPharma, http://www.crystecpharma.com (last accessed 29 September 2011).
70. L. Okavoca, D. Vetchy, A. Franc, M. Rabiskova and B. Kratochvil, Increasing bioavailability of poorly water-soluble drugs by their modification, *Chem. Listy*, 2010, **104**(1), 21.
71. N. Rasenack and B. W. Muller, Micron-size drug particles: common and novel micronization techniques, *Pharm. Dev. Technol.*, 2004, **9**(1), 1.
72. J. C. Chaumeil, Micronization: a method of improving the bioavailability of poorly soluble drugs, *Methods Findings Exp. Clin. Pharmacol*, 1998, **20**(3), 211.
73. R. B. Gupta and J. Shim, *Solubility in Supercritical Carbon Dioxide*, CRC Press, Boca Raton, FL, 2006.
74. K. E. J. Barrett (ed.), *Dispersion Polymerization in Organic Media*, Wiley, New York, 1975.
75. J. M. DeSimone, E. E. Maury, Y. Z. Menceloglu, J. B. McClain, T.J. Romack and J. R. Combes, Dispersion polymerizations in supercritical carbon dioxide, *Science*, 1994, **265**, 356.
76. A. I. Cooper, Polymer synthesis and processing using supercritical carbon dioxide, *J. Mater. Chem.*, 2000, **10**, 207.
77. T. Sarbu, T. Styranec and E. J. Beckman, Non-fluorous polymers with very high solubility in supercritical CO_2 down to low pressures, *Nature*, 2000, **405**, 165.
78. P. Raveendran and S. L. Wallen, *J. Am. Chem. Soc.*, 2002, **124**, 7274.
79. H. Lee, J. W. Pack, W. Wang, K. J. Thurecht and S. M. Howdle, Synthesis and phase behavior of CO_2-soluble hydrocarbon copolymer: poly(vinyl acetate-*alt*-dibutyl maleate), *Macromolecules*, 2010, **43**, 227.
80. N. A. Birkin, N. J. Arrowsmith, E. J. Park, A. P. Richez and S. M. Howdle, *Polym. Chem.*, 2011, **2**, 1293.
81. E. J. Park, A. P. Richez, N. A. Birkin, H. Lee, N. Arrowsmith, K. J. Thurecht and S. M. Howdle, New vinyl ester copolymers as stabilisers for dispersion polymerisation in $scCO_2$, *Polymer*, 2011, **52**, 5403.
82. T. Cosgrove, *Colloid Science: Principles, Methods and Applications*, Wiley-Blackwell, Chichester, 2010.

83. J. Zhang and B. Han, Supercritical CO_2-continuous micro-emulsions and compressed CO_2-expanded reverse microemulsions, *J. Supercrit. Fluids*, 2009, **47**, 531.

84. Y. J. Lee, B. Tan and A. I. Cooper, CO_2-in-water emulsion-templated poly(vinyl alcohol) hydrogels using poly(vinyl acetate)-based surfactants, *Macromolecules*, 2007, **40**, 1955.

85. K. Chen, N. Grant, L. Liang, H. Zhang and B. Tan, Synthesis of CO_2-philic xanthate–oligo(vinyl acetate)-based hydrocarbon surfactants by RAFT polymerization and their applications on preparation of emulsion-templated materials, *Macromolecules*, 2010, **43**, 9355.

86. J. Eastoe, M.J. Hollamby and L. Hudson, Recent advances in nanoparticle synthesis with reversed micelles, *Adv. Colloid Interface Sci.*, 2006, **128**, 5.

87. X. Fan, M.C. McLeod, R.M. Enick and C.B. Roberts, Preparation of silver nanoparticles via reduction of a highly CO_2-soluble hydrocarbon-based metal precursor, *Ind. Eng. Chem. Res.*, 2006, **45**, 3343.

88. M.J. Meziani, P. Pathak, F. Beacham, L.F. Allard and Y.P. Sun, Nanoparticle formation in rapid expansion of water-in-supercritical carbon dioxide microemulsion into liquid solution, *J. Supercrit. Fluids*, 2005, **34**, 91.

89. T. Hasell, K. J. Thurecht, R. D. W. Jones, P. D. Brown and S. M. Howdle, Novel one pot synthesis of silver nanoparticle–polymer composites by RAFT polymerisation in supercritical CO_2, *Chem. Commun.*, 2007, **38**, 3933.

90. D. W. Pu, F. P. Lucien and P. B. Zetterlund, Radical polymerization of CO_2-induced emulsions: a novel route to polymeric nanoparticles, *J. Polym. Sci. Part A: Polym. Chem.*, 2011, **49**, 4307.

91. J. G. Stevens, P. Gómez, R. A. Bourne, T. C. Drage, M. W. George and M. Poliakoff, Could the energy cost of using supercritical fluids be mitigated by using CO_2 from carbon capture and storage (CCS)?, *Green Chem.*, 2011, **13**, 2727.

IV
Materials Related to Energy Conversion, Storage and Distribution

CHAPTER 16

Battery and Fuel Cell Materials

JUSTIN SALMINEN*[1] AND TANJA KALLIO[2]

[1] European Batteries Oy, Karapellontie 11, 02610 Espoo, Finland
[2] Department of Chemistry, Aalto University, 02150 Espoo, Finland
*Email: justin.salminen@europeanbatteries.com

16.1 INTRODUCTION

The number of battery-powered and fuel cell applications is expected to increase rapidly in next few years. Rechargeable batteries and fuel cells are extensively studied for their use as stationary power sources, in electric vehicles (EVs), hybrid electric vehicles (HEVs), marine applications and in industrial vehicles. Fuel cells, in particular, have a high potential for reducing greenhouse gas emissions and could one day partly replace fossil fuel-based power plants and combustion engines in the transportation sector. Both biomass-based liquid and gaseous fuels can be used in fuel cells. Fuel cell technologies usually involve rechargeable batteries and vehicles always require rechargeable batteries even in dual-fuel systems.

The energy capacity required for global electrification of vehicles is huge and it should be noted that in terms of car production, the energy required to make batteries for 3000 electric cars is about

Materials for a Sustainable Future
Edited by Trevor M. Letcher and Janet L. Scott
© The Royal Society of Chemistry 2012
Published by the Royal Society of Chemistry, www.rsc.org

100 MWh. Consequently, an annual production of one million cars would require 100 lithium ion battery factories each with a capacity of 300 MWh.

Battery applications include mature technology related to nickel metal hydride (NiMH) batteries and lead acid batteries and also the new lithium ion batteries. The manufacturing capacity of lithium ion batteries is expanding to meet the increasing demand. Many plants are already in place for the mass production of several lithium ion technologies. Production of lithium iron phosphate (LFP) is growing rapidly and this type of battery will be a strong contender when it comes to choosing a large battery meeting increased safety requirements. Still at an early stage, but with a promising future, is the lithium titanate (LTO) battery. The mass production of fuel cells of all types is still at an embryonic stage.

Present-day lithium ion batteries have limitations, but significant improvements have recently been achieved.[1–4] The main challenges for lithium ion batteries are related to material deterioration, operating temperatures, energy and power output and the lifetime of the batteries. Degradation and lifetime are also an issue with fuel cells in addition to cost reduction. Increased lifetime combined with a higher recycling rate of battery and fuel cell materials is essential for the sustainable energy conversion industry. There are a number of options involving different chemical processes and battery and fuel types. The question really is: which chemical processes and technologies can be scaled up? The limiting factors include the availability of materials, safety, fire hazard, chemical stability and high production costs. Further limitations include the banning of cadmium in batteries by the EU.

Manufacturing processes utilise raw materials and energy. Some of the components used in batteries and fuel cells are highly valuable, such as rare earth metals in NiMH batteries, cobalt in lithium ion batteries and platinum metals in polymer electrolyte fuels cells. Metals such as zinc, nickel, aluminium, copper, manganese, lithium and iron are available in sufficient quantities but the energy required to process them will become higher in the future. This is because, in the future, there will be less easily mineable resources and the types of ores available will be more difficult to process. In addition, the valuable components tend not to be available in large quantities and the mining industry is

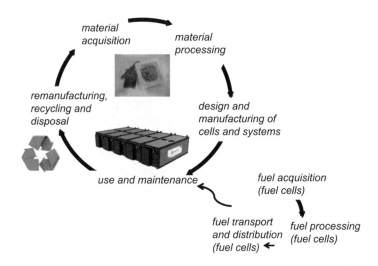

Figure 16.1   Life cycle of battery and fuel cell materials.

forced continually to develop more efficient pyrometallurgical processes and also hydrometallurgical processes that consume less water.

Both large and medium-sizes recycling plants for metals used in the electrochemical and metals processing industry will have to be developed as production volumes increase and recyclable waste becomes available in larger quantities. New innovative recycling processes involving grinding, mechanical separation, leaching, extraction, precipitation and electrochemical methods need to be developed. Recycling is driven by the need to achieve better sustainability and by legislative demands. By improving the durability of materials, longer lifetimes and increased efficiency can be obtained, and by developing more affordable and longer lasting materials, battery and fuel cell systems will become very much more available to the general public. Figure 16.1 shows the lifecycle of batteries and fuel cells.

16.2   RECHARGEABLE LITHIUM ION BATTERY MATERIALS

Batteries are devices that convert stored chemical energy into electricity within a closed system. Electrochemical conversion occurs at two electrodes, the cathode and the anode. In a rechargeable battery (secondary battery), if the external load is replaced with a power supply, the direction of electrons (and metal ions) is reversed and the battery is charged. In a lithium ion battery, lithium ions (Li^+) move from the anode to cathode (in discharging) and from the cathode to anode (in charging). Electrons move in the external circuit in the same direction as the Li^+ ions. The nature of the reactions is dependent on the chemistry of both the electrode and electrolyte–electrode interactions. All chemical reactions are temperature dependent and the most suitable temperature range for existing lithium ion batteries is between 0 to 50 °C, but in many cases it is necessary to discharge in the range between −30 and 60 °C. Keeping the correct operational temperature in a battery system will minimise the unwanted side reactions and early deterioration of the battery materials.

Recently, new materials and chemical processes for lithium ion batteries have been developed. Current research is focused on improving the operating temperature range, lifetime, durability and safety of batteries. The charging of a battery is limited for carbon anodes to temperatures of $>0\,°C$ and a relatively low charging rate. A high charging rate below 0 °C will lower the capacity, *i.e.* the energy content will be smaller.

In reality, no individual cell is used in most applications; instead, battery modules or systems are used. Real applications use clever battery management systems (BMSs); this is an electronic control unit which is able to manage both the operation and the control of the temperature. Liquid or air temperature control can be integrated into the system if needed.[5–8]

There is great emphasis on electrification in the transport sector, replacing part of motor-powered engines with battery-powered applications. There are plans to increase both the energy efficiency and overall need to reduce the consumption of non-renewable liquid fuels. Even more significant applications are dependent on energy storage. Materials required for battery applications require specially made high-quality products.

Lead acid, nickel–metal hydride and lithium ion batteries are the most common rechargeable batteries. Lead acid battery technology is well proven and is more than a century old. However, the lead acid battery has a low gravimetric and volumetric energy density and hence is impractical for use alone in EVs. However, lead acid batteries will always have a market for their own special applications.

Nickel–metal hydride batteries are used in some hybrid vehicles and have been shown to be both reliable and safe. Their downside is their relatively low energy density and low cycle life (*i.e.* number of charging and discharging cycles from full to empty) and relatively high self-discharge rate of up to 10% per month. Furthermore, NiMH batteries consume significant amounts of rare earth metals, which will become a limiting factor in the future as the limited supplies of these metals becomes important.

Rechargeable lithium ion batteries provide cell configurations that operate at over twice the potential of lead acid or NiMH cells. The reactivity of lithium has been a problem, as has been the low cycle life, especially under higher current densities. However, modern lithium-based electrodes provide a much better power density and cycle life and as a result lithium ion cells are being considered for use in vehicles. There are many different types of lithium ion batteries available, with each involving a different chemical process; as can be expected, the best types are being intensively researched.

The anode (negative electrode) of a lithium ion battery is usually graphite or lithium titanate. Graphite-based cells are already in mass production, but lithium titanate is still at the laboratory or pilot scale. The carbon anode material can be manufactured from renewable sources or natural graphite. Many types of carbon conductive additives are also used in both cathodes and anodes. The cathode (positive electrode) is typically lithium–metal oxide or phosphate. Lithium cobalt oxide, $LiCoO_2$, is the oldest type of lithium ion battery and has been produced since 1991 (Sony). Many other types of lithium ion batteries have been developed since that time; they include mixed metal oxides such as $LiCo_{1/3}Ni_{1/3}Mn_{1/3}O_2$ (NCM), $LiMn_2O_4$ (LMO) and $LiNi_{0.8}Co_{0.15}Al_{0.05}O_2$ (NCA). The recently developed $LiFePO_4$ (LFP) is considered to be the most promising lithium ion technology for large-format batteries owing to its long cycle life and its apparent safety. The LFP, LMO and LTO materials are still in the pilot phase, but powder production is rapidly ramping up to the mass production stage.

Table 16.1 shows the production status of the most common lithium ion cathode materials. The lithium ion battery contains

Table 16.1 Properties of different lithium ion cathode materials.

Cathode	Formula	Production status	Environmental impact/fire hazard	Recycling feasibility
LCO	$LiCoO_2$	Mass product	Toxic, flammable	Good
NCM	$LiCo_{1/3}Ni_{1/3}Mn_{1/3}O_2$	Mass product	Toxic, flammable	Good
NCA	$LiNi_{0.8}Co_{0.15}Al_{0.05}O_2$	Mass product	Toxic, flammable	Good
LFP	$LiFePO_4$	Pilot scale/ mass product	Non-flammable	Fair
LMO	$LiMn_2O_4$	Mass product	Flammable	Fair
LTO	$LiTi_2O_5$	Pilot scale	Non-flammable	Fair

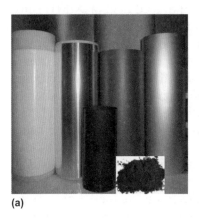 

(a)   (b)

Figure 16.2 (a) Typical materials used in a lithium ion battery (prismatic cell, *i.e.* pouch-type cell). From the left: separator polymer film, aluminium foil, lithium cathode paste coated on aluminium foil, graphite paste coated on copper foil, cover bag material. At the front, LFP powder back roll of copper foil. (b) Slurries and binders. Top right, carbon slurry; bottom left, lithium iron phosphate slurry. Top left shows PvDF and bottom right the SBR binder.

copper and aluminium foils and various amounts of other metals. Most valuable are cobalt, nickel and lithium, followed by manganese and iron. Recovering all the strategic metals such as lithium from the used batteries is vital for battery industry sustainability. Battery recycling also includes plastics, steel parts, aluminium frames and electronic cards – these components can easily be recycled with existing technology.

Figure 16.2 shows typical materials used in a prismatic pouch type cell. Aluminium and copper foil of thickness 10–25 µm are

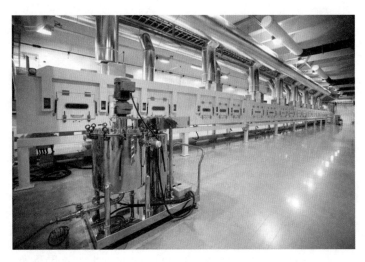

Figure 16.3 Overview of one of the LFP electrode coating lines in the European Batteries plant at Varkaus, Finland. The annual nominal capacity of the plant is 100 MWh.[6]

typically used as the electrode current collector. The separator is a porous single layer or layers of polymeric or ceramic material. The lithium cathode paste is coated on aluminium foil and the graphite paste is coated on copper foil. The cover bag material consists of layers of aluminium, polymer and nylon providing mechanical, chemical and electrical insulation. Aluminium cans or plastic containers are often used. The coated paste consists of active material, solvent (water or organic solvent), conductive additives (different types of carbon particles) and binders such as poly(vinylidene fluoride) (PvDF) and styrene–butadiene copolymer (SBR).

Figure 16.3 shows a coating machine and slurry storage tank. Here the cathode slurry is coated on the aluminium foil and the anode paste on the copper foil with a separate coating machine. The foil is then dried in a long oven, calendared, shaped and dried prior to cell assembly. After putting on the cover bag, the electrolyte is added. The cells are then sealed and sent to formation, ageing and storage sections. Aged cells are assembled into a module that includes a BMS.[6–8] The modules can be combined into larger systems which could include heat exchangers. Figure 16.4 shows the integration from a cell into a system.

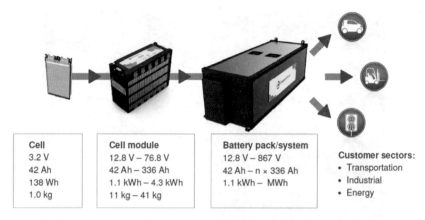

Cell	Cell module	Battery pack/system	Customer sectors:
3.2 V	12.8 V – 76.8 V	12.8 V – 867 V	• Transportation
42 Ah	42 Ah – 336 Ah	42 Ah – n × 336 Ah	• Industrial
138 Wh	1.1 kWh – 4.3 kWh	1.1 kWh – MWh	• Energy
1.0 kg	11 kg – 41 kg		

Figure 16.4   From individual cell to real applications. From the left: recharge-able lithium ion battery cell with a frame, a cell module and battery pack/system. The module and system include a BMS and temperature control. The system also includes a liquid cooling/heating unit.[6,8]

16.3   FUEL CELL COMPONENTS AND RECYCLABILITY

Fuel cells convert chemical energy of hydrogen-rich fuel and oxygen directly into electricity in electrochemical reactions, resulting in low CO_2 emissions, thanks to the inherently highly efficient processes. Unlike batteries, the fuel is stored outside the fuel cell and the fuel cell operates as long as the fuel and oxidant are fed into the system. The fuel flows in the anode, where it reacts, releasing electrons. Meanwhile, oxygen at the cathode consumes the electrons that are transported *via* an external circuit, producing electric power. Ions formed in the reaction move through the electrolyte from one electrode to the other. The reaction mechanisms depend on the system as several fuel cell technologies have been established. Fuel cells are usually classified, on the basis of the electrolyte used, into five different categories as the electrolyte determines the temperature range and offers the reaction environment affecting the selection of electrocatalyst materials. The electrolyte and electrocatalyst materials are the central components of the fuel cell, determining the lifetime and performance along with the selection of the fuel.

In this chapter on the sustainability of fuel cell technology, two promising technologies are discussed: polymer electrolyte fuel cells (PEFCs) and solid oxide fuel cells (SOFCs). Because of the very different operating temperatures set by the properties of the electrolyte, 80 or 120–200 °C for PEFCs compared with 500–700 or 800–1000 °C for SOFCs, these fuel cells have different power ranges, 5 W–400 kW for the PEFCs and 10 kW–100 MW for the SOFCs, and therefore different application areas. PEFC technology is used to provide energy for portable applications, both vehicles and auxiliary power units (APUs) and uninterrupted power production (UPS), whereas SOFCs are used for more continuous power production, *e.g.* domestic combined heat and power (CHP) units, ships, trains and power plants.[9–11]

The materials used for constructing PEFC and SOFC stacks are listed in Table 16.2. The combination of low operating temperature

Table 16.2 Materials of the main components used for constructing PEFCs and SOFCs.

Component	PEFCs	SOFCs
Anode	Pt or Pt alloy on carbon support	Cermet of YSZ and Ni
Cathode	Pt or Pt alloy on carbon support	LaSrMn or LaSr-CoFe perovskites, $LaMnO_3$
Electrolyte	Perfluorosulfonic acid $-(CF_2CF_2)_x-(CFCF_2)-$ $\quad\quad\quad\quad OCF_2CF\ O(CF_2)_2SO_3H$ $\quad\quad\quad\quad\quad\quad CF_3$ Polybenzimidazole doped with phosphoric acid	Yttria-stabilised zirconia (YSZ)
Gaskets sealing material interconnects bipolar plates and flow field plates	PTFE, EPDM, stainless steel, composite materials	Glass, ceramics ($LaCrO_3$), metals (Ni, stainless steel)

and the acidic environment met in PEFCs necessitates the use of noble metal catalyst materials for the fuel oxidation in the anode and the oxygen reduction in the cathode. In a PEFC system, operating with CO-free hydrogen and oxygen, typical platinum loadings of 0.2 and 0.4 mg cm^{-2} are used for the anode and cathode, respectively. Higher loadings at the cathode are required because of retarded oxygen kinetics. Abundant carbon serves as a support material for the noble metal electrocatalyst. Because of durability issues, perfluorinated polymers or aromatic structures are used as the electrolyte. The most commonly used perfluorosulfonic acid membrane requires the presence of liquid water for ion conductivity, thus limiting the operating temperature. Polybenzimidazole (PBI) impregnated with phosphoric acid is often used as an electrolyte at elevated temperatures above the boiling point of water. Other components needed for constructing a PEFC cell include a carbon paper or cloth diffusion layer, which is located between the electrodes and flow field of the bipolar plates. The latter are fabricated either from carbon composites or coated stainless steel. Gaskets made of polymer materials are needed for insulation and the remaining components are usually made of steel.

The US Department of Energy (DoE) estimates that the current cost of a transportation fuel cell system (2010 technology) produced in a high-volume manufacturing plant (500 000 units per year) is $51 per kilowatt. The target of $30 per kilowatt with a catalyst utilisation of 0.2 g(Pt) per kilowatt has been established for 2015. It appears that noble metals will not be replaced by other metals in the near future.[12]

In the case of SOFCs,[10,13] there is a limited choice of materials that can be used for constructing the stack because of thermal compatibility and stability problems. Thanks to more facile electrochemical reactions at higher operating temperatures, transition metals can be used as catalyst material for the SOFCs. However, in selecting the material, the thermal expansion coefficients must be considered in addition to the electrochemical and chemical properties, such as electrocatalytic activity, ionic and electronic conductivity and chemical reactivity of the materials. The most commonly used electrolyte is ZrO_2, in which ~ 8–10 % of Zr has been substituted with Y to increase the oxide ion conduction (Figure 16.5). This YSZ electrolyte is connected to an $LaMnO_3$-based perovskite cathode, the properties of which are adjusted by

Figure 16.5   SOFC single cells. A thin layer of electrolyte lies between the green anode and the black cathode. Reproduced with permission from VTT.

substituting part of the Mn with a rare earth or transition metal such as Sr, Co or Fe to improve the electrocatalytic properties and conductivity and to circumvent problems related to any mismatch in thermal expansion coefficients. The porous anode is most often made of sintering YSZ cement with a percolating nickel phase, which operates as the electrocatalyst while ensuring electrical conductivity. For high-temperature operating SOFCs, the $LaCrO_3$ is doped with alkaline earth metals and is the leading material used as bipolar plates to interconnect the planar SOFCs electrically to build up voltage. It also distributes the gases on the electrodes and separates fuel and oxygen of adjacent cells. For lower operating temperatures, coated steels are preferred for ease of manufacturing as compared with ceramic materials.

After-life reuse, recycling and disposal of materials used in constructing fuel cells have a notably high relative environmental impact when considering the total life cycle which include the low emissions from energy conversion process. Factors necessary for the fuel cell operation such as tight connections between the materials may obstruct recycling and remanufacturing by rendering the separation of the components difficult. As a result, new processes for fuel cell recycling must be introduced or the current processes must be optimised. As metals can be readily reused, they have a higher potential for recycling than other components.[14] Fuel cell hardware recycling includes steel (flow field plates, sealings, interconnections, *etc.*) and noble and transition metals in catalyst materials. Catalysts, in particular, are expected to be recycled; this is already taking place today as valuable metals from battery electrodes and catalytic converters in cars are reused. Composite flow field plates can be either recycled or incinerated. Non-fluorinated membrane materials can be used for energy production but the use of perfluorosulfonic acid membranes may be an issue as municipal incineration is not an option because of the formation of corrosive hydrogen fluoride. However, processes in which the perfluorosulfonated materials can be reused or dissolved after purification are under development and can offer a solution to the problem.[15,16] The recycling of the ceramic components could be an issue because of the chemical and physical changes that occur during the fuel cell operation. This prohibits the reuse of these components and makes it difficult to convert such materials into a reusable form.[10,11]

When evaluating the environmental impact of fuel cells, the raw material choice, fuel utilisation, manufacturing impact and the operation, and also the disposal and recycling of the components, must be considered.

16.4   DURABILITY ISSUES WITH FUEL CELL MATERIALS[10,11,17]

The devices discussed in this chapter have low energy conversion emissions compared with the high emissions and the environmental impact of the processes used in manufacturing the devices. Therefore, extending the lifetime of batteries and fuel cells is important in the light of sustainability. This can be achieved by careful design of the stack, selection of the operating parameters and developing more durable materials. Hence understanding the degradation mechanisms and improving or developing new materials are currently central research trends.

It is generally considered that a fuel cell has reached the end of its life when 10–20% of the voltage, *i.e.* power at a given current density, has been lost compared with the value at the beginning of its life. The targeted lifetimes of fuel cells depend on the application area. It is generally considered that PEFCs for portable applications should tolerate operation times of 3000–5000 h, for vehicles and buses 7000 h and PEFCs and SOFCs for stationary applications 40 000–80 000 h.

To extend the lifetime of a fuel cell, the mechanisms governing the degradation should be understood in order to develop materials and protocols to alleviate or avoid these harmful effects. Because the operating conditions affect the lifetime, fuel cells should be operated under ideal conditions if it is economically feasible. Operating outside the regular parameters of load and thermal cycling accelerates fuel cell degradation. Factors influencing the degradation under steady-state conditions include operating current density, temperature and impurities in the fuel and air.

The activity of the noble metal electrocatalyst in PEFCs decreases during operation because of loss of active electrochemical surface area due to metallic cluster deformation in the electrocatalyst and also because of the oxidation of the carbon support. Cycling increases both metal loss from the catalyst material and corrosion of the carbon support. Moreover, carbon support corrosion can proceed due to the depletion of fuel at the anode during start–stop cycles as the anode potential increases so that water is oxidised when there is not enough fuel for the oxidation process. When using a platinum alloy catalyst, problems may arise as a result of the loss of the other alloying metal and its deposition on the cathode, thus altering the potential and decreasing the membrane conductivity (Figure 16.6). Because of the high potential, the cathode is prone to detrimental corrosion, which can be reduced by operating the stack under load so that the cathode potential remains lower. In addition to the above-mentioned loss of conductivity due to impregnation of the membrane with the

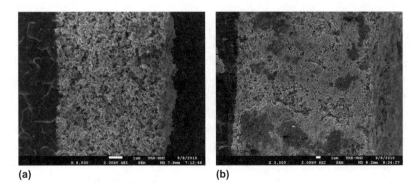

(a)   (b)

Figure 16.6 PtRu supported on a Vulcan anode (a) before and (b) after 200 h of operation in a fuel cell.

corrosion products and other contaminants, the membrane can fail mechanically due to pinhole formation. Other ageing phenomena are related to an increase of contact resistances due to the corrosion and possible delamination of the electrodes and mass transport problems because of degradation of the diffusion layers which distribute the reactants to the electrodes.

In the case of SOFCs, high temperatures combined with the flow of charge carriers and the reactants combined with local moisture induce phase changes. As a result of high temperatures and moisture during the SOFC operation, the interconnecting component tends to corrode as a result of interactions with fuel and oxygen and adjacent components such as the electrodes and electrical contact layers. This is reflected in an increase in surface resistance. Mismatch in thermal expansion, as in the case of barium–calcium–aluminosilicate (BCAS) glass seals and metal interconnecting components, leads to degradation of materials by crack formation and growth. Ageing also proceeds at other surfaces and interphases which inherently tend to change chemical composition as they strive to reach a lower energy level. Therefore, local surface relaxations occur resulting in segregation of constituents and impurities from the components and species from the reactants adsorbs on the surface. This is detrimental to both the electrolyte and the electrodes by reducing electrocatalytic activity and the transfer of the ions formed in the electrochemical reactions. Moreover, sintering and agglomeration change the microstructure of the materials, thus reducing the electrochemically active surface area and lowering performance.

16.5   FUELS FOR FUEL CELLS[9]

The choice of the fuel significantly affects the sustainability of fuel cell technology as alternatives based on utilisation of both fossil fuels and renewable energy sources are available. When choosing the appropriate fuel, not only the electrochemical properties but also issues such as production, processing, storage and transportation of the fuel must be considered. Further, the effect of the fuel on the efficiency and lifetime of the fuel cell should be taken into account when considering the total environmental impact.

The choice of the correct fuel is linked to both the fuel cell type and its application. For portable applications, the fuel must be stored on-board, hence the volume-to-weight ratio of the fuel storage is a crucial issue. Moreover, tolerance of the different fuel cell types to impurities in the fuel also varies. Thanks to the higher operating temperature, the SOFC is more fuel flexible than the PEFC with its relatively low operating temperature. For both cells, sulfur-containing impurities poison the catalyst and must be removed from fuels prior to use. However, whereas CO can be considered as a fuel for the SOFC, tolerance of the PEFC towards CO is limited.

Table 16.3 shows some properties of the most commonly used fuels for PEFCs.

Pure hydrogen is a very good fuel for fuel cells, as it can generate electrical energy with high efficiency and water is its only emission product. However, as hydrogen gas, H_2, does not exist in the Earth's crust, its production, storage and transportation are issues that must be taken into account when evaluating the environmental impact, as each step results in energy losses and emissions. Therefore, H_2 should not be considered as a zero-emission fuel. Hydrogen can be produced at local refuelling sites or centralised locations and transported by road or by water or *via* pipelines to refuelling locations. Hydrogen storage is not a serious technical problem as it can be liquefied, compressed or stored chemically, *e.g.* by absorption in metal hydrides. Furthermore, hydrogen production as a by-product of other process can help in reducing costs. However, it is not yet clear how a safe and reliable hydrogen storage network should be established in cities.

Today, the most economical way to produce hydrogen is by reforming natural gas. Shifting from oil to natural gas can be considered beneficial for the environment as natural gas has a higher hydrogen-to-carbon ratio. On the other hand, hydrogen can be produced by electrolysing water using renewable electrical power sources, such as wind or solar, providing a near zero-emission footprint. Alternatively, renewable biofuels, such as methanol, ethanol or landfill gases, can be used for hydrogen production by either reforming or hydrolysis.

Some hydrogen-rich hydrocarbons can also be directly fed to a fuel cell, resulting in on-site CO_2 emissions in addition to water. Natural gas consists of hydrocarbons with low boiling points, with methane being the main component (44–99 mol%). The overall composition of the natural gas varies with location and season, but

Table 16.3 Fuels used in polymer electrolyte fuel cells. NG is Natural gas.

Property	Hydrogen	Methane	Methanol	Ethanol	Glycerol
Molecular formula	H_2	CH_4	CH_3OH	CH_3CH_2OH	$CH_2OHCHOHCH_2OH$
Freezing point/°C	−259	−183	−98.8	−114	16
Boiling point/°C	−253	−161	64.5	78.3	290
Density/g cm^{-3}	0.07	0.60	0.79	0.78	1.26
Specific heat/ J kg^{-1} K^{-1}	28.8	34.1	76.6	112.4	2400
Flammability limits/vol.%	4–75	4–16	6–36	4–19	Not applicable
Flammability/ toxicity	Very flammable	Very flammable	Flammable, toxic	Flammable	Low hazard
Sources	Fossil: NG Renewable: solar or wind electricity for electrolysis	Fossil: NG Renewable: landfill	Fossil: NG Renewable: biomass	Renewable: biomass	Renewable: biomass

after sulfur removal it can be fed directly into high-temperature fuel cells.

Municipal wastes can also serve as a fuel source as gaseous compounds arising from landfill sites can be collected. The composition varies greatly and landfill gases can be very dilute, containing only ∼50% methane. However, after purification, landfill gases can be used directly for energy conversion in SOFCs.

Fuels produced from renewable biomass, such as methanol, ethanol and glycerol, are attractive candidates. Methanol and ethanol can be produced by fermentation from biomass whereas glycerol is a side product from biodiesel production. However, using land for the production of the biomass reduces the farming area suitable for food production and will no doubt lead to an increase in food prices and could have repercussions related to food shortages. It is noteworthy that processes have been developed to use by-products or leftovers for biofuel production (Figure 16.7).[17] Options include either direct use of ethanol in fuel cells or hydrogen production by steam reforming or electrolysis. Direct use decreases the fuel efficiency and results in on-site emissions, but is a worthy alternative for low-power applications.

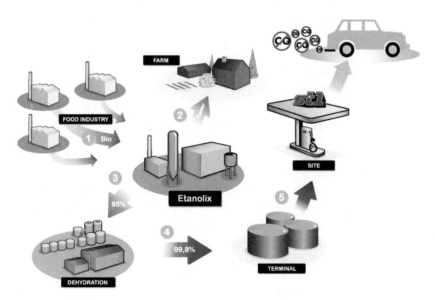

Figure 16.7 Ethanol production from food industry leftovers. Reproduced with permission from ST1.

16.6 DISCUSSION

The global capacity of industrial-scale production of larger lithium ion battery cells and systems may become a limiting factor if current plans for even partial electrification of vehicles or energy storage plans are realised. Lithium cobalt oxide cells benefit from well-established powder and cell production processes and hundreds of millions of cells are produced annually. Currently tens of manufacturers exist for different battery applications, variations, sizes and shapes. The move from small cell sizes to large-format cells has led to new safety issues for cell manufacturers and end users. Novel materials such as lithium iron phosphate (LFP) have been introduced to increase safety in large-format cells. This is gaining popularity and nearing the mass production stage.

The main challenges for lithium ion battery technology are related to the different chemical processes, material deterioration, lifetime, operating temperatures, energy and power output scale-up issues, long-term material supply for some elements and overall costs. Cost targets for lithium ion batteries are around a couple of hundred US dollars per kilowatt-hour, but the present price is several hundred more dollars per kilowatt-hour depending on the size and level of the hardware such as BMS, cooling, heating and connection cables. In the near future, increased volumes will no doubt lower costs.

Batteries are specific in their uses and one type does not fit all purposes. Challenges appear, for example, when individual cells are combined into a larger battery system. In larger combinations, cooling is required to avoid hot spots and deterioration of the lifetime due to overheating. Thermal control is also necessary for safety reasons.

Advanced lithium ion battery systems include the electronic control of BMSs, which is crucial when operating electric vehicles (EVs) and hybrid electric vehicles (HEVs). BMS also prevents both overcharging and heavy discharging of the batteries, thus increasing the operational lifetime of a system.

Major challenges for fuel cells lie in degradation and lifetime issues and overall costs. Fuel cell production volumes are still moderate compared with rechargeable batteries. In addition to niche applications, SOFCs demand for residential combined heat and power production is increasing in Japan and several fuel cell vehicles based on PEFC technology are expected to be brought on to the market in the mid-2010s. If these market introductions take place, production volumes of fuel cells can increase, bringing mass production closer. This will help in achieving the targeted system manufacturing costs of $30 kW^{-1}$ for 500 000 units sold by 2015.

Batteries and also fuel cells can be used for energy conversion in a wide variety of applications, as individual cells can be integrated into larger systems which can be carefully controlled during operation to maintain and optimise the performance and lifetime. The fuel cell technology and its fuel can be selected to comply with its application involving performance, durability, economy and environmental impact of the system.

16.7 CONCLUSION

Metals and metal products and utilisation of fossil fuels play a vital role in the industrial development of batteries and fuel cells. However, for the sustainable use of the Earth's metal resources, the end use and recycling of metals have to be taken into account. Lithium ion batteries have developed rapidly and different chemical processes have been successfully introduced. Common applications for these systems include power sources for cell phones, laptops and other portable devices. Development is currently under way into applications such as energy storage, partly or fully powered electric vehicles, industrial vehicles, lifts, cranes, harbour machines, mining vehicles, boats and submarines. Production of individual cells and battery management system electronics is ramping up globally to large modular solutions. These new applications demand huge amounts of specially made products (copper and aluminium metal foils, electrolytes, lithium metal oxide, separator polymers, binders, graphite, conductive additives, cover bags, tabs and production hardware). Diminishing amounts of easily mineable metal ores influence the long-term material availability. Iron, manganese, lithium and nickel are still relatively abundant but metals such as cobalt and rare earths are becoming progressively scarcer. Because of cost issues, fuel cells are at the pilot-scale level with increasing interest towards such applications as residential combined heat and power systems (CHPs) and vehicle use. If the prediction of the increasing need for fuel cells to power vehicles turns out to be correct, mass production of fuel cells with lower costs will further hasten the market breakthrough.

The driving force behind the growing interest in lithium ion batteries and fuel cells is the desire both to increase energy efficiency and to reduce the consumption of hydrocarbon-based fuels. The deployment of battery systems and the battery industry are growing rapidly while fuel cells are expected to enter the market sometime during the 2010s.

REFERENCES

1. D. Linden and T. B. Reddy (eds), *Handbook of Batteries*, 3rd edn, McGraw-Hill, New York, 2002.
2. M. Winter and R. J. Brodd, What are batteries and fuel cells?, *Chem. Rev.*, 2004, **104**, 4245.
3. J. M. Tarascon and M. Armand, *Nature*, 2001, **414**, 359.
4. M. Yoshio, R. J. Brodd and A. Kozawa (eds), *Lithium Ion Batteries*, 1st edn, Springer, New York, 2009.
5. A. Väyrynen and J. Salminen, Lithium ion battery production, *J. Chem. Thermodyn.*, 2012, **46**, 80.
6. European Batteries, http://www.europeanbatteries.com (last accessed 1 December 2011.).
7. D. Andrea, *Battery Management Systems for Large Lithium-Ion Battery Packs*, Artech House, Norwood, MA, 2010.
8. A.Väyrynen, Control criteria of an electric vehicle battery equalizing system, *Licenciate's thesis*, Helsinki University of Technology, Espoo, 2008 (in Finnish).
9. J. Larminie and A. Dicks, *Fuel Cell System Explained*, 2nd edn, Wiley, Chichester, 2003.
10. M. Gasik (ed.), *Materials for Fuel Cells*, Woodhead Publishing, Cambridge, 2008.
11. W. Vielstich, H. Yokokawa and H. A. Gasteiger (eds), *Handbook of Fuel Cells; Fundamentals, Technologies and Applications*, Vols 5 and 6, Wiley, Chichester, 2009.
12. US Department of Energy, *2010 Fuel Cell Technologies Market Report*, www.eere.energy.gov/hydrogenandfuelcells, June 2011 (last accessed 1 December 2011).
13. W. Vielstich, A. Lamm and H. A. Gasteiger (eds), *Handbook of Fuel Cells; Fundamentals, Technologies and Applications*, Vols 3 and 4, Wiley, Chichester, 2003.

14. C. Handley, N. P. Brandon and R. van der Vorst, Impact of European Union Vehicle Waste Directive on end-of-life options for polymer electrolyte fuel cells, *J. Power Sources*, 2001, **106**, 344.
15. F. Xu, S. Mu and M. Pu, Recycling of memebrane electrode assembly of PEMFC by acid processing, *Int. J. Hydrogen Energy*, 2010, **35**, 2976.
16. S. Grot, Platinum recycle technology development, presented at the DoE Program Review Meeting, Philadelphia, PA, 24 May 2004.
17. R. Borup, J. Meyers, Y. S. Kim, R. Mukundan, N. Garland, D. Myers, M. Wilson, F. Garzon, D. Wood, P. Zelenay, K. Moore, K. Stroh, T. Zawodzinski, J. Boncella, J. E. McGrath, M. Inaba, K. Miyatake, M. Hori, K. Ota, Z. Ogumi, S. Miyata, A. Nishikata, Z. Siroma, Y. Uchimoto, K. Yasuda, K.-I. Kimijima and N. Iwashita, Scientific aspects of polymer electrolyte fuel cell durability and design, *Chem. Rev.*, 2007, **107**, 3904.
18. St1, http://www.st1.eu/index.php?id = 2883 (last accessed 1 December 2011).

CHAPTER 17

Materials for Photovoltaics

IAN FORBES[1] AND LAURENCE M. PETER*[2]

[1] NPAC, School of Computing, Engineering and Information Sciences, Northumbria University, Newcastle upon Tyne NE1 8ST, UK
[2] Department of Chemistry, University of Bath, Bath BA2 7AY, UK
*Email: l.m.peter@bath.co.uk

17.1 INTRODUCTION

The impact on the planet of mankind's insatiable thirst for energy is becoming increasingly evident. According to the *BP Statistical Review of World Energy 2011*,[1] global primary energy consumption in 2010 was equivalent to ~ 12 gigatonnes (Gtoe) of oil, an increase of $\sim 30\%$ compared with a decade earlier (1 toe is equivalent to 42 GJ). About 87% of this energy was generated from carbon-based fuels. The corresponding thermal power output, averaged over a year, is 16 TW. Estimates of future power requirements vary, but most models predict that energy consumption will at least double by 2050.[2,3] Given the potential massive increase in CO_2 release that this could imply, a 'business as usual' scenario for the future is clearly not an option. For this reason, the Head of States Agreement has set an ambitious target of reducing CO_2 emission levels by 80–95% by 2050. Apart from carbon sequestration, the only way to achieve this is to find the additional 16 TW of power from

Materials for a Sustainable Future
Edited by Trevor M. Letcher and Janet L. Scott
© The Royal Society of Chemistry 2012
Published by the Royal Society of Chemistry, www.rsc.org

nuclear or renewable resources. Scenarios considered by the Intergovernmental Panel on Climate Change estimate the potential for power generation by photovoltaics (PV) at around 600–800 GW in 2050, *i.e.* around 2% of the total power required.[4] The International Energy Agency (IEA) 'Vision Scenario'[5] predicts a similar level of power production by PV ($\sim$500 GW), corresponding to nearly 10% of electricity generation. By contrast, a report by EPIA/Greenpeace[6] considers considerably higher figures, between 3 and 4.5 TW. Here the conservative figure of 500 GW will be used as a lower estimate for baseline calculations and estimates. This chapter examines the consequences for resources and sustainability of such large-scale deployment of PV solar cells.

To place these figures in context, let us start with some round figures. Solar cells need sunlight. The total radiant power that the Earth receives from the Sun is around 10^{17} W, *i.e.* several thousand times the estimated total primary power requirement for 2050. However, the average power *density* of sunlight is low and varies considerably depending on location. In Europe, Spain and Italy have particularly high levels of sunshine that are not too different from those in North Africa. Irradiation levels are more modest in the UK but are still usable. The European Commission's map of annual irradiation levels for Europe (Figure 17.1) shows that the southern part of the UK receives more than 50% of the annual amount arriving in North Africa.[7] Given the low power density of sunlight, an important question to answer is: what area would we need to cover with solar cells to generate the 500 GW of electrical power predicted in the IEA scenario? If we take as an example the annual energy input of the Sun for Italy and the south of France of 1500 kWh m^{-2} y^{-1} (equivalent to an average power density of 170 W m^{-2}) and use solar arrays operating at a relatively modest power efficiency of 10%, the area that would be required is around 30 000 km^2, roughly the area of Belgium. If the absorber layer in the cells is 10 μm thick, the required amount of semiconductor material would be 3×10^5 m^3. For silicon cells, which are much thicker, the amount would be at least an order of magnitude higher.

So how far have we progressed towards the target of 500 GW? The current contribution of PV to the total world energy requirement is still very small, with total worldwide installed capacity estimated to exceed 67 GWp in 2011,[8] where GWp stands for

Photovoltaic Solar Electricity Potential in European Countries

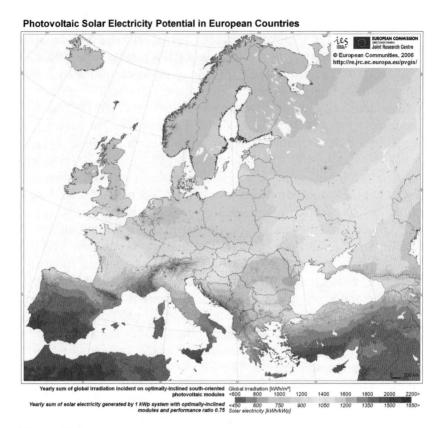

Yearly sum of global irradiation incident on optimally-inclined south-oriented photovoltaic modules · Global irradiation [kWh/m²] <600 800 1000 1200 1400 1600 1800 2000 2200>

Yearly sum of solar electricity generated by 1 kWp system with optimally-inclined modules and performance ratio 0.75 · Solar electricity [kWh/kWp] <450 600 750 900 1050 1200 1350 1500 1650>

Figure 17.1   Yearly sum of global irradiation in Europe. Taken from ref. 7.

gigawatt peak, the unit used for nominal power output based on illumination of the solar cells under standard conditions. These conditions, which are referred to as air mass 1.5 (AM1.5), correspond to an incident illumination power of $1 \, \text{kW m}^{-2}$ with a defined spectral distribution corresponding to a particular angle, 48.19°, of the Sun relative to the Earth's surface. The actual output of this installed capacity averaged over a year will be less than 20% of this nominal figure, *i.e.* below 8 GW [the output fraction of the installed capacity is known as the capacity factor (CF); typical CFs for PV flat plate systems are in the range 10–27%]. So we shall need at least 60 times the current installed capacity by 2050. This is a long way to go but, as Figure 17.2 illustrates, the total installed PV capacity is rising steeply. The annual compound growth rates of installed PV have been as high as 130%, but even if a lower figure

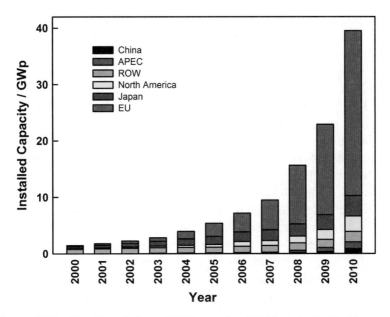

Figure 17.2 Total installed world PV capacity. APEC, Asia–Pacific Economic Cooperation; ROW, rest of the world. Taken from ref. 8.

of 110% is used, then 2 TWp would still be reached before 2050, provided that there are no new barriers. For this reason, consideration of the implications for sustainability of further rapid expansion should already be influencing research directions. The search for new solar cell materials forms an important part of this emerging scenario. Other factors that we need to consider in this chapter include the energy and carbon dioxide balances for PV based on 'cradle to grave' analyses and constraints imposed by mineral scarcity and cost. In order to examine these issues, it is necessary first to look at the different PV technologies that are available and also at newer high-risk technologies that are emerging from the laboratory and moving towards commercialisation.

17.2 EFFICIENCY LIMITS FOR PHOTOVOLTAIC SOLAR ENERGY CONVERSION

PV solar cells work by using light energy to create electronically excited states. In conventional semiconductor solar cells, the optical transition involves the promotion of an electron from the valence band to the conduction band, creating an electron–hole pair. In the case of organic solar cells, an electronically excited molecular state (exciton) is formed first and this subsequently breaks up (dissociates) into an electron–hole pair. In both cases, the minimum photon energy $h\nu$ that is required is given by $h\nu \geq E_g$, where E_g is the bandgap (or energy difference between ground and excited states in the case of organic solar cells). Photovoltaic cells are designed to separate electron–hole pairs efficiently before they can recombine (*i.e.* before excited electrons in the conduction band fall back into the vacancies or holes in the valence band). Readers interested in how this is achieved in different PV systems can find further details in two excellent textbooks.[9,10] The current produced by the cell depends on the fraction of the solar spectrum that it absorbs: the lower the bandgap, the higher is the current. On the other hand, the voltage of the cell depends on the bandgap: the higher the bandgap, the higher is the voltage. These two opposing trends give rise to an optimum bandgap for power conversion (power = current × voltage). This optimum bandgap has been calculated for different illumination conditions by thermodynamic arguments.

The thermodynamic limit for conversion of heat to work is well known as the Carnot limit. For an ideal system consisting of a reservoir at temperature T_1 and a cold sink at temperature T_2, the Carnot efficiency $\eta = 1 - (T_2/T_1)$. The calculation of the limiting thermodynamic efficiency of a solar cell is based on a similar consideration of the detailed thermodynamic balance. Details can be found, for example, in Würfel.[9] The thermodynamic efficiency limit for PV solar energy conversion is known as the Shockley–Queisser limit, after its originators,[11] who calculated it as a function of the bandgap of the semiconductor used in the solar cell. The calculation takes into account that only photons with energies, $h\nu$, higher than E_g are absorbed and that the excess energy $h\nu-E_g$ is lost as heat (vibrational energy). The Shockley–Queisser analysis indicates that the optimum bandgap for terrestrial solar cells is around 1.2 eV, which corresponds to an absorption onset in the near-infrared region at about 1 μm (1000 nm). Under standard AM1.5 conditions for unconcentrated sunlight, the Shockley–Queisser treatment predicts a maximum efficiency of around 30% for 'single-pass' solar cells with bandgaps in the range 1.2–1.4 eV. The thermodynamic limit for cells using concentrated sunlight is

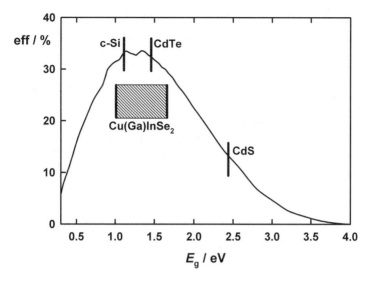

Figure 17.3 Maximum theoretical efficiency (Shockley–Queisser limit) for solar cells under AM1.5 illumination without concentration. Note that the bandgap of the CIGS system can be tuned by controlling the In:Ga ratio.

higher ($\sim$44%) and series connected stacks of cells with different bandgaps have an even higher efficiency limit (up to 60%). Cadmium telluride (CdTe) ($E_g = 1.44$ eV) is a suitable absorber material for single-pass cells. By contrast, the bandgap of CIS (CuInSe$_2$) is too low (1.02 eV), so gallium is added to create a solid solution Cu(In,Ga)Se$_2$ (CIGS) with a bandgap close to the optimum value shown in Figure 17.3. A possible replacement for CIGS is copper zinc tin sulfide (CZTS),[12] which also has a bandgap close to the optimum value. In this case, the bandgap can be tuned to lower values by partial replacement of sulfur by selenium.[13] Research-grade monocrystalline silicon solar cells have already achieved efficiencies close to the Shockley–Queisser limit, whereas thin-film solar cells still have room for improvement.

17.3 OVERVIEW OF CURRENT PV TECHNOLOGY

The current PV market is dominated by silicon solar cells, which come in several different forms: monocrystalline, multicrystalline and amorphous. The market share for crystalline and multi-crystalline silicon solar panels in 2010 was 87%, with the other

13% made up from different types of thin-film solar cells including cadmium telluride (CdTe) (6%), copper indium (gallium) diselenide (CIGS) (2%) and amorphous silicon (5%).[14] Schmidtke[15] has recently given a comprehensive overview of the commercial status of thin-film technologies and materials and the interested reader is referred to this paper for a comprehensive discussion of cell designs. The present chapter focuses on sustainability issues associated with thin-film solar cells utilising compound semiconductors rather than crystalline silicon, since these are generally predicted to take an increasing market share. One of the key advantages of thin-film materials is that they absorb light much more strongly than silicon, so that thin layers of the order of a few microns can be used in the device. However, on the downside they may utilise toxic or rare elements.

17.3.1 Mono-/Multicrystalline Silicon Solar Cells

The basic structure of a state-of-the-art monocrystalline silicon solar cell is shown in Figure 17.4. The device is fabricated from a silicon wafer with a typical thickness of 200 μm. This thickness is required to guarantee efficient absorption of red and near-infrared light, which are more weakly absorbed than blue light by

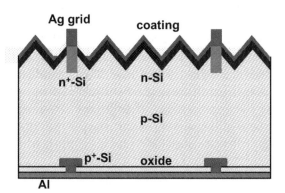

Figure 17.4   Simplified cross-sectional diagram of the structure of a typical monocrystalline silicon solar cell. Light enters the cell through the textured top surface, which is passivated by SiO_2 and coated with an anti-reflection layer. The top contact is a metallic grid and the bottom contact is an aluminium layer separated from the p-Si by a thin layer of SiO_2.

the silicon. The top surface is etched and coated with an anti-reflection layer to minimise reflection losses. Electron–hole pairs created by light absorption in the p-type bulk of the wafer are collected at the selective contacts (heavily doped p-type contacts for holes at the base of the cell and the heavily n-type top layer for electrons). The cell is contacted by a metal grid (usually silver) on the surface of the cell and by an aluminium layer at the base.

The high energy and materials cost of producing silicon wafers have stimulated the growth of silicon cell technologies based on replacing the monocrystalline wafers with polycrystalline silicon produced by the Siemens process in which polycrystalline silicon is grown on high-purity silicon rods by decomposition of tri-chlorosilane at 1150 °C. The lowest cost material is upgraded metallurgical-grade (UMG) silicon, which promises to compete with polycrystalline silicon in the longer term.

17.3.2 CdTe Thin-film Solar Cells

Cadmium telluride-based solar panels are currently the most rapidly expanding thin-film PV technology, with module sales from First Solar reportedly passing the 1 GWp mark in 2009. The technology is particularly interesting since, according to First Solar, it currently offers the lowest production cost of any form of PV: US$0.74 per Wp.[16] The cell design is based on the so-called superstrate config-uration, in which the active materials are deposited on a transparent conducting glass window to allow solar radiation to reach the device junction. The glass sheet is coated with a thin film of a transparent conducting oxide, usually fluorine-doped tin oxide (FTO), and the solar cell is fabricated by sequential deposition of layers of CdS and CdTe, usually by closed-space sublimation, followed by a conduct-ing back contact. The structure of the CdS|CdTe cell is illustrated schematically in Figure 17.5. The record laboratory CdTe cell effi-ciency is 16.7%, with record module efficiencies at 12.8%.[17]

17.3.3 CIGS Thin-film Cells

The efficiency record for laboratory for CIGS cells is now close to 20%.[17] However, commercial production is some way behind CdTe. Most CIGS cells are fabricated by co-evaporation or sput-tering processes involving reaction in a selenium atmosphere.

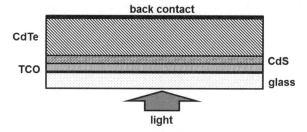

Figure 17.5   Cross-sectional diagram of the structure of a typical CdS|CdTe
solar cell. Light enters the cell through the conducting glass
substrate.

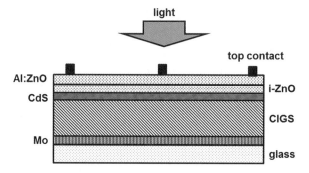

Figure 17.6   Cross-sectional diagram of the structure of a typical CIGS
solar cell. Light enters the cell through the transparent ZnO top
contact.

Record module efficiencies are currently at around 15%.[17] Most
CIGS cells are fabricated in the substrate configuration. This
means that the light enters in the opposite direction compared with
superstrate cells such as CdTe. The substrate in this case normally
consists of a layer of molybdenum (500 nm) sputtered on to a soda
lime glass substrate. The CIGS absorber (1.5–2.0 μm) is then
deposited on the molybdenum layer, followed by a very thin CdS
layer (50 nm). The cell is completed by adding thin transparent
layers of zinc oxide. The top zinc oxide layer is doped with aluminium to make it highly conducting so that it can collect the
current and transfer it to the metallic contact grid. The cell structure is illustrated in Figure 17.6.

17.4 EMERGING PV TECHNOLOGIES

17.4.1 Dye-sensitised Solar Cells

Dye-sensitised solar cells (DSCs) are a type of photoelectrochemical cell in which light is harvested by a layer of dye adsorbed on a high internal surface area of a thin porous oxide film (usually titanium dioxide, TiO_2). The porous oxide layer is deposited on conducting glass from a colloidal paste of nanocrystalline TiO_2 by techniques such as screen printing and the organic components of the paste are burnt off at 450–500 °C, leaving a mesoporous titania film with a porosity of around 50%. Light absorption by the dye molecules adsorbed on the high internal surface area of the mesoporous titania film leads to rapid injection of electrons from the excited state of the dye into the conduction band of the TiO_2 (bandgap 3.2 eV), leaving the dye in its oxidised state. Regeneration of the dye in its original state is effected by electron transfer from the iodide ions present in an electrolyte solution that permeates the mesoporous TiO_2 film. The triiodide formed in the regeneration process diffuses to a second platinised glass electrode, where iodide ions are regenerated by electron transfer. The current validated record cell efficiency is 11%. Further details can be found in a number of reviews.[18,19]

Commercial development of DSC modules is being undertaken by G24i Innovations, a company based in Cardiff. It has developed a roll to roll process for fabrication of flexible DSC modules on titanium foil. In this case, the light enters the DSC through the cathode, which consists of a transparent plastic sheet coated with a transparent conducting oxide (TCO) layer. Also in Wales, Tata and Dyesol are working on a project funded by the Welsh Assembly to develop building-integrated PVs systems based on dye-sensitised solar cells.

17.4.2 Bulk Heterojunction Organic Cells

Progress on thin-film organic photovoltaic (OPV) devices has been rapid in recent years. The cells are based on formation of a 'bulk heterojunction' between a fullerene derivative, which acts as an electron acceptor, and a conducting polymer such as P3HT (poly-3-hexylthiophene), which acts as an electron donor. These two components are dissolved in a solvent such as chlorobenzene and spin-coated on to ITO (indium tin oxide) conducting glass. As the solvent evaporates, the two components appear to phase separate spontaneously into contiguous nano-domains forming the so-called bulk heterojunction.[20] Illumination of the cell creates excitons in the polymer, which diffuse to the interface with the fullerene and inject electrons into the fullerene, creating electron–hole pairs. Electrons are collected by transport through the fullerene phase and the holes by transport through the polymer phase. The polymeric materials used in OPV devices have very high absorption coefficients, so that ultra-thin films (a few hundred nanometres) suffice to harvest light efficiently. Current record cells have achieved 10% efficiency.[17] Commercial development of OPV devices is proceeding rapidly and roll to roll plastic PV technology has the potential for low-cost fabrication, with costs below $1 per Wp.

17.5   EFFICIENCY RECORDS IN THE LABORATORY

The remarkable progress in efficiencies over the last 35 years for a range of different types of solar cells is illustrated by the plot prepared (and periodically updated) by Lawrence Kazmerski (US National Renewable Energy Laboratory). The 2011 version is shown in Figure 17.7.

It can be seen that the thin-film technologies (CdTe and CIGS) are achieving efficiencies that are around two-thirds of the Shockley–Queisser limit. Dye-sensitised and organic solar cells are late starters, but both of these low-cost technologies have reached efficiencies that make them commercially interesting if scale-up and acceptable lifetimes can be demonstrated. An interesting new entry in Kazmerski's plot is CZTSS – copper zinc tin sulfide (selenide).[13,21] This material has attracted widespread attention in the last year or so because it promises to be a viable indium-free substitute for CIGS.

17.6   LIFE CYCLE ANALYSIS, ENERGY BALANCE
AND CO$_2$ FOOTPRINT OF PV

In the context of sustainability, the most important payback time for any PV system is the energy payback time (EPBT). The fabrication, deployment and eventual recycling of a PV system all

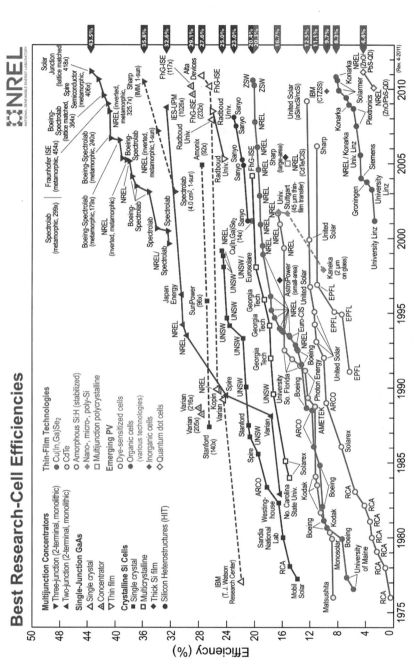

Figure 17.7 Best research solar cell efficiencies: compiled by Lawrence L. Kazmerski, National Renewable Energy Laboratory (NREL), Golden, CO.

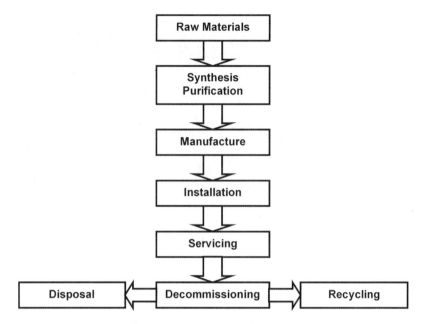

Figure 17.8   Life cycle of a PV array used to calculate the energy payback time
(EPBT, years) and CO_2 emission rate [g (C) $(kWh)^{-1}$].

consume energy. The amount of energy consumed must be calculated for a complete life cycle of the type shown in Figure 17.8. The
same life cycle is used to find the CO_2 emission rate.

The EPBT is defined as

$$\text{EPBT} = \frac{\text{total primary energy requirement of PV system throughout life cycle}}{\text{primary energy saving per year using PV system}}$$

$$(17.1)$$

and the CO_2 emission rate is defined as

$$CO_2 \text{ emission rate} = \frac{\text{total } CO_2 \text{ emission during life cycle}}{\text{annual power generation} \times \text{lifetime}} \quad (17.2)$$

As an example of the EPBT analysis, we can consider a recent
example of the calculation of the EPBT for a CdTe solar module
carried out by the Energy Research Centre of The Netherlands
(ECN) as given by de Wild-Scholten.[22] The primary energy input
derived from the life cycle analysis is 12 236 MJ kWp^{-1}. Based on

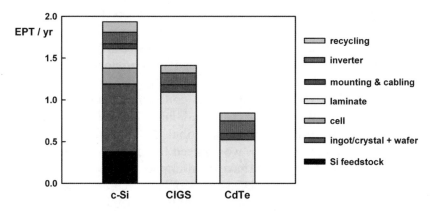

Figure 17.9 Breakdown of EPT for three PV technologies based on data given by Wild-Scholten.[22]

irradiation levels for southern Europe, the annual primary energy saving is 14 535 MJ kW^{-1}. This means that the payback time is $12\,236/14\,535 = 0.84$ years. The energy payback times for other parts of Europe can be obtained using the solar irradiance map in Figure 17.1. The energy payback times calculated by de Wild-Scholten for CIGS modules are higher, at around 1.4 years.

Figure 17.9, taken from the data given by Wild-Scholten, illustrates the relative importance of different the contributions to the EPBT for crystalline silicon, CIGS and CdTe modules. A striking feature of this breakdown in the case of the CIGS and CdTe modules is the high energy cost of the glass/laminate encapsulation of the cells and modules which dominate the EPT. By contrast, the energy input need to produce silicon is the main reason for the higher energy payback for c-Si modules.

It is clear from these calculations that thin-film PV installations begin to make a positive impact on the primary energy requirement from fossil fuels in a short time. The corresponding amelioration of CO_2 emissions is also substantial. Calculations for CdTe modules give CO_2 emission values of 20 g $(kWh)^{-1}$, which should be compared with around 500 g $(kWh)^{-1}$ for conventional electricity generation based on fossil fuels. Hence although PV is certainly not CO_2 free, its carbon footprint is so small compared with conventional power generation from fossil fuels that large-scale deployment of PV can make a substantial contribution to meeting the world energy needs while addressing the problems of climate change.

There have been several reports that consider the energy and environmental implications for various PV technologies that are at various stages of development.[6,23–28] For the newer technologies, there is a shortage of information concerning factors such as equipment capital, running costs and materials usage for large-scale plants. In order to install a global PV capacity of 2 TWp, individual production plants would need to have an annual output of at least $1\,\mathrm{GWp}\ \mathrm{y}^{-1}$ to have any impact. For plants of this size, economies of scale become possible.[29] Crystalline silicon is already produced in large-scale plants and CdTe has a multi-$\mathrm{GWp}\ \mathrm{y}^{-1}$ production capacity.[16,30] CIGS has recently seen the first $\mathrm{GWp}\ \mathrm{y}^{-1}$ production facility become fully operational.[31] Performance and lifetime data for commercial modules are not available for technologies that have received initial investment for commercial production but have not yet been produced on a significant industrial scale (*e.g.* DSCs and OPV devices).

EPBT and life cycle analysis studies involve various assumptions and limitations. Current commercial technologies have all had at least a decade of outdoor and accelerated environmental testing to provide confidence in module lifetime claims. However, newer technologies are only in the early stages of this testing process. At the same time, flexible modules do not have the inherent rigidity or robustness needed for use as building products without additional supporting structures and, notwithstanding their low initial cost, they are likely to require regular replacement. Such issues can often be overlooked when assessing these technologies, but it should be recognised that these factors add to the cost and also the EPBT and CO_2 emission rate values. A recent assessment of OPV devices suggests yearly replacement of modules, which are burned to produce energy and recover any valuable materials.[32] The EPBT and CO_2 emission rate in such a case will depend on the feedstock and manufacturing process, but it is also necessary to consider the energy and CO_2 associated with removal, transport and recycling processes. Regular replacement may allow upgrading of system performance,[25] but a minimum performance improvement is necessary to make it worthwhile in terms of EPBT and CO_2 amelioration. Another resource factor that should be considered is the role and best use of conventional fossil fuel-based energy in PV production technologies. Studies have modelled the benefits of producing PV panels with this energy compared with using it for immediate needs.[33]

In the short term, the scenarios for increased PV market growth to achieve TWp installed capacity will depend to a considerable extent on the existence and form of support mechanisms, such as feed-in-tariffs. As PV approaches parity with retail and wholesale electricity in the medium term and beyond 2020, it will need to maintain growth in a more challenging market environment, where it has to compete against other energy generation technologies.[34,35] Notwithstanding these unknown factors, there will be an increasing need to recycle materials from modules and systems as the installed capacity increases and is replaced. Recycling modules at the end of their operating life can reduce environmental and resource impacts related to PV technologies. Within the period up to 2050, the guaranteed 20–25 year lifetime of currently installed modules will expire and the system inverters will need to be replaced earlier, as these typically have lifetimes of less than 10 years. The need to replace existing capacity will place additional demands on PV production capacity. The volume of modules needing to be replaced will grow exponentially as 2050 approaches and the environmental and energy costs of the recycling process will become clearer. The most effective approach to recycling will depend on a number of factors that include the cost of transportation, the scale and the value (in terms of financial, resources and environmental impact) of the recycling facilities. Some initial work on recycling thin-film modules has indicated that failure to remove the thin-film materials properly results in their incorporation into the recycled glass, reducing its quality.[36] Other work has shown that the benefits will vary but, if optimised, recycling can become an increasingly important factor in resolving cost, resource and environmental constraints on large-scale sustainable growth.[37,38]

17.7 RESOURCE IMPLICATIONS OF LARGE-SCALE DEPLOYMENT OF PV

Each PV technology is based on specific materials such as absorber and window layers that need to be considered separately in the context of large-scale deployment. On the other hand, some materials that are part of both module- and system-level requirements are common to all technologies. For example, glass is used currently as the front transparent protective cover for the majority of modules and for all the highest performing products. Glass–glass modules incorporate an additional sheet of glass. A further common component of most conventional module designs is an aluminium frame. Freestanding PV systems incorporate modules in a frame and support structure, whereas building-integrated PV (BIPV) can utilise the building structure for mechanical support. At the heart of each PV technology is the material used for absorbing the light and, as these materials are critical to the power conversion process, this section considers how resources and production rates may limit the expansion of particular technologies, with particular emphasis on thin-film alternatives to silicon.

The three main inorganic materials that are currently in significant commercial PV production are those based on silicon (single-crystal, multicrystalline and amorphous/micromorph thin film), cadmium telluride (CdTe) and copper indium gallium diselenide (CIGS) technologies. The resource implications for newer thin-film technologies that are beginning to be produced commercially (DSCs and OPV devices) together with new emerging technologies (such as CZTS) will also be considered. DSC technologies are currently dependent on platinum group metals, since they are based on ruthenium dyes and use a platinum-coated electrode. Owing to the absence of any primary production within the EU, these materials are on the EU 'critical raw materials' list.[39–41] Demand for ruthenium was just over 30 t in 2010, with the whole of the platinum group metals amounting to less than 1000 t.[42,43] DSC technologies also use TiO_2, but this is an abundant material: all forms of the mineral amount to a total annual global production of over 16×10^6 t.[44] Even though nanoparticulate TiO_2 is needed for DSCs, the current production is estimated at above 10 000 t in the USA alone.[45] Tin oxide-coated glass is used as the transparent electrodes and although it may be possible to replace the glass with other materials such as transparent polymers in the future, the transparent conducting oxide will still be required. Table 17.1 shows the active and critical materials for the main PV technologies.

Materials that that have a significant associated energy content, and cost and scale with system size, can be considered under two headings: BOM (balance of module) materials such as glass,

Table 17.1 Active and critical materials for PV technology.

PV technology	Si (all)	CdTe	CIGS	CZTS	DSC	OPV
Active material	Si (a-Si)	Cd, Te	Cu, In, Ga, Se, (S)	Cu, Zn, Sn, Se, S	TiO_2, Ru	C_{60} derivatives
Critical components	Ag contacts		Mo	Mo	TCO (Pt, Sn)	TCO (In)

encapsulants and aluminium (module frames), and BOS (balance of systems) components, such as aluminium or other mounting structures and copper (electrical cables). The quantity of module material depends on the technology and module design. Typically, about half of the materials cost of PV modules is associated with inactive components, which are responsible for the majority of a module's mass. The BOS costs for current best practice system designs represent between 40 and 50% of the whole system costs, and the majority of these are associated with mounting structures and cables. Most BOM and BOS materials (glass, aluminium and copper, for instance) do not have significant minerals resource implications. However, each material is associated with significant energy consumption during production. Since the quantity of BOM and BOS materials increases with the area required for a PV system with a given power output, more material is required for low-performance technologies. Energy consumption and costs are also associated with transport of PV arrays when the system is installed and when it is dismantled at the end of its operational life. Technologies with significantly shorter lifetimes (*e.g.* OPV devices) would be expected to have higher transport and dismantling costs for a given total energy output since they need to be replaced more often.

In order to compare the materials resource requirements for the active components of different PV technologies, we need to consider the mass of the relevant elements required to deliver a given electrical power output (Wp). The limit to deployment for a particular technology is ultimately determined by exhaustion of the corresponding material resources. However, determination of the material resource limit is difficult because estimates of the abundance of elements in the Earth's crust do not provide any indication of their distribution there.[39] To be economically viable for exploitation, elements need to be available in sufficient concentration. The term 'resources' is easier to define in terms of exploitable minerals and economic drivers. Other terms used to define the availability of a mineral to be exploited include the 'reserves base' and mineral 'reserves': definitions of these terms are given by the US Geological Survey (USGS).[43] Where information is available, the reserves base can be used as a crude upper limit to the amount of a material that is available. However, other factors, such as a given mineral being a by-product of a more abundant mineral, can become important in determining the supply and impact on the availability for PV. The current world production level of a given material is probably a more useful metric. If PV manufacture were to consume a significant fraction of this production, then additional mining and refining would be needed, which could represent a significant obstacle to development.

Figures 17.10 and 17.11 show the global reserves/reserves base and annual world minerals production for the main PV materials.[46–48] It should be noted that indium, tellurium and gallium are

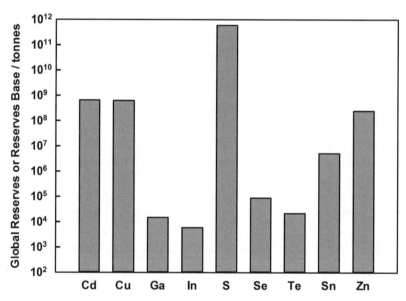

Figure 17.10   Global reserves or reserves base for the elements used in current and emerging inorganic PV technologies based on data from the USGS mineral reports.[43] It should be noted that Indium Corporation estimates the global indium reserves at about 50 000 Mt.

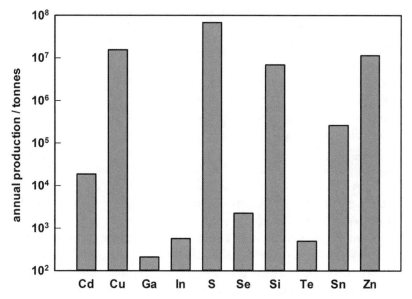

Figure 17.11 Annual global production of source minerals for the elements used in current and emerging flat-plate PV technologies. The variations between the annual production values of the different PV materials are most clearly seen between the elements that are primary (Cu, Zn, Sn, *etc.*) and secondary (Ga, In, Se and Te) products. These figures are reflected in the differences in resources/reserves base shown in Figure 17.10.[44]

produced as by-products of other metals production and their availability is strongly dependent on the markets of the major product. Te is a by-product of copper refining and both indium and gallium are separated during zinc refining, which means that their annual production rates are directly linked to the production rates of the parent materials. A further factor that needs to be considered is the geographical distribution of resources. Strategic concerns regarding resources have been the focus of several reports, including those produced by the European Commission and the US Department of Energy.[40,49] In the context of PV, it is important to note that indium (used for CIGSe and for conducting glass) is considered a critical material in all of these reports. In general, the minerals production data used are from the *World Minerals Production 2006–2010* report by the British Geological Survey[44] and the reserves data are from the USGS 2011 *Mineral Commodity*

Summaries.[43] The figures for ruthenium are from the *Platinum 2011* report produced by Johnson Matthey PLC[42] and the USGS. Fullerene production is not generally reported and is based on estimated US production alone.[45] Fullerenes are the feedstock for the best performing OPV materials and are produced with a yield of about 2%[50] from toluene, which, in turn, originates from the petrochemical industry. Typically they require over 500 MJ(electrical) kg^{-1} and over 5400 MJ(thermal) kg^{-1} to convert the toluene into fullerenes in production,[51] and additional processing into OPV materials increases the energy content further. The figure does not take into account the energy required for toluene manufacture or subsequent processing. Although the amount of material is small within an OPV device, the efficiency and capacity of production will need to increase significantly if these materials are to be capable of meeting multi-TW markets.

In order to calculate the quantities of materials required for each technology, a PV electricity generating capacity of 500 GWp will be used here as a conservative reference scenario for the deployment of PV by 2050. Taking into account realistic capacity factor values, this implies an installed capacity of at least 2 TWp. There have been several of studies of the resource implications of large-scale (TWp) deployment of PV.[46,52] In order to estimate resource requirements, it is necessary to make some assumptions about the likely efficiencies for modules by 2050. Therefore, efficiency values predicted for 2020 in the second edition of the EUPV *Strategic Research Agenda* (2011) (PVSRA)[53] have been considered as an appropriate average for the time period considered. Using these efficiency values together with layer thicknesses and material density values, the total mass of each of the relevant elements can be estimated for 2 TWp installed capacity. Only flat-plate (non-concentrating) and single-junction PV are considered here as the least geographically specific form of PV that is likely to meet and surpass the 2050 multi-TW installed capacity target. However, multijunction PV based on III–V materials will be subject to similar resource issues if used on a large scale.

Table 17.2 shows the target PV module efficiency values identified in PVSRA[53] that have been used for estimating the quantity of materials needed for each technology. To determine the mass of each component material required to meet a 2 TWp installed capacity, the mass of each absorber material must first be

Table 17.2 Data used to calculate the mass of materials used per unit of peak PV power output.

PV technology	Conversion efficiency [2020]/%	Module lifetime [2020]/y	Absorber thickness/ μm	Volume per/m³	Density/ kg m⁻³	Absorber mass per unit peak power/g kW⁻¹
c-Si	21	>35	100	10^{-4}	2330	1110
mc-Si	19	>35	100	10^{-4}	2330	1230
a-Si	14	>30	1	10^{-4}	2330	17
CdTe	16	>30	2	2×10^{-6}	5860	73
CIGS	17	>30	2	2×10^{-6}	5770	68
CZTS	15	>30	2	2×10^{-6}	5170	69
DSC (Ru)	10	>10	10	10^{-5}	–	0.8
OPV (C$_{60}$)	10	>10	0.5	5×10^{-7}	1500	7.5
Ag						80–90

determined per peak watt of installed power. To calculate the power available from each technology, the efficiency values are used together with standard AM1.5 test conditions (STC). Although real systems would operate under much more variable conditions than STC, these standard measurements permit comparisons between technologies and installed capacity values. In order to simplify the estimate of resource requirements, inorganic technologies are considered as having an operating lifetime of 30 years. It is acknowledged that lifetimes may have increased to more than this by 2050 since the c-Si and mc-Si technologies are predicted to have lifetimes of over 35 years before 2030.[6,53] The Ge content in multi-bandgap a-Si devices is not considered here. Germanium is a known critical material and at present multi-bandgap processing increases the cost and energy content of a-Si technologies without increasing the performance to the values needed to compete with rival technologies.

The mass per peak power for each absorber material is determined from its density and the volume required per square metre (*i.e.* the film thickness) together with the percent efficiency (under standard AM1.5 conditions) for the associated PV technology:

$$\text{mass/peak power} \times \frac{\text{thickness} \times \text{density}}{(\text{efficiency}/100) \times 10^3} \tag{17.3}$$

The constituent elements of the PV materials are determined from their relative proportion in terms of atomic mass to the molecular mass, which is applied to the mass per watt figure to yield the elemental mass per unit peak power:

elemental mass/peak power = absorber mass/peak power

$$\times \frac{\text{atomic mass} \times \text{number of atoms of per molecule}}{\text{molecular mass of absorber}}$$

$$(17.4)$$

Table 17.3 shows the relative elemental mass/Wp values for the different active PV materials.

The above calculations were based on a simplified scenario to obtain approximate estimates of the variations in global mineral production that would be required for each technology. Where extra capacity will be needed, it was assumed that this would grow at a rate that matches the PV requirement. However, Figure 17.12 in the next section, which illustrates a possible growth model, indicates that the contribution of CdTe and CIGS to the required growth in annual PV production could become constrained fairly soon by the rate of mineral production. In particular In, Ga, Te and Ru (and Se) are of particular concern because they are by-products in the production of other materials. Ga is present in bauxite at a concentration of 50 ppm. In is present in zinc deposits with an abundance of 1–100 ppm and it is also found in copper, lead and tin ores. Tellurium is a primarily recovered from copper refining, but the production is decreasing due to the introduction of electro-winning processes. The increase in annual production of the parent ores needed to satisfy the materials demands of CIGS and CdTe PV technologies would be between 100 and 500%. For example, if the concentration of Ga in bauxite is 50 ppm, the required increase in primary mineral production would be at least 200% to meet demand for CIGS and would result in aluminium production also needing to increase by this amount. In the case of In present at 1–100 ppm levels in zinc deposits, the production of zinc would have to increase by an amount that would be uneconomic and detrimental to the market for the primary products. In the case of Te for CdTe-based PV, electrolytic refining of Cu would need to increase by ~500%. Even assuming that the resources are available and the market for the primary metal could sustain the increase in primary minerals production, it would

Table 17.3 Elemental masses required for 2 TWP installed PV capacity compared with current annual production.

PV technology	ST conversion efficiency [2020]/%	Elemental mass per peak watt/mg Wp^{-1}	Area required for 2 TWp installed capacity/10^3 km^2	Elemental mass required for 2 TWp installed capacity/kt	Current annual global production [mine/refinery]/kt	Percentage of annual production for 2 TWp installed capacity/%
c-Si	21	1110	9.5	2220	6900	32
mc-Si	19	1230	10.5	2450	6900	36
Ag		80–90		160–180	22.2	720–830
a-Si	14	16.6	14.3	33		
	16		12.5			
CdTe						
Cd		34.4	78	69	18.7	370
Te		39.0		0.5	15 600	
$Cu_{0.7}In_{0.3}GaSe_2$	17		11.8			
Cu		13.4		27	15 500	0.2
In		16.9		34	0.6	5840
Ga		4.4		9	0.2	4220
Se		33.3		67	2.3	2950
$Cu_2ZnSn(S,Se)_4$	15		13.3			
Cu		16.4		33	15 500	0.2
Zn		8.5		17	12 000	0.1
Sn		15.3		31	260	12
S		8.3		17	68 000	0.02
Se		20.4		41	2.3	1800
DSC[a]	10	0.8	20.0			
Ru[a]				1.6×3	0.03	5160
OPV[a]	10	4.2	20.0			
C_{60}-based materials (50% blend)				8.4×3	$0.001–0.04^b$	210–8400

[a] Assuming 3× installation in a 30 year lifetime if not recycled.
[b] Current estimate for the USA only.

represent considerable increases in energy consumption and CO_2 production. Aluminium, copper and zinc and also silicon and glass have significant embodied energy values and PV technology choices will need to take these into account if secondary products require increases in their production capacity. It should be noted that the performance and lifetime of the modules will have a significant effect on these estimates if they vary from the assumed values.

In addition to the active materials represented in the tables and figures above, the quantity of glass that will be needed is estimated using a density of $2500\,kg\,m^{-3}$ and a thickness of 3.2 mm. The figures show that a minimum of between 750 and 1600 Mt would be required, which, if produced equally over the period 2020–2050, would be equivalent to an annual production requirement of between 24 and 52 Mt. Compare this to the global annual production of flat glass in 2009 was ~ 50 Mt, of which between about 64 and 200 kt was used for solar applications.[54] Table 17.4 indicates that if no alternative to the current specifications of module glass can be found, *additional* capacity equivalent to the entire current global flat glass production will be needed. Although the raw material for glass is sand and is readily available, glass production is a high-temperature process and the energy and CO_2 embodied in the material is high.

Alternatives to glass substrates may be used where practical and polymeric materials are often proposed for roll to roll manufacture to reduce costs and increase production rates. Polymer-based materials have also been suggested as future alternatives for the transparent front protection and environmental barrier for PV modules. However, it should be noted that the specific embodied energy and CO_2 values for glass ($15\,MJ\,kg^{-1}$ and $0.85\,kg\,CO_2\,kg^{-1}$), copper and steel are equal to or less than those for current industrial polymeric and plastic materials ($\sim 80\,MJ\,kg^{-1}$ and $\sim 2.5\,kg\,CO_2\,kg^{-1}$), according to Hammond and Jones.[55] By contrast, the volumetric embodied energy of polymeric materials is less than that for these high-density materials. Apart from the potential increase in production rate from roll-to-roll processing, the use of polymeric materials offer less advantages, with respect to embodied energy and CO_2, than expected. Their dependence on crude oil as a feedstock together with a need to use techniques such as organic–inorganic multilayer combinations, using vacuum or CVD processing, to create a suitable moisture barrier,[56–58] and the additional support structures needed to ensure mechanical stability of an array and possible limits

Table 17.4 Calculation of glass requirements and embodied energy for different PV technologies.

PV technology	STC conversion efficiency [2020]/%	Area required for 2 TWp installed capacity/10^3 km^2	Mass of 3.2 mm thick glass required for 2 TWp installed capacity/Gt	Annual production required to meet 2 TWp installed capacity/Mt y^{-1}	Embodied energy/10^{12} J	CO_2/Mt
c-Si	21	9.5	0.76	25	61	650
mc-Si	19	10.5	0.84	27	67	710
a-Si	14	14.3	1.14	37	91	970
CdTe	16	12.5	1.00	32	80	850
Cu(In$_{0.7}$Ga$_{0.3}$)Se$_2$	17	11.8	0.94 (×2 – substrate)	31 (62)	75 (150)	800 (1600)
Cu$_2$Zn(Sn$_{0.5}$S$_{0.5}$Se)$_4$	15	13.3	1.00	32	80	850
DSC	10	20.0	1.60 (×3 over lifetime)	52 (155)	128 (380)	1360 (4080)
OPV	10	20.0	1.60 (×3 over lifetime)	52 (155)	128 (380)	1360 (4080)

to their lifetime, may all reduce the advantages that are perceived for plastics for the medium- to long-term future. Other materials such as silver for contacts for crystalline silicon, indium in the indium tin oxide contacts for OPV and platinum used in the contacts of DSC devices represent potential materials constraints to these technologies. Alternatives will be needed and are being sought, in addition to overcoming the constraints set by the active materials. The environmental impacts due to possible harmful effects of new materials based on dyes and nanoparticles will also need to be assessed because many of these factors are not understood at present.[45,59]

17.8   LOGISTIC GROWTH MODEL FOR ESTIMATION OF RESOURCE REQUIREMENTS FOR PV

In order to estimate the likely requirements for different elements, we need to model the growth of installed PV capacity over the next 40 years or more. This is a difficult task because many factors are involved. The current installed PV capacity is following an exponential growth pattern, but clearly at some stage the installed capacity has to tend towards a constant value. If we consider the simplest logistic growth model, the installed capacity $P(t)$ would follow the expression

$$P(t) = \frac{P_{ss}P_0 e^{rt}}{P_{ss} + P_0(e^{rt} - 1)} \tag{17.5}$$

where P_0 is the initial capacity, P_{ss} is the 'steady-state' installed capacity, t is the elapsed time and r is the growth rate. Figure 17.12 illustrates the type of behaviour predicted by 17.5. The growth rate was chosen to give a reasonable fit to the growth rates in the last decade and the steady-state installed capacity was set at 2 TWp. It can be seen that the model predicts that the installed capacity would already reach the target 2 TWp by 2030, with the predicted annual installation rate peaking at 140 GWp in 2026. More complex growth models can be used (for example, taking into account variations in growth rate dictated by market conditions and subsidies), but here we use the simple logistic growth model to examine the potential demands placed on resources by the increasing installation rate. Figure 17.12 also shows the current annual production rates of the three most critical elements, In, Te and Se, converted to the equivalent installation rates assuming that the

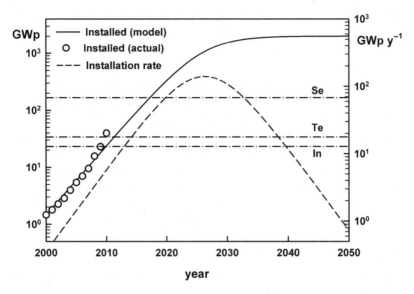

Figure 17.12 Logistic growth model of installed PV capacity calculated for a final capacity of 2 TWp and a growth rate $r = 2.8$ (solid curve). The installed capacity data for the last decade are shown as open circles. The installation rate predicted by the model (dashed curve) peaks at more than 100 GWp y^{-1}. The maximum possible installation rates for CIGS and CdTe thin-film PV corresponding to *total* consumption of In, Se and Te at current production rates are shown as horizontal lines.

entire supply of the element is used for PV. It can be seen that the annual installation rate will soon exceed the equivalent annual production rate of Te and both In and Se production levels are not far behind. This must eventually limit the contribution of CdTe and CIGS to the total installed capacity. It follows that the market share of these two thin-film technologies could *decrease* once the supply of raw materials becomes a critical factor. For example, if we assume that the supply rate of Te and In doubled by 2026 and that 50% of the annual production of the two elements was used by CdTe and CIGSe, respectively, then the market share for CdTe solar cells would be 28% and for CIGSe 12%.

17.9 OUTLOOK AND CONCLUSION

This brief survey of the impact of large-scale terrestrial PV installation on resources has revealed that problems with materials supply will rapidly become acute if the current increase in installation rate continues. The market fraction of existing thin-film technologies based on scarce elements such as indium and tellurium may ultimately be determined by the availability of raw materials. Alternative technologies based on Earth-abundant elements are therefore well worth pursuing now in anticipation of future supply bottlenecks. More sophisticated modelling of the growth in installed PV capacity is also desirable. Improved models should consider economic and environmental factors in addition to the simple supply issues discussed in this chapter. The role of PV in the energy mix for the current millennium is already becoming more important, but it is clear that much more needs to be done to assess the implications of the targets that have been set by governments worldwide. The importance of non-regenerative natural resources in the broader context of a sustainable system of energy supply is highlighted in a paper by Bradshaw and Hamacher[60] that appeared just as this chapter was being submitted.

REFERENCES

1. BP, *Statistical Review: BP Energy Outlook 2030*, http://www.bp. com/sectionbodycopy.do?categoryId=7500&contentId=7068481 (last accessed 7 April 2012).
2. European Commission, *World Energy Technology Outlook – 2050: WETO-H₂*, http://www.ec.europa.eu/research/energy/ pdf/weto-h2_en.pdf (last accessed 7 April 2012).
3. Shell International, *World Energy Consumption - What Might the Future Look Like?*, 2008, http://www.shell.com/home/content/ aboutshell/our_strategy/shell_global_scenarios/?gclid= CMjIzp–8pa4CFQXwzAodliSrPQ (last accessed 7 April 2012).
4. R. P.-M. O. Edenhofer, Y. Sokona, K. Seyboth, P. Matschoss, S. Kadner, T. Zwickel, P. Eickemeier, G. Hansen, S. Schlömer and C. von Stechow (eds), *IPCC Special Report on Renewable Energy Sources and Climate Change Mitigation*, IPCC, Geneva, 2011.
5. International Energy Agency, *Technology Roadmap. Solar Photovoltaic Energy*, 2010, http://www.iea.org/papers/2010/ pv_roadmap.pdf (last accessed 7 April 2012).
6. European Photovoltaic Industry Association and Greenpeace, *Solar Generation 6. Solar Photovoltaic Electricity Empowering*

the World, 2011, http://www.greenpeace.org/international/ Global/international/publications/climate/2011/Final%20Solar Generation%20VI%20full%20report%20lr.pdf (last accessed 7 April 2012).

7. M. Suri, T. A. Huld, E. D. Dunlop and H. A. Ossenbrink, *Solar Energy*, 2007, **81**, 1295.

8. European Photovoltaic Industry Association, *Global Market Outlook for Photovoltaics Until 2014*, 2011, http://www.epia. org/fileadmin/EPIA_docs/public/Global_Market_Outlook_ for_Photovoltaics_until_2014.pdf (last accessed 7 April 2012).

9. P. Würfel and U. Würfel, *Physics of Solar Cells. From Basic Principles to Advanced Concepts*, Wiley-VCH, Weinheim, 2009.

10. J. Nelson, *The Physics of Solar Cells*, Imperial College Press, London, 2003.

11. W. Shockley and H. J. Queisser, *J. Appl. Phys.*, 1961, **32**, 510.

12. H. Katagiri, K. Jimbo, W. S. Maw, K. Oishi, M. Yamazaki, H. Araki and A. Takeuchi, *Thin Solid Films*, 2009, **517**, 2455.

13. D.B. Mitzi, T.K. Todorov, K. Wang and S. Guha, *Solar Energy Mater. Solar Cells*, 2011, **95**, 1421.

14. Four Peaks Technologies, *Solar Cell Central: Solar Markets. Market Share by Technology*. http://www.solarcellcentral.com (last accessed 7 April 2012).

15. J. Schmidtke, *Optics Express*, 2009, **18**, A477.

16. First Solar, *Annual Report 2010, Key Quarterly Financial Data*, http://files.shareholder.com/downloads/FSLR/1699992302x0× 489149/7d51e913-c933-40b8-8cf1-57cf28583eba/Key%20- Quarterly%20Financial%20Data.pdf (last accessed 7 April 2012).

17. M. A. Green, K. Emery, Y. Hishikawa, W. Warta and E. D. Dunlop, *Prog. Photovoltaics*, 2012, **20**, 12.

18. M. Grätzel, *Acc. Chem. Res.*, 2009, **42**, 1788.

19. E. Kalyanasundaram (ed.), *Dye Sensitized Solar Cells*, EPFL Press, Lausanne, 2010.

20. J. Peet, A. J. Heeger and G. C. Bazan, *Accounts of Chemical Research*, 2009, **42**, 1700.

21. T. K. Todorov, K. B. Reuter and D. B. Mitzi, *Adv. Mater.*, 2010, **22**, E156.

22. M. de Wild-Scholten, *Energy Payback Times of PV Modules and Systems*, http://www.apollon-eu.org/Assets/20091218-Energier% C3%BCcklaufzeiten%20f%C3%BCr%20PV-Module%20und%

20Systeme%20-%20deWild%20-%20final.pdf (last accessed 7 April 2012).

23. N. Jungbluth, M. Tuchschmid, M. de Wild-Scholten, *Life Cycle Assessment of Photovoltaics: Update of Ecoinvent Data v2.0*, 2008, http://esu-services.ch/fileadmin/download/jungbluth-2008-LCA-PV-web.pdf (last accessed 7 April 2012).

24. V. M. Fthenakis and H. C. Kim, *Solar Energy*, 2011, **85**, 1609.

25. B. Azzopardi, C. J. M. Emmott, A. Urbina, F. C. Krebs, J. Mutale and J. Nelson, *Energy Environ. Sci.*, 2011, **4**, 3741.

26. SENSE, *SENSE Lifecycle Assessment Results*, SENSE Project (Framework 5 EC Contract No, ENK5-CT-2002-00639), University of Stuttgart, 2008, http://www.sense-eu.net/fileadmin/user_upload/intern/documents/Results_and_Downloads/SENSE_LCA_results.pdf (last accessed 7 April 2012).

27. V. M. Fthenakis, H. C. Kim and E. Alsema, *Environ. Sci. Technol.*, 2008, **42**, 2168.

28. N. H. Reich, E. A. Alsema, W. G. J. H. M. van Sark, W. C. Turkenburg and W. C. Sinke, *Prog. Photovoltaics Res. Appl.*, 2011, **19**, 603.

29. J. E. Trancik and K. Zweibel, in *Conference Record of the 2006 IEEE 4th World Conference on Photovoltaic Energy Conversion*, Vol. 5, Institute of Electrical and Electronics Engineers, Piscataway, NJ, 2006, p. 2490.

30. First Solar, *First Solar Overview*, 2011, http://www.firstsolar.com/~/media/WWW/Files/Downloads/PDF/Document-Library/Company-and-Market-Information/FastFacts_PHX.ashx?la=en (last accessed 7 April 2012).

31. Solar Frontier, *News. Announcement Number 143. Japan's Largest Solar Panel Factory Reaches Full Commercial Operations*, 2011, http://www.solar-frontier.com/news/143 (last accessed 7 April 2012).

32. N. Espinosa, M. Hosel, D. Angmo and F. C. Krebs, *Energy Environ. Sci.*, 2012, **5**, 5117.

33. M. Raugei, P. Frankl, E. Alsema, M. de Wild-Scholten, V. Fthenakis and H. C. Kim, presented at *AIST Symposium Expectations and Advanced Technologies in Renewable Energy, Chiba, Japan*, 2007, http://www.clca.columbia.edu/papers/Raugei_AIST_FINAL.pdf (last accessed 7 April 2012).

34. International Energy Agency, *Deploying Renewables: Best and Future Policy Practice*, 2011, http://www.iea.org/W/bookshop/ add.aspx?id=414 (last accessed 7 April 2012).
35. International Energy Agency, *Renewable Energy: Policy Considerations for Deploying Renewables*, 2011, http://www.iea.org/publications/free_new_Desc.asp?PUBS_ID=2474 (last accessed 7 April 2012).
36. R. Pohl, presented at the 1st International Conference on PV Module Recycling, Berlin, 2010, http://www.epia.org/events/past-events/archives/1st-international-conference-on-pv-module-recycling/1st-international-conference-on-pv-module-recycling-presentations.html
37. J.-K. Choi and V. Fthenakis, *J. Ind. Ecol.*, 2010, **14**, 947.
38. M. Suys, presented at the 1st International Conference on PV Module Recycling, Berlin, 2010, http://www.epia.org/events/past-events/epia-events-within-the-last-year/1st-international-conference-on-pv-module-recycling/1st-international-conference-on-pv-module-recycling-presentations.html.
39. British Geological Survey, *Risk List 2011*, 2011, http://www.bgs.ac.uk/mineralsuk/statistics/riskList.html (last accessed 7 April 2012).
40. Raw Materials Supply Group, European Commission: Enterprise and Industry, *Annex V to the Report of the Ad-hoc Working Group Defining Critical Raw Materials*, 2010, http://ec.europa.eu/enterprise/policies/raw-materials/files/docs/annex-v_en.pdf (last accessed 7 April 2012).
41. Raw Materials Supply Group, European Commission: Enterprise and Industry, *Critical Raw Materials for the UU. Report of the Ad-hoc Working Group Defining Critical Raw Materials*, 2010, http://ec.europa.eu/enterprise/policies/raw-materials/files/docs/report-b_en.pdf (last accessed 7 April 2012).
42. Johnson Matthey, *Platinum 2011 Interim Review*, 2011, http://www.platinum.matthey.com/publications/pgm-market-reviews/archive/platinum-2011-interim-review/ (last accessed 7 April 2012).
43. US Geological Survey, *Mineral Commodity Summaries 2011*, 2011, http://minerals.usgs.gov/minerals/pubs/mcs/2011/mcs2011.pdf (last accessed 7 April 2012).
44. British Geological Survey, *World Minerals Production 2006–10*, 2012, http://www.bgs.ac.uk/mineralsuk/statistics/world-Statistics.html (last accessed 7 April 2012).

45. C. O. Hendren, X. Mesnard, J. Dröge and M. R. Wiesner, *Environ. Sci. Technol.*, 2011, **45**, 2562.
46. A. Feltrin and A. Freundlich, *Renewable Energy*, 2008, **33**, 180.
47. K. Zweibel, in *Thin Film Solar Cells*, ed. J. Poortmans, Wiley, Chichester, 2007, p. 427.
48. C. S. Tao, J. Jiang and M. Tao, *Solar Energy Mater. Solar Cells*, 2011, **95**, 3176.
49. US Department of Energy, *Critical Materials Strategy*, 2010, http://energy.gov/sites/prod/files/piprod/documents/cms_dec_17_full_web.pdf (last accessed 7 April 2012).
50. J. M. Alford, C. Bernal, M. Cates and M. D. Diener, *Carbon*, 2008, **46**, 1623.
51. D. Kushnirand and B. A. Sandén, *J. Ind. Ecol.*, 2008, **12**, 360.
52. P. Rigby, Sustainability aspects for terawatt-scale photovoltaics, presented at Side Event – 26th European Photovoltaic Solar Energy Conference,Hamburg), 2011, http://www.photovoltaic-conference.com/images/stories/26th/3_parallel_events/Sustain abilityAspects/sustainability_event_eupvsec26_06_09_2011_peter_rigby.pdf (last accessed 7 April 2012).
53. P. N. Pearsall and D. A. Jäger-Waldau (eds). *A Strategic Research Agenda for Photovoltaic Solar Energy Technology*, 2nd edn, European Photovoltaic Platform, 2011, Available on the Web at http://www.eupvplatform.org/publications/strategic-research-agenda-implementation-plan.html (last accessed 7 April 2012).
54. Nippon Sheet Glass, *Pilkington and the Flat Glass Industry 2010*, 2010, http://www.pilkington.com/pilkington-information/downloads/pilkington + and + the + flat + glass + industry + 2010.htm (last accessed 7 April 2012).
55. G. P. Hammond and C. I. Jones, *Proc. Instit. Civil Eng. Energy*, 2008, **161**, 87.
56. S. H. Glick, J. A. del Cueto, K. M. Terwilliger, G. J. Jorgensen, J. W. Pankow, B. M. Keyes, L. M. Gedvilas and F. J. Pern, Silicon oxynitride thin film barriers for PV packaging, presented at the DOE Solar Energy Technologies Program Review Meeting, Denver, CO, 2005, paper NREL/CP-520-38959, http://www.nrel.gov/docs/fy06osti/38959.pdf (last accessed 7 April 2012).
57. L. C. Olsen, *Barrier Coatings for Thin Film Solar Cells: Final Subcontract Report NREL/SR-520-47582*, US

Department of Energy, 2010, http://www.osti.gov/bridge/servlets/purl/973968-u7NNkY/ (last accessed 7 April 2012).

58. A. Smith, *SOLAR FAIR 2010 Presentations*, 2010, http://www.solarflair10.com/sf10/sf10-presentations.html (last accessed 7 April 2012).

59. B. Bowerman and V. Fthenakis, *EH&S Analysis of Dye-Sensitized Photovoltaic Solar Cell Production. Final Report: BNL-52640*, Brookhaven National Laboratory, 2001, at http://www.bnl.gov/isd/documents/23176.pdf.

60. A.M. Bradshaw and T. Hamacher, *ChemSusChem*, 2012, **5**, 550.

CHAPTER 18

Materials for Water Splitting

MARK D. SYMES AND LEROY CRONIN*

WestCHEM, School of Chemistry, University of Glasgow, Glasgow
G12 8QQ, UK
*Email: lee.cronin@glasgow.ac.uk

18.1 INTRODUCTION AND SCOPE

In order for there to be any meaningful 'sustainable future' for humanity, there must be sustainable energy generation and storage. In the first instance and of the greatest urgency this means carbon-neutral energy generation and storage. Currently, most of the world's energy needs are met by burning fossil fuels, which has the side effect of increasing atmospheric levels of carbon dioxide (CO_2), with potentially disastrous consequences for the global climate. Furthermore, as energy demand around the world continues to rise, it is to be expected that reserves of fossil fuels will become increasingly expensive to extract and process, making them less attractive as sources of energy. With this in mind, there has been a great upsurge in interest in recent years in the development and deployment of 'alternative' energy sources that do not release carbon into the atmosphere, such as solar power, wind power, tidal power, geothermal power and even the much-maligned nuclear power. Although these sources combined (or in the case of both

Materials for a Sustainable Future
Edited by Trevor M. Letcher and Janet L. Scott
© The Royal Society of Chemistry 2012
Published by the Royal Society of Chemistry, www.rsc.org

nuclear and solar power, potentially alone) could meet the entire world's energy demands both now and as projected for 2050, their suitability as complete replacements for fossil fuels remains in serious doubt.[1] With regard to nuclear power, the costs of both set-up and operation (and more importantly the negative connotations of this technology in the minds of the general public) mean that there is unlikely to be a repetition of the frenzy of nuclear power plant construction that occurred in the 1960s and 1970s. In the case of solar, wind, tidal and geothermal power and the like (hereafter referred to as 'renewables'), the energy produced is almost invariably in the form of electricity. This presents the problem of energy storage for times when the Sun does not shine and the wind does not blow, to say nothing of the need for fuels for applications such as transportation and agriculture. Batteries, compressed air systems, pumped hydro, flywheels and other common alternatives for the storage of electrical energy have limited capacities and often require expensive, bulky or heavy components.[2] For these reasons, the storage of intermittent renewable energy as chemical bonds in a fuel is advantageous, especially for transportation and those applications where high energy densities are required.

Splitting water to give hydrogen and oxygen, with the energy input necessary to achieve this coming from renewables, is increasingly seen as an attractive way to make carbon-neutral fuels for a sustainable energy future [see Equation (18.1)]. Hydrogen itself is an excellent fuel, burning in air to give only energy and water, thus completing the sustainable cycle. Moreover, it is also useful as a feedstock chemical for making more complex molecules.

$$2H_2O \rightarrow O_2 + 2H_2$$
$$\Delta H_{298\ K} = 286\ kJ\ mol^{-1}$$

(18.1)

In this chapter, we focus on *materials* (*i.e.* bulk structures, as opposed to dissolved molecules) that can facilitate the process shown in Equation (18.1), particularly those that can be incorporated into electrochemical and photoelectrochemical cells. We further restrict our discussion to those materials that can be considered as *sustainable*; hence we only consider systems that are composed of Earth-abundant elements. Readers interested in water splitting using precious metal catalysts are directed to several excellent recent reviews in this area and the references therein.[3–6] Likewise, we do not discuss thermochemical approaches to splitting water, as the hazardous, corrosive and high-pressure environments required for such processes do not lend themselves to sustainability.[7] Given these provisos, we discuss in detail some of the latest examples of sustainable approaches to water splitting, looking first at electrochemical systems before moving on to methods for water splitting that aim to mimic photosynthetic processes by using sunlight as the requisite energy input.

18.2   WATER SPLITTING BY ELECTROLYSIS

18.2.1   Fundamentals

As noted previously, an energy input is required to decompose water into hydrogen and oxygen. Inasmuch as this is to be achieved using renewable sources of energy on a large scale and in diverse locations around the world, only two power inputs are practical: (1) direct utilisation of sunlight to effect photocatalytic water splitting and (2) initial generation of electricity from renewable sources, followed by application of this to electrolysis of water. The former direct route is discussed in Section 18.3, and in this section we concentrate on electrolysis. As this method relies on first converting a renewable energy source into electricity before splitting water, such approaches are inherently less efficient than direct solar-to-fuels conversions. On the other hand, *any* renewable source can therefore be used to generate this electricity and so electrolytic water splitting for fuels has the advantage of more input flexibility compared with solar-only systems.

An archetypal water electrolysis cell is shown in Figure 18.1. At the anode, electrons are removed from H_2O to give O_2 and protons. This is known as the oxygen evolution reaction (OER). As the electrons pass in one sense through the external circuit, the protons migrate through the electrolyte solution in the opposite sense towards the cathode. At the cathode, protons and electrons recombine to give hydrogen gas [the hydrogen evolution reaction (HER)]. These two key reactions (OER and HER) are given in Equations (18.2) and (18.3) respectively. Note that the cell in Figure 18.1 is operating under acidic conditions [it is a proton exchange membrane (PEM) electrolyser], where H_2O is the source

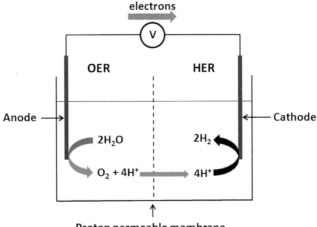

Figure 18.1 General schematic of a PEM water electrolysis cell. Under the influence of an applied potential bias (V), the OER occurs at the anode, generating O_2, protons and electrons. The electrons travel through the external circuit to the cathode, where they combine with the protons liberated in the OER to make H_2. The proton-permeable membranes most commonly used are based on Nafion.

of O_2 and H_3O^+ is the source of H_2. Cells operating under basic conditions are also possible and productive, where O_2 is generated by oxidation of OH^- and H_2O is the source of H_2. In general, these alkaline electrolysers operate under very harsh conditions (around 25–30 wt% KOH or NaOH in water) and tend to use catalysts based on Earth-abundant elements (Ni, Fe, Co, Cu, *etc.*). However, the current densities (proportional to the rate of production of hydrogen and oxygen) that such alkaline cells can reach are only around 100–300 mA cm^{-2} and it is considered that there is little room for significant improvements to these cells. Conversely, PEM electrolysers can achieve current densities that are significantly greater (up to 2000 mA cm^{-2}), but they currently rely on precious metal catalysts. PEM electrolysers are also more efficient than alkaline electrolysers and so the electrical energy costs are lower. Hence a key challenge is to identify Earth-abundant catalysts that are stable under the conditions at which PEM electrolysers operate (very roughly pH 0–9) and that are capable of producing high current densities without the expense of precious metals.[2,8]

$$2H_2O \rightarrow O_2 + 4H^+ + 4e^-$$
$$E_{anodic} = 1.23\ V - 0.059\,(pH)V\ vs\ NHE \qquad (18.2)$$

$$2H^+ + 2e^- \rightarrow H_2$$
$$E_{cathodic} = 0\ V - 0.059(pH)\ V\ vs\ NHE \qquad (18.3)$$

In theory, for both types of cell, a voltage difference between the anode and cathode of just over 1.23 V is required to drive Equations (18.2) and (18.3) at room temperature (*i.e.* 1.23 V is the potential difference at which the forward and reverse reactions proceed at the same rate). In practice, voltage differences of between 1.6 and 2.4 V are required to overcome various kinetic barriers (*e.g.* the electrical resistance of the cell), such that Equations (18.2) and (18.3) can occur at an appreciable rate.

The kinetic barrier to performing Equations (18.2) and (18.3) can be reduced by employing catalysts on the electrodes. To date, the best catalysts found for both the OER and HER are based on precious metals: for the OER, iridium oxide (IrO_2) has the highest activity, whereas platinum metal (Pt) has the greatest catalytic effect on the HER. Clearly, catalysts based on cheaper and more widely available elements are desirable for sustainable water splitting. In recent years, there has also been a move towards cells and catalysts operating at or around neutral pH. The advantages of such mild pH conditions are that cell components need not be engineered to be resistant to the extremely acidic/basic conditions normally employed in electrolytic cells and device housings do not need to meet such stringent safety standards. These both contribute significantly to lowering the cost of such cells, promoting their widespread adoption for domestic and distributed use and, hence, such neutral-pH cells are very much in the spirit of sustainable water splitting. Although sustainable catalysts that are able to operate under such mild conditions are comparatively rare, a few promising candidates exist, a selection of which are reviewed in detail in Sections 18.2.2–18.2.6.

18.2.2 Electrolytic Water Oxidation on Cobalt Oxophosphates (CoPi)

In 2008, Nocera and co-workers reported a cobalt oxide-based catalyst for the OER that self-assembles on the anode when solutions of cobalt(II) salts in neutral phosphate buffer are subjected to a modest anodic bias.[9,10] The composition of the catalyst was determined by a range of structural and analytical techniques and is believed to be based on incomplete Co–oxo cubanes, as illustrated in Figure 18.2.[11,12] Importantly in the context of the current discussion, the catalyst is composed solely of Earth-abundant elements and is able to 'self-repair' at potentials where water oxidation occurs, a crucial requirement for any viable, durable and practical device. Selective oxidation of water to produce oxygen has also been shown to be possible from phosphate-buffered seawater and from untreated river water, which further enhances the low-cost and sustainable credentials of this catalyst.[13]

A proposed mechanism of operation is also shown in Figure 18.2. Initially, soluble Co(II) ions are oxidised to Co(III), forming insoluble Co(III) oxides [possibly with phosphates associated with the Co(III) centres] as a deposit on the surface of the anode. As the anodic potential is increased, some of these Co(III) centres undergo a further oxidation to Co(IV). This is the active state of the catalyst, which oxidises water to O_2 and protons. As this happens, the

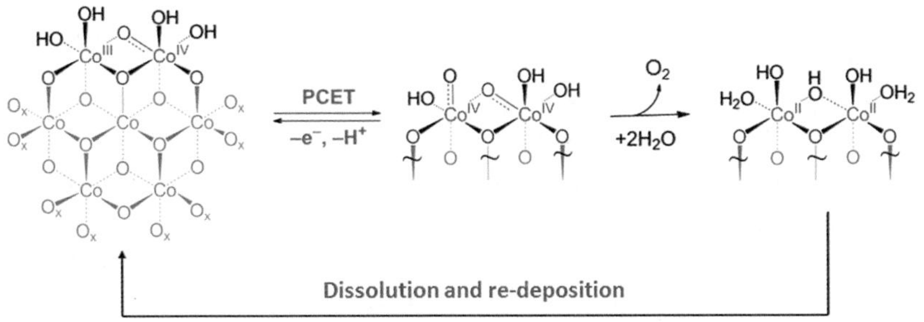

Figure 18.2 The proposed structure and mechanism of operation of the CoPi water oxidation catalyst. In the key step, the mixed Co(III)–Co(IV) resting state of the catalyst undergoes a proton-coupled electron transfer (PCET) event, to give the active Co(IV) oxide state. This reacts with water to liberate O_2, with the Co(IV) centres being reduced to Co(II) in the process. Dissolution and oxidative re-deposition of these Co(II) ions regenerate the catalyst in its resting state, constituting a self-repair mechanism.

Co(IV) centres are reduced to Co(II) ions, which may dissolve into solution and then re-deposit as Co(III) oxides, and it is this dissolution and re-deposition which are postulated to be the mechanism of catalyst re-assembly. Hence the catalyst maintains activity over an extended period of time through functional, rather than structural, stability. The same group also reported a nickel-based analogue of this catalyst that forms in borate buffer and is believed to operate by a similar mechanism.[14,15]

An exciting recent development in this area has been the integration of this CoPi water oxidation catalyst with a catalyst for the HER that is also based on Earth-abundant elements (NiMoZn; see Section 18.2.6). By inserting a buried multi-junction silicon photovoltaic between these two catalyst layers, Nocera and co-workers were able to achieve simultaneous hydrogen and oxygen production from water when these 'artificial leaves' were irradiated with visible light.[16] The principles behind this impressive demonstration of water splitting using sustainable energy and materials inputs are considered in more detail in the subsequent discussion on photochemical approaches to water splitting (Section 18.3).

18.2.3 Electrolytic Water Oxidation on Manganese Oxides

Aside from cobalt–oxos, the most promising non-precious metal candidates for electrolytic water splitting are the various manganese oxides. Mn–oxo cubanes are found in nature in photosystem II, which is responsible for mediating water oxidation in the leaves of plants.[17] In artificial systems, both solely electrocatalytic and photoelectrocatalytic schemes are known and have recently been reviewed.[18] Here, we examine a few contemporary examples in detail, focusing on the mechanisms of action and possible limitations to the use of Mn oxides.

Manganese oxides have been known to effect electrocatalytic water oxidation for some time.[19,20] Over the last few years, however, there has been a resurgence of interest in these materials, with a new emphasis on milder reaction conditions and improved current densities. In 2010, Gorlin and Jaramillo reported a bifunctional Mn-based catalyst able to catalyse both water oxidation and its reverse [the oxygen reduction reaction (ORR)] in 0.1 M KOH solution.[21] The catalyst material was electrodeposited on glassy carbon anodes, before calcination at 480 °C in air for 10 h.

Following this treatment, the Mn–oxo film (which X-ray diffraction and scanning electron microscope data suggested was nanostructured α-Mn$_2$O$_3$) was found to have bifunctional catalytic activity for the OER and ORR comparable to that of precious metals. X-ray photon spectroscopic (XPS) analysis of the active Mn oxide catalyst revealed the Mn centres in the film to be in the +3 oxidation state. However, as already noted, the pH was around 13, which, although significantly less harsh than the conditions prevalent in commercial alkaline electrolysers, is still far from ideal. Some sense of the inherent limitations of such Mn oxide catalysts was provided by Takashima *et al.* in 2012.[22] They noted that manganese oxides are efficient water oxidation catalysts under basic conditions, but that this activity decreases dramatically at neutral pH. Using spectroelectrochemical methods to monitor changes in the composition of Mn oxide catalysts during electrocatalytic water oxidation, Takashima *et al.* observed a remarkable decrease in catalytic activity at pH values below 8. Careful analysis of these results led them to conclude that this decrease in activity occurs at pH values where Mn^{3+} ions start to disproportionate into Mn^{2+} and Mn^{4+}, in the form of soluble Mn(II) salts and insoluble MnO$_2$. Hence, in order to obtain sustained electrocatalytic water oxidation with Mn oxides, the pH should be maintained well above 8 (in practice, at pH 13 or above), at which pH the catalytically active Mn(III) species are stabilised.

Despite these stability arguments, systems relying on milder conditions have been reported recently. Building on work by Harriman *et al.*,[23] Jiao and Frei synthesised nanometre-sized Mn oxide clusters on mesoporous silica scaffolds and found these to be efficient catalysts for water oxidation at room temperature and pH 5.8,[24,25] using the Ru^{2+}(bipyridine)$_3$–persulfate visible light sensitisation system. Likewise, Kurz and co-workers prepared calcium manganese oxides of general formula CaMn$_2$O$_4 \cdot x$H$_2$O and assessed their potential as water oxidation catalysts at pH 4 in the presence of an Ru(bipyridine)$_3$ photosensitiser system.[26]

A final example of electrocatalytic water splitting using Mn oxides, which also has a most interesting history, was reported recently by Spiccia and co-workers.[27] In a previous report, the same group had claimed to observe photo-assisted electrocatalytic water oxidation by a molecular Mn$_4$O$_4$–cubane complex (where the cubane core was supported by coordination to phosphinate ligands) when this molecule was doped into Nafion membranes on the anode.[28] However, subsequent analysis of these films indicated that the Mn–oxo cubanes were decomposing to simpler Mn(II) salts when embedded in the Nafion and that these Mn(II) centres then became oxidised to a mixed Mn(III)–Mn(IV) oxide phase upon application of an anodic bias. It was found that this mixed phase (with a structure similar to that of the mineral birnessite) was the active water oxidation catalyst, reacting with water to form higher oxides such as MnO_2. In support of this theory, simple Mn^{2+} salts and a different Mn coordination compound were embedded in Nafion and found to display similar water oxidation behaviour. This suggests that the original Mn–oxo cubane serves only as a catalyst precursor and not as the active form of the catalyst. To complete the photo-assisted electro-chemical cycle, the MnO_2 generated by water oxidation must be photo-reduced back to Mn(II), with the concomitant release of O_2. The authors suggested that this cycling mechanism may confer some level of self-repairing (or, more accurately, re-assembly) ability on the catalyst, mirroring the work of Nocera and co-workers with CoPi.

18.2.4 Other Electrocatalysts for Water Oxidation

As alluded to in Section 18.2.1, alkaline electrolysers tend to use Ni or Cu anodes, with an oxide coating of metals such as Mn, W or even Ru.[8] However, owing to the potential that PEM electro-lysers show for higher current densities and efficiencies compared with alkaline electrolysers, we consider only one further example in detail here: that of iron-doped nickel oxide.[29] Films of $NiFe_xO_y$ were deposited by reactive sputtering from elemental and alloy targets and assessed as catalysts of the OER at pH 14. Doping with iron was found to reduce the overpotential necessary to evolve oxygen by up to 300 mV compared with undoped nickel oxide, which the authors ascribed to a change in the rate-deter-mining step in the doped films. At modest current densities (around 20 mA cm^{-2}), the films were found to be stable after more than 7000 h of operation. The high stability of this system and its low cost have led to its use in more complicated photocatalytic systems (see Section 18.3.4).

18.2.5 Electrolytic Proton Reduction on Molybdenum Sulfides

The previous sections discussing water oxidation relate to Equation (18.2). The other half of water splitting, that which actually produces hydrogen, involves reduction of the protons produced during water oxidation and is expressed in Equation (18.3). As previously mentioned, the best electrocatalyst for this reaction in acidic media is metallic platinum. However, over the last 5 years or so, molybdenum sulfides have emerged as very promising replacements for proton reduction. In its most thermodynamically stable form, bulk MoS_2 is a rather poor catalyst of the HER.[30] However, nanoparticles of MoS_2 are much more active, as postulated by Chorkendorff and co-workers in 2005,[31] and subsequently confirmed by the same group.[32] These theoretical studies thus facilitated work towards a putative light-driven tandem chemical–solar photocell based on MoS_2 deposited on photo-responsive patterned silicon substrates, which the same group then proposed in 2011 (see Figure 18.3).[33]

Allied to this discovery, Hu and co-workers discovered a simple and mild method for electrodepositing amorphous MoS_x films on the cathode during electrolysis of solutions of a commercial molybdenum salt.[34] These films were then highly active for the HER over a range of pH (0–13), with performance approaching that of Pt in some cases. Using a range of structural techniques, the authors showed that although the amorphous precatalyst deposits could exist in at least two forms (MoS_2 and MoS_3), the active catalyst species was MoS_2. Importantly, the facile nature of the preparation of these catalysts (electrodeposition) is amenable to large-scale manufacture and industrial contexts.

18.2.6 Electrolytic Hydrogen Production Under Basic Conditions

Although originally developed to work under the harsh conditions of commercial alkaline electrolysers (30 wt% KOH), the Earth-abundant alloys studied by Fan *et al.* in the mid-1990s (nickel–molybdenum, nickel–tungsten, cobalt–molybdenum and cobalt–tungsten alloys) show some promise as cathodes for the production of hydrogen from water under milder basic conditions.[35] In particular, those alloys that contained molybdenum were found to display significant reductions in the overpotential

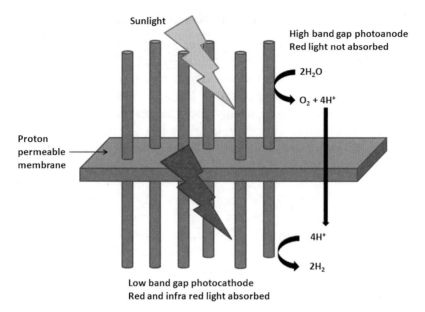

Sunlight

High band gap photoanode
Red light not absorbed

$2H_2O$

$O_2 + 4H^+$

Proton
permeable
membrane

$4H^+$

$2H_2$

Low band gap photocathode
Red and infra red light absorbed

Figure 18.3   The molybdenum sulfide tandem chemical–solar cell proposed by
Chorkendorff and co-workers. Pillars of silicon are first embedded
in a proton-conducting membrane. The blue part of the spectrum
of the incident sunlight is absorbed by the anode and is used for
oxidising water into molecular oxygen and protons. The protons
migrate through the membrane (and the electrons through the
pillars) and are reduced at the cathode side by molybdenum
sulfide clusters adsorbed on the Si pillars, which are in turn excited
by the red part of the spectrum (*i.e.* the light not absorbed by the
photoanode).

necessary to generate hydrogen compared with those alloys that
did not contain any Mo. The best HER catalyst identified from
this work was the CoMo co-deposit, which evinced both high
activity for the HER and high stability. This deposit has been
utilised in light-driven 'artificial leaves' which operate in aqueous
solution at pH 13–14, as described in Section 18.3.4.

In their recent artificial leaf, Nocera and co-workers used a
ternary NiMoZn HER catalyst, which was electrodeposited on the
cathode from a solution of the metals as simple salts.[16] After a
subsequent leaching step in 10 M KOH for 16 h, the electrodeposited alloy was assessed as a catalyst for the HER in 1 M
potassium borate buffer (pH 9.2). Under these conditions, the
NiMoZn alloy was found to generate 50 times more current at a

given potential than a smooth Ni electrode with the same geometric surface area. Subsequent incorporation into a functioning water splitting device proved the effectiveness of this HER catalyst, which importantly is active under mild pH conditions.

18.3 PHOTOCATALYTIC WATER SPLITTING

18.3.1 Basic Concepts

The first demonstration of the photoelectrolysis of water was the TiO_2-based system of Fujishima and Honda in 1972.[36] Even though the hydrogen-evolving side of this reaction set-up used Pt, this example serves as a convenient demonstration of the principles underlying photocatalytic water splitting. The basic concept is outlined in Figure 18.4. In Figure 18.4a, a transparent electrochemical cell is taken and filled with a suitable aqueous electrolyte. A photoanode (in this case rutile TiO_2) is irradiated with simulated sunlight (although due to the bandgap of around 3 eV, only wavelengths less than 415 nm are found to be effective), which causes electrons to be promoted from the valence to the conduction band in the semiconductor. These high-energy electrons can then diffuse to the electrode and thence through the external circuit to the cathode, where they reduce protons to hydrogen [Equation (18.3)]. At the photoanode, the holes thus created in the valence band oxidise water in solution to give oxygen and protons, restoring the semiconductor to its ground state in the process [see Equation (18.2) and Figure 18.4b].

As the potential difference necessary to split water is 1.23 V, light of wavelength < 1000 nm is necessary if photolysis of water is to proceed in such cells unaided. This corresponds to the visible and UV portions of the spectrum. Hence, in order to match the regions of highest solar output, materials that absorb visible light have the greatest potential as light harvesters for applications on the Earth's surface. In reality, however, owing to various kinetic overpotentials and electron transfer-induced losses, the energy requirement for photocatalytic water splitting is generally in the region of 1.6–2.4 eV (corresponding to wavelengths between around 800 and 500 nm) per electron–hole pair generated.[5] This limits the choice of semiconductor materials to those that have a bandgap of at least this magnitude and in addition the conduction band edge and

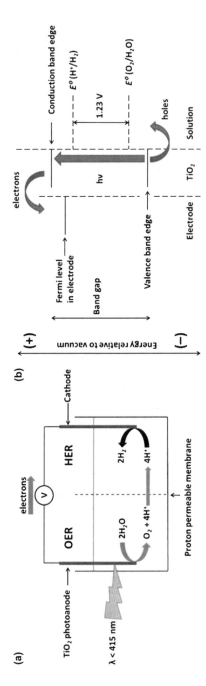

Figure 18.4 The concept of the photocatalytic cell, as exemplified by the cell of Fujishima and Honda. (a) Light of wavelengths below 415 nm is shone upon a TiO_2 photoanode connected to a Pt cathode. In this case, O_2 and H_2 can be evolved in the absence of an applied potential (*i.e.* $V = 0$). (b) Schematic diagram illustrating the relative energy levels of the TiO_2 semiconductor relative to the electrode and the OER and HER in solution. In this instance, the bandgap straddles both the OER potential and the HER potential and so unassisted water splitting should be possible. The energy input to the system is from light (*hv*), which promotes an electron from the valence band to the conduction band. From the conduction band, the electron is able to diffuse energetically downhill into the electrode and thence the external circuit and eventually the cathode. The holes left in the valence band migrate to the semiconductor/solution interface, where they mediate water oxidation.

valence band edge must straddle the electrochemical potentials for the HER and OER if light-driven water splitting is to be possible (see Figure 18.4b), although application of an electrical bias as the electrons pass through the external circuit (assisted photocatalysis) can remove the need for the conduction–valence bandgap to enclose both the HER and OER onset potentials. Moreover, after light absorption has caused electron–hole separation in the semiconductor, these charge carriers must be able to diffuse over sufficiently long distances that they can be collected by the electrode or reach the semiconductor/solution interface. Otherwise, charge recombination will occur within the semiconductor and the energy harvested from the absorbed light will be lost. Charge carrier diffusion distances depend on several factors, including both the nature of the semiconductor material and how it is structured, with this latter consideration itself being a balance of how thick the semiconductor can be (maximising light absorption) before the distances the charge carriers must diffuse become too great. As a further challenge, the efficacy of photocatalytic cells for water splitting is limited by the stability of these semiconductors with respect to the conditions under which they must operate (aqueous solution at variable pH), all of which combine to make photocatalytic water splitting at scale a tremendously difficult problem. Some excellent progress has been made in this field in recent years, however, and we now consider a few selected examples where the greatest emphasis is on sustainability. In the first instance, we examine photoanodes that perform the OER [Equation (18.2)], before moving on to discuss photocathodes for the HER (Section 18.3.3). Finally, in Section 18.3.4, we explore devices that effect overall water splitting using sunlight as the only energy input.

18.3.2 Photocatalytic Water Oxidation

On account of the stringent requirements outlined above, most photoanode materials that have been examined to date are based on the various metal oxides that are stable under the oxidising conditions necessary to evolve oxygen from water.[37] Balancing stability, bandgap size, band edge positions and abundance for widespread use is a key consideration that must be taken into account here. Frequently, photoanodes that meet these requirements must then be integrated with catalysts to bring down the overpotentials for the OER, and most work to date has concentrated on combining photoanodes with precious metal catalysts in order to increase the rate of water oxidation. Hence, to stick to the path of sustainability, we concentrate on those systems that operate with only Earth-abundant catalysts.

Under visible light irradiation, tungsten trioxide (WO_3) is able to oxidise water to oxygen, provided that a suitable electron acceptor is present (the conduction band edge of WO_3 is, however, not quite positive enough relative to the HER potential on the vacuum scale to allow the simultaneous oxidation of water and reduction of protons). A functioning device based on a WO_3 photoanode was first reported by Hodes *et al.* in the mid-1970s.[38] The advantage of this over Fujishima and Honda's system is that WO_3 absorbs more in the visible part of the spectrum than TiO_2 and so there is less need to employ sensitising dyes for light harvesting. That said, however, the visible light absorbance of WO_3 is still far from ideal. In this regard, coloured composite oxides (which clearly absorb more strongly in the visible part of the spectrum than white or off-white TiO_2 and WO_3) are attractive alternatives for photocatalytic water oxidation. Kudo *et al.* demonstrated O_2 evolution from aqueous solutions using a $BiVO_4$ photocatalyst in 1999.[39] The $BiVO_4$ photocatalysts were obtained by reacting layered potassium vanadates (KV_3O_8 and $K_3V_5O_{14}$) with $Bi(NO_3)_3$ at room temperature. Using X-ray diffraction and scanning electron microscopy, the structure was assigned as monoclinic $BiVO_4$, which has a bandgap of $\sim 2.4\,eV$ and characteristic absorptions in the visible (and also UV) portions of the spectrum. The authors used both solely UV (wavelengths between 300 and 380 nm) and visible light (> 420 nm) to photolyse water with $BiVO_4$, in the presence of silver nitrate as an electron acceptor. When irradiated with light at 450 nm, the quantum yield for oxygen evolution on this monoclinic $BiVO_4$ was 9%.

Photoanodes based on a tripartite Mo–Pb–Cr oxide mixture have also shown promise.[40] In this case, oxides of the general formula $PbMo_{1-x}Cr_xO_4$ were synthesised by refluxing the relevant simple salts in aqueous solution, followed by filtration to isolate the products. Under visible light irradiation (> 420 nm), again in the presence of silver nitrate as an electron acceptor, the oxide blend $PbMo_{0.98}Cr_{0.02}O_4$ (with a bandgap of 2.26 eV) was shown to have the highest activity for O_2 evolution from aqueous solution. The work is notable for the significant increase in activity that was obtained with a doping level of Cr of just 2% at Mo sites: the activity of $PbMo_{0.98}Cr_{0.02}O_4$ for water oxidation was found to be one order of magnitude higher than that of undoped $PbMoO_4$ under comparable conditions.

The main hope for sustainable photocatalytic water oxidation at the present time, however, is haematite (α-Fe_2O_3). Haematite has the advantages of being non-toxic, readily available (and hence cheap) and also highly coloured, implying good absorption of visible light. The bandgap of haematite is normally reported as being between 1.9 and 2.2 eV, corresponding to wavelengths between 650 and 560 nm. Indeed, it has been calculated that haematite (as the photoanode in a tandem or dual band gap cell) has the potential to convert 16.8% of incident solar irradiation into hydrogen.[41] The first example of using haematite to split water came from Hardee and Bard in 1976.[42] They made thin films of α-Fe_2O_3 on conducting substrates by chemical vapour deposition and then irradiated these electrodes at > 500 nm. As the bandgap of haematite does not straddle the HER, an additional energy input is required to split water into hydrogen and oxygen under these conditions (see Section 18.3.1), and this was provided by applying an external bias of + 0.8 V (*versus* the saturated calomel electrode) to the cell. Since this initial work, huge efforts have been made to overcome some of the intrinsic drawbacks to the use of haematite as a water oxidation photocatalyst. The chief of these are the large overpotential required to effect photo-assisted water oxidation on haematite, a limitation which will require (preferably sustainable) catalysts to be integrated with haematite photoanodes. In light of this, work aimed at electrodepositing thin layers of the CoPi water oxidation catalyst developed by Nocera and co-workers on α-Fe_2O_3 is of great interest.[43,44] The other main drawback to the use of haematite as a photoanode is its poor efficiency of light harvesting. This has its origin in the high rate of recombination of charge carriers in Fe_2O_3, with the average hole diffusion length being only 2–4 nm.[45] Hence much work has been concentrated in this area on trying to make nanostructured haematite structures to limit charge recombination. Grätzel *et al.* currently leads the way in this field, with a record short-circuit current density of over $3\,mA\,cm^{-2}$ obtained for haematite photoanodes under simulated solar irradiation.[46] However, there is still much room for improvement and a great need for more research in this promising area if practical current densities are to be delivered by haematite-based water splitting cells.

18.3.3 Photocatalytic Hydrogen Evolution

Sustainable materials for the photocatalytic production of hydrogen from aqueous solution are comparatively rare, perhaps because the OER is the rate-determining step in water splitting, hence more research has been directed towards sustainable OER catalysts and photocatalysts than sustainable HER catalysts. This has been exacerbated to some extent by the high efficacy for the HER shown by Pt, even at low loadings. However, as the price of precious metals increases, more readily available alternatives to precious metals and photoelectrode–precious metal combinations are being actively sought.

GaP has been mooted in this context. The material has a bandgap of 2.26 eV (absorbing in the visible part of the spectrum at ~ 550 nm) and its band edges encompass the hydrogen reduction potential (although not quite the OER potential). The p-GaP form is stable in aqueous conditions for numerous rounds of cycling (although not indefinitely) and photocatalytic H_2 generation from water has been demonstrated.[47] However, both the small minority carrier diffusion length inherent in GaP (which makes for inefficient conversion of absorbed light to electric current) and the relative scarcity of gallium in the Earth's crust are serious drawbacks to the widespread adoption of GaP photocathodes.[48]

A more practical alternative for sustainable water splitting photocathodes is p-Si. Although this material has a bandgap of only 1.12 eV, it could be used as the photocathode in dual bandgap water splitting cells (see Section 18.3.4) or in electrically assisted water splitting.[49] Research in this area has shown that planar p-Si photocathodes can be combined with various metal catalysts to reduce the voltage required to produce hydrogen from aqueous solutions, with p-Si decorated with Pt nanoparticles giving a 6% photon-to-hydrogen conversion efficiency (at low level irradiation and a wavelength of 633 nm).[50] Obviously, integration with less rare catalysts is essential for any sustainable system and this has recently been demonstrated by Chorkendorff's group, who combined nanostructured Si with a molybdenum sulfide HER catalyst, as discussed in Section 18.2.5.[33]

18.3.4 Overall Water Splitting in Photocatalytic Cells Using Only Light

There are three basic strategies to achieve overall water splitting to oxygen and hydrogen in photocatalytic cells using only light as the energy input. As outlined in Section 18.3.1, the various criteria that must be met mean that finding a single material with the requisite light harvesting properties, that also has a sufficiently large bandgap (which straddles the potentials for both the OER and the HER) and that has the required stability under operating conditions is a significant challenge. Perhaps surprisingly, then, sustainable materials that can simultaneously produce hydrogen and oxygen from water under irradiation do exist. The greatest drawback to these materials, however, is that in practice the bandgap must be somewhat greater than 3 V (in the UV region of the spectrum), which in turn means that the maximum solar conversion efficiency of such materials is around 2%. $SrTiO_3$ and $KTaO_3$ can both drive unassisted photoelectrolysis of water to hydrogen and oxygen under simulated solar irradiation, although the efficiencies are low ($<1\%$).[51,52]

The second strategy for overall photocatalytic water splitting is to connect a photoanode and a photocathode in series to give a *dual bandgap cell*. In this case, the maximum efficiency of the ensemble is limited to only 50% (as two photons, or one per electrode, have to be absorbed per electron moved), but the advantage of being able to optimise each electrode for its specified reaction more than makes up for this. However, in terms of real-world devices, there are very few examples of systems capable of splitting water to hydrogen and oxygen using only sunlight that do not also rely on precious metals as catalysts. In an early attempt, Nozik reported the use of a p-type GaP photocathode in a back-to-back configuration with a TiO_2 photoanode and obtained an efficiency of under 1%.[53] In addition to this low efficiency, the stability of the GaP photocathode was also a drawback.

The third strategy is to incorporate some sort of photovoltaic device into the cell design to power one or more non-photoactive catalysts. In 1998, Rocheleau and co-workers reported photoelectrochemical solar-to-hydrogen conversion efficiencies as high as 7.8%, using a photocathode consisting of a triple-junction amorphous Si solar array (to provide the electrical driving force for the reaction) covered in a CoMo alloy OER catalyst (see Section 18.2.6), and a separate catalytic anode (based on $NiFe_yO_x$ species, see Section 18.2.4).[54] Building on this work and using very similar materials, Yamada *et al.* developed a water electrolysis cell that could be entirely submerged during operation and that gave a solar-to-hydrogen efficiency of 2.5%.[55] Both of these examples work at fairly strongly alkaline pH (Rocheleau *et al.*'s work was conducted at pH 14 and Yamada *et al.* performed their studies at pH 13). In the light of this, it is instructive to recall the recent results of Nocera and co-workers (Sections 18 2.2 and 18. 2.6), who obtained a solar-to-fuels efficiency of 4.7% using their CoPi-based catalyst system under near-neutral conditions.[16]

18.4   CONCLUSION AND OUTLOOK

The examples in the preceding sections give some sense of the current state-of-the-art in sustainable water splitting. Clearly, much more research is necessary in this area if cost-efficient water splitting driven by renewable energy sources is to become a reality. Indeed, it is a sad fact that it is currently several times cheaper to generate hydrogen by re-forming fossil fuels than it is to generate it by electrolysis (which could potentially be powered by renewable sources),[8] and viable photocatalytic water splitting systems are even less well developed. Above all others, two considerations currently contribute the most to making electrolysis-driven water splitting so much more expensive than re-forming. The first is that electrolysis is a very energy-intensive process. In a true renewable energy future this may not be a great drawback: more energy strikes the surface of the globe every hour in the form of sunlight than the entire human race currently uses in a whole year.[1] Hence there is plenty of effectively free energy available to us if we can harvest it effectively. The other main drawback is the cost of electrolysers. Typically, these are very expensive pieces of hardware on account of the precious metals they require to catalyse the breakdown of water into hydrogen and oxygen. Hence for any sustainable future, catalysts based on cheaper, more abundant elements must be found.

An intriguing possibility from a materials point of view is the application of combinatorial techniques to the screening of vast numbers of metal oxides for their efficacy as water oxidation catalysts. One inspired approach to this challenge is the SHArK (Solar Hydrogen Activity Research Kit) project launched by Parkinson's group at the University of Wyoming, which allows school children to screen potential metal oxide water oxidation candidates and share the results of their research with the community.[56] Schemes such as this have the potential both to generate lots of results very quickly and to engage future generations of scientists from an early age, and are much to be lauded.

Beyond simply splitting water in the laboratory with such systems, we must then move towards functional devices. The artificial leaves reported by Nocera, Rocheleau and Yamada and their coworkers are good places to start, but they are not (yet) workable devices. Current densities must be improved and separation of hydrogen and oxygen must be achieved for safe storage of the hydrogen produced. Nature is able to achieve this latter feat twice over – production of oxygen and reducing equivalents such as NADPH ('natural hydrogen') are separated in *both time and space*.[57] So far, there exist no examples of synthetic systems that are able to accomplish this on any realistic scale, but such technology is sorely needed if cheap, safe and reliable electrolysers are to become part of everyday life.

Finally, there is the issue of whether hydrogen is even a viable fuel for the demands of our society at all. Although it has an excellent mass-to-energy density ratio (Joules stored per kilogram), hydrogen is a gas and must therefore be kept compressed if it is to be stored and used effectively. As hydrogen is also highly flammable, this presents considerable challenges and dramatically increases the real cost of hydrogen as a fuel. Liquid fuels would be much better from this point of view, especially as all the infrastructure necessary for the distribution and utilisation of liquid fossil fuels already exists. Hence the next challenge will be to use the hydrogen generated by sustainable water splitting to make storable liquid fuels, possibly by reacting H_2 with CO_2 captured from the air. Only then will we be able to claim that we have mastered viable and sustainable energy generation and storage.

REFERENCES

1. N. S. Lewis and D. G. Nocera, *Proc. Natl. Acad. Sci. USA*, 2006, **103**, 15729.
2. G. A. Olah, G. K.S. Prakash and A. Goeppert, *J. Am. Chem. Soc.*, 2011, **133**, 12881.
3. A. Kudo and Y. Miseki, *Chem. Soc. Rev.*, 2009, **38**, 253.
4. X. Chen, S. Shen, L. Guo and S. Mao, *Chem. Rev.*, 2010, **110**, 6503.
5. M. G. Walter, E. L. Warren, J. R. McKone, S.W. Boettcher, Q. Mi, E. A. Santori and N. S. Lewis, *Chem. Rev.*, 2010, **110**, 6446.
6. T. R. Cook, D. K. Dogutan, S. Y. Reece, Y. Surendranath, T. S. Teets and D. G. Nocera, *Chem. Rev.*, 2010, **110**, 6474.
7. A. Steinfeld, *Sol. Energy*, 2005, **78**, 603.
8. J. D. Holladay, J. Hu, D. L. King and Y. Wang, *Catal. Today*, 2009, **139**, 244.
9. M. W. Kanan and D. G. Nocera, *Science*, 2008, **321**, 1072.
10. D. A. Lutterman, Y. Surendranath and D. G. Nocera, *J. Am. Chem. Soc.*, 2009, **131**, 3838.
11. M. Risch, V. Khare, I. Zaharieva, L. Gerencser, P. Chernev and H. Dau, *J. Am. Chem. Soc.*, 2009, **131**, 6936.
12. M. W. Kanan, J. Yano, Y. Surendranath, M. Dincă, V. K. Yachandra and D. G. Nocera, *J. Am. Chem. Soc.*, 2010, **132**, 13692.
13. A. J. Esswein, Y Surendranath, S. Y. Reece and D. G. Nocera, *Energy Environ. Sci.*, 2011, **4**, 499.
14. M. Dincă, Y. Surendranath and D. G. Nocera, *Proc. Natl. Acad. Sci. USA*, 2010, **107**, 10337.
15. M. Risch, K. Klingan, J. Heidkamp, D. Ehrenberg, P. Chernev, I. Zaharieva and H. Dau, *Chem. Commun.*, 2011, **47**, 11912.
16. S. Y. Reece, J. A. Hamel, K. Sung, T. D. Jarvi, A. J. Esswein, J. J. H. Pijpers and D. G. Nocera, *Science*, 2011, **334**, 645.
17. J. Barber, *Inorg. Chem.*, 2008, **47**, 1700.
18. M. Wiechen, H.-M. Berends and P. Kurz, *Dalton Trans.*, 2012, **41**, 21.
19. S. Trasatti, *Electrochim. Acta*, 1984, **29**, 1503.
20. M. Morita, C. Iwakura and H. Tamura, *Electrochim. Acta*, 1979, **24**, 357.

21. Y. Gorlin and T. F. Jaramillo, *J. Am. Chem. Soc.*, 2010, **132**, 13612.
22. T. Takashima, K. Hashimoto and R. Nakamura, *J. Am. Chem. Soc.*, 2012, **134**, 1519.
23. A. Harriman, I. J. Pickering, J. M. Thomas and P. A. Christensen, *J. Chem. Soc., Faraday Trans. 1*, 1988, **84**, 2795.
24. Similar methods have recently been used to evolve oxygen from aqueous solutions using nanostructured Co_3O_4 clusters embedded in mesoporous silica, see F. Jiao and H. Frei, *Angew. Chem. Int. Ed.*, 2009, **48**, 1841.
25. F. Jiao and H. Frei, *Chem. Commun.*, 2010, **46**, 2920.
26. M. M. Najafpour, T. Ehrenberg, M. Wiechen and P. Kurz, *Angew. Chem. Int. Ed.*, 2010, **49**, 2233.
27. R. K. Hocking, R. Brimblecombe, L.-Y. Chang, A. Singh, M. H. Cheah, C. Glover, W. H. Casey and L. Spiccia, *Nat. Chem.*, 2011, **3**, 461.
28. R. Brimblecombe, G. F. Swiegers, G. C. Dismukes and L. Spiccia, *Angew. Chem. Int. Ed.*, 2008, **47**, 7335.
29. E. L. Miller and R. E. Rocheleau, *J. Electrochem. Soc.*, 1997, **144**, 3072.
30. W. Jaegermann and H. Tributsch, *Prog. Surf. Sci.*, 1988, **29**, 1.
31. B. Hinnemann, P. G. Moses, J. Bonde, K. P. Jørgensen, J. H. Nielsen, S. Horch, I. Chorkendorff and J. K. Nørskov, *J. Am. Chem. Soc.*, 2005, **127**, 5308.
32. T. F. Jaramillo, K. P. Jørgensen, J. Bonde, J. H. Nielsen, S. Horch and I. Chorkendorff, *Science*, 2007, **317**, 100.
33. Y. Hou, B. L. Abrams, P. C. K. Vesborg, M. E. Björketun, K. Herbst, L. Bech, A. M. Setti, C. D. Damsgaard, T. Pedersen, O. Hansen, J. Rossmeisl, S. Dahl, J. K. Nørskov and I. Chorkendorff, *Nat. Mater.*, 2011, **10**, 434.
34. D. Merki, S. Fierro, H. Vrubel and X. Hu, *Chem. Sci.*, 2011, **2**, 1262.
35. C. Fan, D. L. Piron, A. Sleb and P. Paradis, *J. Electrochem. Soc.*, 1994, **141**, 382.
36. A. Fujishima and K. Honda, *Nature*, 1972, **238**, 37.
37. F. E. Osterloh, *Chem. Mater.*, 2008, **20**, 35.
38. G. Hodes, D. Cahen and J. Manassen, *Nature*, 1976, **260**, 312.
39. A. Kudo, K. Omori and H. Kato, *J. Am. Chem. Soc.*, 1999, **121**, 11459.

40. Y. Shimodaira, H. Kato, H. Kobayashi and A. Kudo, *Bull. Chem. Soc. Jpn.*, 2007, **80**, 885.
41. K. Sivula, F. Le Formal and M. Grätzel, *ChemSusChem.*, 2011, **4**, 432.
42. K. L. Hardee and A. J. Bard, *J. Electrochem. Soc.*, 1976, **123**, 1024.
43. D. K. Zhong, J. Sun, H. Inumaru and D. R. Gamelin, *J. Am. Chem. Soc.*, 2009, **131**, 6086.
44. Deposition of CoPi on WO_3 has also been reported, see J. A. Seabold and K.-S. Choi, *Chem. Mater.*, 2011, **23**, 1105.
45. J. H. Kennedy and K. W. Frese, *J. Electrochem. Soc.*, 1978, **125**, 709.
46. S. D. Tilley, M. Cornuz, K. Sivula and M. Grätzel, *Angew. Chem. Int. Ed.*, 2010, **49**, 6405.
47. R. Memming and G. Schwandt, *Electrochim. Acta*, 1968, **13**, 1299.
48. D. E. Aspnes and A. A. Studna, *Phys. Rev. B*, 1983, **27**, 985.
49. T. W. Hamann and N. S. Lewis, *J. Phys. Chem. B*, 2006, **110**, 22291.
50. R. N. Dominey, N. S. Lewis, J. A. Bruce, D. C. Bookbinder and M. S. Wrighton, *J. Am. Chem. Soc.*, 1982, **104**, 467.
51. F. T. Wagner and G. A. Somorjai, *J. Am. Chem. Soc.*, 1980, **102**, 5494.
52. I. E. Paulauskas, J. E. Katz, G. E. Jellison Jr, N. S. Lewis, L. A. Boatner and G. M. Brown, *J. Electrochem. Soc.*, 2009, **156**, B580.
53. A. J. Nozik, *Appl. Phys. Lett.*, 1976, **29**, 150.
54. R. E. Rocheleau, E. L. Miller and A. Misra, *Energy Fuels*, 1998, **12**, 3.
55. Y. Yamada, N. Matsuki, T. Ohmori, H. Mametsuka, M. Kondo, A. Matsuda and E. Suzuki, *Int. J. Hydrogen Energy*, 2003, **28**, 1167.
56. M. Woodhouse and B. A. Parkinson, *Chem. Mater.*, 2008, **20**, 2495.
57. S. Zein, L. V. Kulik, J. Yano, J. Kern, Y. Pushkar, A. Zouni, V. Yachandra, W. Lubitz, F. Neese and J. Messinger, *Philos. Trans. R. Soc. London B*, 2008, **363**, 1167.

V
Sustainability Related to Materials in the Urban Environment and to Water

CHAPTER 19

Materials Used in Membranes for Water Purification and Recycling

XIAOYING ZHU, KIN-HO WEE AND RENBI BAI*

Department of Civil and Environmental Engineering, Faculty of
Engineering, National University of Singapore, 1 Engineering Drive 2,
Singapore 117576
*Email: ceebairb@nus.edu.sg

19.1 INTRODUCTION TO MEMBRANE TECHNOLOGY

Over the past few decades, membrane technology has increasingly
become the choice for separation technology for many applications. Nowadays, membrane separation is widely incorporated into
many water and wastewater treatment plants to comply with the
more stringent quality regulations in water supply and wastewater
discharge. The main advantages of membrane technology include
the high separation effectiveness, possible elimination of chemical
addition, simple system configuration, short process time and small
footprint.

A number of unit operations are usually employed in conventional
water purification and recycling systems for the removal of components such as ions, charged or neutral molecules, macromolecules,
viruses and bacteria, and fine and coarse particulates that have

Materials for a Sustainable Future
Edited by Trevor M. Letcher and Janet L. Scott
© The Royal Society of Chemistry 2012
Published by the Royal Society of Chemistry, www.rsc.org

different physical and chemical properties through a combination of various physicochemical processes. For instance, ionic or charged contaminants may be removed by ion exchange and adsorption or by coagulation through adding chemicals to form precipitates or settleable particles. Settleable particulates or suspended particles are removed by sedimentation and non-settleable smaller particles by granular bed filtration. Disinfectant is added for the deactivation of bacteria and viruses. The system performance is largely dependent on each of the unit operations and their combination, and the treatment system usually occupies a large area of space.

A number of membrane-based separation techniques have been developed in recent years and they are increasingly replacing many of the conventional water treatment processes. The operating principle of common membrane separation techniques is straightforward: the membrane acts as a physical barrier that selectively allows water and/or other small components to transport through the membrane pores while completely rejecting suspended solids and other substances that have sizes larger than the membrane pore diameter. Depending on the membrane pore size or the size of rejected contaminants, membrane separation for water purification and recycling can be generally divided into more specific processes: microfiltration (MF), ultrafiltration (UF), nanofiltration (NF) and reverse osmosis (RO). Their typical separation functions are shown in Figure 19.1. From MF to RO, the membrane process becomes more effective in rejecting smaller components in water. For example, when the removal of larger particles such as sludge flocs and bacteria in water is intended, a MF or UF membrane process is used. When macromolecules are also to be removed, UF would be needed. Because of the porous structures of the membranes, the separation throughput or productivity of MF and UF is much higher and the operating pressure is much lower (less than 500 kPa). When much smaller substances such as ions and low molecular weight organic molecules need to be removed from water, NF and RO are applied. In NF and RO processes, purification of contaminated water is mainly achieved by diffusion of water molecules through the membranes. The operating pressures required for NF and RO are much higher than those for MF and UF, whereas the permeate flux obtained is much lower. Typical applications of different types of membrane processes in water purification are summarised in Table 19.1.

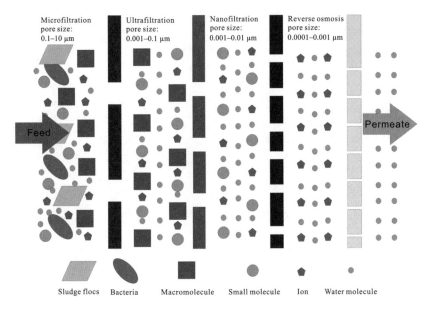

Figure 19.1 Schematic diagrams showing the separation of components in water by different types of membrane filtration systems.

In addition to the membranes retaining components from the feed through the 'size-exclusion' mechanism, another class of membranes that has often found applications in water purification is ion-exchange membranes. These membranes are positively or negatively charged and will interact with the charged components (*i.e.* anions and cations) in the feed. Electrodialysis (ED) is the most established membrane process that uses these ion-exchange membranes for water purification. During an ED process, salt ions from one solution are transported through an ion-exchange membrane to another solution under the influence of an applied electric potential difference.

Many advances made in materials science and engineering have allowed more high-performance materials to be fabricated into various types of membranes for different membrane separation applications in water purification and recycling. Some of these common representative and novel materials that have found use in membrane fabrication are discussed in detail in the following sections.

Table 19.1 Specific breakdown of water purification processes.

Process	MF	UF	NF	RO
Particulates removal (≥ 0.5–$5\,\mu m$)				
• Clarification	√			
• Sludge flocs removal	√			
• Bacteria removal (disinfection)	√	√		
Colloids removal (5–0.01 µm)				
• Viruses removal (disinfection)		√		
• Organic macromolecules removal		√	√	
Ions, dissolved solutes removal (≤ 0.01 µm)				
• Hardness removal (softening)			√	√
• Disinfection by-products removal			√	√
• Ultrapure water production				√

19.2 POLYMERIC MATERIALS FOR MEMBRANE FABRICATION

Organic polymers have dominated commercial applications since the earliest stage of the membrane industry. These polymers offer low-cost fabrication, ease of handling and improved performance in selectivity and permeability. Moreover, they may be chemically derivatised and therefore endowed with different surface properties from a wide selection of synthetic and organic chemistries.

Membranes have various structures based on different fabrication materials and preparation methods. In addition, the prepared membrane's selectivity and permeability are affected by its structure. As shown in Figure 19.2, symmetric membranes have isotropic, homogeneous pore structures without any skin layer. Symmetric polymeric membranes with a dense interior porous structure can provide effective separation. However, these symmetric membranes often have a thickness of 20–200 µm, which gives only low permeation rates. Unfortunately, the direct fabrication of thinner symmetric membranes (e.g., 0.1–1 µm) for

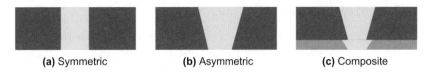

(a) Symmetric   **(b)** Asymmetric   **(c)** Composite

Figure 19.2   Types of membrane structure.

improved permeation and practical applications has been technically challenging. A thin porous membrane film, without a supporting layer, lacks the mechanical strength to sustain the imposed trans-membrane pressure during membrane filtration. To overcome these shortcomings of symmetric membranes, asymmetric membranes that consist of a dense integral skin layer supported by a mechanically durable porous matrix have been researched and developed. They are usually prepared from single polymers through a phase inversion method. In this process, a liquid polymer solution is cast or spun into a membrane and the latter is introduced into a non-solvent where the polymer solution separates into two phases, a solid phase that forms the polymeric matrix of the membrane and a liquid phase that forms the membrane pores.[1] The unique structure of the asymmetric membrane can decouple the interdependency between separation selectivity and membrane permeability, *i.e.* the high permeability is achieved through the use of a supporting porous matrix, and the membrane selectivity can be varied through manipulating the thickness and structure of the thin skin layer. Another recent breakthrough in membrane fabrication was the development of composite membranes with an artificially engineered asymmetric structure. The composite membranes are prepared by various top-layer formation methods such as dip-coating,[2] interfacial polymerisation[3–5] or plasma polymerisation.[6] These methods allow greater flexibility in tailoring the materials and structures used for intended separation objectives. In contrast to asymmetric membranes, the dense skin layer and porous matrix of composite membranes may come from different polymers.

In the following sections, the polymeric materials that are suitable for fabrication of specific types of membranes (MF, UF, NF and RO) are discussed. Subsequently, polymer modification methods that can improve membrane performance or achieve intended separation objectives are introduced.

19.2.1   MF/UF Membrane Materials

MF membranes contain open pores with dimensions in the range of 0.1–10 μm. They are commonly used for separating particles, sludge flocs and bacteria from water. The selectivity of these membranes is determined by the dimensions of the pores relative to the target contaminants to be retained. As a result, the pore formation and pore size control are the major factors during material selection for MF membrane fabrication. In addition, the choice of the membrane material also affects other process performances such as organic adsorption, chemical stability during actual operation and membrane cleaning. Chemical and thermal resistances of the polymeric material become prevailing factors when the membrane stability in harsh, extreme operating conditions needs to be addressed. Another major problem in membrane filtration is flux decline after prolonged usage because of concentration polarisation and fouling. The prevention of membrane fouling and performance reversibility after cleaning during filtration are also dependent on the MF membrane material. A number of synthetic techniques are available for preparing MF membranes, such as sintering, stretching, track-etching and phase inversion. Table 19.2 lists representative polymers which are commonly found in various commercial-grade MF membranes.

Hydrophobic fluoropolymers such as PTFE (polytetrafluoroethylene) and PVDF [poly(vinylidene fluoride)] are often used for MF membranes. PTFE is highly crystalline and exhibits excellent thermal stability. PTFE is chemically inert and insoluble in almost all common solvents. The Poreflon PTFE MF membrane has numerous fine pores of ~0.1 μm on its surface. In terms of strength and chemical resistance, Poreflon exceeds PVDF and other synthetic polymers, allowing easy cleaning of the membrane surface and continuous, stable operation. In addition to its application in the pretreatment of water and wastewater, it is also used in the MBR (membrane bioreactor) system to separate water from sludge flocs and bacteria.[7] However, MF membranes made from PTFE can only be prepared by sintering and stretching, whereas PVDF membranes are made by phase inversion. PVDF also shows good thermal and chemical stabilities, although not as good as PTFE. More importantly, PVDF is soluble in aprotic solvents such as dimethylformamide (DMF), dimethylacetamide (DMAc) and

Table 19.2   Polymers for membrane fabrication.

Material	Chemical structure	Application
Polyethylene (PE)		MF
Polycarbonate (PC)		MF
Polyethylene terephthalate (PET)		MF
Polypropylene (PP)		MF
Polyvinylidene fluoride (PVDF)		MF, UF
Polytetrafluoroethylene (PTFE)		MF
Polyvinyl chloride (PVC)		MF

Table 19.2 (*Continued*)

Material	Chemical structure	Application
Cellulose and cellulose esters (cellulose acetate, cellulose triacetate and cellulose nitrate)		MF, UF, RO
Polyamide (PA)		MF, UF, RO
Polysulfone/Polyarylether sulfone (PES)	R and R' are alkyl groups or benzene rings	MF, UF
Polyetheretherketone (PEEK)		UF
Polyimide/Polyether imide (PEI)		UF
Polyacrylonitrile (PAN)		UF

triethyl phosphate (TEP), and therefore lends itself to easier processing for membrane fabrication. The Kynar PVDF membrane is very chemically inert and it can withstand pH in the range 0–14. This MF membrane provides a barrier to the passage of solids and therefore is capable of removing mineral particles and bacteria, making the filtrate a perfect feed for a reverse osmosis unit or other further treatments.[8]

Another special type of MF membrane with a more precise pore geometry and controllable porosity may be prepared by track-etching polymeric films. Polymers such as PC (polycarbonate) and PET [poly(ethylene terephthalate)] are often used for this purpose because of their outstanding mechanical properties.[9] Polypropylene (PP) is another hydrophobic, thermally and chemically stable polymer that is commonly used as an MF membrane material. These PP membranes may be prepared by melt-spinning, stretching[10] and thermally induced phase inversion.[11] Another class of MF membrane polymers is PA (polyamide), characterised by the amide group. Owing to their outstanding mechanical, thermal and chemical stabilities, aliphatic polyamides such as nylon 6, nylon 6,6 and nylon 4,6 are of great interest in MF membrane fabrication. On the other hand, the aromatic PAs have aromatic groups positioned along their main chains, which considerably reduces the chain flexibility. The aromatic PAs display much better stabilities than aliphatic PAs, hence aromatic PAs are more widely used in making MF, UF and RO membranes.[1] Some other polymers such as PES [poly(ether sulfone)] have also been fabricated into commercial MF products, for example GE's Xpleat filters.[12]

Despite the excellent thermal and chemical stabilities of these hydrophobic polymers, hydrophilic polymers are often favoured as MF membrane materials for water treatment because of their lower adsorption tendencies. The adsorption of organic substances from the feed would lead to a decline in permeation flux or gradual contamination of membrane surface. One of the better known hydrophilic polymers for MF membrane fabrication is cellulose and its derivatives such as cellulose esters, including cellulose acetate, cellulose triacetate, ethylcellulose and cellulose nitrate. These materials are widely used for fabricating a wide range of membranes (MF, UF and RO) for various applications, *e.g.* sludge clarification and seawater desalination. Cellulose is a plant-derived

polysaccharide with a molecular weight varying from 500 000 to
1 500 000 Da. The repeating glucose segment contains three
hydroxyl groups which are highly susceptible towards chemical
reaction, for instance formation of esters (cellulose acetate and
cellulose nitrate) and ethers (ethylcellulose). Cellulose or regenerated cellulose is mainly used for dialysis membranes, cellulose
nitrate and cellulose acetate are usually used for MF or UF
membranes, whereas cellulose triacetate is a common material for
RO membranes used in desalination. Despite their outstanding
membrane properties such as selectivity and permeability, these
cellulose esters suffer limited operating conditions owing to their
susceptibility to thermal, chemical and biological degradation.[1]

UF uses a fine porous membrane to separate bacteria, viruses,
macromolecules and colloids from water. The average pore
diameter of the membrane is in the range of 0.1–0.001 μm.
Hence some MF membrane preparation methods such as sintering,
track-etching and stretching cannot be used to prepare UF
membranes. Most UF membranes are prepared by phase inversion.
Table 19.2 also includes a list of polymers commonly used as
materials for UF membranes.

PES is an important class of polymers for MF and UF membranes fabrication. Owing to their excellent chemical and thermal
stabilities, they are widely used as precursor materials for fabricating UF membranes[13] and as a porous matrix for composite
membranes.[14] PEEK [poly(ether ether ketone)] is another group of
polymers used for UF membrane preparation. They are also very
stable and can only be dissolved in concentrated inorganic acids
such as sulfuric acid and chlorosulfonic acid.[15]

Even though some polymeric materials used for MF membranes
can also be used for fabricating UF membranes, some special
polymers are preferred because of their greater mechanical
strength, which can sustain the higher trans-membrane pressure of
UF. PEI [poly(ether imide)] is extremely strong and also highly
thermally and chemically resistant. Owing to their exceptional
strength, these materials are often exchangeable with glass and
metals in many demanding industrial applications, and are therefore also considered as materials for UF membrane fabrication.[16]

Some other polymers, including cellulose and its esters,[17] PAN
(polyacrylonitrile),[18] aromatic PAs[19] and PVDF are also widely
used to fabricate UF membranes, and especially PVDF is widely

used in commercial UF membranes. The Dow UF double-walled hollow-fibre membrane is fabricated from high-grade PVDF, which has a very small nominal pore diameter that allows for the removal of all particulate matters, bacteria, most viruses and colloids. Despite the small pore diameter, the membrane has a very high porosity, resulting in a flux similar to that of MF, and can effectively replace MF in most cases.[20] Similarly, GE's ZeeWeed UF hollow-fibre membrane produced from PVDF is a pressure-driven barrier to suspended solids, bacteria, viruses, endotoxins and other pathogens, used to produce water with very high purity and low silt density. It can serve as a treatment for surface water, ground water and biologically treated municipal effluent (MBR and tertiary treatment).[21]

19.2.2   NF/RO Membrane Materials

NF is also called loose reverse osmosis or low-pressure reverse osmosis. Typically, NF membranes with pores of ~ 0.001 µm have sodium chloride rejections between 20 and 80% and molecular weight cut-offs for dissolved organic solutes of 200–1000 Da. The materials and methods for NF membrane preparation are similar to those used for RO membranes. Hence the materials for NF membrane fabrication will not be discussed separately, rather they will be mentioned during the discussion of RO materials. RO membranes usually have pores of ~ 0.0001 µm, which are capable of rejecting more than 90% of salt in feed solution.

For NF/RO membranes, the selection of suitable polymeric materials becomes more intricate than that for porous MF and UF membranes, because the chemical composition, spatial alignment and microstructure of the polymers directly determine the separation performance, *i.e.* selectivity and permeability, in addition to the process considerations already mentioned (*e.g.* mechanical strength, fouling, *etc.*).

Cellulose acetate (CA) was the first discovered high-performance NF/RO membrane material. In the late 1950s, Reid and Breton verified that CA RO membranes were capable of separating salt from water, even though the water fluxes obtained were far from practical.[22] In order to improve the RO permeate flux, the Loeb–Sourirajan CA asymmetric membrane was formed with a dense 200 nm thin top layer over a thick microporous supporting layer.

This new membrane produced a permeate flux of at least an order of magnitude higher than the initial symmetric membrane.[23] CA was also used for the preparation of NF membranes.[24]

CA membranes are easy to prepare, mechanically tough and resistant to degradation by common disinfectants such as chlorine and other oxidants. However, the acetate group of CA would be easily hydrolysed in both acidic and alkaline conditions. In addition, CA membranes are prone to microbial fouling. These drawbacks limit the durability and practical feasibility of using CA membranes for NF/RO.[25]

In searching for a substitute for CA with better stability, a few non-cellulosic membrane materials were developed. PA was the first non-cellulosic material that was used for preparing asymmetric RO membranes. For example, DuPont commercialised PA membranes in hollow-fibre form under the trade name Permasep B-9, for application in brackish water desalination. [26] The aromatic PA membranes were also successfully developed by Endoh *et al.*[27] and McKinney and Rhodes,[28] all in hollow-fibre form. Regrettably, these membranes displayed neither better salt rejection nor higher permeate flux than CA membranes. However, their durability, stability and versatility are greater than that of CA membranes.[29] Even though the PA RO membranes could not provide a high filtrate flow rate per unit area; their commercial success was ultimately achieved through highly effective packing of the hollow-fibre modules, which outperformed the CA spiral wound elements in terms of flux per unit module volume. The Permasep membranes were manufactured as B-10 and B-15 for seawater desalination until 2000. The chemical structure of Permasep B-15 is shown in Figure 19.3.[30]

Nevertheless, the amide bonds in PAs are susceptible to degradation by disinfectants such as chlorine (halogens) and ozone. This

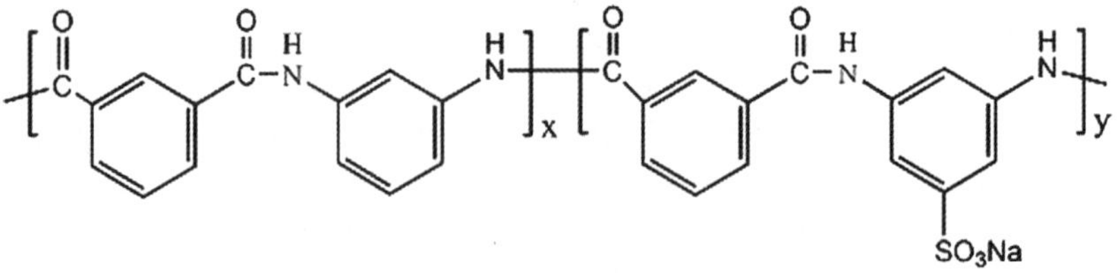

Figure 19.3   Structure of the PA membrane Permasep B-15.[30]

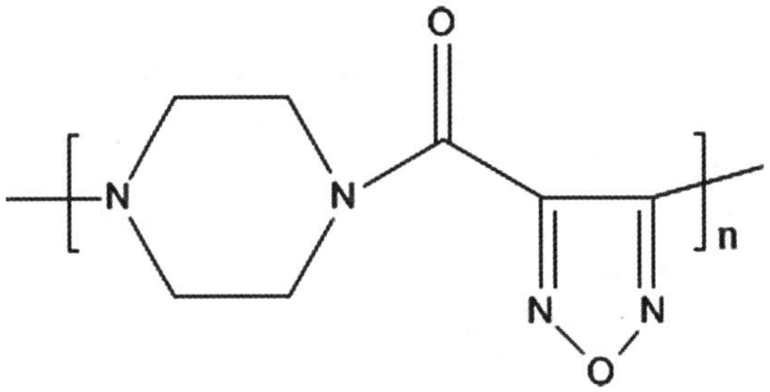

Figure 19.4 Structure of polypiperazinamides.[31]

drawback reduces the operational life span of PA membranes. Subsequently, chlorine-resistant asymmetric membranes based on polypiperazinamides were developed.[31] The reduced presence of amidic hydrogen as shown in Figure 19.4 enhances the chlorine tolerance of these polymers.[32] However, this membrane was not commercialised because of its relatively low salt rejection.[32] Similarly, other polymers such as carboxylated polysulfone[33] and sulfonatedpolysulfone[34] also suffer from inferior salt rejection.

19.2.3 Ion-exchange Membrane Materials

Another class of membranes often found in various water purification applications is ion-exchange membranes or ion-permselective membranes.[35] These membranes were developed and applied across various branches of industry starting from the 1950s, following the invention of ion-exchange resins and its processes for deionisation in the 1930s. The most established membrane process that uses these ion-exchange membranes for water purification is ED.

Based on the sign of charges that these membranes possess in water, the ion-exchange membranes can be categorised as cation- and anion-exchange membranes. The ion-exchange sites distributed throughout the membrane become fully dissociated and electrically charged in the aqueous phase. Unlike their porous counterparts, ion-exchange membranes, which are essentially non-porous, are non-permeable to water, simple ions and salts when hydrostatic pressure is imposed upon them. The separation of ions from brine or saline water is principally achieved by exclusion of ions of the same charge as the fixed ion-exchange sites or ionic groups immobilised on the polymer backbone of the membrane. Cation-exchange membranes contains high volumetric concentrations of negatively charged groups (*e.g.* SO_3^-), which can bind and release cations (*e.g.* Na^+), whereas anion-exchange membranes are able to get rid of anions (*e.g.* Cl^-) through the positively charged groups (*e.g.* NH_4^+). The migration or diffusive transport of mobile species (*e.g.* Na^+, Cl^-) such as free ions and charged molecules is activated only when a voltage or electrical potential difference is applied across the membrane. Typically, the cation- and anion-selective membranes are arranged in an alternating manner, separated by a spacer and stacked into a so-called electrodialysis stack.

The first generation of ion-exchange membranes were fabricated from a finely ground ion-exchange resin and a binder polymer (*e.g.* PVC and PVDF), in which the powdered ion exchangers that serve as ionic transport channels are dispersed and immobilised in the matrix or binder polymer. The most commonly used ion-exchange resins were based on styrene–divinylbenzene, which were either sulfonated or chloromethylated and then aminated to produce cation- and anion-exchange membranes, respectively. The newer generation of ion-exchange membranes with enhanced durability such as perfluorocarbon membranes, *e.g.* Nafion,[36] have been widely studied to assess their use in ED. [37] Synthetic approaches as reviewed in later sections, such as plasma polymerisation and polymer coating that have been used for introducing ion-exchange properties into a pristine matrix or even for modifying the ion permselectivity of commercially available membranes. For instance, Tan *et al.*[38] reported a successful improvement of the permselectivity of the membrane for proton *versus* bivalent cations (Zn^{2+} and Cu^{2+}) by polymerising aniline on the surface of Neosepta[39] cation-exchange membranes.

Another enhancement of the ED process is electrodeionisation (EDI), where the conventional ion-exchanger resins fill up the salt-depleted compartments between membrane stacks in ED to reduce the electrical resistance of the membrane stacks during operationand also to regenerate some of the adjacent ion-exchange resins continuously through splitting of water into protons and hydroxide ions.[40] As a result, this EDI produces deionised water with very low levels of salts, and therefore is often used for ultrapure water production.

19.2.4 Composite Membrane (Top-layer) Formation

In addition to asymmetric membranes, another breakthrough in membrane technology was the invention and subsequent development of composite membranes that eventually led to successful commercialisation. The structure of a composite membrane, which is similar to an asymmetric membrane structure, consists of a thin, dense top layer supported by a porous supporting matrix. However, the two layers of composite membrane can be engineered independently from different polymers with different properties. The advantage of a composite membrane is that the chemical composition and surface properties of the top layer can be manipulated with great degrees of freedom through various top-layer formation methods, to obtain optimal membrane performance with respect to selectivity, permeation rate and chemical and thermal stability or special properties such as antifouling.

19.2.4.1 Dip-coating. Dip-coating is very simple and useful for preparing composite membranes with a very thin but dense top layer. Membranes obtained by this method are usually used in RO, gas separation and pervaporation. The process of dip-coating can be separated into several stages. First, an MF/UF membrane is immersed in the solution which dissolves the coating material for a predefined duration. The membrane is then pulled up after immersion in the precursor polymer solution. A thin polymer layer becomes deposited on the membrane surface and later solidifies after the solvent has been completely removed through evaporation.

Acid polycondensation of low molecular weight hydroxyl-containing compounds through dip-coating is commonly used to prepare thin-film composite membranes. For example, the top layer of the composite RO membrane NS-200 was prepared *via* the reaction of furfuryl alcohol, sulfuric acid and polyoxyethylene.[41] This method was also used to prepare a PEC-1000 thin-film composite membrane by Toray Industries, where 1,3,5-tris(hydroxyethyl)isocyanuric acid was used instead of polyoxyethylene.[42]

19.2.4.2 Interfacial Polymerisation. Interfacial polymerisation deposits a thin, selective polymer layer on top of the supporting porous MF or UF membranes. These membranes with anisotropic morphologies are so-called thin-film or interfacial composite membranes (TFCs). The thin selective layer is formed when a polymerisation reaction occurs between two very reactive but immiscible monomers or mixtures at their boundary of contact or liquid–liquid interface. TFC membranes with a polyamide top layer, which is basically a product from the polycondensation reaction of diacid chloride (or triacid chloride) and diamine, are commonly prepared using this method. Monomers should have two (or more) reactive groups for the polycondensation reaction to occur. For example, if the top surface of a membrane came in contact with an aqueous solution of diamine and was subsequently immersed in diacetyl chloride (or triacetyl chloride) solution in hexane, the polymerisation product, *i.e.* the polyamide, would form spontaneously at the interface.

Interfacial polymerisations of monomeric amines such as aliphatic diamines with trimesoyl chloride were used to prepare TFC membranes. Based on similar chemistry, a range of NF membranes have been commercialised, *e.g.* NF-40 by FilmTec[43] and NTR-7250 by Nitto Denko.[44] Owing to the anionic charged surface due to the presence of carboxylic groups, these membranes can achieve excellent rejection of divalent anions such as sulfate at high flux and are more chlorine tolerant than earlier membranes.

TFC membranes were also prepared by interfacial polymerisation of monomeric aromatic amines with trimesoyl chloride. The FT-30 RO membrane, which has an all-aromatic structure based on the reaction of phenylenediamine and trimesoyl chloride, provided exceptional salt rejection and very high fluxes.[45] The aromatic PA structure of FT-30 exhibits excellent mechanical, thermal and chemical stabilities. Although not completely free from chlorine attack, FT-30 shows a moderate tolerance to chlorine and can withstand very low-level exposure to chlorine for prolonged periods or exposure to ppm levels for a few days.[46]

19.2.4.3 Plasma Polymerisation. Another method of fabricating the top layer of TFC membranes is *via* plasma polymerisation. In order to initiate the polymerisation, gaseous or liquid monomers, which contain reactive vinyl groups, are activated or fragmented

Table 19.3 Common monomers for plasma polymerisation.

Compound	Structure
Thiophene	
Pyridine	
Acrylonitrile	
Furan	
Styrene	
Acetylene	$H-C{\equiv}C-H$
2-Methyloxazoline	
Tetramethyldisiloxane	$H-\overset{\overset{\textstyle CH_3}{|}}{Si}-O-\overset{\overset{\textstyle CH_3}{|}}{\underset{\underset{\textstyle CH_3}{|}}{Si}}-H$

by the energy of the gas plasma. Subsequently, the monomeric and polymeric radicals propagate and then graft on to the surface of a porous support, *e.g.* polysulfone support, to form a thin polymeric film. As can be seen in Table 19.3, many functional monomers are readily available for plasma polymerisation, with the one precondition that the monomer must be a small ionizable species as it must be able to enter the plasma state.

The permeability characteristics of plasma polymers deposited on porous substrates are different from those found in the usual polymer films. The characteristics depend on a deposition and polymerisation mechanism.Polymers formed from this process are often cross-linked and branched due to the multiple propagating species present in the plasma. Plasma polymers as membranes for separation of oxygen and nitrogen, ethanol and water and water vapour permeation have all been studied.[47] The application of plasma polymerised thin films as RO membranes has also attracted considerable attention.[48] Further research has shown that varying the monomers of the membrane also offers other properties, such as chlorine resistance. [49]

19.3 POLYMERIC MATERIAL MODIFICATIONS

Most polymers described earlier cannot be directly fabricated into asymmetric or functional MF/UF membrane products without proper conditioning or chemical modification. In addition, pristine membranes prepared from conventional polymeric materials are usually prone to many process-related problems such as membrane fouling. These technological gaps led to many research efforts aimed at utilising and improving various synthetic or modification methods for either enhanced separation performance or longer operating life span. These modification methods can be classified into bulk modifications and surface modifications.

19.3.1 Bulk Modifications

To modify the bulk properties of a polymer for membrane fabrication, the following modification methods were developed. After modification, the novel polymers or copolymers would have membrane selectivity and/or permeability which are remarkably different from those of the unmodified polymers.

19.3.1.1 Copolymerisations. A number of copolymers were developed from block or graft copolymerisation of more than one of the monomers that are usually used for producing homopolymers. The copolymer is intended to enhance the separation properties of the homopolymer-based membranes and other properties such as mechanical strength and morphology. These novel copolymers, once incorporated into membranes, can also bring new applications to the membranes.

PEI and related polymers, synthesised from aromatic monomers, have rigid-chain structures resulting in low permeability. Hence PEI usually can only be used for UF membrane fabrication. To overcome the low permeability of membranes prepared from PEI, block copolymers incorporated with flexible segments such as poly(ethylene oxide) for the preparation of MF membranes,[50] sulfonated diamines[51] and ether linkages[52] for proton exchange membranes have been studied. The incorporation of the flexible segments increases the solubility, making the materials easier to process and achieving higher membrane permeability.

PVDF is widely used in MF and UF membrane preparation. Attempts have been made to endow PVDF membranes with new properties or functions through copolymerisation methods. pH-sensitive MF membrane was prepared from PVDF copolymer synthesised *via* thermally induced graft copolymerisation with acrylic acid (AAc), the permeate flux of which could change reversibly in response to pH variation of the feed solution. [53] In addition, PVDF copolymer produced *via* thermally induced graft copolymerisation with *N*-isopropylacrylamide (NIPAAM) made the membrane permeate flux temperature sensitive.[54] Membrane morphology can also be adjusted through copolymerisation. For example, the pore size and distribution of an MF membrane prepared from graft copolymers synthesised *via* living radical graft polymerisation of poly(ethylene glycol) methyl ether methacrylate (PEGMA) with PVDF in the reversible addition–fragmentation chain transfer (RAFT)-mediated process, could be changed with the graft concentration and the density of graft points.[55] The PVDF membrane ishighly prone to fouling, due to PVDF's hydrophobicity. PVDF copolymers produced by grafting copolymerisation of hydrophilic poly(ethylene glycol) with PVDF could improve the membrane's hydrophilicity to resist membrane fouling.[56]

19.3.1.2 Blending.

Compared with other modification methods, polymer blending is preferred in membrane preparation owing to its simplicity and reproducibility. The prepared membranes' properties can be enhanced by blending in functional materials with various characteristics.

Multifunctional membranes can be produced through polymer blending. In conventional membrane treatment, removal of free ions such as metal ions is only achievable through desalting with RO membranes. However, MF membranes that are capable of selectively sequestrating heavy metal ions through adsorption can be fabricated by using polymer blending, without resorting to an energy-demanding RO process. Bai's group reported the use of chitosan (CS) and cellulose acetate (CA) blend solutions to prepare highly porous adsorptive hollow-fibre membranes that could be applied for the removal of heavy metals such as copper and lead ions from aqueous solutions during membrane filtration.[57-60]

Hydrophilisation of the membrane surface, widely considered as an important strategy for preventing membrane fouling and improving permeate flux, can also be achieved through blending. PVDF was blended with amphiphilic polymers to produce membranes with reliable hydrophilicity and protein resistance.[61,62] Poly(ether sulfone) (PES) was also blended with novel branched amphiphilic polymers to improve the fouling-resistant ability and UF membrane performance.[63,64]

Further, biocides can be blended with matrix membrane materials to give the membrane an anti-biofouling property. One of the examples was to introduce on a membrane the biocide and slow-releasing materials that are able to induce inactivation of microbes upon contact.[65] In addition, NF membranes were impregnated with silver nanoparticles to prevent biofouling.[66]

19.3.2 Surface Modifications

In addition to modifying the bulk materials prior to membrane fabrication, local and targeted modification of membrane surface only is also widely applied. Membrane fouling is a major obstacle to the widespread use of membrane technology. During filtration, the membrane surface contacting with the incoming feed directly is exposed to abuse by and accumulation of organic foulants, corrosive media and waterborne particles. Therefore, the maintenance of a clean membrane is crucial for ensuring steady and sustainable membrane filtration. Membranes with inherent antifouling or self-cleaning properties can be prepared through surface modification. Because most starting materials for commercially available membranes are hydrophobic in nature, the common practice in membrane research and development is to improve the surface hydrophilicity of membranes, as a high affinity of the membrane surface for water is capable of reducing the membrane fouling by waterborne hydrophobic foulants. In contrast to bulk modification, this method only alters the chemical properties of surfaces subjected to the modification process and therefore is more economic.

19.3.2.1 Surface Adsorption and Coating. Surface adsorption and coating are popular methods to enhance membrane surface hydrophilicity with little or no chemical reaction. For example, various polymers and surfactants were adsorbed on PS membranes to increase their hydrophilicity for resisting protein fouling.[67–69] Hydrophilisation has also been achieved by coating the membrane surface with more hydrophilic compounds. It has been reported that stable performance over time can be attributed to a hydrophilic poly(vinyl alcohol) (PVA) coating on the surfaces of conventional hydrophobic RO or NF membranes[70] and UF membranes.[71] However, the surface layer prepared through this approach is physically bound to the membrane surface and therefore could easily leach away and eventually diminish after prolonged usage of membranes.

19.3.2.2 Surface Reaction and Grafting. To overcome the low stability of physical coatings, another effective method to enhance membrane surface hydrophilicity is the use of chemicals such as strong acids/bases or high-energy irradiation sources such as plasma and UV to alter permanently the surface properties of membranes. Through optimising the modification chemistry, hydrophilic functional groups can be grafted on to membrane surfaces to boost hydrophilicity. For example, a polyvinylbutyral UF membrane was treated with hydrochloric acid to resist BSA fouling,[72] a PAN UF membrane was reacted with organic bases (ethanolamine, triethylamine) and inorganic bases (NaOH, KOH) to prevent BSA fouling,[73] PS membranes were treated with CO_2 plasma,[74] N_2 plasma[75] and O_2 plasma[76] to inhibit protein fouling and NOM fouling of PES and sulfonated PS membranes was significantly reduced after UV treatment of the membrane surface.[77]

Unlike graft copolymerisation described earlier, surface grafting covalently attaches functional monomers or polymers on the membrane surface only, *via* either free radical-, photochemical-, radiation-, redox- or plasma-induced grafting. A variety of functional monomers are available for preparing multifunctional membranes. To improve membrane hydrophilicity, hydrophilic functional monomers can be grafted on to the membrane surfaces. For example, a layer of polymer brushes is formed by gas plasma surface activation followed by free radical graft polymerisation using methacrylic acid or acrylamide monomers on the surface of conventional PA TFC membranes to reduce effectively the adhesion of foulants.[78] Hydrophilic monomers were also grafted on to PES UF membranes *via* UV-induced graft polymerisation to render the membrane surfaces more hydrophilic and less prone to fouling.[14] A charged membrane surface can be fabricated from a neutral, hydrophobic polymer through grafting either positively or negatively charged monomers on to the membrane surfaces. For example, copolymer brushes were grafted on to the surface of a PP membrane through UV-induced polymerisation of two oppositely charged monomers to prevent fouling by proteins and bacteria.[79] In addition, a PP membrane was grafted with thiol groups to immobilise silver stably, which could provide long-term anti-biofouling performance.[80]

19.4 INORGANIC MATERIALS FOR MEMBRANE FABRICATION

Inorganic membranes for water purification are usually prepared from metallic oxides (alumina, titania and zirconia)and zeolite. Even though the manufacturing cost of inorganic membranes is much higher than that of polymeric membranes, inorganic membranes are commonly preferred in separation processes that involve harsh, extreme operating conditions where polymeric membranes would degrade rapidly, *e.g.* high operating temperatures, radioactive/ heavily contaminated feeds and highly reactive environments.[81]

Generally, inorganic ceramic membranes consist of a macroporous support layer and a meso- or microporous active layer. The macroporous support layer is prepared through sintering ceramic powders treated by extrusion or slip-casting. Subsequently, a thin layer with an average pore diameter of 0.2–1 μm that coincides with typical MF pore sizes is applied on the support layer through suspension coating. For smaller pores, alternative pore-forming processes such as the sol–gel process are widely employed to produce ceramic UF membranes. In the case of the sol–gel process, polymeric silica molecules are deposited on the support system consisting of a 1 μm thick mesoporous γ-Al$_2$O$_3$, over a macroporous α-Al$_2$O$_3$, of a desired thickness.[82,83] The silica layer is then calcined at 400 °C to fuse into an active separation layer with a thickness of 50–100 nm. Ceramic membranes prepared by the sol–gel process usually have pore sizes in the range 2–4 nm, which are suitable for UF.

Zeolites are another popular precursor material for the fabrication of inorganic membranes. Zeolites are crystalline microporous aluminosilicates, built up of three-dimensional networks of SiO_4 and AlO_4. The zeolite structure contains open channels with different sizes that can be used for molecular sieving. Composite membranes with a zeolite top layer have been developed. Zeolites as a top layer provide a number of advantages, including the following: (1) the pore size of the active separation layer can be adjusted by selecting the appropriate type of zeolite; (2) the hydrophilicity of the membrane can be adjusted by introducing a different type of zeolite [*e.g.*, zeolite LTA (type A) is very hydrophilic, but silicalite-1 is very hydrophobic]; and (3) the intrinsic catalytic properties of the zeolite are conveniently available for making catalytic membrane reactors.[84]

REFERENCES

1. M. Mulder, *Basic Principles of Membrane Technology*, Kluwer, Dordrecht, 1991.
2. H. Yanagishita, D. Kitamoto, K. Haraya, T. Nakane, T. Okada, H. Matsuda, Y. Idemoto and N. Koura, *J. Membr. Sci.*, 2001, **188**, 165.
3. L.-C. Li, B.-G. Wang, H.-M. Tan, T.-L. Chen and J.-P. Xu, *J. Membr. Sci.*, 2006, **269**, 84.
4. V. Freger, *Langmuir*, 2003, **19**, 4791.
5. B.-H. Jeong, E. M. V. Hoek, Y. Yan, A. Subramani, X. Huang, G. Hurwitz, A. K. Ghosh and A. Jawor, *J. Membr. Sci.*, 2007, **294**, 1.
6. R. Q. Kou, Z. K. Xu, H. T. Deng, Z. M. Liu, P. Seta and Y. Y. Xu, *Langmuir*, 2003, **19**, 6869.
7. Sumitomo Electric Interconnect Products, *Poreflon*® *Overview*, http://www.seipusa.com/products/W0200501.asp (last accesed October 2011).
8. Arkema, *Kynar*® *and Kynar Flex*® *PVDF*, http://www.arkema-inc.com/kynar/page.cfm?pag = 1118 (last accesed October 2011).

9. P. Apel, *Radiat. Meas.*, 2001, **34**, 559.

10. J.-J. Kim, T.-S. Jang, Y.-D. Kwon, U. Y. Kim and S. S. Kim, *J. Membr. Sci.*, 1994, **93**, 209.

11. D. R. Lloyd, K. E. Kinzer and H. S. Tseng, *J. Membr. Sci.*, 1990, **52**, 239.

12. General Electric, *GE Water and Power: Water & Process Technologies*, http://www.gewater.com (last accesed October 2011).

13. B. K. Chaturvedi, A. K. Ghosh, V. Ramachandhran, M. K. Trivedi, M. S. Hanra and B. M. Misra, *Desalination*, 2001, **133**, 31.

14. J. Pieracci, J. V. Crivello and G. Belfort, *J. Membr. Sci.*, 1999, **156**, 223.

15. T. Shimoda and H. Hachiya, Process for preparing a polyether ether ketone membrane, *US Patent*, 5 997 741, 1999.

16. Z.-L. Xu, T.-S. Chung and Y. Huang, *J. Appl. Polym. Sci.*, 1999, **74**, 2220.

17. R. Tuccelli and P. V. McGrath, Cellulosic ultrafiltration membrane, *US Patent*, 5 522 991, 1996.

18. K. Nouzaki, M. Nagata, J. Arai, Y. Idemoto, N. Koura, H. Yanagishita, H. Negishi, D. Kitamoto, T. Ikegami and K. Haraya, *Desalination*, 2002, **144**, 53.

19. J. Hosch and E. Staude, *J. Membr. Sci.*, 1996, **121**, 71.

20. Dow Chemical, *Dow Ultrafiltration*, http://www.dowwaterand process.com/products/uf/index.htm (last accesed October 2011).

21. General Electric, *Ultrafiltration: Effective Solids Separation and Pathogen Removal*, http://www.gewater.com/products/ equipment/mf_uf_mbr/uf.jsp (last accesed October 2011).

22. C. E. Reid and E. J. Breton, *J. Appl. Polym. Sci.*, 1959, **1**, 133.

23. S. Loeb, in *Synthetic Membranes, ACS Symposium Series*, Vol. 153, American Chemical Society,Washington, DC, 1981, pp. 1–9.

24. R. Haddada, E. Ferjani, M. S. Roudesli and A. Deratani, *Desalination*, 2004, **167**, 403.

25. K. J. Edgar, C. M. Buchanan, J. S. Debenham, P. A. Rundquist, B. D. Seiler, M. C. Shelton and D. Tindall, *Prog. Polym. Sci.*, 2001, **26**, 1605.

26. H. H. Hoehn and J. W. Richter, Aromatic polyimide, polyester and polyamide separation membranes, *US Patent Application*, RE30351, 1980.

27. R. Endoh, T. Tanaka, M. Kurihara and K. Ikeda, *Desalination*, 1977, **21**, 35.
28. R. McKinney Jr and J. Rhodes, *Macromolecules*, 1971, **4**, 633.
29. J. Beasley, *Desalination*, 1977, **22**, 181.
30. H. Hoehn, in *Materials Science of Synthetic Membranes, ACS Symposium Series*, Vol. 269, American Chemical Society, Washington, DC, 1985, pp. 81–98.
31. L. Credali, G. Baruzzi and V. Guidotti, Reverse osmosis anisotropic membranes based on polypiperazine amide, *US Patent*, 4 129 559, 1978.
32. P. Parrini, *Desalination*, 1983, **48**, 67.
33. M. D. Guiver, A. Y. Tremblay and C. M. Tam, Method of manufacturing a reverse osmosis membrane and the membrane so produced. *US Patent*, 4 894 159, 1990.
34. C. Brousse, R. Chapurlat and J. P. Quentin, *Desalination*, 1976, **18**, 137.
35. H. Strathmann, *Ion-exchange Membrane Separation Processes*, Elsevier, Amsterdam, 2004.
36. DuPont, *Fuel Cells: DuPont™ Nafion® Membranes and Dispersions*, http://www2.dupont.com/FuelCells/en_US/products/nafion.html (last accesed October 2011).
37. J. D. Norton and M. F. Buehler, *Sep. Sci. Technol.*, 1994, **29**, 1553.
38. S. Tan, A. Laforgue and D. Bélanger, in *Advanced Materials for Membrane Separations, ACS Symposium Series*, Vol. 876, American Chemical Society, Washington, DC, 2004, pp. 311–323.
39. Astom, *NEOSEPTA*, http://www.astom-corp.jp/en/en-main2-neosepta.html (last accesed October 2011).
40. Z. Matějka, *J. Appl. Chem. Biotechnol.*, 1971, **21**, 117.
41. J. E. Cadotte, Reverse osmosis membrane, *US Patent*, 3 926 798, 1975.
42. M. Kurihara, N. Kanamaru, N. Harumiya, K. Yoshimura and S. Hagiwara, *Desalination*, 1980, **32**, 13.
43. J. E. Cadotte, Interfacially synthesized reverse osmosis membrane, *US Patent*, 4 277 344, 1981.
44. Y. Kamiyama, N. Yoshioka, K. Matsui and K. Nakagome, *Desalination*, 1984, **51**, 79.
45. R. E. Larson, J. E. Cadotte and R. J. Petersen, *Desalination*, 1981, **38**, 473.

46. J. Glater, M. R. Zachariah, S. B. McCray and J. W. McCutchan, *Desalination*, 1983, **48**, 1.
47. N. Inagaki, *Plasma Surface Modification and Plasma Polymerization*, Technomic Publishing, Lancaster, PA, 1996.
48. H. K. Yasuda, *Plasma Polymerization and Plasma Treatment*, Wiley, New York, 1984.
49. H. K. Yasuda, in *Plasma Polymerization and Plasma Interactions with Polymeric Materials: Proceedings of the Symposium on Plasma Polymerization and Plasma Interactions with Polymeric Materials, held at the ACS 199th National Meeting in Boston, MA, April 1990*, Wiley, New York, 1990.
50. G. C. Eastmond, M. Gibas, W. F. Pacynko and J. Paprotny, *J. Membr. Sci.*, 2002, **207**, 29.
51. B. R. Einsla, Y.-T. Hong, Y. Seung Kim, F. Wang, N. Gunduz and J. E. McGrath, *J. Polym. Sci., Part A: Polym. Chem.*, 2004, **42**, 862.
52. H.-S. Lee, A. S. Badami, A. Roy and J. E. McGrath, *J. Polym. Sci., Part A: Polym. Chem.*, 2007, **45**, 4879.
53. L. Ying, P. Wang, E. T. Kang and K. G. Neoh, *Macromolecules*, 2001, **35**, 673.
54. L. Ying, E. T. Kang and K. G. Neoh, *Langmuir*, 2002, **18**, 6416.
55. Y. Chen, L. Ying, W. Yu, E. T. Kang and K. G. Neoh, *Macromolecules*, 2003, **36**, 9451.
56. P. Wang, K. L. Tan, E. T. Kang and K. G. Neoh, *J. Mater. Chem.*, 2001, **11**, 783.
57. C. X. Liu and R. B. Bai, *J. Membr. Sci.*, 2006, **284**, 313.
58. C. X. Liu and R. B. Bai, *J. Membr. Sci.*, 2005, **267**, 68.
59. C. X. Liu and R. B. Bai, *J. Membr. Sci.*, 2006, **279**, 336.
60. W. Han, C. X. Liu and R. B. Bai, *J. Membr. Sci*, 2007, **302**, 150.
61. Y.-H. Zhao, B.-K. Zhu, L. Kong and Y.-Y. Xu, *Langmuir*, 2007, **23**, 5779.
62. Y.-H. Zhao, Y.-L. Qian, B.-K. Zhu and Y.-Y. Xu, *J. Membr. Sci.*, 2008, **310**, 567.
63. Y.-Q. Wang, Y.-L. Su, X.-L. Ma, Q. Sun and Z.-Y. Jiang, *J. Membr. Sci*, 2006, **283**, 440.
64. Y.-Q. Wang, Y.-L. Su, Q. Sun, X.-L. Ma and Z.-Y. Jiang, *J. Membr. Sci.*, 2006, **286**, 228.
65. G. Golomb and A. Shpigelman, *J. Biomed. Mater. Res., Part A*, 1991, **25**, 937.

66. K. Zodrow, L. Brunet, S. Mahendra, D. Li, A. Zhang, Q. L. Li and P. J. J. Alvarez, *Water Res.*, 2009, **43**, 715.
67. L. E. S. Brink and D. J. Romijn, *Desalination*, 1990, **78**, 209.
68. A. G. Fane, C. J. D. Fell and K. J. Kim, *Desalination*, 1985, **53**, 37.
69. K. J. Kim, A. G. Fane and C. J. D. Fell, *Desalination*, 1988, **70**, 229.
70. C. Y. Tang, Y.-N. Kwon and J. O. Leckie, *Desalination*, 2009, **242**, 149.
71. X. F. Wang, D. F. Fang, K. Yoon, B. S. Hsiao and B. Chu, *J. Membr. Sci*, 2006, **278**, 261.
72. X. Ma, Q. Sun, Y. Su, Y. Wang and Z. Jiang, *Sep. Purif. Technol.*, 2007, **54**, 220.
73. H. R. Lohokare, S. C. Kumbharkar, Y. S. Bhole and U. K. Kharul, *J. Appl. Polym. Sci.*, 2006, **101**, 4378.
74. I. Gancarz, G. Poźniak and M. Bryjak, *Eur. Polym. J.*, 1999, **35**, 1419.
75. I. Gancarz, G. Poźniak and M. Bryjak, *Eur. Polym. J.*, 2000, **36**, 1563.
76. K. S. Kim, K. H. Lee, K. Cho and C. E. Park, *J. Membr. Sci.*, 2002, **199**, 135.
77. J. E. Kilduff, S. Mattaraj, J. P. Pieracci and G. Belfort, *Desalination*, 2000, **132**, 133.
78. N. H. Lin, M.-M Ki, G. T Lewis and Y Cohen, *J. Mater. Chem.*, **20**, 4642.
79. Y. H. Zhao, X. Y. Zhu, K. H. Wee and R. B. Bai, *J. Phys. Chem. B*, 2010, **114**, 2422.
80. X. Y. Zhu, PhD dissertation, National University of Singapore, 2012.
81. C. A. M. Siskens, Chapter 13 Applications of ceramic membranes in liquid filtration, In: A.J. Burggraaf and L. Cot, Editor(s), *Membrane Science and Technology*, Elsevier, 1996, **4**, 619–639.
82. A. Julbe, C. Guizard, A. Larbot, L. Cot and A. Giroir-Fendler, *J. Membr. Sci.*, 1993, **77**, 137.
83. C. J. Brinker, A. J. Hurd, P. R. Schunk, G. C. Frye and C. S. Ashley, *J. Non-Cryst. Solids*, 1992, **147–148**, 424.
84. C. Baerlocher, L. B. McCusker and D. Olson, *Atlas of Zeolite Framework Types*, 6th edn, Elsevier, Amsterdam, 2007.

CHAPTER 20

Glass and New Technologies

ANDREAS KAFIZAS AND IVAN P. PARKIN*

Department of Chemistry, University College London, 20 Gordon Street, London WC1H 0AJ, UK
*Email: i.p.parkin@ucl.ac.uk

20.1 INTRODUCTION

Glass is one of the most important materials in the modern world, used in a wide range of applications from architectural glazing and high-precision optics to food containers. It is well known that the qualities of glass, such as colour and thermal expansion, can be modified by the addition of various minerals during manufacture. However, improving the properties of glass through the application of thin films has become today's mode of choice in functional glass production. The attraction of such coatings is their ability to alter surface chemical/physical properties, while preserving the bulk qualities of glass.[1]

Glass for use in windows is predominately made by the float glass process, which produces a continuous ribbon of highly uniform glass. For standard glass windows, the required size is cut from the ribbon. However, the glass can also be shaped to form curved windows for automotive, aerospace and other industries. The float glass process, first discovered and commercialised by

Materials for a Sustainable Future
Edited by Trevor M. Letcher and Janet L. Scott
© The Royal Society of Chemistry 2012
Published by the Royal Society of Chemistry, www.rsc.org

Pilkington, is today used to manufacture over 90% of the world's flat glass. This involves melting the premixed raw materials of glass in a furnace at 1500 °C and then feeding this molten liquid into a tin bath at a constant rate under an inert nitrogen gas atmosphere.[2] By maintaining the temperature at 1000 °C, irregularities and defects are avoided. After a sufficient time, the glass is cooled to around 600 °C and passed through rollers that thin the glass. In a typical plant, this fully automated process is operated continuously for up to 15 years, producing ~6000 km of glass per year.[3] To improve the functions of glass, thin-film coatings are applied. Most functional coatings are applied in tandem with the on-line production of float glass, typically while the ribbon is still hot.

Architectural glazing is perhaps the most active area of coating innovation and also the fastest growing commercially. With the significance of global warming being appreciated and rising fuel prices, there has been increased pressure on glass construction companies to produce more energy-efficient glass. One such type of coating that has achieved unparalleled success is low-emissivity coatings that retain the heat generated within a building.[4] Other coatings include selfcleaning,[5] 'intelligent' colour changing[6] and heat-reflective coatings for buildings in warm climates.[7] However, the largest sector in which functional coatings are applied is the glass bottle industry, in which anti-friction coatings are applied to increase resistance and strength. Functional coatings have also been applied to automotive vehicle windows, tiles, mirrors, optics, glass fibres and displays.[8]

The primary focus of this book is to look at sustainability. There is no problem with the world running out of the precursor materials to make window glass. Glass is manufactured from materials such as silica (sand) that are for particle purposes virtually limitless on the Earth. However, a lot of energy goes into the glass-making process with furnace temperatures in the region of 1200–1400 °C and a typical float glass plant having the same energy demand as a city of ~100 000 inhabitants. Despite this, glass is the 'ultimate in recyclable' material as it can be reused to make further glass. In this reusable process, the broken recycled glass is called 'cullet' and in fact not only is incorporated in new glass but also acts to reduce the furnace temperature by acting as a catalytic former around which new glass will form. By so doing, it reduces the furnace temperature by around 200–400 °C with a saving of ~20% in energy use. Coated glass products for energy conservation purposes have one of the shortest payback times in terms of the energy saving they provide against the extra energy required to make the coating. Within 1 year of use, a coated window has conserved more energy than was required in making the coating.

In this chapter, we focus on existing, developing and next-generation window coatings. We focus on how they are produced and also the underlying scientific principles on how they work. In addition, recent scientific research into new avenues for improving the functions of glass are presented together with their potential future impact. These new products illustrate how employing the appropriate coating can reduce energy usage within homes and buildings, hence leading to a more sustainable, lower energy economy in the future.

20.2   LOW-EMISSIVITY COATINGS

Since the development and production of multiple glazed heat-insulating glass units in the 1950s, a significant ecological drive to improve the heat retention of such systems has followed.[4] Although such systems, consisting of two or more panes of glass separated by vacuum or gas in-fill layers that inhibit kinetic thermal loss (molecules in air colliding with the interior window pane), a significant level of heat can be lost through radiative means (photons). This led to the development of thin-layer materials with strong radiative reflection coatings on the inner vacuum side of such multi-glazed units. To date, systems based on silver thin-film layers have achieved the greatest success and form the market standard (Figure 20.1).

Low *emissivity* (ε) or, more accurately, low *thermal emissivity* is a measure of the quality of a material to reflect radiant thermal energy. This is scaled from 0 to 1, where a perfect heat reflector would have an emissivity of 0. At ambient temperatures (~25 °C), a blackbody radiates thermal energy through the emission of photons in the infrared region of the electromagnetic spectrum. Therefore, a measure of a material's *thermal emissivity* for architectural applications is in fact an indirect measure of the material's infrared reflectivity. Over the past 50 years, a range of low-emissivity materials for the primary use of being coated on the

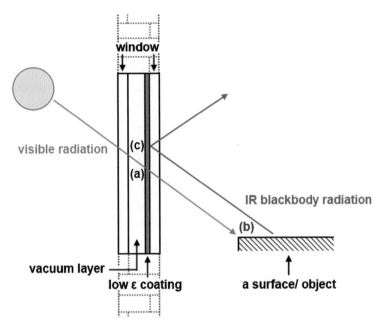

Figure 20.1 A diagram representing the high radiative heat retention of low-emissivity (ε) window coatings. Given the high visible light transparency of the coating, sunlight can pass through into the building (a). This light is absorbed by the contents of the room and re-emitted as blackbody radiation in the infrared (IR) region (b). Given the high reflectivity of the low-emissivity (ε) coating, any stray IR radiation is reflected back into the room and the heat is retained.

inner side of double-glazed glass units have been investigated.[9] These layers would serve to keep any heat generated in a building (infrared radiation) by reflecting it back inside. In addition to their infrared-reflecting properties, such materials require a high transparency in the visible region, allowing natural light into the building and creating a greater aesthetic appeal. In combining the two properties of high visible light transparency and high infrared reflectivity, these layers can significantly reduce building heating costs, where sunlight can enter and warm the building and the low emissivity coating retains any blackbody heat (infrared) generated inside (Figure 20.1). Additionally, any radiative heat generated internally through central heating systems, *etc.*, is also retained.

Given this specific functionality, low-emissivity coatings have been deemed *solar control* coatings.

Since the mid-1980s, Pilkington NSG have commercialised a low-emissivity product under the brand name K Glass. This product is produced in-line during the hot-end float glass process. A typical coating is dual layered, consisting of a heat-absorbing layer and a low-emissivity layer.[10] In their patented processes, Pilkington produce sodium-doped tungsten oxide ($Na:WO_3$) 80–200 nm thick heat-absorbing layers from the atmospheric pressure chemical vapour deposition (APCVD) of tungsten hexafluoride (WF_6) and sodium vapour. By next introducing tin precursors in ethylene–carbon dioxide precursor mixtures at 600–750 °C with a fluorine source such as fluoroacetic acid or hydrogen fluoride, a low emissivity fluorine-doped SnO_2 ($F:SnO_2$) layer is formed on top (200–500 nm thick).[11] However, the technical values and the coating quality of K Glass (roughly 80% solar reflectance) are not as high as those displayed by coatings based on layers of silver (up to 97% solar reflectance) and it has only received regional market acceptance.

Silver layer coatings have been produced since the early 1980s through a magnetron sputtering process and have had the most prominent market success of any solar control coating, primarily owing to their high transparency and neutral colour. Today, more than 100 in-line magnetron coaters are installed worldwide, producing more than 200 million m^2 of silver-based low-emissivity glass. Most silver-based coatings today are multi-layered systems that consist primarily of either a zinc oxide (ZnO_2) or SnO_2 seed layer (~ 10 nm thick) that aids crystalline silver layer growth on top followed by a robust sub-oxidic titanium oxide (TiO_x) interface layer (~ 3 nm thick) that prevents oxidation of the silver layer and increases the lifetime of the coating.[12] By European and US legislation, all new buildings are obliged to use solar control coatings. Such coatings are today provided as standard in all flat glass, representing a global market value of more than US$30 billion.[13]

20.3   INTELLIGENT WINDOW COATINGS

Intelligent window coatings describe materials that have the ability to alter their optical properties when a change in condition occurs.

There are three main branches of such intelligent coatings: electrochromic (change with electrical flow),[14] thermochromic (change with temperature)[6] and photochromic (change with lighting conditions).[15] In this section, we discuss each type of intelligent window in turn.

20.3.1 Electrochromic Windows

Electrochromic materials can reversibly change colour in response to a change in voltage altering the material's optical properties and thus allowing control over the amount and type of light that can pass through. In electrochromic windows based on tungsten(VI) oxide (WO_3), the material changes its opacity between a coloured translucent state (dark blue) and a transparent state (yellow tinge). A burst of electricity is required to induce this colour change, but once the change has been effected, no further current is needed to maintain the particular shade that has been reached.

Such windows, termed *smart glass*, are used in a wide variety of devices that rely on rapid, reversible changes in colour. Today, the main use of smart glass is for internal partitions in an office environment, where many companies now enjoy the ability to switch screens and doors from clear to private. A similar use is found in the healthcare industry for patient privacy, replacing traditional blind systems, that are often difficult to clean, with a flat and more easily cleanable surface. Smart glass has also been applied in exhibitions and displays, for both revealing and protecting new products and museum artwork. It is also used in television studios, passenger–driver separator windows in trains and glass roofing in automobiles.[16]

The properties of an electrochromic device as a whole are largely dictated by the electrochromic layer. To create a successful device, the electrochromic colour change must be reversible over a large number of cycles, without degradation or loss of performance. How many cycles are necessary depends on the application; a pane of window glass would be expected to last over 20 years and survive wide variations in temperature and humidity. More innovative smart glass design coming on to the consumer market, with extended life cycling (~ 20 years) without degradation or loss of performance, has driven competition and made smart glass

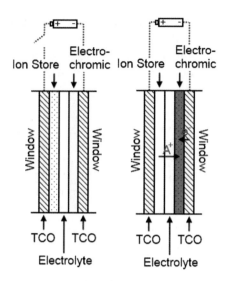

Figure 20.2 Images showing the arrangement for a typical WO_3 smart glass window consisting of seven layers, (a) before and (b) after a sufficient current is passed through the circuit to induce a colour change. Ions (A^+) move from the ion store and intercalate into the WO_3 layer where electrons serve to reduce W^{6+} ions to W^{5+}, turning the material from pale yellow to deep blue in colour.

increasingly affordable, with the market forecast to reach US$418 million by 2020.

A typical smart glass window consists of seven layers (Figure 20.2). The working layers act as an electrochemical cell in which the electrochromic material, normally tungsten oxide (WO_3), is housed, usually as a working electrode.[17] Vanadium(V) oxide and molybdenum oxides have also been investigated as electrochromic layers; however, WO_3 shows better visible light transparency in its transparent state coupled with a large difference in transmission between the bleached and coloured states. This layer is connected to an ion storage layer (either H^+ or Li^+) *via* an electrolyte. This is sandwiched between two TCO layers, normally made from indium-doped SnO_2, that acts as a transparent medium in which the current can be fed. Upon applying a current, ions (A^+) are pushed from the ion storage layer into the electrochromic layer. If the potential difference applied exceeds the activation barrier, ions are intercalated within the WO_3 electrolyte layer. This forms a partially reduced A_x:WO_3 layer containing W^{5+} and W^{6+} ions. The W^{5+}

ions absorb visible light strongly through polaron transitions, turning the material dark blue in appearance. The overall reaction can be written as follows:

$$WO_3 + xA^+ + xe^- \rightleftharpoons A_xWO_3 \equiv A_x(W^{6+}, xW^{5+})O_3 \qquad (20.1)$$

As thin films of WO_3 are unstable in acid–base solutions, a permeable protection layer, such as organic polymers or hydrated tantalum oxide (Ta_2O_5), is required if a protic electrolyte is used. The mobility of ions into and out of the WO_3 layer dictates the time taken for a colour switch to occur. Electrochromic layers with high surface areas are used when a rapid colour switch is required. As a high-crystallinity layer reduces the mobility of ions within the material, reducing the electrochromic effect, amorphous tungsten oxide (α-WO_3) films are favoured, typically formed through physical vapour deposition (PVD) sputtering processes.

Smart windows based on an organic polymer dispersed liquid crystal (PDLC) electrochromic layer have also been commercialised. When no current is applied through the PDLC, the crystals remain randomly oriented, highly scattering light and appearing a translucent milky white. Upon applying a sufficient current, the crystals become ordered and the window turns transparent. Although having some success in privacy control, the product has had only modest success in comparison with WO_3-based electrochromics as it requires a constant current to maintain its transparent state.

However, the electrochromic smart mirror has found the most prominent market success, having been applied in the rear- and side-view mirrors of high-end automobiles.[18] This device consists of two lenses that sandwich an electrochromic gel layer, which normally contains WO_3 as the active electrochromic, propylene carbonate as the electrolyte and viologen salt as the ion source. The inside sides of the lenses are coated with a TCO layer (normally F:SnO_2) and the deepest lens has a reflective mirror coating (normally an Al:Ti alloy). The device uses a forward sensor which measures the outside ambient light and a rearward sensor to look for glare. When dark enough, it sends current to the electrostatic gel, darkening it a rate which is related to the level of ambient darkness and rearward glare. When the outside ambient light is strong, a reverse current is passed through the gel to make it clear once more. Worldwide annual sales of smart mirrors currently far exceed $1 billion, more than double the market share for smart windows.[19]

20.3.2 Thermochromic Windows

A thermochromic material changes its optical properties with a change in temperature. One potential application of such a material would be in a window solar control coating that changes its transmittance and reflectance properties in order to regulate the temperature within a room. Such windows would be suited to geographical regions where the temperature may change dramatically during a day or over the year. Suitable materials should become more reflective in the infrared region when the room becomes too hot; thereby better blocking the infrared portion (heat) of sunlight from entering. A thermochromic window would therefore act as an advanced solar control coating that actively responds to a user's needs. From assessing the solar spectrum at sea level (Figure 20.3), it can be seen that a substantial portion of sunlight, in terms of the number of photons, lies in the infrared

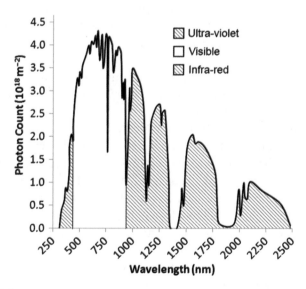

Figure 20.3   The solar spectrum at sea level in terms of the number of photons (10^{18} m^{-2}) sectioned into ultraviolet, visible and infrared.

region. This highlights the importance of controlling the level of heat that can enter a building, potentially leading to energy savings from reduced air conditioning costs.

The most popularly studied material for use as a thermochromic solar control coating is vanadium dioxide (VO_2) as it possesses the lowest temperature metal–semiconductor phase transition of any material. Upon exceeding its transition temperature, VO_2 changes phase from monoclinic (semiconducting) to tetragonal (metallic), as shown in Figure 20.4. In its low-temperature monoclinic state the structure contains homopolar V^{4+}–V^{4+} bonds that prevent electrical conduction, making the material more infrared transparent.[6] However, in its high-temperature state, these homopolar bonds are broken, causing an increase in electrical conduction and infrared reflectivity. This phase transition is completely reversible in thin films of the material and can be cycled an extended number of times without damage or functional loss.

The transition temperature of 68 °C was clearly too high to be useful in regulating room temperature. Nevertheless, this problem was overcome through W doping of VO_2, where roughly a 2 at.% doping level reduced the transition temperature to 25 °C. This was

(a)   (b)

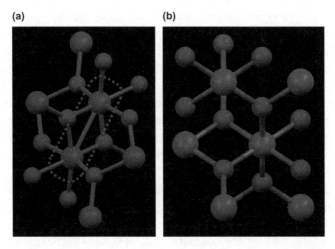

Figure 20.4 Arrangement of atoms in vanadium dioxide (vanadium = purple, oxygen = red) (a) before (monoclinic – semiconducting) and (b) after (tetragonal – metallic) undergoing a metal–semiconductor phase transition at 68 °C. The homopolar bonds of the low-temperature monoclinic state are highlighted.

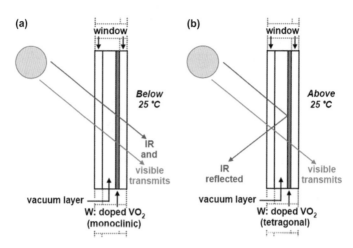

Figure 20.5   Diagrams showing how a thermochromic window based on W: VO$_2$ (2% doping level) would function (a) before and (b) after its metal–semiconductor phase transition at 25 °C.

achieved through the APCVD of VCl$_4$, WCl$_6$ and H$_2$O on glass at 550 °C,[20] a process that could simply be introduced in-line during the hot-end production of float glass. A diagram of how this material might be used as a thermochromic window is shown in Figure 20.5. Figure 20.5(a) shows the case where the enclosed room is below 25 °C. Therefore, the thermochromic window layer is in its semiconducting state and allows significant portions of both visible and infrared frequencies through, thus better allowing the room to heat up. Figure 20.5(b) shows the case where the room temperature exceeds 25 °C. This induces the thermochromic window layer to undergo a phase transition to its metallic state. While retaining a degree of visible light transmittance, the material now reflects infrared light more strongly and inhibits the Sun from heating the room.

However, neither VO$_2$ nor its modified derivatives have been commercialised in thermochromic windows. This is due to the yellow–brown appearance of VO$_2$ that makes it highly undesirable as a window coating for cosmetic reasons. An alternative patented thermochromic material product based on N-substituted acrylamide polymers called Cloud Gel has also been highly researched, but is yet to make mass production.[21] Although this material displayed excellent infrared reflectance properties, altering from roughly 0 to 90% infrared reflectance when increased in temperature from 27 to 31 °C, the originally colourless polymer changes to a clear white, obscuring visibility. Although it has been suggested that Cloud Gel could find prominent success in sky-lighting, this is yet to materialise even after more than 30 years of research and heavy investment. Nevertheless, a novel thermochromic window product, not based on either VO_2 or Cloud Gel, called Pleotint SRT, will soon reach the consumer markets.

20.3.3 Photochromic Glass

The only successful commercial application of photochromic glass materials to date has been in eyeglass lenses. These lenses darken on exposure to ultraviolet radiation, protecting the eye from high-energy light, and lighten to their original state on removal from ultraviolet exposure. In glass lenses, the active photochromic material most often used is AgCl, which reversibly darkens upon ultraviolet exposure by disintegrating into elemental chlorine and silver. In a plastic lens, the active photochromic material is organic and generally of the spirooxazine or naphthopyran class. Looking at the spirooxazine family as an example, the spiro form of an oxazine is a colourless leuco dye where the conjugated system of the oxazine and another aromatic part of the molecule is separated by an sp^3-hybridised 'spiro' carbon (Figure 20.6).

After exposure to UV irradiation, the bond between the spiro carbon and the oxazine breaks and opens the ring. The spiro carbon becomes planar and the aromatic group rotates, aligning its π-orbitals with the rest of the molecule and forming a light-absorbent

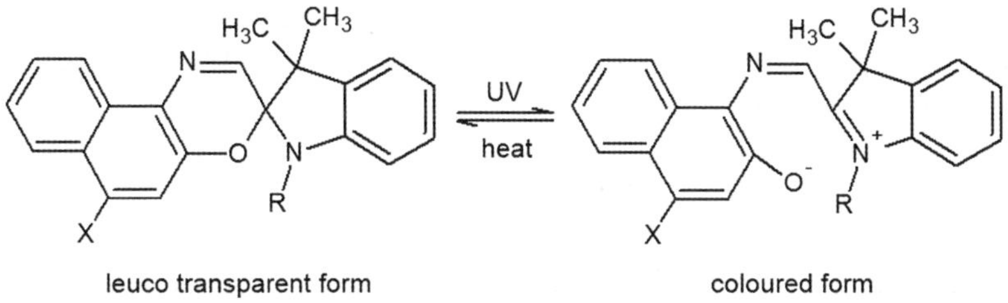

leuco transparent form   coloured form

Figure 20.6 Spirooxazine, an organic class of photochromic material used in transition eyeglass lenses, before (transparent) and after (coloured) exposure to UV radiation.

π-conjugated system. When the UV light is removed, the molecules gradually relax to their ground state, the spiro carbon–oxygen bond re-forms and the molecule returns to its colourless state. The most successful product in this genre has been the Transitions lens that consists of an impact-resistant polycarbonate that houses the active photochromic agent and today accounts for at least US$100 million in annual sales worldwide. Nevertheless, the lifetime of such lenses is affected by the level of UV exposure they endure, where, like most organic dyes, they are susceptible to degradation by oxygen and free radicals. The focus of Transitions lens innovation has been on increasing the mechanical and free radical resistance of the polymer matrices that house the active photochromic, prolonging their lifetime and increasing performance.

The development of photochromic windows has been hindered in comparison with photochromic lenses. This has mainly been due to the cost of scaling up the processes involved in lens production for windows; however, other factors include poor scratch resistance and UV exposure lifetime. Although photochromic windows would prevent the overheating of a building in summer, its fundamental drawback is that it would prevent heat from entering in winter. Nevertheless, some new designs based on a combination of photovoltaic and electrochromic cells have been presented as 'solutions' to producing photochromic windows. One such solution involves mixing $Li:WO_3$ within a TiO_2-based dye-sensitised solar cell. Although not being a direct photochromic material, the photovoltaic serves to create a voltage by the absorption of light that is sufficient to induce the intercalation of Li^+ in WO_3 and a change in window colour. A working prototype darkens visible light transmittance from 60 to 4% upon light exposure. When light is removed, a reversal occurs where electrons, primarily from WO_3, flow to I^{3-} ions present in the electrolyte. This reverse reaction is typically quite slow but can be accelerated by adding a catalyst such as platinum.

20.4   ANTI-FRICTION COATINGS

Anti-friction coatings on glass containers have revolutionised the manufacture of bottle products. By creating smoother containers, the speed at which the bottles can be moved around the plant can be increased. In addition, such anti-friction coatings also serve to improve the mechanical strength of the containers. Therefore, such bottles are less liable to fracture. In combining these two functions, the production rate and efficiency of bottled glass manufacture has increased.

Bottle containers are typically formed from moulding molten glass and then applying compressed air to form the cavity. In most instances two forms of coating are applied to these containers, one at the hot end and the other at the cold end of manufacture. At the hot end, a hard ceramic coating of either a ~ 10 nm thick TiO_2 or SnO_2 layer is formed from atmospheric pressure CVD at temperatures ranging from 450 to 600 °C. Such coatings must be deposited quickly (0.1–0.5 s time frame). In the formation of SnO_2 layers, tin tetrachloride ($SnCl_4$) and monobutyltin trichloride ($BuSnCl_3$) precursors dominate whereas for TiO_2 layers titanium tetrachloride ($TiCl_4$) and titanium isopropoxide [$Ti(O^iPr)_4$] are most common. The cold-end coating is typically a polyethylene wax that is spray coated from a water-based emulsion. This coating is ~ 50 nm thick and is applied to the glass when cooler at 135–165 °C. This makes the glass slippery, increasing its scratch resistance and reduces the sticking of neighbouring containers whilst being transported around the conveyor. Efforts have been made to avoid the use of a hot-end CVD coating and instead have a single cold-end coating stage alone; however, such products generally show inferior mechanical and handling properties due to poorer coating adhesion.

Although this mature technology now competes with the sale of plastic alternatives, it is perceived as the 'premium' quality packaging format, accounting for more than 180 billion units per annum and with revenues of at least $22 billion per annum in the USA alone.[22]

20.5   SELF-CLEANING COATINGS

Self-cleaning windows have become a highly desirable feature in architecture and have been successfully commercialised by a number of companies. Currently, all commercial self-cleaning windows are coated with a thin transparent layer of anatase titanium dioxide (TiO_2), a coating which acts to clean the window in sunlight through two distinct properties; photocatalysis and photoinduced hydrophilicity (PIH).[23] The mechanism of photocatalysis is the

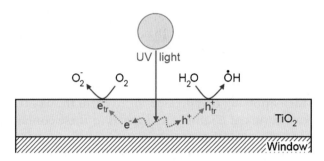

Figure 20.7 Summary of the photocatalytic processes that occur on TiO₂ thin films that generate highly reactive species and break down organic dirt and grime.

driving force for chemically breaking down organic material that comes into contact with the surface of the film (Figure 20.7). This process first requires a photoexcitation, caused by absorption of a photon greater in energy than the material bandgap (~ 3.2 eV), where electrons are excited from the valence band into the conduction band, leaving a concomitant hole behind. These photogenerated electrons and holes can travel freely through the material and migrate to the lower energy surface of the film. These electrons and holes are stabilised by falling into shallow traps and react on contact with donor molecules from the air. Trapped electrons preferentially react with dioxygen (O_2) and trapped holes with water (H_2O). This yields highly reactive superoxide (O_2^-) and radical ($H^\bullet$) species that serve to degrade any organic material in contact with the film to carbon dioxide (CO_2), H_2O and mineral acids. The efficacy of TiO₂ in destroying a range of organic media including bacteria, cancer cells and viruses has also been demonstrated.

The mechanism of PIH is a driving force that causes 'sheet' wetting of the surface. Analogous to photocatalysis, this process requires a photoexcitation. However, upon migration and trapping of electrons and holes, a different reaction can occur. In the case of PIH, holes abstract oxygen at the surface of the film between Ti–O–Ti bridges. This allows water in the vicinity of the film surface to insert within these holes, forming surface hydroxyl bonds. Several hydroxylated domains form on the surface of TiO₂, increasing the surface polarity and water attraction force. This encourages water droplets that hit the surface of the film to spread to superhydrophilic contact angles ($\theta < 5°$), washing away dirt and

Figure 20.8 Series of side-on black and white camera images of (a) a water droplet after being cast on (b) a non-UV-irradiated and (c) a UV-irradiated TiO_2 thin film clearly demonstrating the transition to superhydrophilicity.

grime. An example of this phenomenon is shown in Figure 20.8, where a water droplet is cast on to a TiO_2 thin film before and after UV irradiation.

Both photocatalysis and PIH processes work in tandem upon photoexcitation, which requires high-energy photons from the UV region. Given that only a small proportion of the solar spectrum falls in the UV region (Figure 20.3), the efficiency of this self-cleaning process on TiO_2 is low, with quantum yields as low as 0.04% having been observed in commercially available samples.[24] Nevertheless, TiO_2 has become the material of choice for self-cleaning window applications because of its high biological and chemical inertness, mechanical robustness, durability to extended photocatalytic cycling, low cost and high photocatalytic activity. TiO_2 can also be deposited as a thin film relatively easily, compatible with a hot-end float glass production line.

Although TiO_2 can exist in several polymorphs (anatase, rutile and brookite), the anatase phase has been consistently applied in thin-film self-cleaning coatings owing to its higher photocatalysis. In its commercial production, APCVD has been the method of choice for depositing a thin layer of this photocatalyst at the hot end of their in-line production of float glass. Pilkington NSG, the first company to produce self-cleaning windows commercially in 2001, use $TiCl_4$ and oxygen-containing precursors to deposit an ~ 10–20 nm thick layer of anatase TiO_2 during the cooling stage of their window manufacture over the 650–670 °C temperature range in their production of Activ self-cleaning glass. Pittsburgh Plate Glass (PPG), in their production of SunClean, similarly deposit an ~ 20 nm thick layer of TiO_2 with a titanium isopropoxide precursor at 650 °C. Saint-Gobain in their production of Bioclean, first deposit a 15 nm thick layer of TiO_2 but also add a 2 nm thick Al-doped SiO_2 (8 at.%) layer at the cold end by magnetron sputtering for increased hydrophilicity. Cardinal Glass, in their formation of Neat Glass, apply solely a cold-end magnetron sputtering stage in their formation of a 10 nm thick layer of TiO_2. Although self-cleaning windows currently hold only a 0.2% share of the total glass market in Europe, sales are expected to increase significantly in the next few years with consumer 'green awareness'.

Self-cleaning tiles that utilise the photocatalytic power of TiO_2 have also been produced. Such tiles, called Hydrotect, are produced through a sol–gel coating and subsequent calcination process by TOTO. They have been applied in a variety of environments, from the walls of building to the walls of public urinals. In order to promote better green awareness, TOTO has shown that a Hydrotect-coated household (150 m^2) could decompose the same amount of NO_x produced by 12 cars driving 30 km per day and purify the same amount of air as a forest area the size of four tennis courts on an average day.

20.6 WATER-REPELLENT WINDOWS

The most popular consumer requirement for a water-repellent coating has been in the automotive industry for car windscreens that increase visibility in wet weather.[25] Several companies have developed a range of products for tackling this problem. One product, produced by Nanoprotect under the name Nanoprotect NG, contains nanoparticles dispersed within an ethanol solution that is sprayed onto the window layer. Through a self-assembly process, the nanoparticles form a transparent and hydrophobic monolayer that is also chemical and abrasion resistant for up to 2 years of usage. To increase the durability of nanoparticle-based hydrophobic coatings, Volvo developed and recently patented a process that encases the nanoparticles within a thermosetting acrylic polymer. Silicon dioxide (SiO_2) nanoparticles (5–100 nm in diameter) are first coated with a hydrophobic polydimethylsiloxane layer before being cast within the acrylic. The windscreens are then coated with this paste and cured at 175 °C to form the robust hydrophobic coating.

Although the use of nanoparticles within hydrophobic coatings has become increasingly popular, alternative methods based on polymer chemistry alone have had more prominent commercial success. For instance, DFI International applies a two-step procedure in which silicone-based polymers are first cross-linked to form a branched structure that is then capped with hydrophobic chains in the production of their Diamond Fusion coating. Learning their lessons from Nature, Saint-Gobain have also developed a robust polymer-based hydrophobic coating, with microstructure similar to that of the lotus leaf, famed for its cleanliness through its ability to repel water. Mimicking this effect required several stages in their production of Hydrophobic Glass. First, the windscreen is coated with a 100 nm layer of SiO_2 through magnetron sputtering. A 15 nm thick layer of Ag metal is then deposited by sputtering through a metal mask containing an array of nanometre-sized circular holes. After etching the entire surface, an acid treatment stage removes the unwanted Ag template layer, yielding a microstructure similar to that of the lotus leaf. A monolayer of a fluorinated chloro/alkylsilane is then chemically bound to this surface, forming a highly robust and hydrophobic surface with <4% haze and more than 92% visible light transmittance, and forms a 165° contact angle with water.[26]

20.7 ANTI-FOGGING COATINGS

There have been several approaches to solving the problem of fogging from water mist that occurs on windows and mirrors.[27] One commercialised solution has involved spraying surfaces with a thin layer of a transparent wax from an aerosol propellant. However, such simple methods provide only a temporary solution to the problem of fogging as the waxes can be easily smudged and removed when handled. More permanent solutions involve a curing stage, such as the product marketed by Hydrocover, where surfaces are coated with a thin layer (10–25 μm) of melamine resin and then heat treated at 125 °C. Although this method provides a permanent solution when coating plastics, as the melamine becomes chemically bound to the surface, problems arise when coating glass. To counteract this problem, companies such as TOTO have applied inorganic layers. This is achieved through vacuum sputtering a base layer (∼200 nm thick) of TiO_2 and a porous 15 nm thick SiO_2

above it. The anti-fog function of this product is due to the photo-induced hydrophilic (PIH) properties of TiO_2. Upon ultra-bandgap irradiance, the surface of the TiO_2 layer becomes inundated with hydroxyl groups that encourage the even spreading of water across its surface. The porous SiO_2 top layer serves to increase the amount of water in contact with the hydrophilic TiO_2 sub-layer and increase the effect of water spreading. Both layers, working in tandem, inhibit water droplets that land on the surface of the glass/mirror from beading and causing haze. Nevertheless, such a product does not function well indoors, as the UV irradiation required to excite the TiO_2 layer is scarcely available. However, when used outdoors, the TiO_2 layer can reach its hydrophilic state more readily, remaining in this anti-fogging state for several days. To counteract such problems associated with indoor use of TiO_2, Laroche *et al.* developed a method for depositing permanent anti-fogging layers without the need for TiO_2. This patented method first involves the cold plasma treatment of the window to form nucleophilic surface groups that covalently bond to a polyanhydride polymer. A hydrophilic poly(vinyl alcohol) (PVA) layer is then chemically bound to this polymer surface layer, providing the desired anti-fogging properties.[28]

20.8 PRIVACY COATINGS

Privacy coatings on glass are typically applied after the window pane has been installed. They have found prominence in the automobile industry for tinting car windows, where the entire sector generates more than US$250 million worldwide per year. A typical privacy coating is made from polyester that is stretched to a desired thickness and heat treated at around 200 °C. By varying the thickness or side chains attached to the polyester, the deepness of tint can be adjusted. To increase the lifetime of the polyester, UV stabilisers such as benzophenone are added prior to its curing. In addition, silica particles ranging from 1 to 10 μm in diameter can also be added to control the turbidity and gloss of the final product. An adhesive resin is then added to one side of the plastic so that it can later be stuck to the desired pane.

An alternative approach has involved depositing a thin layer of titanium nitride (TiN), and such a product called Insulatir is distributed by Johnson Laminating and Coating. As TiN is so highly absorbing in the visible region, only a thin layer is required. It can also be applied in tandem with the float glass process at the hot end through atmospheric pressure CVD. This is typically achieved using either a single source such as $Ti[N(CH_3)_2]_4$ or $TiCl_4$ and NH_3 dual sources at around 650 °C. Given the strongly black colouration of the material, to meet the strict transparency requirements for automotive glass only a layer 20–60 nm thick is required to tint the window sufficiently. TiN is highly infrared heat reflective, making the product more desirable for automobiles used in hot climates. More importantly, TiN is highly scratch and weather resistant, so much so that it is used to coat machine tools such as drill-heads and milling cutters – increasing the lifetime of the coated pane by a factor of three or more. In addition, windows coated with this hard ceramic never fade or discolour, the main drawback of traditional polymer-based privacy coatings.

20.9 THIN-FILM PHOTOVOLTAIC WINDOWS

Photovoltaic devices convert photoenergy from sunlight into electrical energy. The market for such devices is expanding rapidly with global revenues greater than \$80 billion. Given their high commercial success, a great deal of investment has-been put into improving such devices. There are two main approaches to improving photovoltaics, improving either their photoelectrical conversion efficiency or the wattage produced to production cost ratio. Although conversion efficiencies have always been considerably lower in thin-film photovoltaics (10–16%) compared with their polycrystalline silicon counterparts (30–35%), their wattage:cost efficiencies have typically been stronger.[29] As electricity can be generated more cheaply using thin-film photovoltaics, their interest and scope have increased. One recently commercialised application has been in transparent thin-film photovoltaics for generating electricity in windows.[30] Several companies, including Konarta, SCHOTT, Danish Solar Energy and Pythagoras Solar, are producing such integrated window units that can generate around 5 W m^{-2} K^{-1} while retaining up to 10% transmittance in the visible region. A typical device consists of a multi-layered structure, as shown in Figure 20.9. Starting from the window surface, a thin TCO layer (nanometre scale) such as $F:SnO_2$ is first deposited. A thicker amorphous silicon (α-Si) layer

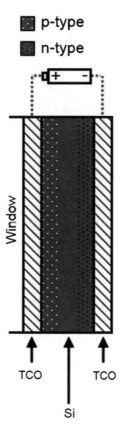

Figure 20.9　A representation of a typical transparent thin-film photovoltaic device for generating electricity from windows. The key represents the regions within the silicon (Si) layer that consist of n/p-type doping.

is then deposited on top of this (micrometre scale); however, specific dopants are added to create a graded film from p- to n-type doping. Conventional p-type dopants include gallium and aluminium, and conventional n-type dopants include phosphorus and arsenic. Another thin TCO layer is then deposited on top of this and connected in circuit with its superimposed partner layer. All layers can be deposited by CVD, making their integration within traditional float glass production much simpler. Amorphous silicon has a bandgap of 1.7 eV (730 nm), thereby forming photo-generated electron–hole pairs from most of the visible range. By introducing an n/p-type dopant gradient, each species is driven in opposite directions, electrons travelling to the p-type layer and holes to the n-type layer. The electrons fall into the first TCO layer and travel around the circuit to the second TCO layer, generating a current. The electrons then fall back into the holes directed initially towards the n-type layer. Transparent organic solar cells, such as those sold by Konarta, work in a similar fashion to such inorganic modules. By dispersing a suitable aromatic species, such as perylene, within an electrolyte, solar photons can promote electrons located in the highest occupied molecular orbital (HOMO) of the organic into their lowest unoccupied molecular orbital (LUMO) – a pπ to pπ^* transition. The electrons are then ferried around a circuit, generating a current, before falling back into their original HOMO. In producing such transparent organic modules, Konarta have already generated more than US$100 million in revenue under the brand name Power Plastic.

20.10   CONCLUSION

Glass production and coating technologies account for nearly $100 billion in worldwide revenue each year. It is an ever-growing and highly competitive market that continually invokes innovation in function, design and application. This technology provides cost-effective solutions to tackle a variety of problems, including pollution, insulation and energy generation, while retaining resilience and aesthetic appeal.

On looking back at the historical development of coating technology, one can see how the intricacy of devices has continually increased with multi-layered hetero-junction devices becoming the standard. Continued investment in glass coating research has increased the technical know-how in device function from the nanoscale to the bulk. This has spurred a plethora of new production methods with increased nanostructural control and nanoparticulate application. Current levels of investment and market growth in this sector can only promise greater innovation and a promising future for window technologies.

REFERENCES

1. A. Arnaud, *J. Non-Cryst. Solids*, 1997, **218**, 12.
2. L.A.B. Pilkington, *Proc. R. Soc. London, Ser. A*, 1969, **314**, 1.

3. B. J. Savaete, *From Flat Glass to Float Glass Industry: 40 Years of Major Change*, http://verrehistoire.typepad.com/histoire_entreprises/files/textebernardsavate.pdf, 2007 (last accessed 26 April 2012).

4. S. Chaudhuri, D. Bhattacharyya, A. B. Maity and A. K. Pal, *Mater. Sci. Forum*, 1997, **246**, 181.

5. I. P. Parkin and R. G. Palgrave, *J. Mater. Chem.*, 2005, **15**, 1689.

6. T. D. Manning, I. P. Parkin, M. E.. Pemble, D. Sheel and D. Vernardou, *Chem. Mater.*, 2004, **16**, 744.

7. A. Synnefa, M. Santamouri and I. Livada, *Solar Energy*, 2005, **80**, 981.

8. I. P. Parkin and R. G. Palgrave, in *Chemical Vapour Deposition*, ed. A. C. Jones and M. L. Hitchman, Royal Society of Chemistry, Cambridge, 2008.

9. H. J. Gläser, *Appl. Opt.*, 2008, **47**, C193.

10. J. M. Gallego and J. R. Siddle, Coated glass, *US Patent*, 6 048 621, 1997.

11. M. S. Jenkins, A. F. Simpson and D. A. Porter, Coatings on glass, *US Patent*, 4 828 880, 1987.

12. F. H. Hart, Low emissivity coatings on transparent substrates, *US Patent*, 4 462 883, 1983.

13. Freedonia, *World Flat Glass: Industry Study with Forecasts for 2014 and 2019*, Study 2715, Freedonia Group, Cleveland, OH, 2011.

14. C. G. Granqvist, *Solar Energy Mater., Solar Cells*, 2000, **60**, 201.

15. D. Levy, *Chem. Mater.*, 1997, **9**, 2666.

16. Wikipedia, *Smart Glass*, http://en.wikipedia.org/wiki/Smart_glass, 2012 (last accessed 8 April 2012).

17. G. A. Niklasson and C. G. Granqvist, *J. Mater. Chem.*, 2007, **17**, 127.

18. D. R. Rosseinsky and R. J. Mortimer, *Adv. Mater.*, 2001, **13**, 783.

19. BCC Research, *Smart Glass: Technologies and Global Markets*, Report VM065A, BCC Research, Wellesley, MA, 2009.

20. C. S. Blackman, C. Piccirillo, R. B. Binions and I. P. Parkin, *Thin Solid Films*, 2009, **517**, 4565.

21. D. Chahroudi, *History of Cloud Gel*, http://suntekllp.com/34.php, 2011 (last accessed 26 April 2012).

22. IBISWorld, *Glass Product Manufacturing in the US – Industry Risk Rating Report*, NAICS 32721, IBISWorld, Los Angeles, 2011.
23. A. Fujishima, X. Zhang and D. Tryk, *Surf. Sci. Rep.*, 2008, **63**, 515.
24. A. Mills, A. Hill, M. Crow and S. Hodgen, *J. Appl. Electrochem.*, 2005, **35**, 641.
25. A. Nakajima, K. Hashimoto and T. Watanabe, *Monatsh. Chem./Chem. Monthly*, 2001, **132**, 31.
26. C. Gandon, C. Marzolin, B. Rogier, B. and E. Royer, Transparent textured substrate and methods for obtaining same, *US Patent*, 7 959 815, 2009.
27. J. A. Howarter and J. P. Youngblood, *Macromol. Rapid Commun.*, 2008, **29**, 455.
28. G. Laroche, D. Mantovani, D. and P. Chevalier, Process for producing anti-fog coating, *US Patent Application*, 2009/0246513, 2009.
29. A. Shah, P. Torres, P. Tscharner, N. Wyrsch and H. Keppner, *Science*, 1999, **285**, 692.
30. K. E. Park, K. E. Kang, H. I. Kim, G. J. Yu and J. T. Kim, *Energy*, 2010, **35**, 2681.

CHAPTER 21

Sustainable Materials in Building and Architecture

LLEWELLYN VAN WYK,* JOE MAPIRAVANA AND
NAA LAMKAI AMPOFO-ANTI

Built Environment Unit, CSIR, Pretoria, Gauteng 0001, South Africa
*Email: lvwyk@csir.co.za

21.1 INTRODUCTION

Most buildings up to the early nineteenth century, with the notable exception of prestigious public buildings, were constructed from easily available local materials. In the nineteenth century, developments in transport modes enabled the movement of heavy and bulky materials and opened up the era of prefabricated elements and product catalogues. At the same time, new materials were invented. Notwithstanding this, timber and timber-derived products, masonry units of clay and cement, concrete, steel, aluminium and glass remained and still remain the dominant materials used in construction. Recent developments in the twentieth century have included plastics in a number of forms, with the most recent material development including biocomposites for construction.

Construction remains one of the most intensive material consumers: 40% of materials manufactured end up in buildings.[1]

Materials for a Sustainable Future
Edited by Trevor M. Letcher and Janet L. Scott
© The Royal Society of Chemistry 2012
Published by the Royal Society of Chemistry, www.rsc.org

Material demand is still dominated by the building sector, especially with respect to cement and fired clay.[2]

21.2 GLOBAL INDUSTRY TRENDS

The construction sector works with many other parties, including built environment consultants, material manufacturers and suppliers and specialist suppliers and contractors. The global construction industry grew from US$4.6 trillion ($4.6 \times 10^{12}$) in 2006 to US$7.5 trillion in 2010.[3] The bulk of construction expenditure is occurring in Asia in general and China specifically.[4] It is estimated that construction output will grow by 70% to US$12.7 trillion in 2020 and will account for 14.6% of world industrial output.[3]

The total number of construction workers worldwide exceeded 111 million in 2001, with most employed in low- and middle-income countries.[5] In 2001, construction accounted for 7% of total employment with 75% of all construction workers found in developing countries: typically over 90% of workers are employed in micro firms with less than 10 persons.[6]

The industry relies on close relationships with related industries specialising in building materials, including cement, plastics, wood, steel and glass, and also innovative new construction products and systems offering higher performance characteristics, especially with regard to strength, thermal and low-maintenance properties. Growth in the world market for building materials is forecast by one analyst to reach US$706.7 billion ($706.7 \times 10^9$) in 2015, although another analyst forecasts that material value will reach US$752.7 billion by 2013.[7,8] The primary factors driving future material growth include increased investment in the residential and commercial sectors, increased spending by governments in infrastructure, improving liquidity in the financial markets, a softening of interest rates and ongoing industrialisation in developing economies.[7] The building material manufacturing and supply sector is diverse and makes a significant contribution to the industrial base of countries. The sector includes a diverse range of manufacturers (including cement, steel, aluminium, glass and clay) and suppliers (wholesale and do-it-yourself), in addition to providing a large market to goods manufacturers including the furniture, paint, plumbing and electrical industries.

The building material manufacturing and supply sector is also in a state of transformation, driven in large part by regulations

requiring environmentally friendly or 'green' materials. The green
building market was worth ~$60 billion in 2009 in the USA
alone.[9] The push for green products has spurred interest in bio-
based materials, including woven floor coverings, bamboo and
cork, and recycled materials, including recycled steel, aluminium,
bricks and the inclusion of waste materials such as fly-ash in concrete. In addition, green building has raised the performance
requirements for fittings and fixtures such as low-flow taps and
energy-efficient fittings and fixtures, resulting in analysts forecast-
ing double-digit growth every year to 2013.[9] Some new emerging
technologies, such as dynamic smart windows (that can shift from a
clear to a darkened state to provide shading from the Sun) and
advanced insulation, such as aerogels and phase-change materials,
are forecast to grow into a US$829 million industry by 2020.[10] The
construction sector provides an ideal test bed for scientific innovation in many ways: the risks are lower compared with the
automobile or aeronautical industries and the scale is significantly
larger. Innovation need not be limited to new material either;
innovation often offers more opportunities in the manner in which
conventional materials are used or combined. Designers, contractors and material manufacturers and suppliers have to give
consideration to both material usage and building regulations that
are likely to emerge in a world dominated by calls for the reduction
of carbon emissions. Building regulations are likely to embark on a
pathway of increasing building performance requirements, especially with regard to energy and water efficiency and reduced
environmental impact. Net-zero buildings and smart cities, both of
which react to resource consumption and occupant comfort, are
emerging concepts that will provide a fertile ground for innovation
in building materials.

21.3 BACKGROUND AND CONTEXT TO SUSTAINABLE BUILDING AND CONSTRUCTION

The global pressures exerted on finite natural resources through
'modern'[†] consumption and production patterns and the concomitant depletion of non-renewable resources occupy a central
position in the developmental strategies of governments and

[†] The term 'modern' is used to refer to the period commencing with the industrial revolution.

non-governmental organisations. These strategies flow substantially from the *Habitat Agenda* and *Agenda 21*.[‡]

Buildings are significant consumers of raw materials.[11] The environmental capital they contain is enormous:

- 50% of all resources globally go into construction.
- 45% of energy generated is used to heat, light and ventilate buildings and 5% to construct them.
- 40% of water used globally is for sanitation and other uses in buildings.
- 60% of prime agricultural land lost to farming is used for building purposes.
- 70% of global timber products end up in building construction.
- 40% of CO_2 releases come from building construction and operation.
- 40% of total wastes result from building and demolition activities.

Chapter 7 of *Agenda 21* promotes sustainable construction industry activities and seeks to adopt policies and technologies to enable the construction sector to meet human settlement development goals while avoiding harmful side effects on human health and the biosphere. Among the identified activities are strengthening the indigenous building materials industry based on inputs of locally available natural resources and enhancing the utilisation of local materials by expanding the technology platform.

A broader life-cycle approach – where the full impact of the building across its lifetime is understood – is therefore required as a starting point for 'sustainable building and construction' (SBC), a term used in *Agenda 21*.[12] The main challenges that emerge to achieve sustainable construction industry activities are[12]

- Promoting energy efficiency (energy-saving measures; extensive retrofit programmes; transport aspects; use of renewable energies).

[‡]The importance of *Agenda 21* lies in the confirmation by the General Assembly of the United Nations in 1997 that *Agenda 21* remains the fundamental programme of action for achieving sustainable development.

- Reducing consumption of high-quality drinking water (relying on rainwater/grey water; reducing domestic consumption with water management systems; waterless sanitation systems and use of drought resistant plants).
- Selecting materials based on environmental performance (use of renewable materials; reduction of the use of natural resources; recycling).
- Contributing to a sustainable urban development model (efficient use of land; design for long service life; the longevity of buildings through adaptability and flexibility; converting existing buildings in lieu of building new; refurbishment and technology upgrading; sustainable management of facilities; prevention of urban decay and reduction of sprawl; contributing to employment creation; cultural heritage preservation and social cohesion promotion).
- Contribution to poverty reduction.
- Healthy and safe working environments.

Sustainable material use is thus predicated on the replacement of future flows of conventional building materials with 'green' materials. From an environmental perspective, 'green' materials would be those materials with the least 'embodied effects', where the word *embodied* refers to attribution or allocation in an accounting sense as opposed to true physical embodiment. In the building community, the tendency is to refer only to 'embodied energy'.[13] However, when investigated in a life cycle assessment (LCA) study, all the extractions from and releases to Nature are *embodied* effects and there are also embodied effects associated with the making and moving of energy itself as depicted in Table 21.1.

21.4   BULK CONSTRUCTION MATERIAL RESOURCE RESERVES

The raw materials that make up the bulk of materials used in construction in most cases are also those resources that are also the most plentiful on the planet. Steel is made from iron ore: iron is the most plentiful element on Earth, forming most of the outer and inner core, and it is also the fourth most common element in the Earth's crust. Notwithstanding this, most of the iron ore is bound in silicate and, more rarely, carbonate minerals.

Table 21.1 Embodied effects typically investigated in an LCA.

Inputs (extractions from Nature)	Outputs (releases to the environment)
Energy	Acidification
Land	Climate change (global warming)
Materials	Eutrophication
Water	Ecotoxocity
	Human toxicity
	Photochemical oxidant formation
	Stratospheric ozone depletion

The thermodynamic barriers to separating pure iron from these minerals are formidable and energy intensive: iron ore melts at about 1375 °C. The World Steel Association forecasts that steel use in 2011 will amount to 1359 Mt (1359 million tonnes) and reach a new record of 1441 Mt 2012. Although iron ore reserves seem vast, some commentators suggest that reserves will run out in 64 years based on an extremely conservative extrapolation of 2% per annum growth. Cement is made by heating limestone with small quantities of other materials to 1450 °C in kilns. Cement is the basic ingredient in concrete, mortar and grout. Although individual cement companies may face shortages, cement raw materials, especially limestone, are geologically widespread and abundant and overall shortages are not likely in the future.[14] World production in 2003 was 1860 Mt, of which the bulk (880 Mt) comes from China. Clay is a naturally occurring aluminium silicate composed primarily of fine-grained materials. Clay bricks are moulded and fired at temperatures between 1000 and 1300 °C depending on the type and colour of brick required. World mine production increased from 9.66 Mt in 2009 to 10 Mt in 2010. Resources of all clays are thought to be extremely large. Aluminium is a silvery white member of the boron group of chemical elements. It is the most abundant metal in the Earth's crust. Aluminium oxide has a melting point of over 2000 °C and pure cryolite, used to separate aluminium from oxygen in the oxide ores, has a melting point of 1012 °C. World smelter production increased from 37.3 Mt in 2009 to 41.4 Mt in 2010, with the largest contribution coming from China. It is believed that the world reserves of bauxite are sufficient to meet global demand well into the future.[15]

21.5   THREATS FACING THE BULK CONSTRUCTION
MATERIAL SECTOR

The life cycle assessment (LCA) concept has emerged as one of the most appropriate tools for assessing product-related environmental impacts and for supporting an effective integration of environmental concerns into economic activity. The LCA procedure investigates the whole product life cycle, from the acquisition of raw materials, through manufacture of the product, transportation and distribution, use and maintenance to disposal of the used product at the end-of-life based on multiple environmental effect categories (Table 25.2).[16] Although the main driver for LCA is sustainable development, the methodology does not as yet incorporate criteria for measuring the social and economic dimensions. The three major environmental areas of protection (AoPs) of interest to society, namely human health, ecological health and natural resources, are considered in a comprehensive manner, however. The LCA procedure is standardised under the ISO 14040 sub-series *Environmental Management Life Cycle Assessment.*

While resource depletion in the short to medium term would therefore appear not be a threat, a number of related threats nonetheless exist. There are two categories of embodied effects associated with the building material life cycle, namely outdoor and indoor environmental effects. The outdoor environmental effects are caused by the chemical content of building materials. They can occur during any of the five generic life cycle stages (Table 21.2) and may affect any of the three AoPs. LCA measures the external embodied effects of products on the basis of the effect categories listed in Table 21.1. The indoor environmental effects manifest as human health effects such as asthma or sick building syndrome. The human health effects result from indoor air concentrations of chemical toxins off-gassed from various indoor sources, including interior finishing materials.

21.5.1   Embodied Energy

The greatest challenge presented to construction materials comes from the energy required to manufacture the materials. The dual crises created by the current energy scenario – energy scarcity and greenhouse gas (GHG) emissions – are likely to drive the

Table 21.2   Generic life cycle stages of construction products.

Material extraction and processing	Construction material fabrication	On-site construction	Facility operation and maintenance	End-of-life
Forest products, coal and petroleum, plastic products, natural gas, iron, copper, zinc, aluminium, sand, stone, limestone	Timber products, fibreboard, cellulose products, steel, aluminium, appliances, wire, paint, solvents, plate glass, carpets	Foundation and site earthworks, concrete pouring, structural framing, roofing, mechanical and electrical systems, painting and cleanup	Space and water heating, space cooling, appliances, lighting, facility improvement and maintenance materials	Stripping reusable materials, knock-down, site clearing, disposal

development of new materials that are less energy dependent. Coal, oil and gas still remain the main sources of energy. At current global production levels, proven global coal reserves are estimated to last a further 119 years.[16] In contrast, proven global oil and gas reserves are equivalent to 46 and 63 years, respectively, at current production levels. Over 62% of oil and 64% of gas reserves are located in the Middle East and Russia. Forecasts for peak coal – the point at which the maximum global production is reached – is somewhere between 2010 and 2048.[17] Collective projections generally predict that global peak coal production may occur sometime between 2010 and 2025 at 30% above current production in the worst-case scenario.[18] Peak oil, the point at which maximum petroleum is extracted, is projected by some to occur around 2020, whereas others believe that peak oil has already been reached. Notwithstanding the known supply, the impact of geo-political pressures will significantly influence availability and create supply uncertainties in the future.

21.5.2   Embodied Water

Water is a significant component in the production of construction materials and electricity. Hence water shortages pose a significant threat to the production of construction materials. Peak water has been put forward as a concept to help understand growing constraints on the availability, quality and use of fresh water resources. There is a growing view that peak water has already been reached: if present trends continue, 1.8 billion people will be living with absolute water scarcity by 2025 and two-thirds of the world's population could be subject to water stress.

21.5.3 Embodied Toxicity

Many building material production processes contribute to the widespread dispersion of toxins in the environment. Ceramic tile production has recently been linked to air pollutants that contribute to human health and ecological health effects.[19] Cement kilns contribute to airborne emissions of sulfur dioxide, nitrous oxide, mercury, dioxins, furans and particulate matter.[20] Poly(vinyl chloride) (PVC) is associated with a uniquely wide and potent range of emissions throughout its life cycle, including volatile organic compounds (VOCs), phthalates, heavy metals, halogenated flame retardants (HFRs) and perfluorocarbons (PFCs).[21] A study analysed the environmental impact of materials for the six environmental impact types within the 'Human health' damage category for 2000 and also three impact types within the 'Ecosystem quality' damage category.[22] Within the 'Human health' category, the impact types 'Carcinogenic', 'Winter smog' and 'Climate change' contribute much more to the human health damage category than the other three impact types, namely 'Summer smog', 'Radiation' and 'Ozone layer depletion'. In the 'Carcinogenic' category, steel, paper and paperboard and zinc contributed the most; in the 'Winter smog' category, paper and paperboard, cement, steel and lime contributed the most; and in the 'Climate change' category, cement and steel contributed the most. In the 'Ecosystem quality' damage category, zinc and steel contributed the most to ecotoxicity.

With regard to building applications, non-residential and residential buildings ranked second and third, respectively, in terms of environmental impact. The environmental impact of the application areas reflects the demand for one or more bulk materials. The building sector uses a large amount of steel, cement and fired clay: cement ranks high on the environmental impacts for summer smog and climate change, resulting in a significant impact of the building sector on these human health impacts; steel in the building sector causes carcinogenic effects; and roads and infrastructure are largely responsible for climate change effects because of the use of cement and steel. The materials included in the study accounted for roughly one-quarter of the total environmental impact caused by human activities in the EU. Materials have a considerable impact on 'Carcinogenics' (24%), 'Climate change' (16%), 'Ecotoxicity' (39%) and 'Fossil fuel depletion' (28%). The largest part of the environmental problems related to materials (in particular carcinogenics, winter smog, climate change and fossil fuel depletion) is due to the production of the main bulk materials steel, cement, plastics and paper and paperboard. These therefore constitute areas where clean technologies are needed most.

21.6 A FIVE-PILLARED APPROACH TO CONSTRUCTION MATERIALS

Having regard for the challenge posed in Chapter 4(G) of *Agenda 21*, a five-pillared strategy, based on optimising existing environmentally sound technologies and the research and development of new environmentally sound technologies, is proposed.[23] The five pillars, which are not mutually exclusive, are to:

- optimise existing technologies;
- integrate environmentally sound fringe technologies into mainstream technologies;
- accelerate hybrid technologies into mainstream technologies;
- develop biotechnology applications for construction; and
- develop nanotechnology applications for construction.

21.6.1 Optimise Existing Technologies

Optimising the use of mainstream technologies in the construction sector could assist in improving the sustainability of the sector. Some of the strategies that would help include the following:

- *Promoting dimensional coordination* – The adoption of a set of standard metric-based modular dimensions would dramatically reduce the waste produced by the construction sector.
- *Off-site production and assembly* – Reducing the amount of wet work done on-site will greatly reduce wastage.

- *On-site installation techniques* – The greater use of pre-assembled installations will reduce the wastage generated on-site and improve the accuracy and the quality of work.
- *Design technology* – More careful consideration of detailing, especially with regard to the sealing of joints in buildings, will reduce the heat gains and losses through leakage. Techniques include the forming of barriers between materials that are exposed to different temperatures and the contact with other more isolated materials.
- *Recycling* – The recycling and reuse of industrial and agricultural waste materials to widen the range of building materials, reduce building product costs, ecological reduction of the CO_2 footprint, as alternative new pozzolanic binders, as cement replacement, for glass ceramic manufacture and for composite manufacture continue to advance. Some countries are looking at toxicity challenges associated with the use of wastes such as phosphor gypsum, municipal incinerator ashes and some coal combustion ashes. Solutions such as source reduction of toxins, oxidative destruction of persistent organics and reduction of leachability of toxic elements by pinning them down in glass and glass ceramic matrices have been proffered.

21.6.2 Integrate Environmentally Sound Fringe Technologies into Mainstream Technologies

There are a number of so-called fringe technologies that are used infrequently in the construction sector, but could contribute significantly to sustainable consumption and production patterns in the sector. There would be considerable potential to increase building efficiencies if construction relied less on traditional brick construction. Contemporary drywall construction offers far greater thermal and acoustic properties than traditional brick walling, in addition to offering a higher standard of finish. In addition, the construction of masonry walls, particular in multi-storeyed buildings, places the workforce at considerable health and safety risks. Great strides have been made in façade engineering: enclosing systems that are essentially prefabricated and installed on-site once the superstructure of the building allows it. These systems also offer greater speed of construction with higher performance standards due to the extensive research and development inputs into jointing and fixing details. Where the use of brick is desirable (*e.g.* to promote labour-intensive construction for job creation purposes), the design and construction should actively pursue an energy conservation strategy that relies on bricks' inherent thermal mass to provide a comfortable base temperature from which active heating or cooling can top up the system as required. This will require building considerably thicker walls, with fewer openings.

21.6.3 Accelerate Hybrid Technologies into Mainstream Technologies

Hybrid technologies combine two or more technologies into making a building material or building component. A structurally insulated panel (SIP) is a prime example where two outer skins are bonded to a central insulation core to make up a single building component. Hybrid technologies can also be combined with conventional technologies: constructing the outer skin of a building with a high-performance SIP and building an inner skin of masonry construction creates a highly insulated building but having the benefits of thermal mass internally.

21.6.4 Develop Biotechnology Applications

Biotechnology applications in construction explore the use of bio-based and biodegradable materials, such as agricultural crops (plant and animal), as the basic raw material for construction products and materials. Current research in this field is examining the use of agricultural crops for insulation, paints, floor and wall coverings, geotextiles, thatch, reinforcement, boards, biocomposites, glues and mortar.[24] Basic science studies on developing, understanding, testing and applying fundamental principles of reinforcement to fibre/matrix engineering to enhance toughening and strengthening effects between matrix and reinforcement particles and/or fibres have been developed. Fundamental studies of polymer–polymer, polymer–fibre and polymer–particle interfaces have produced plausible theories on interfacial fracture mechanics that are based on thermodynamics. Basic science research on composites is looking at issues such as developing computational techniques to determine the extent of stress concentration due to dynamic loading, studying the effect of surface treatment of natural

fibres on mechanical properties of natural fibre-reinforced composites, modelling and measurement of residual stresses in composites due to differential thermal expansion between fibre and
matrix, modelling dynamic stress concentration on particulate or
discontinuous fibre-reinforced composites and modelling of
environmental degradation of composites based on chemical link
density (degree of entanglement for polymers) and cohesive force
variation to predict durability on the basis of short-term exposure
tests.[25] Significant strides have been made in these areas and fire
resistance and retarding remain the challenges. The effect of fibre
content on the mechanical and thermal characteristics of hemp
fibre-reinforced 1-pentene–polypropylene copolymer composites
has led to the optimisation of the composites in South Africa.

Fibre-reinforced foamed concrete technology, strain-hardened
cement-based composite technology and foamed polyolefin-rein-
forced natural fibre composite technology have also been developed in South Africa. Globally, product development research
produced a plethora of cement/concrete matrix composites, polymer matrix composites and bio-based composites. Monolithic
roofs and structural I-beams have been produced from fibre-reinforced resin matrix soya bean-based materials. Design guidelines
and procedures exist for applying fibre-reinforced composites that
take into account the lack of ductility by allowing higher safety
factors than for steel-reinforced concrete design and have environmental exposure reduction factors for various fibre-reinforced
polymer systems. Product development from recycled fibre-optic
waste, poly(ethylene terephthalate), polyolefinic polymer wastes
(including: polypropylene, polyethylene and polystyrene) have
been developed. Ongoing composite projects involve reducing the
cost of fibre reinforcement and related technological operations,
addressing repairability, fire resistance, durability and environmental concerns associated with composites.

Biomimetics – the imitation of Nature – studies the ways in
which Nature addresses problems, such as the adhesive capability
of a mussel. Biomimetics aims to replicate the processes of Nature:
just as a plant bends towards or away from the Sun (phototropism)
or a leaf opens and closes in response to environmental conditions,
so a building will continuously respond to internal and external
environmental conditions and patterns of use. Intelligent buildings
make use of chemical science, like its organic counterpart, to

promote life. Plants make use of carbon dioxide and water and, through the process of photosynthesis, build up complex substances, releasing oxygen as a by-product. The use of biomimetics may lead to new adhesives that enable buildings to be glued together, instead of relying on the variable adhesive qualities of mortar. This technology requires the collaborative knowledge of biologists, physicists, chemists and engineers working in integrated teams.

21.6.5 Develop Nanotechnology Applications

The value of Nature is only recently being appreciated as environmental capital. The attraction of nanotechnology is its ability to package atoms in a controlled way and package any unwanted atoms for recycling or return to their source. Current nanotechnology research focuses on the design, modelling, synthesis and characterisation of a range of nanomaterials with specific properties for a range of applications. Nanolayer deposition techniques and technologies for the production of single-layer carbon nanotubes are being developed. Other research interests include the purification and functionalisation of carbon nanotubes, the development of polymer nanocomposites for barrier resistance flame retardancy and investigating nanostructures and properties of nanocomposites. Characterisation of nanoparticles takes centre stage. Ultra-high-performance concrete and rapid strength development concrete repair nanomaterials and high-performance biodegradable nanocomposites have been developed. The most widely studied nanomaterial is TiO_2. Basic science studies on the mechanism of its photocatalytic behaviour have been carried out and are now well understood. The optimal incorporation of TiO_2 into building materials such as window panes, paints, cementitious materials and tiles has led to the development of nanoscale sol–gel coating or painting of carrier substrates such as metals, ceramics, glass, polymers, concrete and wood to achieve photoactive layers a few microns thick. Investigations have shown that porosity, concrete aggregate size, cement to aggregate ratio, curing age, TiO_2 content and its degradation affect the photocatalytic behaviour in cementitious materials. Carbonation of cement will reduce the photocatalytic activity of TiO_2-containing concrete as the TiO_2 particles are closed from light. Titania solar cells are said to convert sunlight directly into electricity through a process similar to photosynthesis. Unlike silicon-based solar cells, they are said to perform well even in low light/shade and to give consistent performance over a wide temperature range. Substrates do not have to be transparent. Thus titania solar cells can be vertically mounted façade panels, power-generating windows of integral components of buildings.

Worldwide, nano silica fume is added to Portland cement-based products to enhance strength, abrasion resistance and durability. It is also commonly used for polymer reinforcement and increases hardness, stiffness, weatherability and fire retardant properties. Other nanoparticles of alumina, titania, fullerenes, carbon nanotubes and nanoclay also increase the stiffness, strength and toughness of epoxy resin polymers by inducing energy-absorbing toughening mechanisms such as crack deflection, plastic deformation of the resin matrix, crack pinning, crack blunting and microcracking of the resin matrix. Global emphasis on nanomaterial manufacturing is currently on titania, fullerenes, nanotubes and fibres. Research is required to quantify any health risks that are associated with exposure to nanomaterials.

The result of this development in materials science is that new and smart materials are being invented that can swell and flex, repel paint, repair themselves, adapt to the environment and capture and store the energy of the Sun. Structures built from new materials have the ability to provide information to their occupants and be more environmentally sustainable. New materials such as nanotubes that are 100 times stronger than steel and one-fifth lighter are likely to lead to fundamental new applications in energy, materials and transportation and to improve productivity significantly.[29] Nanotechnology can produce efficient solar cells that are as cheap as newspaper and as tough as asphalt, in fact tough enough to resurface roads so that they become solar energy collectors.[26] Nanotechnology can produce fuels from solar energy, air and water. Consumption of these fuels in nanomechanical systems will return these exact components to the air, along with some additional water vapour. The fuels are made and consumed in a cycle that produces no net pollution. In the process, it obviates the need for fossil fuels, thereby reducing carbon dioxide levels – and global warming. Molecular manufacturing does not require foaming activities, so the release of chlorofluorocarbons (CFCs) used in foaming plastics has ceased. Waste, including sewage and toxic waste, is made of harmless atoms arranged into noxious molecules: nanotechnology can convert these wastes into harmless forms and reconstitute them into renewable materials. Products can be made extremely durable, thereby obviating the need for disposal or be made genuinely biodegradable, designed to decompose after use, leaving humus and mineral grit. Nanotechnology can purify the soil in the same manner that living organisms clean the environment: particular organisms or groups of organisms after all manage the ecosystem. At the core of molecular manufacturing or nano-technology[§] is a kind of technology that many believe will replace the technological system of our current industrial world.[27] Molecular nanotechnology is defined as 'thorough, inexpensive control of the structure of matter based on molecule-by-molecule control of products and by-products'.[26] Throughout human history, production has relied on relatively crude technology that cuts, grinds, melts, stirs, bakes, sprays and etches to make the materials that societies require to produce the products they use. Nanotechnology, by contrast, constructs materials from the smallest molecular level up.

Carbon fibre also holds great promise. Carbon, along with air and water, makes up the polymers of wool and polyester and of wood and nylon. Composite materials are increasingly used in industry in applications ranging from aircraft fuselages to Formula One Grand Prix racing cars and to tennis rackets because they are strong, lightweight and easily moulded into an almost endless variety of shapes. Carbon fibre is well suited to structural applications as it is extremely strong in tension – up to five times stronger than steel. It is already being used in bridge design and other civil engineering projects. Applying inexpensive solar energy, carbon dioxide can be removed directly from the air, producing oxygen and glossy graphite pebbles as a by-product. Extracting carbon from the air to reduce its level back to pre-industrial levels will remove over 300 Gt (300×10^9 tonnes) from the atmosphere and yield enough material to construct large houses for 10 billion people and have 95% left over.[26] Atmospheric waste is therefore

---

[§]Nanotechnology like 'modern industry' describes a huge range of technologies that have the precise arrangement of molecules at their core.

capable of providing an abundant source of structural materials for the foreseeable future.

21.7 TRENDS

It is highly unlikely that there will be a reduction in or a major shift away from conventional bulk materials in the short to medium term. Some new trends will emerge, however, most likely in insulation materials (in response to energy efficiency building regulations), in a shift away from ceramic products, in a shift away from zinc and copper use in piping towards PVC and other plastics and in an uptake of recycled materials most notably in concrete (aggregate substitution), steel and aluminium.

21.7.1 Trends in Conventional Bulk Construction Materials

The partial to total replacement of Portland cement by various cement extenders has been attracting the attention of researchers throughout the world. The scope of extenders investigated, developed and used is wide, including fly ash, rice husk ash, maize cob ash, nanostructured zonolite, nanoclay cement binder (montmorillonite), metakaolin, limestone and/or dolomite fines, stronger macro-defect-free glycerol-plasticised PVA–calcium aluminate (Secar 71) and calcium aluminate–phenol resin cements. Innovative cement development for total replacement of Portland cement using geopolymers is increasing. New carbon-negative magnesium-based cement that absorbs 0.6 t of CO_2 per tonne of cement on hardening has been discovered. Researchers have also been looking at the basic science of pozzolanic materials for a better understanding of the underlying principles. Various cement standards have been reviewed, taking account of numerous blending possibilities that are coming up. Concrete research in the world has incorporated both basic and applied science aspects. A wider range of concretes have been developed, performance evaluated and applied globally. The world's tallest building (828 m high) in Dubai used high-performance concrete of mean strength >100 MPa. Depolluting, self-cleaning and photocatalytic features have also been incorporated into concrete. South Africa leads in textile concrete research and has come up with novel geotextiles for mine tunnel wall and road stabilisation. The country also leads the world in the application of continuously reinforced thin concrete road pavement emplacement and testing and in the production and use of zero-waste, low-cost modular concrete block building systems.

The use of advanced characterisation techniques such as mercury intrusion porosimetry, oxygen permeability, gas adsorption and electron microscopy techniques for the characterisation of materials has allowed the rapid development of building materials. The use of composite panels rather than bricks in construction is increasing. The recycling and reuse of recycled materials for bricks and concrete blocks is also growing. There have also been advances in the use of alternative concrete block production processes such as autoclaving for rapid strength development.

Research on the use of adobe or earth has been wider and has even included basic science, characterisation and selection of clays for unstabilised and stabilised earth building products. Researchers have looked at the manufacture of pozzolana for the extension of Portland cement from clays using flash calcination of kaolinitic clays and unfired natural fibre reinforcement of mud bricks and their characterisation. Rammed earth construction has been extensively and successfully used for two- to three-storey buildings, although the technology had not made headway elsewhere in spite of some of its environmental benefits.

Some work has been done on the evaluation of timber frame-built structures, mechanical characterisation of timber and development of formaldehyde-free polyurethane glues for laminated timber production. Transgenic trees with decreased or altered lignin content are being genetically engineered to reduce the consumption of energy and chemicals for delignification and refining of wood chips. Substitution of chemical preservatives and conventional chemical glues has been achieved through the application of fungal cultures and isolated enzymes as preservatives and adhesion promoters that require less energy to produce than chemicals. Innovative water glass-bonded wool–wooden fibre composites and bio-glue adhesive-free board materials have also been developed. Owing to proper preservation, the durability of modern wooden structures now exceeds 25 years and matches that of steel and concrete. Modular, panel and timber frame structures are now industrially manufactured – significantly lowering production costs and delivery lead times.

Higher productivity of ceramic tiles and sanitary ware has been achieved by technological advances in processing that introduced fast firing, battery pressure casting and robotic glazing systems. Large 800 × 800 mm ceramic floor tiles are now available on the market thanks to research and development. Photoactive self-cleaning and depolluting ceramic tiles and façade systems are also available.

Glass research came up with strong, durable and insulating structural glass–mica composites, insulating foam glasses and glass ceramics from the recycling of glass cullet, vitrification and devi-trification of fly ash and slag waste. The toughening of glass was achieved by molten salt diffusion, lamination with stiffer and stronger interlayer materials being developed. Glass has remained the major substrate for silicon solar panels. Glass-fibre research has led to the development of zirconia-coated fibres that can reinforce concrete matrices without reacting with the basic cement. Stable, durable glasses and glass fibres that do not devitrify have been developed.

The development of manufactured high specific strength light-weight steel-framed buildings and with a variety of greener clad-ding and infill materials is catching on. Steel-encased polymer concrete, steel fibre-reinforced concrete, steel fibre-reinforced fly ash concrete, steel fibre-reinforced metakaolin-blended concrete and high-strength steel fibre-reinforced concrete were developed and characterised. Research on the recrystallisation behaviour of low-carbon hot-rolled steel strip has led to the production of stiffer steel strip from cold working of hot-rolled strip that originally contained low sulfur levels. Lower Young's modulus steel strip was also developed from higher sulfur content hot-rolled steel strip. Corrosion-resistant nitrogen-alloyed stainless steels that offer the best combination of strength and toughness of any material known to date have also been developed in South Africa. Techniques to negate nitrogen losses during welding of stainless steel to retain strength and toughness and stabilise the welding arc and reduce spatter have been developed. Basic science evaluations of the effect of concrete strength and reinforcement on the toughness of rein-forced concrete beams, the effect of aspect ratio and volume frac-tion of steel fibres on mechanical properties of steel fibre-reinforced concrete, strength reliability of steel–polypropylene hybrid fibre-reinforced concrete, structural performance evaluation of steel fibre-reinforced concrete, fatigue of steel cable-stayed highway bridges and fatigue and fracture performance of cold-drawn wires for prestressed concrete have been carried out. New non-ferrous metal products are being developed. These include a superplastic Al–Mg–Cu alloy with 700 and 25% elongation at 550 and 20 °C, respectively, AlMn alloy sheeting for a helium-cooled thorium high-temperature reactor cooling tower and light-weight high specific strength hybrid carbon fibre-reinforced plastic (CFRP)–aluminium laminates. Short- and long-term corrosion rate tests on zinc, copper, aluminium, mild steel, stainless steel and Corten steel have been carried out at various exposure locations. The information obtained is used for long-term planning in terms of material selection, cost of corrosion, corrosion protection, durability, design, maintenance and life cycle cost analysis. Aluminium alloys now face serious competition as construction materials from a range of polymeric composites and glass-reinforced plastic materials that have property advantages. Internationally there is more extensive use of recycled rather than virgin aluminium and copper products.

There has been increased and wider use of industrial minerals and rocks such as granite, basalt, basalt tuffs, limestone, dolomite, slate, dolerite, expanded perlite, pumice, exfoliated vermiculite, diatomite, bloating clays, dunite, forsterite and norite globally for aggregates, as dimension stone and for the production of construction materials.

Refractories research work appears to take place largely in the private sector. Globally, the use of unshaped monolithic refractories has increased and the high alumina cement content of the castables has been dramatically reduced – to zero in some cases. The trend has also been towards the use of purer higher refractoriness materials and unfired refractories that are cured and sintered *in situ*.

21.7.2 Trends in Bio-based Construction Materials

There is, however, great potential for crops and by-products of animal husbandry to help to reduce the environmental impact of construction. In addition to environmental benefits in production, many of the products that can be produced using renewable resources offer environmental benefits in use. The use of agricultural crops would furthermore replace part of this material and energy consumption with lower energy and renewably sourced alternatives. In addition, wastes from agricultural crops lend themselves to safe and easy disposal with little or no environmental damage and possibly some benefits. Renewable plant materials can contribute to cutting greenhouse gases in many further ways where they substitute for fossil-based materials. The demand for industrial non-food agricultural crops (plant) and their products in Europe is generally at a very early stage. Plants can synthesise an immense range of compounds. As 'cell factories' they contain structures that can be used by the physical, chemical and biochemical sciences to produce useful materials such as fibres, starch, oils, solvents, dyes, resins, proteins, speciality chemicals and pharmaceuticals. Physical and chemical sciences can combine to produce new applications including fully bio-based composites such as boards in which the fibre component is made from hemp, flax or timber and the resin binder from rapeseed oil rather than the more commonly used synthetic chemical resins. Some renewable products have distinct functional advantages such as biodegradability and absence of toxicity.

The automotive industry has historically experimented with natural fibre composites – or biocomposites. It used them in 1941 to manufacture a prototype car (Henry Ford, 1941) with bodywork manufactured from flax, hemp and other organic materials. Contemporary automobile manufacturers are using jute-based materials for door panels and hemp for dashboards (Daimler Chrysler and Mercedes Benz E and C class models). Current interest is primarily focused on the use of bast fibres, namely hemp and flax, and certain sub-tropical fibres such as jute and kenaf. What is of particular importance with regard to applications within the automotive industry is the ability to mould complicated three-dimensional shapes: this has a specific importance for future products in the construction industry. The UK participates in the work of IENICA (Interactive European Network for Industrial Crops and their Applications) through the National Non-Food Crops Centre (NNFCC).[28] The NNFCC has identified significant potential markets for renewable raw materials in four main areas:

- chemicals – polymers and plastics, dyes, paints and pigments;
- speciality chemicals – adhesives, agrochemicals, personal specialised organics;

- industrial fibres – paper and board, composites, textile fibres;
- industrial oils – two-cycle oils, transmission fluids, lubricants.

The application of agricultural crops (animal) in the manufacture of products is not as advanced as that of crops. The use of animal crops, particularly wastes, although featuring extensively in vernacular construction, has limited application in contemporary urban construction. There are a number of reasons for this:

- *Collection* – The dispersion of animal residues makes the collection difficult, certainly in terms of the volume required.
- *Health risks* – Collecting certain animal crops, such as skins and hair, poses distinct associated health risks as the raw material must generally be sourced from abattoirs.
- *Marketability* – Animal sources suffer a distinct marketing disadvantage in terms of social acceptance.

Notwithstanding these difficulties, the use of animal fibres (hair) offers huge potential and this has therefore been included in the general fibre section.

The application of agricultural crops in construction is related to the type of crop, although it appears that fibre is the most suitable in the short term. For the purpose of this chapter, crops cover all materials deliberately planted/reared on farms.

21.7.2.1 Fibre Crops. Plant fibres have a wide range of uses and are becoming an important component in the search for sustainable production and consumption patterns. There are a number of different fibres found in plants, including bast fibres, leaf fibres, seed fibres, fruit fibres and wood fibres. Fibre crops are generally divided into two categories: long and short fibre crops. Long fibre crops include flax and hemp. Flax has been used traditionally to produce high-value long fibre material for the textile industry. Development work is ongoing in this field and offers a significant body of knowledge for applications in the construction sector. Hemp has a wide range of potential uses and has established market niches using fibre, pith, seeds and seed oils.[30] Unfortunately, the presence of the psychotropic agent THC (Δ^9-tetrahydrocannabinol) in fibre hemp creates a significant policing problem generally. Work is in progress in the UK to minimise the THC content and to develop visual and other simple field diagnosis tests for these types. Short fibre crops include cereal straws, *Miscanthus* grass, canary grass and short rotation coppice. Cereal straws are by-products of the cereal industry and have been used for some time by the board and paper industries as sources of fibre. *Miscanthus* is a highly productive C_4 grass species. Its primary market is as a biofuel, although there are a number of alternative markets for its short fibre components. Some difficulties are being experienced in overseas markets with its nomenclature and taxonomy that are undermining its applicability for future development work, which will need to be further explored for relevance within the South African context. Reed canary grass has been developed in Scandinavia as a source of short fibre for paper pulp. It is a lower productive alternative to *Miscanthus* and is ideally suited to low temperatures and poor soil conditions. Short rotation coppice is willow and poplar coppiced in a 3 yearly cycle. Although it too is used as a biofuel, it also offers a limited range of alternative uses.

21.7.2.2 Carbohydrates. Although all plants contain carbohydrate in one form or another, of primary interest are those in which it is stored for later extraction. The crops that are starch producing include maize, wheat, potatoes, barley, oats, peas and, in the longer term, quinola. With regard to the physical, chemical and genetic modification of starch, wheat offers one of the largest potential sources of renewable raw materials for industry, with products ranging from biodegradable plastics to adhesives. The paper and board industries are among the largest consumers of starch and derivatives in the non-food sector. Most of the starch is used to reinforce surface fibres and provide a smooth finish. Starch is also used in three major types of plastic: loaded products, where the starch represents a very small portion of material (10%), polymer mixtures and thermoplastic starches (produced by extrusion).[31] One building application where starch is already used is in the manufacture of paint. It is argued that starch-based paints are as economic as synthetic coatings and could have novel properties. Starch-based paints are as strong, shiny and as liquid as synthetic paints and, although they are biodegradable, do not rot after application.

21.7.2.3 Proteins. Plant proteins are an important source of raw material for both food and non-food uses, although it is mainly the structural or storage proteins such as collagen, keratin and seed proteins that are of interest for product applications. In agriculture they are sourced from seeds (soybean and pea proteins, wheat gluten) and tubers (potatoes). In addition, protein end products take the form of fibres, films and adhesives. The production of protein-based plant products for non-food products may be based upon fractionation of existing crops (wheat, barley), by current or newly developed processes (peas, soybean) or by extracting protein from existing low-value by-products (rapeseed meal). Soybean and wheat protein currently offer the most important resource for protein production, with pea proteins and cottonseed proteins emerging as viable additions. Proteins are used in the manufacture of paper coatings, plywood adhesives, floor coatings, textiles and plastics.

21.7.2.4 Oils. Oils crops are principally grown for food uses, although a significant proportion is used in non-food applications in the EU. Oilseed rape and soya bean cultivation dominate oil crop cultivation within the EU, although little of the soya bean production is used in non-food products. Despite considerable research activity having taken place in many European countries, significant further investigation will be required to refine and quantify specific commercial opportunities and to develop field productive types. Oil crops around the world include castor, coconut, cottonseed, groundnut, linseed, olive, palm, rape, safflower, sesame, soya and sunflower. Two species, Carmbe and Pot Marigold, are already showing commercial promise. Oilseed rape has been successfully used in the Reichstag building in Berlin as a biofuel to generate electricity. The processing potential for ricinoleic acid derived from castor in construction products includes coatings, flashing agent for pigment, polyurethane, nylon (nylon 11 and nylon 6,10), alkyd resin, polyester resin, plasticiser, hardener for epoxy resin, rust inhibitor, caulking material, adhesive, paint additive, potting and encapsulating material for electrical insulation, structural adhesive, flooring, elastomer sealant, waterproofing material, stabiliser for PVC, durable rapid dry coating and insulating varnish.[32]

Vegetable oils are currently already in use for the manufacture of

- gloss (oil-based) paints;
- oil-modified alkyl resins which form reticulated films with variable drying times and are used extensively in paints and varnishes;
- oleoresinous varnishes.

The majority of oil-modified resins are based on soya oil, although sunflower is a potential direct replacement. Linseed, tung and soya oils are most commonly used in paints as they contain high levels of polyunsaturated fatty acids that aid drying. Calendula oil offers an alternative to tung oil. Linoleum – a well-known building product – is made from linseed oil, resins, wood, cork powder, calcium, vegetable pigments and hessian (jute). This material is regaining popularity as an environmentally friendly product and works well in high-tech contexts requiring a static-free environment. The solvent market is a significant market and stands to benefit from the environmental advantages offered by the use of oil crops (most solvents are very environmentally unfriendly and subject to increasingly stringent environmental regulation). Oilseed-based solvents, by contrast, are fully biodegradable, non-toxic, odourless and do not contain VOCs or PAHs (polycyclic aromatic hydrocarbons). Solvents are used extensively in the construction sector in products such as paints, varnishes, metal treatments, wood treatments and for treating textiles.

21.7.2.5 Other Crops. Other crops that offer potential use in the manufacture of construction products are natural latex, cork, bamboo, lavender, clary sage and thatch. Many of these crops have been used traditionally in vernacular architecture and many are regaining popularity owing to their environmentally friendly characteristics, both in manufacture and in use.

21.7.3 Agricultural Crop Products

From the above, it is clear that there are a number of construction products that can be made from an alternative resource such as agricultural crops. For the purposes of this chapter, a construction product is any component that is used in the creation of an immovable asset, both buildings and construction works.

The product areas with the greatest current impacts (in terms of quantities of agricultural material used) are[33]

- insulation
- paints
- floor coverings/internal finishes
- geotextiles/soil stabilisation
- thatch

the product areas that have limited current impact but offer significant future potential are

- boards
- reinforcement in blocks
- adhesives
- sealing materials
- reinforcement in polymer composites
- dyes
- structural members

and the product areas that are beginning to emerge following recent research, and which hold significant promise for the future, are

- Plastics from bio-resins
- Polymer composites from bio-resins.

The use of crop-based materials for the applications mentioned above depends upon a number of factors, including the potential performance of the materials, their sustainable consumerism, the 'green' movement and global climate change. Environmental and consumer considerations, and also cost drivers of crop-derived products as sources of sustainable raw materials, will continue to drive the utilisation of such materials in the future.

21.8 CONCLUSION

It is critical that the construction industry in general, and construction material manufacturers in particular, identify first the environmental bottlenecks related to current and future material production and consumption, and second the technological opportunities to mitigate those environmental problems. The construction industry needs to concentrate on those environmental impacts of materials that are caused during material production (extraction, energy and water use) and end-of-life management (waste management and recycling). In some cases, the in-use phase may dominate the overall environmental impacts of product life cycles because of a continuous supply of energy and/or materials during use, *e.g.* buildings (operation, maintenance, repair).

The unsustainable contemporary patterns of consumption and production are an outcome of a technology arising from the industrial revolution. It is therefore unlikely that the continuing use of this technology will do any more than perpetuate those patterns. However, reconciling cutting-edge technology with innovative consumption and production patterns offers a platform for achieving real progress towards the achievement of sustainable construction. Efforts to fix our current industrial system will be undermined by its inherent dependence on resource extraction. Almost everything we make requires something to be extracted and consumed from the thin layer of the Earth that supports life and the wastes to be disposed of back into this layer, often to remain in their waste state for hundreds of years. Pollution is the consequence of the application of low-level technology.

Sustainable building and construction activities will not be achieved automatically or through writing best practice based on current thinking. At best, this is a short-term palliative, albeit a necessary one. However, understanding and maximising the benefits of these new materials will require a significant commitment to research and the integration of multi-disciplinary skills. This technology relies upon the collaborative knowledge of physicists, chemists, biologists, multi-disciplinary engineers and computer programmers. Since technologies such as molecular manufacturing fall between traditional disciplines such as molecular scientists and mechanical engineers, new disciplines will need to emerge to fill the gap. Up until now, the articulated alternatives have been dismissed as expensive, untested and obstructive.

Cutting-edge technologies offer a clean, efficient and sustainable alternative that will yield better quality, lower cost, greater safety and restored environments and will be economically sustainable.

REFERENCES

1. B. Edwards, *Rough Guide to Sustainability*, RIBA Publications, London, 2002, pp. 5–6.
2. D. Phylipsen, M. Kerssemeeckers, K. Blok, M. Patel and J. de Beer, *Assessing the Environmental Potential of Clean Material Technologies*, Report EUR 20515 EN, Joint Research Centre, European Communities, Brussels, 2002, p. 12.
3. Global Construction Perspectives and Oxford Economics, *Global Construction 2020*, Global Construction Perspectives and Oxford Economics, London, 2010, p. 6.
4. Global Construction Perspectives and Oxford Economics, *Global Construction 2020*, Global Construction Perspectives and Oxford Economics, London, 2010, p. 8.
5. ILO, *The Construction Industry in the Twenty-first Century: Its Image, Employment Prospects and Skills Requirements*, International Labour Organization, Geneva, 2001, p. 6.
6. ILO, *The Construction Industry in the Twenty-first Century: Its Image, Employment Prospects and Skills Requirements*, International Labour Organization, Geneva, 2001, p. 23.
7. GIA, *World Building Materials Market to Reach US$706.7 Billion by 2015*, Report by Global Industry Analysts, http://www.StrategyR.com, 2011 (last accessed 28 October 2011).
8. Datamonitor, *Construction Materials: Global Industry Guide*, http://www.datamonitor.com, 2009 (last accessed 28 October 2011).
9. QFinance, *Construction and Building Materials Industry*, http://www.qfinance.com/sector-profiles/construction-and-building-materials, 2011 (last accessed 28 October 2011).
10. Climate Control, *Advanced Materials to Lead Global Construction Market*, http://www.climatecontrolme.com/en/2011/08/, 2011 (last accessed 28 October 2011).
11. B. Edwards, *Rough Guide to Sustainability*, RIBA Publications, London, 2002, p. 10.
12. United Nations, *Agenda 21*, UN Department of Economic and Social Affairs: Division for Sustainable Development, 1992,

Chapter 7(G), p. 70, http://www.un.org/esa/dsd/agenda21/ (last accessed 28 October 2011).

13. W. Trusty and S. Horst, LCA tools around the world in 2005 White Paper: Life Cycle Assessment and Sustainability. Available at: http://www.bdcnetwork.com/2005-white-paper-life-cycle-assessment-and-sustainability, 2005 (last accessed 7 May 2012).

14. USGS, *Mineral Commodity Survey 2004*, U.S. Geological Survey, online http://minerals.usgs.gov/minerals/pubs/mcs, 2004 (last accessed 28 October 2011).

15. USGS, *Mineral Commodity Survey 2011*, U.S. Geological Survey, http://minerals.usgs.gov/minerals/pubs/mcs (last accessed 28 October 2011).

16. G. Keoleian, S. Blanchard and P. Repper, Life-cycle energy, costs and strategies for improving a single family house, *Applic. Implement.*, 2001, **4**, 135.

17. World Coal Association, *Coal Statistics*, www.worldcoal.org/resources/coal-statistics/ (last accessed 26 May 2011).

18. University of Newcastle, Australia, *Research Forecasts World Coal Production Could Peak as Soon as 2010*, October 28, 2009, http://www.newcastle.edu.au/news/2009/10/28/research-fore casts-world-coal-production-could-peak-as-soon-as-2010.html, 2009 (last accessed 28 October 2011).

19. N. Tikul and P. Srichandr, Assessing the environmental impact of ceramic tile production in Thailand, *J. Ceram. Soc. Jpn.*, 2010, **118**, 887.

20. Groundwork. Analysis and write-up of burning alternative fuels in cement kilns. Available at: http://www.docstoc.com/docs/73288986/Cement-Kilns—groundWork-Report, 2006 (last accessed 7 May 2012).

21. Healthy Building Network, *Screening the Toxics out of Building Materials*, http://www.healthybuilding.net/ (last accessed 1 December 2011).

22. D. Phylipsen, M. Kerssemeeckers, K. Blok, M. Patel and J. de Beer, *Assessing the Environmental Potential of Clean Material Technologies*, Report EUR 20515 EN, Joint Research Centre, European Communities, Brussels, 2002, pp. 16–19.

23. L. van Wyk, *Changing Consumption and Production Patterns in the Construction Industry*, CSIR Technical Report, CSIR/BE/CON/IR/2006/0026/B, CSIR, Pretoria, 2006, p. 17.

24. L. Van Wyk, *The Use of Agricultural Crops, Industrial Waste and Recycled Materials in Construction*, CSIR Technical Report, CSIR/PDP TH/2004/020, CSIR, Pretoria, 2005, pp. 10–16.
25. J. Mapiravana, *Building Materials Research and Development Priorities,* CSIR Technical Report, CSIR/BE/CON/IR/2007/006/B, CSIR, Pretoria, 2009, pp. 106–111.
26. E. Drexler, C. Petersen with G. Pergamint, *Unbounding the Future: the Nanotechnology Revolution*, William Morrow, New York, 1991, p. 2.
27. E. Drexler, C. Petersen with G. Pergamint, *Unbounding the Future: the Nanotechnology Revolution*, William Morrow, New York, 1991, p. 7.
28. L. van Wyk, *The Use of Agricultural Crops, Industrial Waste and Recycled Materials in Construction*, CSIR Technical Report, CSIR/PDP TH/2004/020, CSIR, Pretoria, 2005, p. 11.
29. E. Drexler and C. Petersen with G. Pergamint, *Unbounding the Future: the Nanotechnology Revolution*. William Morrow, New York, 1991, p. 134.
30. IENICA, *Summary Report for European Union – Fibre Crops*, Interactive European Network for Industrial Crops and their Applications, Brussels, 2000, p. 29.
31. IENICA, *Summary Report for European Union – Carbohydrate Crops*, Interactive European Network for Industrial Crops and their Applications, Brussels, 2000, p. 52.
32. IENICA, *Summary Report for European Union – Oil Crops*, Interactive European Network for Industrial Crops and their Applications, Brussels, 2000, p. 9.
33. L. van Wyk, *The Use of Agricultural Crops, Industrial Waste and Recycled Materials in Construction*, CSIR Technical Report, CSIR/PDP TH/2004/020, CSIR, Pretoria, 2005, p. 16.

CHAPTER 22

Biomass in Composite Materials

MARIANNE LABET,[1,2,†] KAZI M. ZAKIR HOSSAIN,[3]
IFTY AHMED[3] AND WIM THIELEMANS*[1,2]

[1] Process and Environmental Research Division, Faculty of
Engineering, University of Nottingham, University Park, Nottingham
NG7 2RD, UK; [2] School of Chemistry, University of Nottingham,
University Park, Nottingham NG7 2RD, UK; [3] Division of Materials,
Mechanics and Structures, Faculty of Engineering, University of
Nottingham, University Park, Nottingham NG7 2RD, UK
*Email: wim.thielemans@nottingham.ac.uk

22.1 INTRODUCTION

Polymers and plastics are ubiquitous in our daily lives and contribute substantially to increased comfort, food safety and even
lower energy consumption through light-weight materials and
improved insulation. Polymers are macromolecules which can be
produced by connecting (forming chemical linkages between)
small molecules called monomers. Most man-made polymers are
relatively simple as they are generally formed using a limited
number of different monomers (usually one or two, generally less
than four). They can be either thermoplastic when they are formed

---

†Present address: Materials and Engineering Research Institute, Sheffield Hallam University,
Howard Street, Sheffield S1 1WB, UK.

Materials for a Sustainable Future
Edited by Trevor M. Letcher and Janet L. Scott
© The Royal Society of Chemistry 2012
Published by the Royal Society of Chemistry, www.rsc.org

of independent chains held together by physical interactions or thermosetting when they form a three-dimensional chemically connected network.

Nature continuously produces a variety of polymers and monomers through its biological activity. The most common, wood, is produced in quantities of over 100 Gt (100 billion tonnes) each year and has found wide uses as a construction material. Wood is actually made up of cellulose, hemicellulose and lignin, three polymers, in addition to waxes, pectins and proteins. Over its history, mankind has used natural materials to construct shelters and produce weapons and tools, and also clothing. As technology has developed, so too has the amount of modification of the naturally occurring materials to improve performance or to fit better the needs that they address.[1] Up to the first half of the twentieth century, a wide range of materials had been studied, resulting in major breakthroughs in the use of bio-based polymers. Examples can be found in dying of natural fibres, leather tanning, vulcanisation of rubber and derivatisation of cellulose for a variety of uses (*e.g.* smokeless gunpowder and thermoplastic polymers).

The advent of the coal and petroleum industry at the beginning of the twentieth century was founded due to considerable advances in coal-based chemistry and the reliable and relatively inexpensive cracking of crude oil, resulting in a vast supply of monomeric substances that could easily be turned into polymers on an industrial scale. As new lightweight and cheap polymers were produced in great quantities, new uses were constantly being discovered that fitted the polymer properties. To increase the potential fields of application and the range of attainable properties, polymers were blended and reinforced using glass fibres, carbon fibres and, more recently, natural fibres, creating composite materials.[2]

In recent years, there has been much interest in moving away from petroleum-based materials and increasing the renewable content of polymers and composite materials. The most commonly cited reasons for this trend are (i) the limited availability of fossil materials, (ii) insecurity in terms of the cost of these materials, (iii) geographic location of fossil materials, (iv) increasing energy demands for extraction of remaining fossil deposits, (v) the understanding that biorefineries can be economically sustainable as proven by pilot plants and (vi) a growing awareness by both governments and the wider public of environmental issues and concerns.

Composite materials, or composites, are a combination of two or more immiscible materials, with improved properties compared with the individual constituents.[2] They are constituted of a continuous phase called the matrix and immiscible additives which can be reinforcements or fillers. Reinforcements improve the mechanical properties of the composite, whereas fillers are added to reduce weight, change colour or reduce production costs without affecting the mechanical properties. When preparing composites, the interface between the different phases is very important as it will dictate stress transfer upon deformation and thus the final mechanical properties. As a result, biomass utilisation in composite materials can focus on the three separate elements: the matrix, the additives (reinforcement and/or filler) and the interface, which can be improved by the use of biomass-derived compatibilising agents. Wood is the quintessential example, with lignin constituting the polymer matrix, cellulose the reinforcement and hemicellulose the compatibilising interfacial agent. In this chapter, we focus on biomass-derived polymers for use as the matrix and biomass-derived reinforcements. Biomass-derived compatibilisers are mentioned in the discussion of renewable polymers and composite materials.

It is important to point out, however, that the use of mineral fillers or reinforcement materials should not be looked upon negatively. Most natural composite materials are made up of biopolymers and minerals, with the stiffest materials such as tooth enamel and sea shells containing little organic material.[3] Flexible materials such as wood are essentially completely organic composites whereas materials requiring both stiffness and toughness contain both mineral and organic matter. Bone material, for example, is essentially a hydroxyapatite nanoplatelet-reinforced collagen composite.[4]

22.2 BIO-BASED POLYMERS

There are a variety of polymers derived from renewable sources, all of which could be used as the polymer matrix phase in composite materials. A brief overview of these polymers is presented with references to detailed reviews for a more detailed description. Similarly to petroleum-based polymers, bio-based polymers can be reinforced with natural, synthetic and mineral materials to alter their properties and achieve a wide range of toughness, stiffness, thermal conductivity, acoustic insulation and dimensional stability.

22.2.1 Polysaccharides

Polysaccharides are the most abundant organic materials on Earth and encompass cellulose, starch and chitin, amongst others such as hemicellulose.[5] Whereas cellulose and chitin are structural polymers (cellulose is used as the structural agent in plants and chitin is found in the exoskeleton of insects and arthropods such as crustaceans), starch is used in Nature as an energy storage material. The common base of these naturally occurring polymers is D-glucose, which Nature adapted to various extents to accommodate different environments and obtain different properties. These variations result in varying crystallinity, solubility and ease of chemical modification, so that they may span a wide field of potential uses. Modification of polysaccharides is often performed to render the polymer more hydrophobic and to improve its solubility in organic solvents. The large amount of hydroxyl groups offers a direct and straightforward modification opportunity.

Cellulose is the most abundantly available organic material on Earth. Natural fibres (constituted of cellulose, lignin, hemicellulose and a variety of pectins and waxes) are commonly used as a reinforcement agent for composite materials. The hydrophilicity and water absorption of natural fibres reduce the dimensional stability of composites. Therefore, surface modifications are commonly used to improve the dimensional stability by making the fibre surface less hydrophilic, which may also improve the interfacial compatibility between polymers and the natural fibres. As cellulose is remarkably insoluble due to the significant hydrogen bonding present between polymer chains, pure cellulose is generally used in a modified form (commonly etherified or esterified). Cellulose modification requires hydrogen-disrupting solvents such as dimethylacetamide–LiCl, various ionic liquids or a tetrabutylammonium fluoride–dimethyl sulfoxide (TBAF–DMSO) mixture.[6] Virtually any imaginable chemical modification of cellulose has been reported, although there has been only limited commercial exploitation of most of these modifications.[5,7] However, cellulose esters have been around as important commercial polymers for nearly a century. The most common ones include cellulose acetate (CA), cellulose acetate butyrate (CAB), cellulose acetate propionate (CAP) and nitrocellulose (or smokeless gunpowder). Cellulose esters have found applications in coatings, controlled release formulations, membranes, optical films, heavy metal sorbents and chromatography materials.[8] They have also been widely investigated in blends with other polymers, as binders in natural fibre composites, paper, explosives and as filler particles in wax compositions for metal casting.[8] Oxidised cellulose has found wide use in medical applications such as absorbable haemostatic scaffolding materials, postsurgical adhesion prevention layers and as a carrier material for agricultural, cosmetic and pharmaceutical products. Recent efforts appear to have focused more on the creation of well-defined cellulose surfaces and surface modification of cellulose.[9,10]

In addition to its use as foodstuff, starch has found uses as a biodegradable polymer in the form of thermoplastic starch (TPS). The starch grains and their semicrystalline structure are disrupted by a combination of thermal treatment and the addition of one or more plasticisers (generally water and/or multifunctional alcohols). The major drawback of these polymers is their hydrophilicity, reducing its dimensional stability and mechanical properties, and an ageing effect making the polymer more brittle with time. Blending with a hydrophobic polymer can improve some of these properties and composite systems and a variety of polymers such as polycaprolactone, polyhydroxyalkanoate (PHA), polylactic acid (PLA), polyester amides and lignin and also natural fibres and polysaccharide micro- and nanoparticles have been reported, permitting the formation of soft and rigid materials.[11-14] However, this is not always required. Novamont, a leading Italian manufacturer of bioplastic resins, produces a compostable polymer based on cornstarch, Mater-Bi, which is used for shopping bags, netting for fruit and produce and packaging.[15,16] Various classes of Mater-Bi exist with various properties and biodegradation profiles. The production of starch foams for shipping fill material and as a precursor for mesoporous materials is also under investigation.[17,18] Just like cellulose, virtually any modification of starch has been reported, even though commercial applications as structural materials are limited.

The extreme insolubility of chitin limits its use and it is not widely used as found in Nature. However, deacetylation of chitin results in chitosan, a very promising material since it is readily soluble in mildly acidic media and various common organic solvents. Chemical modification of chitin through its abundant hydroxyl groups is also used to improve its processability.[19] Since chitin and chitosan are biodegradable, biocompatible and non-toxic materials and show low immunogenicity and also antibacterial activity, they have been widely investigated for use in biomedical applications.[20] The most developed applications for chitin- and chitosan-based materials are in tissue engineering. To achieve the required mechanical properties, modified chitin and chitosan have been reinforced with biomaterials such as hydroxyapatite and bioactive glasses, which, depending on their concentration, can bind or not bind to soft tissue and/or bone and show variable resorption rates.[20,21] Chitosan has also been combined with a variety of minerals (*e.g.* montmorillonite, kaolin and magnetite) and both synthetic and natural polymers [*e.g.* poly(vinyl chloride), poly(vinyl alcohol), cellulose and calcium alginate] to prepare adsorbents for organics and heavy metals.[22] Hydrophilic chitosan has also been blended successfully with hydrophobic polycaprolactone (PCL) (25, 50 and 75% PCL) using an acetic acid solution.[23] Owing to the low melting point of PCL, membranes were processed into films in a water-bath at 25, 37 and 55 °C or in an oven at 55 °C. Membranes were also solvent annealed using chloroform vapour. Tensile properties were analysed in either dry or wet conditions at 25 and 37 °C and 50:50 blends processed at 55 °C revealed a significant improvement in mechanical properties and also support for cellular activity relative to chitosan.

Alginates are hydrophilic anionic polysaccharides consisting of alternating blocks of β-D-mannuronic acid and α-L-guluronic acid residues with C5 carrying a carboxylate and connected through (1–4)-linkages.[24] They can be extracted from marine brown algae and soil bacteria. Cell adhesion does not readily occur in the unmodified form,[25] and alginates have been investigated as biomaterials since 1947.[26] Alginate composites have been produced by combining alginates with magnetic inorganic nanoparticles and in some instances also additives such as activated carbon and phosphinic acid derivatives, which can be used for metal and organics remediation.[27–31] Composites made from alginate fibres covered with chitosan have been proposed as advanced wound dressing materials as they showed improved tensile properties.[32] Composites of alginates with other biocompatible materials can be expected to improve the alginate properties and extend their applicability beyond what is currently feasible.

Hemicellulose (obtained from wood) and ulvan (a polysaccharide derived from seaweed) are other polysaccharides available in larger quantities. Hemicellulose has found uses in films and coatings, polyelectrolytes and rheology modifiers. However, materials with hemicelluloses as the major constituent are not expected to be developed as a significant amount of research has been devoted to them without much success.[5] Although ulvan solutions are generally of low viscosity, they can be made to self-assemble and gel. Industrial use of ulvan is currently virtually non-existent but its unique chemical and physical properties make it very attractive for a variety of applications in the medical, pharmaceutical and agricultural domains.[33] In the field of materials, ulvan has been shown to intercalate into clay, opening up an application in the field of clay nanocomposites.[34]

Polysaccharides have also found their way into nanoscience. The amorphous sections of semicrystalline polysaccharides (cellulose, starch and chitin) can be hydrolysed to release the nanosized crystalline domains. The monocrystalline particles have nanometric sizes with starch nanoparticles having a platelet-like shape and cellulose and chitin having a rod-like structure. Mechanical disintegration can also lead to nanoscopic flexible fibres. Their use as nanoreinforcements has already received considerable attention,[14,35–37] and this is discussed later.

22.2.2 Lignin

Lignin is a complex polyphenolic polymer found in all vascular plants and is the second most abundant polymer in Nature.[38] It is obtained industrially as a by-product of the production of cellulose pulp at an annual rate of about 70 Mt (70 million tonnes). However, only an estimated 2% of this is used as a chemical product, either directly or after modification. The remaining 98% is burned to recover its energy.[39] Depending on the extraction process, industrial lignins can have a low molecular weight and can be insoluble in water (soda lignin), have a low to high molecular weight (1–150 kDa), 4–8% sulfur content and be dispersible in water (lignosulfonates), or are recovered as a highly pure material with medium molecular weights (2.5–39 kDa) that is water insoluble (Kraft lignin, the predominant industrial lignin).

Lignosulfonates are used widely as dispersants and binders.[40] Soda lignins have found limited use as a replacement for phenol in phenol–formaldehyde resins, as a component in animal feed and as a dispersant.[39] Industrial uses of Kraft lignin are limited to dispersants and emulsifiers and to some extent as a source for low molecular weight aromatic compounds, and a significant amount of academic work has been conducted to unlock the potential of this promising compound in polymeric materials.[41,42] Especially the blending of lignin with thermoplastic and thermosetting polymers is commonly regarded as a promising pathway.[43–45] Lignin has been used in blends with renewable polymers such as proteins, starch, PHAs, polylactides and cross-linked triglycerides, and also with petroleum-based polyolefins, epoxy and phenol–formaldehyde resins and polyesters.[41] As lignin is generally poorly soluble in the polymer, a composite material with lignin as filler is obtained. In general, the mechanical properties of lignin–polymer blends either show a small improvement at low lignin contents followed by degradation of properties at higher contents, or no mechanical property improvement at all. However, lignin does appear to protect polymers from oxidative degradation due to stabilisation of radicals, also evident is its ability to slow free radical polymerisation reactions.[46,47] The use of the large number of hydroxyl groups present on Kraft lignin makes it an obvious candidate for targeted chemical modification where blending with polymers becomes easier,[48,49] for grafting with polymers to generate a thermo-processable material,[50–52] or as a copolymer in, for example, polyurethane resins.[45] These lignin-containing polyurethanes are sometimes described as composites even though the lignin is expected to be incorporated into the polymer backbone, making it rather like a copolymer. To our knowledge, no peer-reviewed work has been published on lignin-containing polyurethanes reinforced with natural fibres, even though some relevant conference proceedings exist. In these systems, lignin can be expected to improve the fibre–polymer interactions, as has been shown for plant oil-based polymers containing lignin.[53] *In situ* polymerisation of lignin by enzymes (phenol oxidases) results in a bioadhesive which can be used in the fabrication of particle boards under normal manufacturing conditions.[52] Lignin has also found applications in controlled release granules, where it can be blended with a variety of other biodegradable polymers (synthetic or natural) to control the release of pesticides or herbicides.[54]

Furthermore, recent advances in understanding lignin bio-synthesis may offer the potential to engineer the lignin structure to render lignin extraction less energy intensive and to promote the formation of lignin structures desirable for polymer and composite fabrication.[55,56]

22.2.3 Plant Oils

Plant oils, or triglycerides, consist of three fatty acid arms connected to a glycerol centre by an ester linkage. The fatty acid arms vary in length and chemical functionality depending on the plant from which it is extracted.[57] The most common fatty acids are 14–22 carbons long with 0–3 double bonds per fatty acid unit. Some naturally epoxidised triglycerides also exist. Triglyceride oils have been used extensively to produce coatings, plasticisers, lubricants, agrochemicals and inks.[58,59] They have also been used as toughening agents and interpenetrating networks that improve the fracture properties of thermosetting polymers.[60] The double bond functionality can also be chemically modified to an epoxy and subsequently to acrylates, secondary alcohols or maleates. Epoxidised plant oils can be used directly as a comonomer in epoxy resins or after further modification.[57,61] Hydroxylated plant oils have been used in polyurethanes, and acrylated and maleated triglycerides can be used as additives or as comonomers in unsaturated polyesters or vinyl esters.[57,62] The conversion of plant oil triglycerides into difunctional monomers to produce linear polymers has also been described and they have be converted into building blocks for polyamides (production of nylon 11), polyesters, polyacrylates and polymethacrylates.[61]

Since the creation of transgenic plants allows the preferential production of specific triglycerides in plants, it should be possible to extend this framework towards ever more useful monomers.[63] The production of highly epoxidised oils by plants would provide an interesting starting material that could be used directly or with minimal chemical transformations. Given that plant oil extraction is already well developed and optimised industrially, this is a very promising route to obtain functional monomers in much larger volumes.

The use of plant oil-based polymers in composite materials is largely concentrated on thermosetting plant oil-based polymers with two general approaches taken: the double bonds of triglycerides are either converted to epoxide or hydroxyl groups and used in polyurethanes after blending with polyamines or isocyanates, respectively, or they are modified with terminal double bond functionalities (*e.g.* acrylate, methacrylate, maleate) capable of free radical polymerisation.[57,64–66] Free radical polymerisation is generally performed after blending with low molecular weight species such as styrene or methyl methacrylate to reduce viscosity.[57,66,67] Some work has attempted to replace these low molecular weight species with fatty acid methyl ester-based molecules to increase the renewable content of the final polymer, but the properties were significantly poorer than when styrene was used.[68] It is also possible to prepare alkyd resins from plant oil triglycerides. However, this requires partial hydrolysis of the ester linkages between the fatty acid chain and the glycerol centre through alcoholysis with glycerol to form mono- and diglycerides, which are then cross-linked with a cyclic anhydride.[64] Alkyd resins are easy to prepare, tend to be inexpensive and are in general also biodegradable to a greater extent due to the hydrolysable ester linkages used in the cross-linking reaction. Polyurethanes have also been prepared by blending mono- and diglycerides with isocyanates.[64]

Plant oil triglyceride-based polymers have been widely studied as a matrix phase in composite materials with both natural and synthetic fibres as structural materials.[66,69–73] Lignin has also been used as a filler but the mechanical properties of the composites were rather poor.[44] Silver nanoparticle-containing composites have been prepared by silver reduction achieved by the radicals generated from the free radical initiator in a partial glycerides–styrene matrix resulting in antibacterial materials.[74]

22.2.4 Polyhydroxyalkanoates

Polyhydroxyalkanoates (PHAs) are polyesters of 3-, 4-, 5- and 6-hydroxy acids which are synthesised naturally by a large variety of bacteria.[75] PHAs are impermeable to water and air and therefore suitable to be used as the materials for bottles, films and fibres.[75] The use of PHAs as a low commodity polymer is inhibited by its relatively high cost (5–10 times the cost of polypropylene).[76] Variations in monomer composition have a significant effect on the mechanical properties, with poly-3-hydroxybutyrate being relatively hard and brittle, whereas poly(3-hydroxybutyrate-co-3-hydroxyvalerate) improves on the brittleness. Inclusion of longer chain monomers results in materials with similar properties to polypropylene.[75] To improve the economics of large-scale PHA production, attention has been diverted towards its production in plants. Much work has been performed with reported production of up to 40% polymer based on dry shoot weight (in transgenic *Arabidopsis thaliana*, the resulting dwarf plant could no longer produce seeds). Oil rape plant production of up to 8% dry weight remained viable and it did produce seeds.[75] There still is a large potential to improve production efficiency and keep the plants viable at the same time. Commercial production of PHAs is carried out by Biomer in Germany and Metabolix in the USA company. The industrial production route still follows the bacterial pathway, but Metabolix are also investigating the production of PHAs in switchgrass, a perennial grass that thrives on land of marginal use for other crops.[75]

The use of PHA in composite materials has largely focused on tissue engineering applications, making use of their biodegradability, biocompatibility and also piezoelectric properties, which can be used to stimulate bone growth and promote wound healing.[77,78] Investigations on PHA for medical applications have also been helped by US FDA approval of poly(4-hydroxybutyrate) for wound sutures.[79] In recent work, a drug delivery composite obtained by blending natural rubber with poly(3-hydroxybutyrate)-co-poly(3-hydroxyvalerate), showing good adhesion between the two phases and a reduction in drug release rate with increasing rubber content, was proposed.[80] Most of the work on PHA composites has focused on short chain length PHAs (monomer with less than six carbon atoms) reinforced with hydroxyapatite, wollastonite and bioactive glasses. Some limited work on medium chain length PHAs (mcl-PHAs) such as poly (3-hydroxybutyrate)-co-poly(3-hydroxyhexanoate) with hydroxyapatite and poly(3-hydroxyoctanoate) with carbon nanotubes, as these mcl-PHAs are the only ones currently available in significant quantities.[78] However, this field offers great development potential given the large variety of available bioceramics such as alumina, zirconia and calcium phosphates and the potential to prepare mcl-PHA composites with appropriate mechanical properties for tissue engineering and graft scaffold applications.

22.2.5 Polylactic Acid

Polylactic acid (PLA) is one of the biggest success stories of bio-based polymers. It was first developed by Dow Chemical in the 1950s, but its high cost limited its use to specialised medical devices such as sutures and soft tissue implants.[16] Lactic acid, the monomer used to produce PLA, is obtained through fermentation of glucose, with conversion reaching 90%.[81] PLA is generally formed by ring-opening polymerisation of lactide, a circular dimer of lactic acid, resulting in high molecular weight polymers. Recent advances in fermentation have significantly reduced the production cost of PLA, currently around the same price as poly(ethylene ter-ephthalate) (PET), commonly used in drink bottles and the polymer that it most closely resembles in terms of properties and potential applications. Unlike PET, PLA is also compostable. NatureWorks, an independently managed business unit of Cargill in the USA, produces PLA from dextrose maize sugar on an industrial scale and has an annual production capacity of ~136 kt.[15] Greenhouse gas emissions are said to be reduced by 80–90% with ~65% less use of fossil fuels than for traditional plastics. PLA has found wide use in films, bottles, labels, disposable cups and serviceware, *etc.* Toyota is also developing PLA production from starch-rich sweet potatoes, containing 40–50% more starch than does corn. The starch is converted to lactic acid, which is then polymerised to PLA. The first use of Eco-Plastic, the Toyota PLA, was in 2003 in commercial vehicles.[15] Toyota is targeting the production of 20 Mt of PLA by 2020. A recently reported advance over the currently employed production pathway (fermentation followed by a chemical poly-merisation step) used a modified enzyme to polymerise lactic acid directly, allowing for a one-step production of PLA.[82] This will further reduce the production cost of PLA and improve even further its environmentally friendly properties.

To reduce the cost of PLA-based products and also the brittleness of PLA, blending with other polymers is common.[83,84] PLA–starch blends are very sensitive to the composition and the use of plasticisers is generally required to reduce the brittleness of the overall composite. PLA has been combined with various petroleum-based polymers such as poly(ethylene oxide), poly-(propylene oxide) and poly(vinyl acetate).[85-87] The use of epox-idised plant oils has also been reported but its toughening effect was rather poor.[88] Due to PLA's environmental credentials, it has been widely studied and continues to be, in composite materials with natural fibre reinforcements.[89-93] Nearly all natural fibres have been studied as reinforcements for PLA. The major hurdles to industrial uptake are fibre property variability, processing diffi-culties and brittleness of untoughened PLA.

22.2.6 Natural Rubber

Rubber is a polymer of isoprene and is currently the most widely used polymer derived from a natural source.[16] The properties of natural rubber are reviewed in detail elsewhere.[94,95] The excellent properties exhibited by natural rubber (malleability, elasticity, heat dissipation, resistance to impact and abrasion) have not been matched by synthetic rubber owing to the presence of naturally present property-enhancing secondary components such as pro-teins, lipids, carbohydrates and minerals which are ill-characterised. All commercial natural rubber is harvested from the Para rubber tree (*Hevea brasiliensis*), one of the most genetically restricted crops grown on commercial plantations largely located in South-East Asia. The reliance on a single species to produce vast amounts of a natural resource poses significant danger to the supply due to potential fatal diseases. Therefore, other crops such as the Russian dandelion and the Mexican shrub guayule are under investigation for commercial rubber production.[96] The Russian dandelion, which accumulates rubber in lactifers in the roots, is a very promising candidate since it produces rubber with sig-nificantly higher molecular weight than guayule and the *Hevea* tree (2 MDa) and it has already shown its potential during World War II, when its rubber was used for motor tyres in various countries, including the USA and the UK. The Russian dandelion also accumulates 25–40% root dry weight of inulin, a fructose-based sugar, which can be used for bioethanol production. Guayule rubber is considered hypoallergenic compared with *Hevea*-extracted rubber owing to a lower concentration of proteins and the absence of reaction between the proteins present and *Hevea* immuno-globulins.[97] Extraction of guayule rubber is more difficult as it cannot be tapped, resulting in higher production costs, coupled with one- to two-thirds lower rubber production per acre compared with *Hevea*. Guayule rubber is currently a more specialised rubber product and Yulex (Chandler, AZ, USA) plans to couple guayule rubber production with biofuel production, improving the process economics.[16]

The use of rubber in composite materials is widespread owing to the need of rubber tyres for the transport industry. Tyres can be made of synthetic rubber, natural rubber or a blend and also contains fabric, metal wire and various chemicals to improve performance, heat dissipation, UV stability and mechanical integrity. Carbon black is the most widely used filler in rubber materials, but cheaper and more environmentally friendlier inorganic particles such as silica have also been reported.[98,99] Strong interaction between the filler and the rubber matrix is paramount to achieve the desired mechanical reinforcement as it reduces particle aggregation during processing and improves stress transfer. As inorganic fillers tend to have a reduced affinity for rubber compared with carbon black, the use of compatibilisers is usually required for optimal performance.[100,101] Polyisoprene has also been used in composites with carbon nanotubes for pressure and gas sensors, showing that the elastomeric properties of rubber can have important applications when combined with appropriate fillers.[102] Natural rubber has also been reinforced with natural fibres, with the reinforcing fibres restricting mobility of the rubber chains and improvements to the rubber–fibre interface with the use of coupling agents.[103,104]

22.3 REINFORCEMENTS

There has been huge interest in the use of natural fibres in fields as varied as automobiles, textiles, food packaging, biomedical and tissue engineering. The United Nations General Assembly declared 2009 the International Year of Natural Fibres to raise awareness of and promote demand for sustainably produced natural fibres. For thousands of years, people all over the world have used fibres from plants and animals to make cloth, string and paper, to strengthen building materials and as structural materials in buildings. Roughly 30 Mt of natural fibres are produced annually worldwide, providing a large supply of renewable fibrous materials.

Around 25 Mt of cotton are produced each year. China, the USA, Pakistan, India, Uzbekistan, Turkey and Brazil are the major producers, with more than 80 countries, many in Africa, recording some production. Approximately 2.3–2.8 Mt of jute are produced each year, with India producing 60% of the world's jute and Bangladesh being the other major producer. Other countries, including Myanmar and Nepal, produce much smaller quantities. Other fibres are produced in lower quantities such as flax (~ 1.9 Mt), hemp (~ 50 kt), kenaf (~ 1 Mt) and sisal (~ 250 kt) (values taken from http://www.fao.org). In addition to the common lignocellulosic fibres, other potential sources for renewable reinforcing fibres include wool with an annual production of around 2 Mt and chicken feathers produced by the poultry industry as waste at a rate of 1.4 Mt y^{-1}. The rising cost of oil, together with a greater awareness of the environment and advances in technology, are creating an increasing market for natural fibres which are finding inroads into commercial applications in composite materials.

22.3.1   Fibre Reinforcement

Fibre reinforcement of polymeric matrices has been widely used to improve both the mechanical and thermomechanical properties of polymers owing to their excellent load transfer capacity and the light-weight structures that can be formed. The mechanical properties of reinforcing fibres, which can be natural or synthetic, can have a significant influence on the overall properties of the composites produced. Table 22.1 presents the density, tensile and elongation properties of some of the fibre types used as reinforcement.

Successful reinforcement is achieved through chemical and/or mechanical coupling between the fibres/fillers and matrix so that the load is shared between the fibres and the matrix.[2,106] Natural reinforcements such as cellulose and starch have a strong tendency for self-association owing to their strongly interacting surface hydroxyl groups.[107] This poses problems during processing to

Table 22.1 Density and mechanical properties of some natural and synthetic reinforcing fibres.[105]

Fibre	Density/g cm^{-3}	Tensile strength/MPa	Tensile modulus/GPa	Elongation/%
Cotton	1.5–1.6	400	5.5–12.6	7.0–8.0
Jute	1.3	373–773	26.5	1.5–1.8
Flax	1.5	500–1500	27.6	2.7–3.2
Hemp	1.47	690	70	2–4
Kenaf	1.45	930	53	1.6
Ramie	–	400–938	61.4–128	3.6–3.8
Sisal	1.5	511–635	9.4–22	2.0–2.5
Coir	1.2	593	4.6–6	30
E-glass	2.5	2 000–3500	70	0.5
Aramid	1.4	3 000–3150	63–67	3.3–3.7
Carbon	1.4	4 000	230–240	1.4–1.8

attain individualisation and can also aid in the formation of stabilising percolated networks. The properties of composites prepared by fibre-reinforcing polymers are largely dependent on the fibre architecture, such as aspect ratio, fibre mechanical properties, volume fraction and alignment of fibres within the matrix. The compounding processes are also an important factor controlling the properties of composite materials as they affect the fibre orientation, individualisation in the case of short fibres and fibre aspect ratio. The orientation and distribution of short fibres in the matrix are, relatively, more difficult to control, to measure and to characterise, whereas long fibres offer better control of the distribution of fibres in the matrix, resulting in improved composite properties. Despite this, industrial and large-scale production of composite materials has so far been restricted to short fibre-reinforced matrices using injection moulding and extrusion processes. These methods can easily be turned into continuous processing whereas long fibre composite manufacturing with natural fibres generally requires significant manual labour and longer batch processing times. The cost of manual labour and longer processing times is inhibitive for low-cost applications at which natural fibre-reinforced composites are currently targeted. However, compression and injection moulding processes are still not properly optimised and variability in composite properties can easily occur.

Short natural fibres have attracted much attention as reinforcements in polymers in order to produce biodegradable

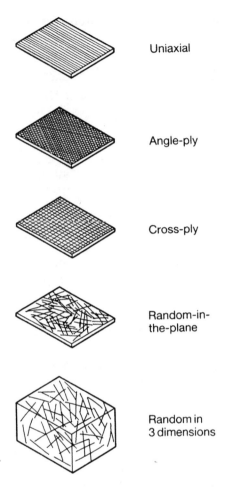

Uniaxial

Angle-ply

Cross-ply

Random-in-
the-plane

Random in
3 dimensions

Figure 22.1 Patterns of fibre reinforcement in polymers.[2] Reprinted by permission of Oxford University Press.

composites.[108–110] Natural fibres, isolated from plants, animals and also some minerals sources, have been used aligned or randomly distributed in polymer matrices either in the same plane or in three dimensions[2] (see Figure 22.1). Aligned reinforcing fibres show strong reinforcement in the longitudinal direction of the fibre, with low reinforcement capability in the transversal direction. Random three-dimensional fibre orientation results in equal reinforcement in all directions. The mode of their reinforcement has also

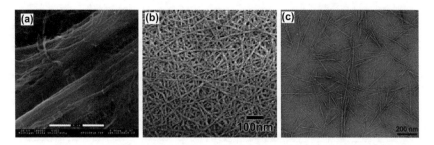

Figure 22.2 Structure and appearance of cellulose fibres by electron microscopy: (a) ESEM image of cellulose microfibres;[215] (b) FE-SEM
image of cellulose nanofibres;[216] (c) TEM image of cellulose
nanowhiskers.[172]
Sources: (a) Reprinted from ref. 215, Copyright 2008, with permission from Elsevier; (b) reprinted with permission from ref. 216,
Copyright 2007 American Chemical Society; (c) reprinted from ref.
172, Figure 1, with permission from Springer Science + Business
Media.

been studied extensively through experimental and theoretical
investigations.[35,111]

22.3.1.1 Macro-, Micro- and Nano-reinforcement. Polymers
have been reinforced with macroscopic and also micro- and
nanofibres. Microfibres have diameters in the micrometre range
whereas nanofibres have at least one dimension on the
nanometre scale (see Figure 22.2). Micro- and nanofibres derived
from various natural sources have received much attention for
composite production owing to their good mechanical properties
and high reinforcement capability at low loadings.[112,113]
Improved mechanical properties stem from the reduction of
defects with reduction in size, while the improved reinforcement
capability arises from a higher surface-to-volume ratio and
smaller distance between reinforcing fibres with decreasing fibre
size for the same fibre loading. The fibre diameter and aspect
ratio have also been reported to have a great influence on the
mechanical properties of the fibre itself (see Figure 22.3).

An added advantage of nanoscale reinforcement is that the
product tends to be more transparent than micro- and macro-
reinforced composites as the fibres are smaller than the wavelength
of visible light. This is certainly an advantage for packaging

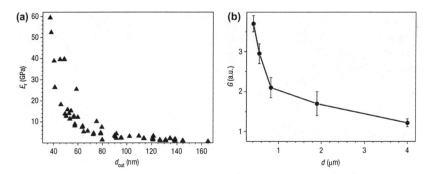

Figure 22.3   Dependence of modulus on fibre diameter: (a) elastic modulus *versus* diameter of polypyrrole nanofibre; (b) relative shear modulus *versus* diameter of electrospun polystyrene fibre.[214] Reprinted with permission from Macmillan Publishers, *Nature Nanotechnology*, copyright 2007.

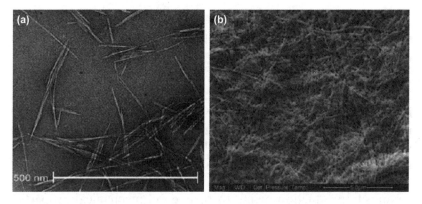

Figure 22.4   (a) TEM image of cellulose nanowhiskers from sisal and (b) an SEM image of cellulose microfibres from sisal.[115] Adapted and reprinted with permission from ref. 115, Copyright 2009 American Chemical Society.

materials where consumers want to be able to see the products they buy. For example, comparing pea starch nanocrystals and native pea starch in a poly(vinyl alcohol) matrix, 10 wt% nanocrystals led to an increase in light transmittance from 69 to 91%, an increase in the tensile strength from 35 to 40 MPa, an increase in the elongation at break from 579 to 734% and a decrease in the moisture uptake from 73 to 71% in comparison with a reinforcement with 10 wt% microsized starch granules.[114]

However, on comparing nanowhiskers and microfibrillated cellulose (MFC) derived from sisal (Figure 22.4) in a PCL matrix and using octadecyl isocyanate as a fibre grafting agent to improve compatibilisation, MFC showed a significantly higher reinforcing capability (60% *versus* 15% increase in tensile modulus for the same amount of reinforcing fibres).[115] The MFC allowed better inter-fibre entanglement compared with the nanoscale reinforcements, resulting in a better reinforcing performance. This clearly shows that fibre properties and interfacial compatibility alone are not sufficient to predict composite performance.

Since it is virtually impossible to cover the vast amount of work that has been done on natural fibre-reinforced composites, some representative examples are given that illustrate the potential of this technology and also some potential hurdles. The two most important materials used as fillers/reinforcements in natural materials are cellulose (as lignocellulosic fibres) and starch (as granules). These are covered in some detail, followed by some interesting work on other types of fibres.

22.3.2 Cellulosic Fibres

Cellulosic fibres, sometimes called lignocellulosic fibres, are a composite constituted of cellulose, lignin and hemicelluloses obtained from renewable natural sources, such as cotton, jute, flax and hemp (see Table 22.1). The exact composition depends on the natural source and even the environmental conditions. Cellulose fibres from various origins (*e.g.* hardwood,[116–121] softwood,[117,118,122] bagasse,[121] nutshell,[121] kenaf,[123] pineapple leaf,[124] cotton,[118] jute,[125] sisal[126]) have been widely used as reinforcements in polymer composites. These fibres have been used to reinforce a large variety of polymer matrices [*e.g.* polyethylene (PE),[117,119,126,127] poly(vinyl chloride) (PVC),[118,120,121] polypropylene (PP)[116,122,123] and plant oil-based thermosets[69,71]].

As is obvious from this small sample, an enormous amount of work has been performed over the last couple of decades, which is impossible to review within this limited space. In the following we mention some work that shows important aspects of natural fibre composites and refer the interested reader to an array of review papers published in the recent literature.[89,92,128–133]

The interface between the polymer matrix and the natural fibre is critical in stress transfer between the polymer and the reinforcing fibre. As a result, a large amount of work has been focused on improving the interface by modifying the fibre surface with functionalities which are more compatible with the polymer or which can even chemically react with the polymer. These strategies include etherification, esterification, siloxane and urethane formation, 'click' chemistry and adsorption of lignin or carboxylic waxes.[53,134–141] An added benefit which may result is reduced water absorption by the hydrophilic fibres resulting in an increased dimensional stability. Mercerisation (alkali treatment) is also a common procedure as it results in a roughened fibre surface, improving mechanical polymer–fibre interlocking. It also reduces fibre shrinkage and lowers the microfibrillar angle, which results in improved fibre mechanical properties.[142,143] However, it may also reduce the mechanical properties of the natural fibres.[103]

Between 1985 and 1993, Kokta, Daneault and co-workers investigated wood fibres of different origins (aspen, birch, spruce and balsam fir) in different forms: chemithermomechanical pulp (CTMP)[117,119] thermomechanical pulp (TMP),[117] Kraft pulp,[118] bleached pulp[118] and sawdust.[118,120,121] Cellulose derived from CTMP pulp (from hardwood) was better in terms of mechanical properties than cellulose derived from other kinds of pulp.[117,118] This was probably due to the higher aspect ratio of the fibrils resulting from the treatments used. HDPE-based composites performed better than LDPE-based composites.[119] To improve the compatibility of the fibres with the polymer matrix, the fibres were generally treated before incorporation into the polymer matrix. The treatments investigated included impregnation[117] or encapsulation[118] with polymer, coating with latex[120] and grafting with silanes,[117] with isocyanates,[118,119,121] with carboxylic acids[120] and with acid anhydrides,[120,121] with grafting found to be more effective than coating. Among grafting agents, isocyanates[121] and linoleic acid[120] were found to perform the best. Composites reinforced with wood fibres presented mechanical properties either comparable to or even better than those of composites reinforced with glass fibres[117,119] or mica[119,121] and had a lower cost.[119] As a logical continuation of their work on the chemical modification of cellulose fibres to increase their compatibility with polymer matrices, Kokta, Daneault and co-workers also investigated grafting polymer to the wood fibre chain, resulting in aspen-g-PE or aspen-g-PVC copolymer composites.[117,120]

In 1996, Hedenberg and Gatenholm converted LDPE and cellulose waste into a composite and introduced ketone, carboxy and hydroxyperoxy groups at the surface of LDPE by ozone treatment.[127] The ketone and carboxy groups were able to hydrogen bond with the cellulose fibres, whereas hydroxyperoxy groups could decompose during processing and subsequently form covalent bonds with the cellulosic fibres. In 1997, Devi *et al.* reported on composites made of pineapple fibres as reinforcement for a polyester resin.[124] The ideal fibre length and loading were found to be 30 mm and 30%, respectively, whilst treating the fibres with silane improved the mechanical properties of the composite. Pineapple leaf fibres were also found to increase the mechanical properties better than other cellulosic fibres. The same year, Trejo-O'Reilly *et al.* investigated the chemical surface modification of cellulose from different sources for use in composites.[144] The grafting agents they used carried anhydride [*e.g.* alkenylsuccinic anhydrides, poly(styrene-co-maleic anhydride)] or isocyanate groups [*e.g.* poly(styrene-co-3-isopropenyl-α,α'-dimethylbenzyl isocyanate)]. In 1999, Kazayawoko *et al.* synthesised composites with ester bonds between the cellulose fibres (bleached Kraft cellulose and unbleached TMP) and a maleated PP polymer matrix.[145] More recent work by Bengtsson *et al.* confirmed the improved mechanical properties of such a system by extruding maleated PP with different cellulose pulps to prepare composites with up to 60% cellulose.[122] TMP reaction with the maleated polymer was found not to occur, indicating that insufficient hydroxyl groups exist on the unbleached pulp surface for reaction. Arbelaiz *et al.*[146] also investigated the mechanical properties of flax fibres modified with 5% maleic anhydride–polypropylene (MAPP) copolymer for PP reinforcement. The SEM images of untreated, MAPP-treated flax–PP and E-glass–PP composites are represented in Figure 22.5b–d. The untreated flax fibre–PP composites revealed a gap between the fibre and matrix, suggesting poor fibre–polymer adhesion. MAPP-treated flax–PP composites, however, showed an enhanced fibre–matrix interface. The E-glass fibre–PP matrix also displayed poor interfacial adhesion due to the polar surface of the glass fibre. Composites containing ∼30 wt% fibres were reported to have tensile strengths of 27 and 37 MPa for untreated and

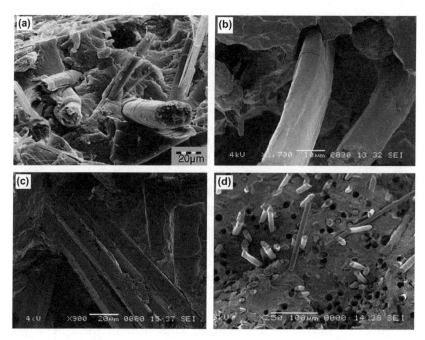

Figure 22.5   SEM micrographs of fracture surface of composites: (a) flax–
PLA;[90] (b) untreated flax–PP;[146] (c) MAPP modified flax–PP;[146]
(d) E-glass–PP.[146] Reprinted from (a) ref. 90, Copyright 2008, and
(b–d) ref. 146, Copyright 2003, with permission from Elsevier.

MAPP-treated flax–PP composites, respectively, compared with
32.4 MPa for PP alone. The tensile modulus values for flax and
MAPP-treated flax fibre–PP composites were ~1.6 and ~2.1 GPa,
respectively, compared with ~2.2 GPa for the E-glass fibre–PP
composites with a similar fibre content. Other natural fibres, such
as kenaf, coir, sisal, hemp and jute fibres, have also been widely
used to reinforce polypropylene (PP). For example, in 40 wt%
composites, coir-reinforced PP composites revealed the lowest and
hemp-reinforced PP composites the highest mechanical properties
(Figure 22.6).[147] The lower strength of coir composites compared
with other natural fibre-reinforced composites was attributed to a
lower cellulose content and higher microfibrillar angle. For comparison, the tensile strength, tensile modulus, flexural strength and
flexural modulus properties of E-glass fibre–PP composites were
reported to be 88.6 MPa, 6.2 GPa, 4.38 GPa and 60 MPa,
respectively. The mechanical properties of the natural fibre–PP

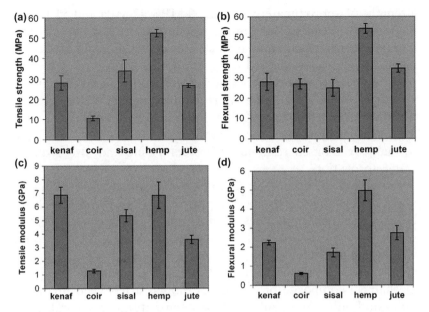

Figure 22.6 (a) Tensile strength, (b) flexural strength, (c) tensile modulus and
(d) flexural modulus of natural fibre-reinforced PP composites.[147]
Reprinted from ref. 147, Copyright 2003, with permission from
Elsevier.

composites were found to be comparable to the corresponding
properties of E-glass fibre–PP composites.[147–149] However, some
mechanical properties, such as the tensile and flexural modulus of
natural fibre–PP composites, were higher than those of E-glass
fibre–PP composites, suggesting that these fibres could potentially
replace E-glass fibre, especially where very high load-bearing
properties are not required.

The properties of jute fibre reinforced PP composites were
investigated by Liu and Dai,[150] who reported the interfacial shear
strength (IFSS), flexural strength, tensile strength and impact
strength of untreated jute–PP composites as 3.49 MPa, 39.1 MPa,
28.4 MPa and 65.0 J m^{-1}, respectively. However, surface modification of the jute fibres using 2% NaOH (mercerisation) and
maleic anhydride-grafted polypropylene (MAPP) revealed a significant improvement in fibre–matrix adhesion, which as a result
also improved the mechanical properties. These treated jute fibre
mat–PP composites revealed IFSS, flexural, tensile and impact

strengths of approximately 9 MPa, 59 MPa, 42 MPa and 54 J m^{-1}, respectively. However, the impact strength of treated fibre–matrix composites decreased compared with untreated fibre–matrix composite, which was suggested to be due to the surface modification resulting in increased rigidity of the fibres with reduced toughness.

Mitra *et al.* used jute fibres impregnated with different precondensates based on phenol–formaldehyde resins.[125] This treatment improved the mechanical properties of the formaldehyde-based resin–jute fibre composite and also reduced water absorption.

A significant amount of completely biomass-derived cellulosic fibre composites have also been reported.[130] These composites have bio-based polymer matrices such as plant oil triglyceride-based resins,[69,71,151–153] PLA,[91,154,155] soy protein,[142,156] PHA[157–159] and wheat gluten.[160] Compared with petroleum-based polymer matrices, bio-based polymers are not a magical solution for good mechanical composite properties. Although it has been predicted that good adhesion could occur between natural fibres and more hydrophilic polymers which display groups capable of hydrogen bonding, this is not obvious from work using bio-based polymers such as PLA.[90,161,162] For example, Bax and Müssig reviewed PLA–flax composites and compared their properties with those of PLA–rayon fibre composites.[90] Flax-reinforced composites were comparable to their rayon reinforced counterparts in modulus and strength, but performed poorly with respect to impact properties, similar to the results of other work using flax and jute fibres. Despite the relatively poor adhesion between the polymer matrix and the fibres (Figure 22.5a), which usually is beneficial for impact, a decrease in impact strength was observed upon fibre reinforcement. In general, bio-based polymer composites encounter the same issues, and control over the fibre–polymer interface and improvement of the inherent fibre properties remain important fields of research.[53,134–141]

Cellulose reinforcements have also arrived in the micro- and nano-reinforcement area. Treatment of pure cellulose fibres by enzymatic, mechanical and/or concentrated acid results in a reduction of the cellulose fibre size. There are two main methods for obtaining micro/nano-sized cellulose from macroscopic cellulose fibres. A method described first by Turbak *et al.* uses mechanical disintegration as the main driver, resulting in long, somewhat flexible nanofibrils.[113] A second method relies on acid hydrolysis using inorganic acids, first reported by Nickerson and Habrle,[163] which results in highly crystalline ribbon-like nanowhiskers. If the disintegration is reduced in time, micro-sized fibres are obtained.

In 1995, Favier *et al.* were the first to mention the use of cellulose whiskers as reinforcements in composites.[111] Since then, an enormous amount of work has appeared on nanocellulose-reinforced composites. One of the main reasons is the high stiffness of nanocrystalline cellulose; the Young's modulus of crystalline cellulose along the chain axis was calculated to be 167.5 GPa for native cellulose and 162.1 GPa for regenerated cellulose, which was found to be consistent with X-ray data (120–140 GPa for native cellulose and 110 GPa for regenerated cellulose).[164] However, the Young's modulus was seen to be significantly affected by the intramolecular hydrogen bonding along the chain axis, resulting in a reduction of the tensile properties.

For example, cellulose nanocrystals from sources such as tunicin,[111,165–167] cotton,[168] softwood,[168] flax,[169] hemp,[169] bacteria,[168,170] sugar beet[171] and sisal[172]) have been used as reinforcements in polymer matrices such as poly(styrene-co-butyl acetate),[173,174] PHA, plasticised starch,[166,168] pectin,[168] epoxy resin,[169] poly(vinyl acetate) (PVAc),[172] poly(ethylene oxide) (PEO),[170] poly(vinyl alcohol) (PVA)[171] and PLA.[175] Layer-by-layer assembly processes have also been used to make layered composites for mechanical applications,[176] which can also be used in sensor development.[177–179]

As a sign of the amount of activity, several review papers appear every year,[35,36,180–184] and it is therefore not our intention to attempt to cover all developments in the limited space here. In the original work by Favier *et al.*,[111] a matrix resulting from the copolymerisation of styrene (33–35 wt%) with butyl acetate (65–66 wt%) and a small amount of acrylic acid was used. When reinforced with whiskers made from tunicin (*Microcosmus fulcatus*), the drop in the shear modulus of the composite after the glass transition temperature was dramatically reduced, in particular when the whisker content was above the percolation threshold; this is due to the large number of hydrogen bonds between nanocrystals, which results in a network. This has since been reproduced with a variety of polymer matrices and cellulose nanowhiskers from a variety of cellulose sources. The most pronounced effects are seen for soft polymers with low glass transition temperatures. The cellulose nanowhiskers mostly are obtained by acid hydrolysis with sulfuric acid and, as a result, display a negative surface charge capable of stabilising their dispersion. Most composites are thus formed by water evaporation from an aqueous nano-whisker dispersion and either a water-soluble polymer or a polymer latex.[36,180,183] Common industrial processing techniques such as extrusion and injection moulding result in significant nanowhisker aggregation, which reduces the nanoscale reinforcing effect. Major breakthroughs will therefore need to come from the development of reliable processing techniques for nanocomposites with the retention of mechanical property improvement on an industrial scale. Some progress has been made by using cellulose nanowhiskers grafted with polymeric chains,[185] infusion of preformed nanowhisker networks with dissolved polymer chains[186] and electrospinning of nanowhiskers with polymers to form fibres.[187] In another straightforward approach, coating of PLA films with a cellulose nanofibre film reduced the oxygen permeability significantly.[188]

22.3.3 Starch Composites

Starch granules of various origins (*e.g.* amylomaize,[189,190] corn,[191–196] pea,[114] potato,[197] sago,[198] rice,[195,196,199,200] waxy maize[189] and wheat[201]) have been studied as potential reinforcements in polymer composites. A wide variety of matrices, including PE,[191] poly-caprolactone (PCL),[189,190,194,198,202] polyvalerolactone (PVL),[202] poly(3-hydroxybutyrate-co-3-hydroxyvalerate) (PHBV),[189] PVA,[114,193,197] poly(propylene carbonate) (PPC),[192] PLA,[199,201] polyacrylamide,[195,200] poly(acrylic acid),[196,200] poly(methacrylic acid)[200] and polyacrylonitrile,[200] have been investigated.

In 2000, Odusanya *et al.* prepared starch–PCL composites with up to 60 wt% starch in the PCL matrix before phase inversion occurred.[198] Adhesion at the starch–PCL interface was very poor, resulting in reduced tensile strength. In the same year, Avella and co-workers also prepared PCL–starch composites.[190] To improve the compatibility between starch and PCL, they used a low molecular weight PCL modified with pyromellitic anhydride as a compatibiliser. The use of the compatibiliser improved the performance of the materials without being detrimental to its biodegradability.

Biodegradability of starch–PE composites was also investigated by Wool *et al.*[191] They found that for a starch fraction smaller than the percolation threshold, the degradation of starch mainly occurred by microbial invasion. Further work on starch PCL composites was subsequently reported by Wu.[203] To improve compatibility, PCL was modified using maleic anhydride.

Siddaramaiah *et al.*[193] later investigated PVA–starch composites. Owing to the possibility of hydrogen bonding between the hydroxyl groups of starch and the hydroxyl groups of PVA, the composites did not suffer from adhesion problems at the interface. As a result, the mechanical properties of the composite were maintained on increasing the starch content. The main drawback of this composite is its poor biodegradability; a 50:50 PVA–starch blend does not lose more than 40% of its weight, even after 3 months in compost mud, because PVA is a water-soluble polymer that can only be degraded in water in the presence of specific microorganisms.

Ge *et al.* reported in 2004 that incorporating starch granules in a PPC matrix led to greater stiffness and tensile strength, while also significantly improving the thermal stability of the material.[192] In the same year, Zhang and Sun grafted PLA to the surface of starch using maleic anhydride as the grafting agent, resulting in an improvement of the mechanical properties compared with the pure physical composites.[201]

Similarly to cellulose, starch has also entered the nanoparticle era since acid hydrolysis of starch granules results in the formation of nanosized platelets.[37] In contrast to cellulose nanowhiskers, only a few articles have reported the use of starch nanocrystals. Nonetheless, starch nanocrystals from pea,[114,204] potato[205] and waxy maize[206–208] have been produced to be used as reinforcements in nanocomposites. The matrices investigated were poly(styrene-co-butyl acrylate),[205] poly(β-hydroxyoctanoate) (PHO),[204] natural rubber,[207] pullulan[208] and PVA.[114]

22.4 OTHER NATURAL FIBRES–REINFORCEMENTS

Chitosan fibres have been incorporated within PLLA–β-tricalcium phosphate (β-TCP) (PLLA: Poly-L-lactide) as a reinforcing agent to prepare scaffolds with sufficient compressive strength and adequate degradation rate for bone tissue engineering materials.[209] The chitosan fibres reinforced PLLA–β-TCP scaffolds with 36.2

and 50.6% porosity showed 40.0 and 20.2% increases in compressive strength, respectively, compared with control scaffolds (11.48 MPa) without chitosan fibres. The interconnected porous structure formed by degradation of chitosan fibres after immersion in simulated body fluid (SBF) for 24 days exhibited a favourable structure for new bone growth and the compressive strength was calculated to be 5.28 MPa after degradation. Furthermore, a layer of bone-like apatite formed on the surface of scaffold composites after degradation, which was also confirmed and indicated good biological activity.

The development of a novel composite structure combining microparticles and nanofibres was investigated using an innovative combination obtained by melt extruding a particulate composite reinforced with a chitosan nanofibre mesh (0.05 wt%) produced via electrospinning.[210] The reinforced microfibres were analysed by SEM and showed considerable alignment of the chitosan nanofibres along the longitudinal axis of the microfibre composite structure due to the melt flow during extrusion. The tensile properties revealed that the introduction of the nanofibre reinforcement increased the tensile modulus by up to 70% compared with microfibres without reinforcement (175.6 MPa). The presence of chitosan nanofibres in the composite enhanced the water uptake by up to 24%, which was suggested to be due to the high surface area of nanofibres facilitating higher water intake and the hydrophilicity of chitosan polymers. This type of composite shows great potential

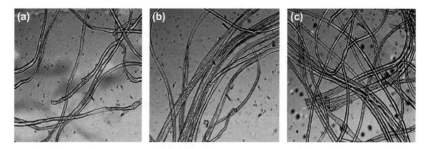

Figure 22.7 Optical micrographs ($\times 400$) of A1 alginate fibres treated with hydrolysed C_3 chitosan: (a) 0 wt% (*i.e.* untreated), (b) 2 wt% and (c) 5 wt%.[211] Reprinted from ref. 211, Copyright 2004, with permission from Elsevier.

for various biomedical applications owing to their good mechanical properties and enhanced degradability.

Chitosans (unhydrolysed and hydrolysed) were used to modify sodium alginate fibres (Figure 22.7) and the tensile properties (percentage elongation and tenacity) of chitosan–alginate fibres were analysed to evaluate their compatibility for potential biomedical applications, especially in wound dressings.[211] Alginate fibres modified by unhydrolysed chitosan (0–6 wt%) showed a reduction in tensile properties. The unhydrolysed chitosan did not reinforce as it was only coated on the surface of alginate fibre rather than penetrating it. However, incorporation of hydrolysed chitosan (7–25 wt%) in the alginate fibre revealed significant reinforcement and showed enhanced tensile properties compared with unhydrolysed chitosan–alginate fibres. Antibacterial properties of hydrolysed chitosan–alginate fibres were also confirmed in terms of timely and effective bacterial reduction with initial use. However, a slow release or leaching of hydrolysed chitosan fibres supported the sustained released of antibacterial components.

Chitin-based fibres are generally prepared not by disintegration of native chitin, but through chemical modification of chitin to render it soluble, followed by spinning.[212] Dibutyrylchitin fibres, for example, have been investigated in composites with chitosan glycolate for wound healing applications and were found to be effective in restoring the subcutaneous architecture.[213]

22.5 CONCLUSION

There is an enormous potential for the use of biomass in composite materials. The biomass content can vary from 100% down to negligible amounts. Although a vast amount of work has been done, improvements in fibre-reinforcement compatibility, consistency in performance and large-scale processing capability are paramount to ensure that biomass-containing composite materials continue to increase their market share.

REFERENCES

1. R. Ulber, D. Sell and T. Hirth (eds), *Renewable Raw Materials: New Feedstocks for the Chemical Industry*, Wiley-VCH, Weinheim, 2011.

2. N. G. McCrum, C. P. Buckley and C. B. Bucknall, *Principles of Polymer Engineering*, 2nd edn, Oxford Science Publications, Oxford, 1997, p. 254.
3. J. W. C. Dunlop and P. Fratzl, *Annu. Rev. Mater. Res.*, 2010, **40**, 1.
4. P. Fratzl, H. S. Gupta, E. P. Paschalis and P. Roschger, *J. Mater. Chem.*, 2004, **14**, 2115.
5. A. Gandini, *Macromolecules*, 2008, **41**, 9491.
6. T. Heinze and K. Petzold, Chapter 16 - Cellulose chemistry: Novel products and synthesis paths, in *Monomers, Polymers and Composites from Renewable Resources*, ed. M. N. Belgacem and A. Gandini, Elsevier, Amsterdam, 2008, pp. 343–368.
7. T. Heinze and T. Liebert, *Prog. Polym. Sci.*, 2001, **26**, 1689.
8. K. J. Edgar, C. M. Buchanan, J. S. Debenham, P. A. Rundquist, B. D. Seiler, M. C. Shelton and D. Tindall, *Prog. Polym. Sci.*, 2001, **26**, 1605.
9. M. Roman, in *Model Cellulosic Surfaces, ACS Symposuium Series*, Vol. 1019, ed. M. Roman, American Chemical Society, Washington, DC, 2009, p. 3.
10. C. S. R. Freire and A. Gandini, *Cellul. Chem. Technol.*, 2006, **40**, 691.
11. L. Averous, *J. Macromol. Sci. Polym. Rev.*, 2004, **C44**, 231.
12. L. Yu, K. Dean and L. Li, *Prog. Polym. Sci.*, 2006, **31**, 576.
13. I. Siro and D. Plackett, *Cellulose*, 2010, **17**, 459.
14. A. Dufresne, *Molecules*, 2010, **15**, 4111.
15. R. Stewart, *Plast. Eng.*, 2007, **63**, 25.
16. B. P. Mooney, *Biochem. J.*, 2009, **418**, 219.
17. J. L. Willett and R. L. Shogren, *Polymer*, 2002, **43**, 5935.
18. V. Budarin, J. H. Clark, J. J. E. Hardy, R. Luque, K. Milkowski, S. J. Tavener and A. J. Wilson, *Angew. Chem. Int. Ed.*, 2006, **45**, 3782.
19. K. T. Shalumon, N. S. Binulal, N. Selvamurugan, S. V. Nair, D. Menon, T. Furuike, H. Tamura and R. Jayakumar, *Carbohydr. Polym.*, 2009, **77**, 863.
20. R. Jayakumar, D. Menon, K. Manzoor, S. V. Nair and H. Tamura, *Carbohydr. Polym.*, 2010, **82**, 227.
21. M. Dash, F. Chiellini, R. M. Ottenbrite and E. Chiellini, *Prog. Polym. Sci.*, 2011, **36**, 981.
22. W. S. Wan Ngah, L. C. Teong and M. A. K. M. Hanafiah, *Carbohydr. Polym.*, 2011, **83**, 1446.

23. A. Sarasam and S. V. Madihally, *Biomaterials*, 2005, **26**, 5500.
24. A. D. Baldwin and K. L. Kiick, *Biopolymers*, 2010, **94**, 128.
25. A. D. Augst, H. J. Kong and D. J. Mooney, *Macromol. Biosci.*, 2006, **6**, 623.
26. M. G. Blaine, *Ann. Surg.*, 1947, **125**, 102.
27. A. I. Zouboulis and I. A. Katsoyiannis, *Ind. Eng. Chem. Res.*, 2002, **41**, 6149.
28. V. Rocher, J.-M. Siaugue, V. Cabuil and A. Bee, *Water Res.*, 2008, **42**, 1290.
29. A.-F. Ngomsik, A. Bee, J.-M. Siaugue, D. Talbot, V. Cabuil and G. Cote, *J. Hazard. Mater.*, 2009, **166**, 1043.
30. A. F. Ngomsik, A. Bee, M. Draye, G. Cote and V. Cabuil, *C. R. Chim.*, 2005, **8**, 963.
31. X. Zhao, L. Lv, B. C. Pan, W. M. Zhang, S. J. Zhang and Q. X. Zhang, *Chem. Eng. J.*, 2011, **170**, 381.
32. H. Tamura, Y. Tsuruta and S. Tokura, *Mater. Sci. Eng. C*, 2002, **20**, 143.
33. M. Lahaye and A. Robic, *Biomacromolecules*, 2007, **8**, 1765.
34. H. Demais, J. Brendle, H. Le Deit, A. L. Laza, L. Lurton and D. Brault, *Eur. Patent WO 2006020075*, 2006.
35. S. J. Eichhorn, A. Dufresne, M. Aranguren, N. E. Marcovich, J. R. Capadona, S. J. Rowan, C. Weder, W. Thielemans, M. Roman, S. Renneckar, W. Gindl, S. Veigel, J. Keckes, H. Yano, K. Abe, M. Nogi, A. N. Nakagaito, A. Mangalam, J. Simonsen, A. S. Benight, A. Bismarck, L. A. Berglund and T. Peijs, *J. Mater. Sci.*, 2010, **45**, 1.
36. S. J. Eichhorn, *Soft Matter*, 2011, **7**, 303.
37. D. Le Corre, J. Bras and A. Dufresne, *Biomacromolecules*, 2010, **11**, 1139.
38. G. Gellerstedt and G. Henriksson, Chapter 9 - Lignins: major sources, structure and properties, in *Monomers, Polymers and Composites from Renewable Resources*, ed. M. N. Belgacem and A. Gandini, Elsevier, Amsterdam, 2008, pp. 201–224.
39. J. Lora, Chapter 10 - Industrial commercial lignins: sources, properties and applications, in *Monomers, Polymers and Composites from Renewable Resources*, ed. M. N. Belgacem and A. Gandini, Elsevier, Amsterdam, 2008, pp. 225–242.

40. J. D. Gargulak and S. E. Lebo, in *Lignin: Historical, Biological and Materials Perspectives, ACS Symposuium Series*, Vol. 742, ed. W. G. Glasser, R. A. Northey and T. P. Schultz, American Chemical Society, Washington, DC, 2000, p. 304.

41. W. O. S. Doherty, P. Mousavioun and C. M. Fellows, *Ind. Crops Prod.*, 2011, **33**, 259.

42. W. G. Glasser, in *Classification of Lignin According to Chemical and Molecular Structure, ACS Symposuium Series*, Vol. 742, ed. W. G. Glasser, R. A. Northey and T. P. Schultz, American Chemical Society, Washington, DC, 2000, p. 216.

43. D. Fan, P. R. Chang, N. Lin, J. H. Yu and J. Huang, *Iran. Polym. J.*, 2011, **20**, 3.

44. W. Thielemans, E. Can, S. S. Morye and R. P. Wool, *J. Appl. Polym. Sci.*, 2002, **83**, 323.

45. A. Gandini and M. N. Belgacem, Chapter 11 - Lignins as composites of macromolecular materials, in *Monomers, Polymers and Composites 'from Renewable Resources*, ed. M. N. Belgacem and A. Gandini, Elsevier, Amsterdam, 2008, pp. 243–272.

46. N. A. Rizk, A. Nagaty and O. Y. Mansour, *Acta Polym.*, 1984, **35**, 61.

47. B. Kosikova, J. Labaj, D. Slamenova, E. Slavikova and A. Gregorova, *Novel Environmentally Friendly Use of Lignin Biomass Component*, Nova Science, Hauppauge, NY, 2006.

48. A. V. Maldhure, A. R. Chaudhari and J. D. Ekhe, *J. Thermal Anal. Calorim.*, 2011, **103**, 625.

49. W. Thielemans and R. P. Wool, *Biomacromolecules*, 2005, **6**, 1895.

50. J. J. Meister, D. R. Patil and H. Channell, *J. Appl. Polym. Sci.*, 1984, **29**, 3457.

51. E. Chiellini and R. Solaro (eds), *Biodegradable Polymers and Plastics*, Kluwer Academic/Plenum Publishers, New York, 2003.

52. A. Huttermann, C. Mai and A. Kharazipour, *Appl. Microbiol. Biotechnol.*, 2001, **55**, 387.

53. W. Thielemans and R. P. Wool, *Composites, Part A*, 2004, **35A**, 327.

54. M. Fernandez-Perez, M. Villafranca-Sanchez, F. Flores-Cespedes and I. Daza-Fernandez, *Carbohydr. Polym.*, 2011, **83**, 1672.

55. R. Vanholme, K. Morreel, J. Ralph and W. Boerjan, *Curr. Opin. Plant Biol.*, 2008, **11**, 278.
56. G. Neutelings, *Plant Sci.*, 2011, **181**, 379.
57. R. P. Wool, S. N. Khot, J. J. LaScala, S. P. Bunker, J. Lu, W. Thielemans, E. Can, S. S. Morye and G. I. Williams, in *Advancing Sustainability Through Green Chemistry and Engineering, ACS Symposuium Series*, Vol. 823, ed. R. L. Lankey and P. T. Anastas, American Chemical Society, Washington, DC, 2002, pp. 177–204.
58. A. Cunningham and A. Yapp, *US Patent 3,827,993*, 1974.
59. G. C. Force and F. S. Starr, *US Patent 4,740,367*, 1988.
60. L. Barrett, L. Sperling and C. Murphy, *J. Am. Oil Chem. Soc.*, 1993, **70**, 523.
61. M. A. R. Meier, J. O. Metzger and U. S. Schubert, *Chem. Soc. Rev.*, 2007, **36**, 1788.
62. Y. Lu and R. C. Larock, *ChemSusChem*, 2009, **2**, 136.
63. C. K. Williams and M. A. Hillmyer, *Polym. Rev.*, 2008, **48**, 1.
64. F. S. Guner, Y. Yagci and A. T. Erciyes, *Prog. Polym. Sci.*, 2006, **31**, 633.
65. H. P. Zhao, J. F. Zhang, X. S. Sun and D. H. Hua, *J. Appl. Polym. Sci.*, 2008, **110**, 647.
66. S. N. Khot, J. J. Lascala, E. Can, S. S. Morye, G. I. Williams, G. R. Palmese, S. H. Kusefoglu and R. P. Wool, *J. Appl. Polym. Sci.*, 2001, **82**, 703.
67. Y. T. Li, L. Y. Fu, S. F. Lai, X. C. Cai and L. T. Yang, *Eur. J. Lipid Sci. Technol.*, 2010, **112**, 511.
68. A. Campanella, J. J. La Scala and R. P. Wool, *Polym. Eng. Sci.*, 2009, **49**, 2384.
69. R. P. Wool, S. N. Khot, J. J. LaScala, S. P. Bunker, J. Lu, W. Thielemans, E. Can, S. S. Morye and G. I. Williams, in *Advancing Sustainability Through Green Chemistry and Engineering, ACS Symposuium Series*, Vol. 823, ed. R. L. Lankey and P. T. Anastas, American Chemical Society, Washington, DC, 2002, pp. 205–224.
70. K. Adekunle, D. Akesson and M. Skrifvars, *J. Appl. Polym. Sci.*, 2010, **115**, 3137.
71. G. I. Williams and R. P. Wool, *Appl. Compos. Mater.*, 2000, **7**, 421.
72. A. O'Donnell, M. A. Dweib and R. P. Wool, *Compos. Sci. Technol.*, 2004, **64**, 1135.

73. C. K. Hong and R. P. Wool, *J. Appl. Polym. Sci.*, 2005, **95**, 1524.
74. O. Eksik, A. T. Erciyes and Y. Yagci, *J. Macromol. Sci. Part A: Pure Appl. Chem.*, 2008, **45**, 698.
75. S. Philip, T. Keshavarz and I. Roy, *J. Chem. Technol. Biotechnol.*, 2007, **82**, 233.
76. E. Rezzonico, L. Moire and Y. Poirier, *Phytochem. Rev.*, 2002, **1**, 87.
77. S. K. Misra, S. P. Valappil, I. Roy and A. R. Boccaccini, *Biomacromolecules*, 2006, **7**, 2249.
78. R. Rai, T. Keshavarz, J. A. Roether, A. R. Boccaccini and I. Roy, *Mater. Sci. Eng. R: Rep.*, 2011, **72**, 29.
79. G.-Q. Chen and Q. Wu, *Biomaterials*, 2005, **26**, 6565.
80. J. F. J. Coelho, J. R. Gois, A. C. Fonseca and M. H. Gil, *J. Appl. Polym. Sci.*, 2010, **116**, 718.
81. R. Auras, B. Harte and S. Selke, *Macromol. Biosci.*, 2004, **4**, 835.
82. S. Taguchi, M. Yamada, K. I. Matsumoto, K. Tajima, Y. Satoh, M. Munekata, K. Ohno, K. Kohda, T. Shimamura, H. Kambe and S. Obata, *Proc. Natl. Acad. Sci. USA*, 2008, **105**, 17323.
83. K. M. Nampoothiri, N. R. Nair and R. P. John, *Bioresour. Technol.*, 2010, **101**, 8493.
84. H. Liu and J. Zhang, *J. Polym. Sci. Part B: Polym. Phys.*, 2011, **49**, 1051.
85. A. J. Nijenhuis, E. Colstee, D. W. Grijpma and A. J. Pennings, *Polymer*, 1996, **37**, 5849.
86. A. M. Gajria, V. Davé, R. A. Gross and S. P. McCarthy, *Polymer*, 1996, **37**, 437.
87. S. McCarthy and X. Song, *J. Appl. Med. Polym.*, 2002, **6**, 64.
88. F. Ali, Y.-W. Chang, S. C. Kang and J. Y. Yoon, *Polym. Bull.*, 2009, **62**, 91.
89. J. K. Pandey, S. H. Ahn, C. S. Lee, A. K. Mohanty and M. Misra, *Macromol. Mater. Eng.*, 2010, **295**, 975.
90. B. Bax and J. Müssig, *Compos. Sci. Technol.*, 2008, **68**, 1601.
91. T. Mukherjee and N. Kao, *J. Polym. Environ.*, 2011, **19**, 714.
92. H. M. Akil, M. F. Omar, A. A. M. Mazuki, S. Safiee, Z. A. M. Ishak and A. Abu Bakar, *Mater. Des.*, 2011, **32**, 4107.
93. C. Nyambo, A. K. Mohanty and M. Misra, *Biomacromolecules*, 2010, **11**, 1654.

94. M. B. Rodgers, D. S. Tracey and W. H. Waddell, *Rubber World*, 2005, **232**, 32.
95. M. B. Rodgers, D. S. Tracey and W. H. Waddell, *Rubber World*, 2005, **232**, 41.
96. J. B. van Beilen and Y. Poirier, *Crit. Rev. Biotechnol.*, 2007, **27**, 217.
97. D. J. Siler, K. Cornish and R. G. Hamilton, *J. Allergy Clin. Immunol.*, 1996, **98**, 895.
98. J. B. Donnet, R. C. Bansal and M. J. Wang, *Carbon Black: Science and Technology*, CRC Press, Boca Raton, FL, 1993.
99. A. Voet, J. C. Morawski and J. B. Donnet, *Rubber Chem. Technol.*, 1977, **50**, 342.
100. J. L. Valentín, I. Mora-Barrantes, J. Carretero-González, M. A. López-Manchado, P. Sotta, D. R. Long and K. Saalwaächter, *Macromolecules*, 2009, **43**, 334.
101. Y. Nakamura, H. Honda, A. Harada, S. Fujii and K. Nagata, *J. Appl. Polym. Sci.*, 2009, **113**, 1507.
102. M. Knite, K. Ozols, J. Zavickis, V. Tupureina, I. Klemenoks and R. Orlovs, *J. Nanosci. Nanotechnol.*, 2009, **9**, 3587.
103. M. Jacob, J. Jose, S. Jose, K. T. Varughese and S. Thomas, *J. Appl. Polym. Sci.*, 2010, **117**, 614.
104. S. Joseph, S. P. Appukuttan, J. M. Kenny, D. Puglia, S. Thomas and K. Joseph, *J. Appl. Polym. Sci.*, 2010, **117**, 1298.
105. J. Holbery and D. Houston, *JOM J. Miner. Met. Mater. Soc.*, 2006, **58**, 80.
106. D. Hull and T. W. Clyne, in *An Introduction to Composite Materials*, 2nd ed., Cambridge University Press, Cambridge, 1996, p. 6.
107. M. M. de Souza Lima and R. Borsali, *Macromol. Rapid Commun.*, 2004, **25**, 771.
108. W. G. Glasser, B. K. McCartney and G. Samaranayake, *Biotechnol. Prog.*, 1994, **10**, 214.
109. C. H. Park, Y. K. Kang and S. S. Im, *J. Appl.Polym. Sci.*, 2004, **94**, 248.
110. H. Suh, K. Duckett and G. Bhat, *Text. Res. J.*, 1996, **66**, 230.
111. V. Favier, H. Chanzy and J. Y. Cavaille, *Macromolecules*, 1995, **28**, 6365.
112. F. W. Herrick, R. L. Casebier, J. K. Hamilton and K. R. Sandberg, *J. Appl. Polym. Sci. Appl. Polym. Sci.*, 1983, **37**, 797.

113. A. F. Turbak, F. W. Snyder and K. R. Sandberg, *J. Appl. Polym. Sci. Appl. Polym. Symp.*, 1983, **37**, 813.

114. Y. Chen, X. Cao, P. R. Chang and M. A. Huneault, *Carbohydr. Polym.*, 2008, **73**, 8.

115. G. Siqueira, J. Bras and A. Dufresne, *Biomacromolecules*, 2009, **10**, 425.

116. P. Bataille, L. Ricard and S. Sapieha, *Polym. Compos.*, 1989, **10**, 103.

117. A. D. Beshay, B. V. Kokta and C. Daneault, *Polym. Compos.*, 1985, **6**, 261.

118. D. Maldas, B. V. Kokta and C. Daneault, *J. Vinyl Technol.*, 1989, **11**, 90.

119. R. G. Raj, B. V. Kokta, D. Maldas and C. Daneault, *Polym. Compos.*, 1988, **9**, 404.

120. B. V. Kokta, D. Maldas, C. Daneault and P. Béland, *Polym. Compos.*, 1990, **11**, 84.

121. D. Maldas and B. V. Kokta, *J. Vinyl Technol.*, 1993, **15**, 38.

122. M. Bengtsson, M. L. Baillif and K. Oksman, *Composites Part A: Appl. Sci. Manuf.*, 2007, **38**, 1922.

123. A. R. Sanadi, D. F. Caulfield, R. E. Jacobson and R. M. Rowell, *Ind. Eng. Chem. Res.*, 1995, **34**, 1889.

124. L. U. Devi, S. S. Bhagawan and S. Thomas, *J. Appl. Polym. Sci.*, 1997, **64**, 1739.

125. B. C. Mitra, R. K. Basak and M. Sarkar, *J. Appl. Polym. Sci.*, 1998, **67**, 1093.

126. K. Joseph, S. Thomas and C. Pavithran, *Compos. Sci. Technol.*, 1995, **53**, 99.

127. P. Hedenberg and P. Gatenholm, *J. Appl. Polym. Sci.*, 1996, **60**, 2377.

128. F. P. La Mantia and M. Morreale, *Composites Part A: Appl. Sci. Manuf.*, 2011, **42**, 579.

129. A. Alavudeen, M. Thiruchitrambalam, N. Venkateshwaran and A. Athijayamani, *Rev. Adv. Mater. Sci.*, 2011, **27**, 146.

130. J. J. Blaker, K. Y. Lee and A. Bismarck, *J Biobased Mater. Bioenergy*, 2011, **5**, 1.

131. S. Shinoj, R. Visvanathan, S. Panigrahi and M. Kochubabu, *Ind. Crops Prod.*, 2011, **33**, 7.

132. J. Summerscales, N. P. J. Dissanayake, A. S. Virk and W. Hall, *Composites Part A: Appl. Sci. Manuf.*, 2010, **41**, 1329.

133. K. G. Satyanarayana, G. G. C. Arizaga and F. Wypych, *Prog. Polym. Sci.*, 2009, **34**, 982.

134. R. T. Woodhams, G. Thomas and D. K. Rodgers, *Polym. Eng. Sci.*, 1984, **24**, 1166.

135. M. N. Belgacem and A. Gandini, *Surface Modification of Cellulose Fibres*, Elsevier, Amsterdam, 2008.

136. M. N. Belgacem and A. Gandini, *Natural Fibre-Surface Modification and Characterisation*, Old City Publishing, Philadelphia, PA, 2009.

137. W. Thielemans and R. P. Wool, *Polym. Compos.*, 2005, **26**, 695.

138. W. H. Binder and R. Sachsenhofer, *Macromol. Rapid Commun.*, 2007, **28**, 15.

139. E. H. B. Ly, J. Bras, P. Sadocco, M. N. Belgacem, A. Dufresne and W. Thielemans, *Mater. Chem. Phys.*, 2010, **120**, 438.

140. O. Paquet, M. Krouit, J. Bras, W. Thielemans and M. N. Belgacem, *Acta Mater.*, 2010, **58**, 792.

141. A. N. Frone, S. Berlioz, J. F. Chailan, D. M. Panaitescu and D. Donescu, *Polym. Compos.*, 2011, **32**, 976.

142. J. T. Kim and A. N. Netravali, *Composites Part A: Appl. Sci. Manuf.*, 2010, **41**, 1245.

143. Y. Li, Y. W. Mai and L. Ye, *Compos. Sci. Technol.*, 2000, **60**, 2037.

144. J.-A. Trejo-O'Reilly, J.-Y. Cavaille and A. Gandini, *Cellulose*, 1997, **4**, 305.

145. M. Kazayawoko, J. J. Balatinecz and L. M. Matuana, *J. Mater. Sci.*, 1999, **34**, 6189.

146. A. Arbelaiz, B. Fernández, G. Cantero, R. Llano-Ponte, A. Valea and I. Mondragon, *Composites Part A: Appl. Sci. Manuf*, 2005, **36**, 1637.

147. P. Wambua, J. Ivens and I. Verpoest, *Compos. Sci. Technol.*, 2003, **63**, 1259.

148. N.-J. Lee and J. Jang, *Compos. Sci. Technol.*, 2000, **60**, 209.

149. N.-J. Lee and J. Jang, *Composites Part A: Appl. Sci. Manuf.*, 1999, **30**, 815.

150. X. Y. Liu and G. C. Dai, *eXPRESS Polym. Lett.*, 2007, **1**, 299.

151. J. D. Espinoza-Perez, C. Λ. Ulven and D. P. Wiesenborn, *Trans. ASABE*, 2010, **53**, 1167.

152. K. Adekunle, D. Akesson and M. Skrifvars, *J. Appl. Polym. Sci.*, 2010, **116**, 1759.
153. D. Akesson, M. Skrifvars and P. Walkenstrom, *J. Appl. Polym. Sci.*, 2009, **114**, 2502.
154. N. Graupner, A. S. Herrmann and J. Mussig, *Composites Part A: Appl. Sci. Manuf.*, 2009, **40**, 810.
155. C. S. Wu, *Polym. Degrad. Stab.*, 2009, **94**, 1076.
156. N. Reddy and Y. Q. Yang, *Ind. Crops Prod.*, 2011, **33**, 35.
157. K. C. Batista, D. A. K. Silva, L. A. F. Coelho, S. H. Pezzin and A. P. T. Pezzin, *J. Polym. Environ.*, 2010, **18**, 346.
158. J. D. Macedo, M. F. Costa, M. I. B. Tavares and R. Thire, *Polym. Eng. Sci.*, 2010, **50**, 1466.
159. L. Jiang, F. Chen, J. Qian, J. J. Huang, M. Wolcott, L. S. Liu and J. W. Zhang, *Ind. Eng. Chem. Res.*, 2010, **49**, 572.
160. N. Reddy and Y. Q. Yang, *Polym. Int.*, 2011, **60**, 711.
161. D. Plackett, T. Løgstrup Andersen, W. Batsberg Pedersen and L. Nielsen, *Compos. Sci. Technol.*, 2003, **63**, 1287.
162. K. Oksman, M. Skrifvars and J. F. Selin, *Compos. Sci. Technol.*, 2003, **63**, 1317.
163. R. F. Nickerson and J. A. Habrle, *J. Ind. Eng. Chem.*, 1947, **39**, 1507.
164. K. Tashiro and M. Kobayashi, *Polymer*, 1991, **32**, 1516.
165. A. Dufresne, M. B. Kellerhals and B. Witholt, *Macromolecules*, 1999, **32**, 7396.
166. M. N. Anglès and A. Dufresne, *Macromolecules*, 2000, **33**, 8344.
167. A. P. Mathew, W. Thielemans and A. Dufresne, *J. Appl. Polym. Sci.*, 2008, **109**, 4065.
168. W. J. Orts, J. Shey, S. H. Imam, G. M. Glenn, M. E. Guttman and J.-F. Revol, *J. Polym. Environ.*, 2005, **13**, 301.
169. R. Kohler and K. Nebel, *Macromol. Symp.*, 2006, **244**, 97.
170. E. E. Brown and M.-P. G. Laborie, *Biomacromolecules*, 2007, **8**, 3074.
171. J. Leitner, B. Hinterstoisser, M. Wastyn, J. Keckes and W. Gindl, *Cellulose*, 2007, **14**, 419.
172. N. L. G. de Rodriguez, W. Thielemans and A. Dufresne, *Cellulose*, 2006, **13**, 261.
173. V. Favier, G. R. Canova, J. Y. Cavaillé, H. Chanzy, A. Dufresne and C. Gauthier, *Polym. Adv. Technol.*, 1995, **6**, 351.

174. P. Hajji, J. Y. Cavaillé, V. Favier, C. Gauthier and G. Vigier, *Polym. Compos.*, 1996, **17**, 612.

175. K. Hossain, I. Ahmed, A. Parsons, C. Scotchford, G. Walker, W. Thielemans and C. Rudd, *J. Mater. Sci.*, 2012, **47**, 2675.

176. J. P. de Mesquita, C. L. Donnici and F. V. Pereira, *Biomacromolecules*, 2010, **11**, 473.

177. M. J. Bonne, K. J. Edler, J. G. Buchanan, D. Wolverson, E. Psillakis, M. Helton, W. Thielemans and F. Marken, *J. Phys. Chem. C*, 2008, **112**, 2660.

178. M. J. Bonne, E. Galbraith, T. D. James, M. J. Wasbrough, K. J. Edler, A. T. A. Jenkins, M. Helton, A. McKee, W. Thielemans, E. Psillakis and F. Marken, *J. Mater. Chem.*, 2010, **20**, 588.

179. K. Tsourounaki, M. J. Bonne, W. Thielemans, E. Psillakis, M. Helton, A. McKee and F. Marken, *Electroanalysis*, 2008, **20**, 2395.

180. R. J. Moon, A. Martini, J. Nairn, J. Simonsen and J. Youngblood, *Chem. Soc. Rev.*, 2011, **40**, 3941.

181. Y. Habibi, L. A. Lucia and O. J. Rojas, *Chem. Rev.*, 2010, **110**, 3479.

182. D. Klemm, F. Kramer, S. Moritz, T. Lindström, M. Ankerfors, D. Gray and A. Dorris, *Angew. Chem. Int. Ed.*, 2011, **50**, 5438.

183. A. Dufresne, *Can. J. Chem. Rev. Can. Chim.*, 2008, **86**, 484.

184. I. Siró and D. Plackett, *Cellulose*, 2010, **17**, 459.

185. G. Chen, A. Dufresne, J. Huang and P. R. Chang, *Macromol. Mater. Eng.*, 2009, **294**, 59.

186. J. R. Capadona, O. Van Den Berg, L. A. Capadona, M. Schroeter, S. J. Rowan, D. J. Tyler and C. Weder, *Nat. Nanotechnol.*, 2007, **2**, 765.

187. W.-I. Park, M. Kang, H.-S. Kim and H.-J. Jin, *Macromol. Symp.*, 2007, **249–250**, 289.

188. A. Isogai, T. Saito and H. Fukuzumi, *Nanoscale*, 2011, **3**, 71.

189. M. F. Koenig and S. J. Huang, *Polymer*, 1995, **36**, 1877.

190. M. Avella, M. E. Errico, P. Laurienzo, E. Martuscelli, M. Raimo and R. Rimedio, *Polymer*, 2000, **41**, 3875.

191. R. P. Wool, D. Raghavan, G. C. Wagner and S. Billieux, *J. Appl. Polym. Sci.*, 2000, **77**, 1643.

192. X. C. Ge, X. H. Li, Q. Zhu, L. Li and Y. Z. Meng, *Polym. Eng. Sci.*, 2004, **44**, 2134.

193. B. Siddaramaiah, B. Raj and R. Somashekar, *J. Appl. Polym. Sci.*, 2004, **91**, 630.

194. P. Dubois, M. Krishnan and R. Narayan, *Polymer*, 1999, **40**, 3091.

195. A. Hebeish, M. H. El-Rafie, A. Higazy and M. Ramadan, *Starch Stärke*, 1996, **48**, 175.

196. A. Hebeish, M. H. El-Rafie, A. Higazy and M. A. Ramadan, *Starch Stärke*, 1992, **44**, 101.

197. B. Ramaraj, *J. Appl. Polym. Sci.*, 2007, **103**, 1127.

198. O. S. Odusanya, U. S. Ishiaku, B. M. N. Azemi, B. D. M. Manan and H. W. Kammer, *Polym. Eng. Sci.*, 2000, **40**, 1298.

199. G. H. Yew, A. M. M. Yusof, Z. A. M. Ishak and U. S. Ishiaku, *Polym. Degrad. Stab.*, 2005, **90**, 488.

200. A. Hebeish, E. El-Alfy and A. Bayazeed, *Starch Stärke*, 1988, **40**, 191.

201. J.-F. Zhang and X. Sun, *Biomacromolecules*, 2004, **5**, 1446.

202. D. Rutot, P. Degée, R. Narayan and P. Dubois, *Compos. Interface*, 2000, **7**, 215.

203. C.-S. Wu, *Polym. Degrad. Stab.*, 2003, **80**, 127.

204. D. Dubief, E. Samain and A. Dufresne, *Macromolecules*, 1999, **32**, 5765.

205. A. Dufresne, J.-Y. Cavaillé and W. Helbert, *Macromolecules*, 1996, **29**, 7624.

206. H. Angellier, J.-L. Putaux, S. Molina-Boisseau, D. Dupeyre and A. Dufresne, *Macromol. Symp.*, 2005, **221**, 95.

207. H. Angellier, S. Molina-Boisseau and A. Dufresne, *Macromol. Symp.*, 2006, **233**, 132.

208. E. Kristo and C. G. Biliaderis, *Carbohydr. Polym.*, 2007, **68**, 146.

209. J. Wang, L. Qu, X. Meng, J. Gao, H. Li and G. Wen, *Biomed. Mater.*, 2008, **3**, 25004.

210. E. D. Pinho, A. Martins, J. V. Araújo, R. L. Reis and N. M. Neves, *Acta Biomater.*, 2009, **5**, 1104.

211. C. J. Knill, J. F. Kennedy, J. Mistry, M. Miraftab, G. Smart, M. R. Groocock and H. J. Williams, *Carbohydr. Polym.*, 2004, **55**, 65.

212. R. Jayakumar, M. Prabaharan, P. T. Sudheesh Kumar, S. V. Nair and H. Tamura, *Biotechnol. Adv.*, 29, 322.

213. M. Mattioli-Belmonte, A. Zizzi, G. Lucarini, F. Giantomassi, G. Biagini, G. Tucci, F. Orlando, M. Provinciali, F. Carezzi and P. Morganti, *J. Bioact. Compat. Polym.*, 2007, **22**, 525.
214. A. Arinstein, M. Burman, O. Gendelman and E. Zussman, *Nat. Nanotechnol.*, 2007, **2**, 59.
215. J. Lu, T. Wang and L. T. Drzal, *Composites Part A: Appl. Sci. Manuf.*, 2008, **39**, 738.
216. K. Abe, S. Iwamoto and H. Yano, *Biomacromolecules*, 2007, **8**, 3276.

Subject Index

References to tables and charts are in bold type